Animal Genetics: Theories and Applications

Animal Genetics:
Theories and Applications

Editor: Dominic Fasso

R CALLISTO REFERENCE

www.callistoreference.com

Callisto Reference,
118-35 Queens Blvd., Suite 400,
Forest Hills, NY 11375, USA

Visit us on the World Wide Web at:
www.callistoreference.com

ISBN: 978-1-64116-101-5 (Hardback)

Cataloging-in-Publication Data

Animal genetics : theories and applications / edited by Dominic Fasso.
    p. cm.
Includes bibliographical references and index.
ISBN 978-1-64116-101-5
1. Animal genetics. 2. Genetics. 3. Veterinary genetics. I. Fasso, Dominic.
QH432 .A55 2019
591.35--dc23

Table of Contents

Preface

Genetics is the study of genes, variation and heredity. The fundamental aspects of this discipline are trait inheritance and molecular inheritance. The study of animal genetics is concerned with the development of methodologies to improve the genetic makeup of domestic and farm animals for the purposes of increasing disease resistance and stimulating inheritance of desired characteristics. The study and enhancement of genomes of economically crucial animals is important in the modern scenario to maximize productivity. Humans depend on animals for a variety of products like meat, egg, milk, fur, leather, etc. This book studies, analyzes and upholds the pillars of animal genetics and its utmost significance in modern times. It includes contributions of experts and scientists, which will provide innovative insights into this field. It aims to serve as a resource guide for students and experts alike and contribute to the growth of the discipline.

The researches compiled throughout the book are authentic and of high quality, combining several disciplines and from very diverse regions from around the world. Drawing on the contributions of many researchers from diverse countries, the book's objective is to provide the readers with the latest achievements in the area of research. This book will surely be a source of knowledge to all interested and researching the field.

In the end, I would like to express my deep sense of gratitude to all the authors for meeting the set deadlines in completing and submitting their research chapters. I would also like to thank the publisher for the support offered to us throughout the course of the book. Finally, I extend my sincere thanks to my family for being a constant source of inspiration and encouragement.

Editor

Genomic regions associated with host response to porcine reproductive and respiratory syndrome vaccination and co-infection in nursery pigs

Jenelle R. Dunkelberger[1,2], Nick V. L. Serão[1], Ziqing Weng[1,3], Emily H. Waide[1,4], Megan C. Niederwerder[5], Maureen A. Kerrigan[5], Joan K. Lunney[6], Raymond R. R. Rowland[5] and Jack C. M. Dekkers[1*]

Abstract

Background: The WUR1000125 (**WUR**) single nucleotide polymorphism (**SNP**) can be used as a genetic marker for host response to porcine reproductive and respiratory syndrome (**PRRS**), PRRS vaccination, and co-infection with porcine circovirus type 2b (**PCV2b**). Objectives of this study were to identify genomic regions other than WUR associated with host response to PRRS vaccination and PRRSV/PCV2b co-infection and regions with a different effect on host response to co-infection, depending on previous vaccination for PRRS.

Methods: Commercial crossbred nursery pigs were pre-selected for WUR genotype (n = 171 AA and 198 AB pigs) where B is the dominant and favorable allele. Half of the pigs were vaccinated for PRRS and 4 weeks later, all pigs were co-infected with PRRS virus and PCV2b. Average daily gain (**ADG**) and viral load (**VL**) were quantified post vaccination (**Post Vx**) and post co-infection (**Post Co-X**). Single-SNP genome-wide association analyses were then conducted to identify genomic regions associated with response to vaccination and co-infection.

Results: Multiple SNPs near the major histocompatibility complex were significantly associated with PCV2b VL ($-log_{10}P \geq 5.5$), regardless of prior vaccination for PRRS. Several SNPs were also significantly associated with ADG Post Vx and Post Co-X. SNPs with a different effect on ADG, depending on prior vaccination for PRRS, were identified Post Vx ($-log_{10}P = 5.6$) and Post Co-X ($-log_{10}P = 5.5$). No SNPs were significantly associated with vaccination VL ($-log_{10}P \leq 4.7$) or PRRS VL ($-log_{10}P \leq 4.3$). Genes near SNPs associated with vaccination VL, PRRS VL, and PCV2b VL were enriched ($P \leq 0.01$) for immune-related pathways and genes near SNPs associated with ADG were enriched for metabolism pathways ($P \leq 0.04$). SNPs associated with vaccination VL, PRRS VL, and PCV2b VL showed overrepresentation of health QTL identified in previous studies and SNPs associated with ADG Post Vx of Non-Vx pigs showed overrepresentation of growth QTL.

Conclusions: Multiple genomic regions were associated with PCV2b VL and ADG Post Vx and Post Co-X. Different SNPs were associated with ADG, depending on previous vaccination for PRRS. Results of functional annotation analyses and novel approaches of using previously-reported QTL support the identified regions.

Keywords: Functional annotation, Genome-wide association study, PCV2, PRRS, Quantitative trait locus, Swine, WUR10000125

* Correspondence: jdekkers@iastate.edu
[1]Department of Animal Science, Iowa State University, Ames, IA 50011, USA
Full list of author information is available at the end of the article

Background

Guanylate binding protein 5 (*GBP5*), located on *Sus scrofa* chromosome (**SSC**) 4, was identified as a major gene for host response to porcine reproductive and respiratory syndrome (**PRRS**) [1, 2]. Since the causative mutation for *GBP5* does not appear on commercial genotyping platforms, the single nucleotide polymorphism (**SNP**) WUR10000125 (**WUR**), which is in complete linkage disequilibrium with the putative causative mutation [2–4] can be used as a genetic marker for this mutation.

Since the identification of this quantitative trait locus (**QTL**), the effect of WUR has been associated with host response to PRRS virus (**PRRSV**)-infection following infection with two different PRRSV isolates [3–6], PRRS vaccination [7], and co-infection with PRRSV and porcine circovirus type 2b (**PCV2b**) [7]. In addition, WUR was associated with PCV2b viral load (**VL**) following PRRSV/PCV2b co-infection for pigs previously vaccinated for PRRS, but not for non-vaccinated pigs [7]. Other genomic regions, including regions on SSC7 and SSC12 have been associated with PCV2b VL following experimental infection with PCV2b only [8].

The first objective of this study was to identify genomic regions other than WUR that are associated with host response to PRRS vaccination and co-infection with PRRSV and PCV2b. A second objective was to identify regions with a different effect on host response to PRRSV/PCV2b-infection, depending on whether pigs were previously vaccinated for PRRS. The final objective was to assess the biological relevance of genomic regions associated with each trait to provide support for, and assign biological function to, these statistically-associated regions.

Methods

Experiments involving animals and virus were performed in accordance with the Federation of Animal Science Societies Guide for the Care and Use of Agricultural Animals in Research and Teaching, the USDA Animal Welfare Act and Animal Welfare Regulations, or according to the National Institutes of Health's Guide for the Care and Use of Laboratory Animals, and were approved by the Kansas State University and Iowa State University institutional animal care and use committees and institutional biosafety committees. Animals were humanely euthanized by pentobarbital overdose following the American Veterinary Medical Association (AVMA) guidelines for the euthanasia of animals, and all efforts were made to minimize suffering.

Animals

A detailed description of the animals used for this study is in Dunkelberger et al. [7]. Briefly, commercial Large White x Landrace crossbred barrows from two experimental co-infection trials (trial 1 n = 199; trial 2 n = 197) were used. Pigs originated from the same, high health multiplier farm and were pre-selected based on WUR marker genotype: approximately half for the AA genotype (n = 184) and the other half for the AB genotype (n = 212), where the B allele is the favorable and dominant allele [1]. Pigs were shipped to Kansas State University at weaning (between 18 and 28 days of age) and sorted into one of 2 rooms upon arrival. Within each room, pens were balanced according to WUR genotype and pigs were given 3-4 days to acclimate to their new surroundings before all pigs within one room were vaccinated with a 2-mL dose of a commercial PRRS MLV vaccine (Ingelvac PRRS®, Boehringer Ingelheim Vetmedica Inc., St. Joseph, MO). Four weeks later, all pigs were co-infected with field strains of PRRSV and PCV2b [9]. Pigs were followed for the next 42 days, after which all surviving pigs were humanely euthanized using an intravenous injection of pentobarbital sodium and tissue was collected for genotyping. Body weights were recorded weekly throughout the vaccination (–28 to 0 days post-infection [**dpi**]) and co-infection (0 to 42 dpi) periods on both vaccinated (**Vx**) and non-vaccinated (**Non-Vx**) pigs. Serum samples were collected on Vx pigs following vaccination (**Post Vx**) at –28, –24, –21, –17, –14, and –7 dpi and on both Vx and Non-Vx pigs following co-infection (**Post Co-X**) at 0, 4, 7, 11, 14, 21, 28, 35, and 42 dpi. Serum samples were used to quantify PRRSV and PCV2b viremia using real-time polymerase chain reaction according to Niederwerder et al. [9].

Traits

A detailed description of the traits analyzed for this study, including descriptive statistics for all traits analyzed, is in Dunkelberger et al. [7]. Briefly, 89 Non-Vx AA, 106 Non-Vx AB, 95 Vx AA, and 106 Vx AB pigs were used for all analyses. ADG was calculated as the regression of body weight on dpi using body weight data from –28 to 0 and 0 to 42 dpi for ADG Post Vx and Post Co-X, respectively. Vaccination VL, PRRS VL, and PCV2b VL were calculated for each individual as the area under the curve of $\log_{10}$-transformed viremia from –28 to 0, 0 to 21, and 0 to 42 dpi, respectively.

Genotype data

Ear tissue was used to genotype pigs from trials 1 and 2 using the GeneSeek-Neogen PorcineSNP80 BeadChip (GeneSeek, Igenity; Lincoln, NE). Quality control of genotype data was performed in the following steps: 1) fixed SNPs were removed, 2) SNP genotypes with a gene call score lower than 0.3 were set to missing, and 3) SNPs missing more than 15% of genotypes were removed. Final genotyping rate, calculated as the percent of SNP genotypes called out of the total number of genotypes, was 99.43%.

For all analyses presented, a genomic relationship matrix (**GRM**) was used to account for relationships among the 369 individuals with both genotypes and phenotypes and was constructed according to VanRaden [10] by centering and scaling genotypes for all individuals across the 61,729 SNPs that remained after quality control.

Genome-wide association analyses (GWAS)
All GWAS were performed using single-SNP analyses with ASReml 4.0 [11].

Univariate GWAS
The following univariate animal model was used to test the effect of SNP genotype on vaccination VL by fitting the effect of one SNP at a time, with animal, litter, and pen (trial) fitted as random effects to account for genetic, common environmental, and random environmental effects, respectively:

Model [1]

$$Y_{ijklmn} = \mu + Trial_j + WUR_k + SNP_l$$
$$+ \beta_1 * WtVx_i + \beta_2 * VxAge_i + \beta_3 * PCV2_{0_i}$$
$$+ Animal_i + Litter_m + Pen_{n(j)} + e_{ijklmn}$$

where Y_{ijklm} = is the observed phenotype; $Trial_j$ = fixed effect of the jth trial (1 or 2); WUR_k = fixed effect of WUR SNP genotype (AA or AB); SNP = fixed effect of the lth genotype (AA, AB, or BB); β_p = partial regression coefficient for the covariate weight at −28 dpi (WtVx) ($P = 1$), age at −28 dpi (VxAge) ($P = 2$), and PCV2b viremia at 0 dpi ($P = 3$); $Animal_i$ = random animal genetic effect of the ith individual, with a variance-covariance structure proportional to the GRM with the assumption $\sim N(0, G\sigma_a^2)$; Litter = random litter effect (123 levels), assumed to be $\sim N(0, I\sigma_l^2)$; and Pen = random effect of pen nested within trial (40 levels), assumed to be $\sim N(0, I\sigma_p^2)$. Interaction effects of trial with each fixed effect were fitted but removed since they were not significant ($P > 0.10$).

Including WUR genotype in the model as a fixed effect allowed for the identification of regions other than WUR associated with host response to PRRS vaccination, while simultaneously accounting for the effect of this marker. Level of PCV2b viremia at 0 dpi was fitted as a covariate because 37 (24 Non-Vx and 13 Vx) pigs had non-zero PCV2b viremia titers at 0 dpi, suggesting that they were exposed to PCV2b prior to entering the facility, likely from their mothers; all phenotypes were adjusted to PCV2_0 = 0 to model the situation that all pigs were negative for PCV2b at 0 dpi.

Bivariate GWAS
To identify the effect of vaccination on the effect of genomic regions for host response to co-infection, bivariate animal models were used to analyze ADG Post Vx, PRRS VL, PCV2b VL, and ADG Post Co-X of Vx versus Non-Vx groups as two separate traits, as described in Dunkelberger et al. [7]. For each bivariate GWAS, the following model was used to test the effect of each SNP averaged across vaccination status (**main effect**) and the interaction of SNP genotype with vaccination status (**interaction effect**):

Model [2]

$$\begin{bmatrix} y_N \\ y_V \end{bmatrix} = \begin{bmatrix} X_N & 0 \\ 0 & X_V \end{bmatrix}\begin{bmatrix} b_N \\ b_V \end{bmatrix} + \begin{bmatrix} W_{iN} & 0 \\ 0 & W_{iV} \end{bmatrix}\begin{bmatrix} g_{iN} \\ g_{iV} \end{bmatrix} + \begin{bmatrix} Z_{\alpha_N} & 0 \\ 0 & Z_{\alpha_V} \end{bmatrix}\begin{bmatrix} \alpha_N \\ \alpha_V \end{bmatrix}$$
$$+ \begin{bmatrix} Z_{l_N} & 0 \\ 0 & Z_{l_V} \end{bmatrix}\begin{bmatrix} l_N \\ l_V \end{bmatrix} + \begin{bmatrix} Z_{p_N} & 0 \\ 0 & Z_{p_V} \end{bmatrix}\begin{bmatrix} p_N \\ p_V \end{bmatrix} + \begin{bmatrix} e_N \\ e_V \end{bmatrix}$$

where subscripts N and V represent a trait recorded on Non-Vx and Vx pigs, respectively; $y_N(y_V)$ = vector of phenotypes; $X_N(X_V)$ = design matrix of fixed effects (same as for Model [1]); $b_N(b_V)$ = vector of solutions for fixed effects; $W_{iN}(W_{iV})$ = matrix of fixed genotype effects for the ith SNP genotype and $g_{iN}(g_{iV})$ = corresponding vector of solutions; $Z_{\alpha_N}(Z_{\alpha_V})$ = design matrix of random genetic effects and $\alpha_N(\alpha_V)$ = corresponding vector of solutions; $Z_{1_N}(Z_{1_V})$ = design matrix of random litter effects and $l_N(l_V)$ = corresponding vector of solutions; $Z_{p_N}(Z_{p_V})$ = design matrix of random pen effects and $p_N(p_V)$ = corresponding vector of solutions. The covariance matrix of random animal genetic, litter, pen, and residual effects was specified as follows:

$$Var\begin{bmatrix} \alpha_N \\ \alpha_V \\ l_N \\ l_V \\ p_N \\ p_V \\ e_N \\ e_V \end{bmatrix} = \begin{bmatrix} G_N\sigma_{a_N}^2 & G_{NV}\sigma_{a_{N,V}} & 0 & 0 & 0 & 0 & 0 & 0 \\ G_{NV}\sigma_{a_{N,V}} & G_V\sigma_{a_V}^2 & 0 & 0 & 0 & 0 & 0 & 0 \\ 0 & 0 & I\sigma_{l_N}^2 & I\sigma_{l_{N,V}} & 0 & 0 & 0 & 0 \\ 0 & 0 & I\sigma_{l_{N,V}} & I\sigma_{l_V}^2 & 0 & 0 & 0 & 0 \\ 0 & 0 & 0 & 0 & I\sigma_{p_N}^2 & 0 & 0 & 0 \\ 0 & 0 & 0 & 0 & 0 & I\sigma_{p_V}^2 & 0 & 0 \\ 0 & 0 & 0 & 0 & 0 & 0 & I\sigma_{e_N}^2 & 0 \\ 0 & 0 & 0 & 0 & 0 & 0 & 0 & I\sigma_{e_V}^2 \end{bmatrix}$$

where **G** represents the GRM for the Non-Vx(**N**) and Vx(**V**) individuals and **I** represents the identity matrix. For all analyses, covariances were allowed between traits for the animal genetic effect and litter effect. However, covariances were constrained to 0 for pen effects and residuals since Vx and Non-Vx pigs were allocated to different rooms and no pig had records for both traits.

Correction for multiple testing

Multiple test correction was performed to determine the appropriate significance threshold to interpret GWAS results. First, principal component analysis was used to determine the number of independent tests (i.e. SNPs) per chromosome as the number of principal components required to capture 99.5% of the variation [6, 12] using the *princomp* function of R software [13]. Results are presented in Additional file 1: Table S1. Since the number of individuals used for analyses serves as an upper bound for the number of principal components that can be identified, chromosomes were divided into segments so that for each segment, the number of SNPs was fewer than the total number of individuals (i.e. $n < 369$). The number of independent tests per chromosome segment was then summed to calculate the total number of independent tests across the genome. Unmapped SNPs were not included in this calculation. Bonferroni correction was then applied using the following equation:

$$p\text{-value} = \frac{\alpha}{\text{number of independent tests across the genome}}$$

where $\alpha = 0.10$ and the number of independent tests for the whole genome = 26, 272. The resulting genome-wise significance $-\log_{10}p$-value threshold was 5.4.

Pathway analyses

Pathway analyses were conducted to identify protein pathways that were enriched for genes near SNPs associated with each trait according to the procedure described by Waide et al. [6]. First, lists of SNPs at three $-\log_{10}$ p-value thresholds (2, 2.5, and 3, referred to as SNP lists **T2**, **T2.5**, and **T3**, respectively) for the combined additive and dominance effect (2-df test) of each SNP were generated for each trait. Next, 1-megabase (**Mb**) windows were constructed for each SNP, which spanned 0.5 Mb on either side of each SNP. Then, for each SNP, Ensembl IDs (http://www.ensembl.org) [14] were obtained for all genes contained within or flanking either side of the 1-Mb SNP window. These Ensembl IDs were then used to perform a statistical overrepresentation analysis using the gene ontology (**GO**) Slim feature of the PANTHER software [15]. Compared to using the entire GO term database, GO Slim uses a limited set of GO terms (550 versus 45,237 total annotations) to provide a more general list of protein pathways that map to gene IDs (http://www.pantherdb.org) [16]. A custom background gene list was used, which consisted of all genes that mapped to 0.5 Mb on either side of each of the 61,729 SNPs used for GWAS. *P*-values for statistical overrepresentation of protein pathways were corrected by PANTHER software using the Bonferroni multiple testing correction method. The number of SNPs and

genes corresponding to each SNP list for each trait are shown in Additional file 1: Tables S2 and S3.

QTL test

A second type of analysis, hereafter referred to as the **QTL Test**, was also used to provide evidence for regions associated with each trait in this study. A master list of previously reported QTL for all traits in the pig genome was downloaded from the QTL database for pigs at *animalgenome.org* [17]. In the database, each QTL entry is assigned to one of the following trait categories: exterior, health, meat and carcass, production, and reproduction. In total, 16,032 QTL entries were downloaded. Filtering of the master list was then performed according to the following steps: duplicate QTL entries (i.e. entries for the same trait, location, and publication) were removed, QTL for the same trait from the same study with overlapping positions were concatenated, QTL spanning more than 5 Mb were removed, and all ADG, body weight, and growth-related QTL, which belong to the "production" category, were re-assigned to a new category entitled "growth". After filtering, 9892 unique QTL entries remained.

The filtered master QTL list was then used to construct a list of all QTL that mapped to each SNP. A QTL was mapped to a SNP if the QTL was either entirely contained within or entirely overlapped the 1-Mb SNP window. This list was used to determine the number of health and growth QTL that mapped to SNPs associated with VL and ADG traits, respectively. These lists were then used to assess overrepresentation of health or growth QTL mapping to SNPs for each list.

Using QTL_I as the trait QTL (i.e. QTL for health or growth) mapped to one of the three SNP lists for a trait (i.e. T2, T2.5 or T3), QTL_T as the list of all QTL (of all QTL types) mapped to one of the three SNP lists, and QTL_G as the list of all QTL (for all QTL types) throughout the genome, the null hypothesis (H_0) for this test was that the ratio $P_1 = QTL_I{:}QTL_T$ was equal to the ratio $P_2 = QTL_I{:}QTL_G$. The alternative hypothesis (H_a) was that $P_1 > P_2$.

P-values for the QTL Test were calculated as the probability of identifying QTL_I or more trait QTL for a given SNP list under H_0 using the binomial distribution, as follows:

$$p\text{-value} = \sum_{x=k}^{K} \binom{K}{x} P^x (1-P)^{K-x}$$

where $k = QTL_I$; $K = QTL_T$; and $P = P_2$. Analyses were performed separately for health and growth QTL for each SNP list.

SNP test

A third test, hereafter referred to as the **SNP Test**, was also performed to assess overrepresentation of health or

Genomic regions associated with host response to porcine reproductive and respiratory syndrome...

5

growth QTL, but using slightly different information than the QTL Test. Using SNP_I as the number of unique SNPs within a health or growth QTL (i.e. QTL_I), SNP_T as the number of SNPs in each SNP list (i.e. T2, T2.5, or T3), and SNP_G the number of SNPs across the genome (i.e. 61,729), the H_0 for this test was that $P_3 = SNP_I:SNP_T$ was equal to $P_4 = SNP_I:SNP_G$ (H_0: $P_3 = P_4$). The alternative hypothesis (H_a) was that $P_3 > P_4$. P-values for the SNP Test were also calculated using the above binomial distribution but using $k = SNP_I$, $K = SNP_T$, and $P = P_4$.

Results
GWAS post vaccination
Vaccination viral load
No SNPs were significantly associated with vaccination VL (maximum $-log_{10}P = 4.65$) (Fig. 1a).

Average daily gain
Five total SNPs, located on SSC9 at 52 Mb (i.e. 9_52), 11_53, and from 12_5 to 12_6 (Fig. 1b) were significantly ($-log_{10}P \geq 5.6$) associated with ADG Post Vx for Vx pigs (Table 1). A different set of SNPs, including SNPs located at 15_129 and 5_5 (Fig. 1b) were significantly ($-log_{10}P \geq 5.4$) associated with ADG Post Vx of Non-Vx pigs (Table 1).

Seven SNPs had a significant ($-log_{10}P \geq 5.6$) main effect on ADG Post Vx (Table 2). These SNPs were located at 6_108, 7_81 to 7_82, 17_32, and 18_4 (Fig. 2). One of the only two SNPs with a significant interaction effect identified in this study was also associated with ADG Post Vx ($-log_{10}P = 5.6$) (Fig. 2), and was located at 17_57 (Table 2). However, the effect of this SNP on ADG Post Vx was not significant when analyzed for Vx ($-log_{10}P = 5.1$) or Non-Vx ($-log_{10}P = 1.0$) pigs separately (Fig. 1b).

GWAS post co-infection
Viral load
Genomic regions associated with host response to co-infection are presented in Fig. 3. One SNP, located at 7_20 (Fig. 3b) was significantly ($-log_{10}P = 6.1$) associated with PCV2b VL of Non-Vx pigs (Table 1). No SNPs were significantly associated with PCV2b VL of Vx pigs ($-log_{10}P \leq 5.1$)(Fig. 3b), PRRS VL of Vx pigs ($-log_{10}P \leq 4.3$)(Fig. 3a), or PRRS VL of Non-Vx pigs ($-log_{10}P \leq 4.3$) (Fig. 3a).

SNPs with a significant main effect on host response to co-infection were identified for some traits Post Co-X (Fig. 4). SNPs associated with main and interaction effects of SNP for PRRS VL, PCV2b VL, and ADG are presented in Fig. 4a, b, and c, respectively. Four SNPs had significant ($-log_{10}P \geq 5.5$) main effects on PCV2b VL (Table 2), all of which were located on SSC7 at 19, 20, 32, and 41 Mb (Fig. 4b). Of these SNPs, the effect of SNP H3GA0020199 (7_20) was significant for PCV2b VL of Non-Vx pigs ($-log_{10}P = 6.1$), but not Vx pigs ($-log_{10}P = 4.8$) (Fig. 3b). The interaction effect of SNP genotype by VxStatus was not significant for any SNP for analysis of PCV2b VL ($-log_{10}P \leq 4.4$) (Fig. 3b). No SNPs had a significant ($-log_{10}P \leq 4.6$) main or interaction effect on PRRS VL Post Co-X (Fig. 4a).

Average daily gain
Three SNPs, located at 1_47 and 14_61 to 14_62 (Fig. 3c) were significantly ($-log_{10}P \geq 5.5$) associated with ADG Post Co-X of Vx pigs (Table 1). SNPs at 7_27 and 15_140 (Fig. 3c) were significantly ($-log_{10}P \geq 5.4$) associated with ADG Post Co-X of Non-Vx pigs (Table 1).

Three SNPs had a significant ($-log_{10}P \geq 6.0$) main effect on ADG Post Co-X (Table 2). These SNPs were located at 1_47, 7_24, and 15_140 (Fig. 4c). Of these SNPs, MARC0021766

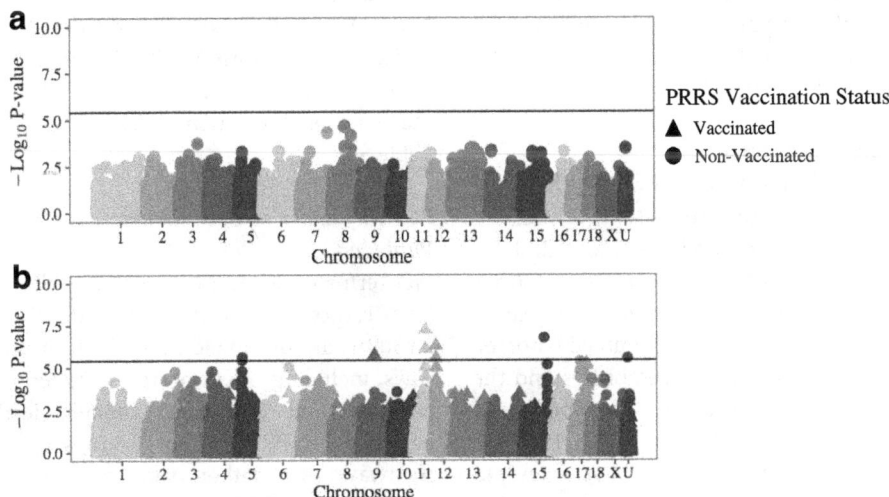

Fig. 1 Significance of GWAS results for vaccination VL (**a**) and ADG (**b**) following vaccination

Table 1 SNPs associated with host response to PRRS vaccination and PRRSV/PCV2b co-infection

Infection Period	Trait	VxStatus[a]	SNP Name	Chromosome	Mb[b]	P-value[c]
Post Vaccination	Vaccination VL[d]	Vx	–	–	–	–
	ADG[e]	Vx	ALGA0052956	9	52	5.7
			WU_10.2_11_53143619	11	53	7.2
			ALGA0062289	11	53	6.1
			ASGA0083776	12	5	6.3
			ISU10000072	12	6	5.6
		Non-Vx	WU_10.2_5_5635354	5	5	5.6
			WU_10.2_5_5693454	5	5	5.4
			MARC0034977	15	129	6.8
Post Co-Infection	PRRS VL	Vx	–	–	–	–
		Non-Vx	–	–	–	–
	PCV2b VL	Vx	–	–	–	–
		Non-Vx	H3GA0020199	7	20	6.1
	ADG	Vx	MARC0021766	1	47	6.9
			ALGA0077929	14	61	5.5
			H3GA0040428	14	62	5.5
		Non-Vx	H3GA0020408	7	27	5.8
			WU_10.2_15_140171163	15	140	5.4

[a]VxStatus, vaccination status: Pigs were either vaccinated (**Vx**) or not (**Non-Vx**) for PRRS prior to co-infection of PRRSV with PCV2b 28 days later
[b]Mb, megabase
[c]P-value: -log$_{10}$p-value
[d]VL: calculated as the area under the curve from −28 to 0, 0 to 21, or 0 to 42 dpi for vaccination VL, PRRS VL, and PCV2b VL, respectively
[e]ADG: calculated as the regression of body weight on dpi post vaccination and post PRRSV/PCV2b co-infection

(1_47) was also significant for ADG Post Co-X of Vx pigs ($-log_{10}P$ = 6.9), but not Non-Vx pigs ($-log_{10}P$ = 0.5) (Fig. 3c). Conversely, WU_10.2_15_140171163 (15_140) was significantly ($-log_{10}P$ = 5.4) associated with ADG Post Co-X for Non-Vx pigs, but not Vx pigs ($-log_{10}P$ = 1.0) (Fig. 3c).

The second SNP (WU_10.2_4_6084304) with a significant ($-log_{10}P$ = 5.5) interaction effect of SNP genotype by VxStatus identified in this study was associated with ADG Post Co-X (Table 2) located at 4_6 (Fig. 4c). However, the effect of this SNP was not significant for ADG Post Co-X when analyzed separately for Vx ($-log_{10}P$ = 5.2) or Non-Vx ($-log_{10}P$ = 0.5) pigs (Fig. 3c).

Pathway analyses post vaccination
For each trait, only GO terms that were significantly (P < 0.05 following Bonferroni correction) associated with genes near SNPs for each SNP list (i.e. T2, T2.5, and T3) are presented. These significant associations include several protein pathways that were significantly underrepresented, which are presented but will not be discussed. The number of mapped Ensembl IDs and the total number of Ensembl IDs (i.e. mapped and unmapped) are presented in Additional file 1: Table S2 for each trait analyzed separately by VxStatus and in Additional file 1: Table S3 for the main and interaction effects of SNP genotype by VxStatus.

Vaccination viral load
GO terms enriched for genes near SNPs for host response to PRRS vaccination are presented in Table 3. Genes near SNPs associated with vaccination VL were enriched for pathways related to cell proliferation, response to stimuli/stress, behavior, cell movement, and cell signaling.

Average daily gain
No pathways were significantly enriched for genes near SNPs associated with ADG of Vx or Non-Vx pigs (Table 3) or for genes near SNPs associated with ADG Post Vx for the main or interaction effects of SNP (Additional file 1: Table S4).

Pathway analyses post co-infection
Viral load
GO terms enriched for genes near SNPs associated with host response to PRRSV and PCV2b co-infection by VxStatus are presented in Table 4. Immune-related pathways, including the G-protein coupled receptor signaling pathway and chromatin organization pathway were enriched for PRRS VL of Vx pigs. None of these same pathways, but rather, sensory-related pathways, were enriched for PRRS VL of Non-Vx pigs. Chromatin organization was enriched for genes near SNPs for

Table 2 SNPs with significant main/interaction effects with vaccination status following PRRS vaccination and PRRSV/PCV2b co-infection

Infection Period	Trait	Effect[a]	SNP Name	Chromosome	Mb[b]	P-value[c]
Post Vaccination	ADG[d]	Main	ALGA0036437	6	108	7.7
			ALGA0107326	7	81	6.1
			MARC0056209	7	81	5.6
			ALGA0042683	7	82	5.6
			WU_10.2_17_32849954	17	32	6.9
			WU_10.2_18_4027654	18	4	6.3
			WU_10.2_18_4233046	18	4	6.1
		Interaction	ASGA0077518	17	57	5.6
Post Co-Infection	PRRS VL[e]	Main	–	–	–	–
		Interaction	–	–	–	–
	PCV2b VL	Main	DRGA0007276	7	19	6.2
			H3GA0020199	7	20	8.7
			ASGA0032282	7	32	5.5
			MARC0060135	7	41	7.3
		Interaction	–	–	–	–
	ADG	Main	MARC0021766	1	47	7.1
			MARC0059955	7	24	7.7
			WU_10.2_15_140171163	15	140	6.0
		Interaction	WU_10.2_4_6084304	4	6	5.5

[a]Effect: The effect of SNP across groups vaccinated, or not, for PRRS (**main**) versus the effect of SNP interacting with vaccination status (**interaction**)
[b]Mb, megabase
[c]P-value: -log$_{10}$ p-value
[d]ADG: calculated as the regression of body weight on dpi post PRRS vaccination and post PRRSV/PCV2b co-infection
[e]VL: calculated as the area under the curve from −28 to 0, 0 to 21, or 0 to 42 dpi for vaccination VL, PRRS VL, and PCV2b VL, respectively

PCV2b VL of Vx and Non-Vx pigs. Additional enriched pathways for Vx pigs included cellular defense response, cell-cell adhesion, and nucleobase-containing compound metabolic process.

Genes near SNPs associated with the interaction effect of SNP genotype by VxStatus were enriched for the response to stimulus and chromatin organization pathways for PRRS VL (Additional file 1: Table S5). No pathways were significantly enriched for SNPs associated with the main effect of SNP for PRRS VL. The chromatin organization, response to interferon (**IFN**)-γ, and primary metabolic process pathways were significantly enriched for genes near SNPs associated with the main effect of SNP for PCV2b VL (Additional file 1: Table S6). Only the transport pathway was enriched for the interaction effect of SNP genotype by VxStatus for PCV2b VL (Additional file 1: Table S6).

Average daily gain

Consistent with results for PRRS VL of Vx pigs and PCV2b VL of both Vx and Non-Vx pigs, genes near SNPs were significantly enriched for chromatin organization for ADG of Vx and Non-Vx pigs Post Co-X (Table 4). Another significantly enriched pathway for ADG Post Co-X of Vx pigs included the cholesterol metabolic process pathway (Table 4).

Only the chromatin organization pathway was significantly enriched for genes near SNPs associated with the main effect of SNP for ADG Post Co-X (Additional file 1: Table S7). This same pathway was enriched for the

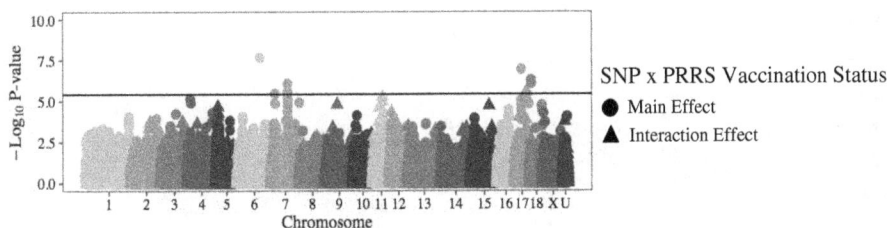

Fig. 2 Significance of GWAS results for main/interaction effects for ADG following vaccination

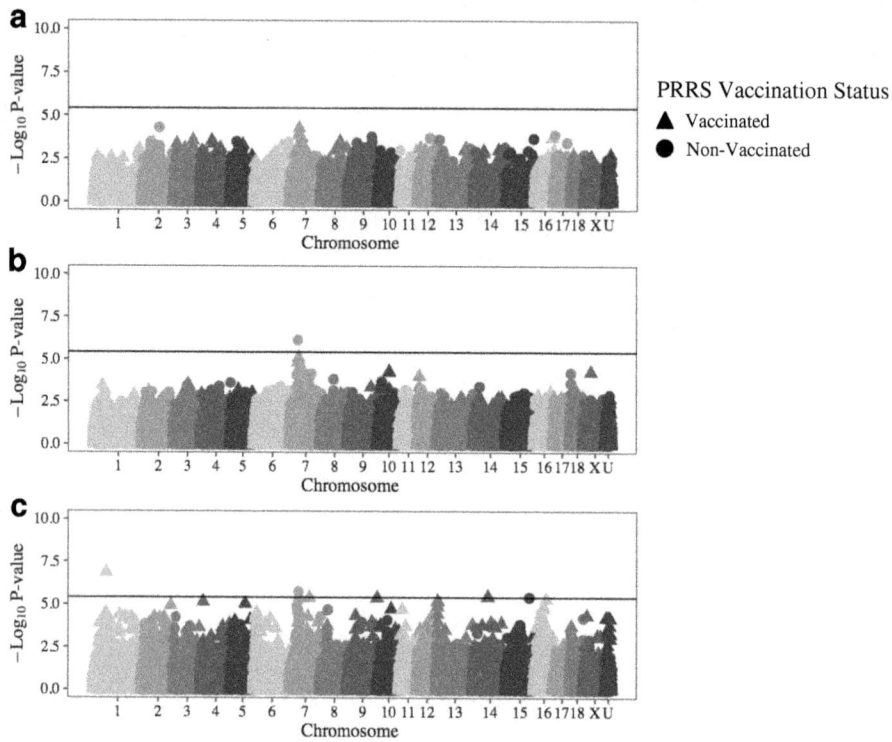

Fig. 3 GWAS results for PRRS VL (**a**), PCV2b VL (**b**), and ADG (**c**) following co-infection

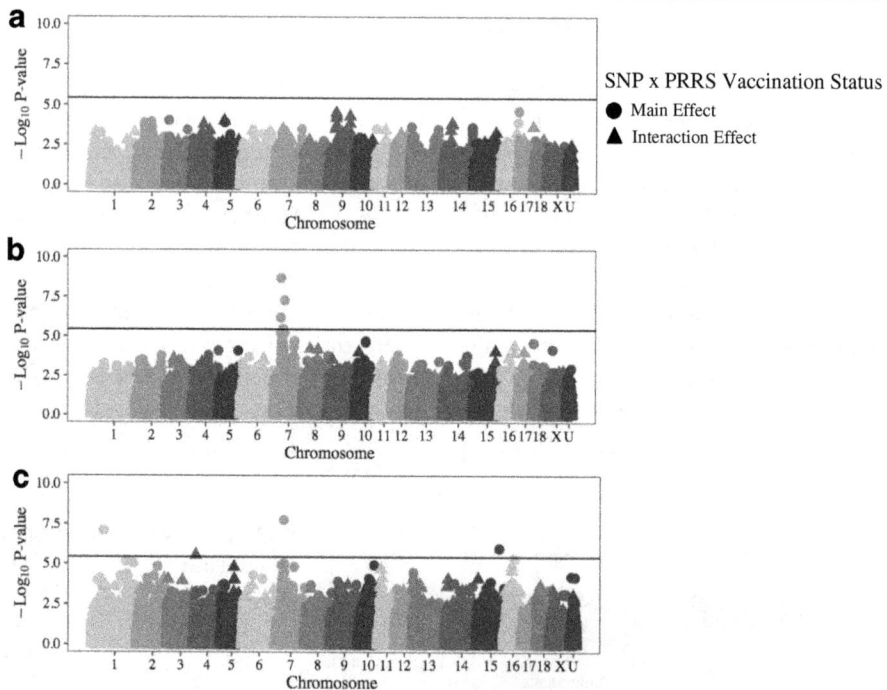

Fig. 4 Main/interaction effects for PRRS VL (**a**), PCV2b VL (**b**), and ADG (**c**) following co-infection

Table 3 Significantly enriched GO terms for genes near SNPs associated with traits following PRRS vaccination

Trait	SNP List[a]	GO term	Fold change	P-value[b]
Vaccination VL	2	Sensory perception of chemical stimulus	0.53	4.9E-2
	2.5	Behavior	7.38	4.0E-3
		Response to biotic stimulus	5.98	2.2E-3
		Cell proliferation	3.55	2.4E-2
	3	Behavior	23.09	7.1E-7
		Response to biotic stimulus	18.72	5.4E-8
		Cell proliferation	11.12	5.7E-8
		Locomotion	9.30	1.6E-4
		Cytokine-mediated signaling pathway	7.07	4.4E-4
		Cell surface receptor signaling pathway	2.29	3.5E-2
		Signal transduction	1.96	2.2E-2
		Response to external stimulus	5.82	2.4E-3
		Cellular component movement	4.87	7.4E-4
		Response to stress	3.40	4.8E-3
ADG Non-Vx[c]	2	Sensory perception of smell	0.53	1.7E-2
	2.5	–	–	–
	3	–	–	–
ADG Vx	2	G-protein coupled receptor signaling pathway	0.61	2.4E-2
		Cell surface receptor signaling pathway	0.68	7.4E-3
		Sensory perception of smell	< 0.20	1.2E-17
		Sensory perception of chemical stimulus	0.40	2.4E-8
		Sensory perception	0.55	2.6E-5
		Neurological system process	0.69	3.0E-3
		System process	0.73	2.1E-2
	2.5	Sensory perception of smell	< 0.20	7.5E-9
		Sensory perception of chemical stimulus	0.45	2.4E-2
	3	Sensory perception of smell	< 0.20	4.4E-3

[a]SNP List: Lists of SNPs from the GWAS with a –$\log_{10}$ p-value above 2, 2.5 or 3
[b]P-value: Bonferroni corrected p-value
[c]Non-Vx, Non-Vaccinated: Pigs were either vaccinated (Vx) or not (Non-Vx) for PRRS

Table 4 Significantly enriched GO terms for genes near SNPs associated with traits following PRRSV/PCV2b co-infection

Trait	SNP List[a]	GO term	Fold change	P-value[b]
PRRS VL Non-Vx[c]	2	Sensory perception of smell	1.67	3.5E-3
	2.5	Sensory perception	1.85	4.3E-2
	3	–	–	–
PRRS VL Vx	2	Cellular protein modification process	0.53	1.2E-2
		Immune response	0.47	4.9E-2
	2.5	Chromatin organization	2.80	2.7E-2
		G-protein coupled receptor signaling pathway	1.82	2.7E-2
	3	–	–	–
PCV2b VL Non-Vx	2	Sensory perception of chemical stimulus	0.50	2.9E-4
		Sensory perception	0.63	2.5E-2
	2.5	Chromatin organization	2.82	2.4E-2
	3	–	–	–
PCV2b VL Vx	2	Cellular defense response	2.61	1.0E-2
		Cell-cell adhesion	2.26	4.7E-3
		Sensory perception of chemical stimulus	0.54	1.4E-2
		Sensory perception	0.61	4.3E-2
	2.5	Cellular defense response	4.38	1.4E-3
		Chromatin organization	3.75	3.2E-4
		Nucleobase-containing compound metabolic process	1.47	3.6E-2
	3	Cellular defense response	8.52	3.3E-4
ADG Non-Vx	2	Chromatin organization	2.07	3.9E-2
	2.5	Chromatin organization	3.14	1.4E-3
	3	Chromatin organization	4.75	8.5E-6
		Nucleobase-containing compound metabolic process	1.53	3.6E-2
		Primary metabolic process	1.35	3.7E-2
ADG Vx	2	Chromatin organization	1.86	2.5E-2
		JAK-STAT cascade	< 0.20	4.1E-2
	2.5	Cholesterol metabolic process	3.54	4.9E-2
	3	Chromatin organization	2.73	1.1E-2
		Sensory perception of chemical stimulus	0.31	9.8E-4

[a]SNP List: Lists of SNPs from the GWAS with a –$\log_{10}$ p-value above 2, 2.5 or 3
[b]P-value: Bonferroni corrected p-value
[c]Non-Vx, Non-Vaccinated: Pigs were either vaccinated (Vx) or not (Non-Vx) for PRRS

interaction effect of SNP genotype by VxStatus, in addition to lipid transport and localization (Additional file 1: Table S7).

QTL test

The QTL Test was performed to assess overrepresentation of health or growth QTL for SNPs associated with VL or ADG, respectively. Results for analyses of each trait by VxStatus are presented in Table 5 and Additional file 1: Table S8 and S9. Results for analyses of the main and interaction effects of SNP for each trait are presented in Tables 6 and 7.

Viral load

Results indicated that SNPs associated with vaccination VL were significantly ($P = 6.5E-14$) overrepresented for health QTL for T2 (Table 5). SNPs associated with the main effect of SNP genotype on PRRS VL were significantly ($P = 4.6E-7$) overrepresented for health QTL for T3 (Table 6). The latter result was driven by significant ($P = 7.9E-7$) overrepresentation of health QTL for SNPs associated with PRRS VL of Vx pigs for T3 (Additional file 1: Table S8). SNPs associated with the interaction effect of SNP genotype by VxStatus on PRRS VL were also significantly ($P = 0.02$) overrepresented for health QTL for T2 (Table 6). This result was driven by significant overrepresentation of health QTL for SNPs associated with PRRS VL of both Vx ($P = 6.7E-12$) and Non-Vx ($P = 3.3E-18$) pigs for T2 (Additional file 1: Table S8).

All three lists of SNPs associated with the main effect of SNP on PCV2b VL showed significant ($P = 0.03$, $2.3E-8$, and $1.5E-8$) overrepresentation of health QTL (Table 6), driven by significant overrepresentation of health QTL for all three lists of SNPs for PCV2b VL for Vx ($P = 6.7E-5$ to $6.0E-4$) and Non-Vx ($P = 1.7E-9$ to $4.0E-4$) pigs (Additional file 1: Table S8).

Average daily gain post vaccination

Results indicated little evidence of overrepresentation of growth QTL for SNPs associated with ADG Post Vx of Vx ($P \geq 0.15$) or Non-Vx ($P \geq 0.04$) pigs (Additional file 1: Table S9). However, SNPs associated with the main effect of SNP on ADG Post Vx were significantly overrepresented for growth QTL for T2.5 ($P = 0.01$) and T3 ($P = 0.03$) (Table 7).

Average daily gain post co-infection

SNPs associated with ADG Post Co-X were not significantly overrepresented for growth QTL for Vx ($P \geq 0.36$) or Non-Vx ($P \geq 0.74$) pigs (Additional file 1: Table S9), or for the main ($P \geq 0.32$) or interaction ($P \geq 0.12$) effects of SNP by VxStatus (Table 7).

SNP test

The SNP Test was conducted to assess overrepresentation of health or growth QTL based on the number of unique SNPs mapping to trait QTL. Results for analysis of each trait by VxStatus are presented in Table 8 and Additional file 1: Table S10 and S11 and results for the main and interaction effects of SNP by VxStatus are presented in Tables 9 and 10.

Viral load

Consistent with results for the QTL Test, SNPs associated with vaccination VL were significantly ($P = 2.2E-5$) overrepresented for health QTL for T2 (Table 8). A tendency ($P = 0.06$) for significant overrepresentation of health QTL was also detected for T2.5, but not T3 ($P = 0.26$) (Table 8). For analysis of the main and interaction effects of SNP genotype by VxStatus on PRRS and PCV2b VL, the same SNP lists that showed significant overrepresentation of health QTL for the QTL Test also showed significant overrepresentation of health QTL for the SNP Test (Table 9). An exception was for SNPs associated with the main effect of SNP genotype on PRRS VL for T3, which was not significant ($P = 0.88$) for the SNP Test (Table 9). SNPs associated with PRRS VL of Vx pigs, PCV2b VL of Non-Vx pigs, and PCV2b VL of Vx pigs were significantly ($P = 7.0E-12$ to $7.0E-3$; $P = 1.1E-4$ to 0.02; $P = 0.04$ to 0.01) overrepresented for health QTL (Additional file 1: Table S10). One exception was T3 for PCV2b VL of Vx pigs, which did not show significant ($P = 0.26$) overrepresentation of health QTL (Additional file 1: Table S10).

Table 5 QTL Test results for SNPs associated with vaccination VL

Trait	SNP List[a]	# SNPs above threshold	# health QTL in region[b]	Total # QTL in region[c]	P-value
Vaccination VL	2	392	290	1092	6.5E-14
	2.5	124	37	367	0.99
	3	51	16	100	0.69

[a]SNP List: Lists of SNPs from the GWAS with a $-log_{10}p$-value above 2, 2.5 or 3
[b]The total number of health QTL in the genome (after filtering) is 1732
[c]The total number of QTL in the genome (i.e. across all QTL types and after filtering) is 9892

Table 6 QTL Test results for the main/interaction effects of SNPs associated with PRRS and PCV2b VL

Trait	Effect[a]	SNP List[b]	# of SNPs above threshold	# of health QTL in region[c]	Total # of QTL in region[d]	P-value
PRRS VL	Main	2	528	189	1046	0.33
		2.5	168	64	382	0.67
		3	37	34	83	4.6E-7
	Interaction	2	516	246	1239	0.02
		2.5	159	79	442	0.44
		3	42	2	97	0.99
PCV2b VL	Main	2	823	372	1943	0.03
		2.5	350	260	1076	2.3E-8
		3	160	162	609	1.5E-8
	Interaction	2	520	182	1140	0.92
		2.5	156	42	399	0.99
		3	44	9	126	0.99

[a]Effect: The effect of SNP across groups vaccinated, or not, for PRRS (main) or the effect of SNP interacting with PRRS vaccination status (interaction)
[b]SNP List: Lists of SNPs from the GWAS with a −log$_{10}$ p-value above 2, 2.5 or 3
[c]The total number of health QTL in the genome (after filtering) is 1732
[d]The total number of QTL in the genome (i.e. across all QTL types and after filtering) is 9892

Average daily gain post vaccination

SNPs associated with the main effect of SNP on ADG Post Vx showed significant overrepresentation of growth QTL for T2 ($P = 0.01$) and T2.5 ($P = 7.1E-6$)(Table 10). SNPs associated with the interaction effect of SNP genotype by VxStatus on ADG Post Vx also showed significant overrepresentation of growth QTL for T2.5 ($P = 0.05$) and T3 ($P = 2.8E-6$) (Table 10). For T2, overrepresentation of growth QTL for SNPs associated with the main effect of SNP was driven by significant overrepresentation of growth QTL for Non-Vx pigs for T2 ($P = 2.0E-4$)(Additional file 1: Table S11). Overrepresentation of growth QTL for SNPs associated with the interaction effect of SNP was driven by significant overrepresentation of growth QTL for Non-Vx pigs for T3 ($P = 8.7E-6$) (Additional file 1: Table S11).

Average daily gain post co-infection

SNPs associated with the main effect ($P \geq 0.11$) and interaction effect ($P \geq 0.44$) of SNP on ADG Post Co-X were not significantly overrepresented for any of the SNP lists (Table 10). Similarly, growth QTL were not significantly overrepresented for SNPs associated with ADG of Vx ($P \geq 0.29$) or Non-Vx ($P \geq 0.53$) pigs Post Co-X (Additional file 1: Table S11).

Table 7 QTL Test results for the main/interaction effects of SNPs associated with ADG

Infection Period	Effect[a]	SNP List[b]	# of SNPs above threshold	# of growth QTL in region[c]	Total # QTL in region[d]	P-value
Post Vaccination	Main	2	902	85	1633	0.32
		2.5	365	45	643	0.01
		3	141	23	300	0.03
	Interaction	2	687	62	1323	0.68
		2.5	251	20	443	0.69
		3	110	11	230	0.58
Post Co-Infection	Main	2	870	107	2069	0.32
		2.5	364	55	1065	0.38
		3	176	23	502	0.67
	Interaction	2	682	77	1519	0.42
		2.5	291	29	536	0.33
		3	120	20	305	0.12

[a]Effect: The effect of SNP across groups vaccinated, or not, for PRRS (main) or the effect of SNP interacting with PRRS vaccination status (interaction)
[b]SNP List: Lists of SNPs from the GWAS with a −log$_{10}$ p-value above 2, 2.5 or 3
[c]The total number of growth QTL in the genome (after filtering) is 488
[d]The total number of QTL in the genome (i.e. across all QTL types after filtering) is 9892

Table 8 SNP Test results for SNPs associated with PRRS vaccination VL

Trait	SNP List[a]	# SNPs mapping to health QTL in SNP list[b]	# SNPs in SNP list[c]	P-value
Vaccination VL	2	125	392	2.2E-5
	2.5	36	124	0.06
	3	14	51	0.26

[a]SNP List: Lists of SNPs from the GWAS with a $-\log_{10}$ p-value above 2, 2.5 or 3
[b]The number of unique SNPs mapping to health QTL in the genome is 14,063
[c]The total number of SNPs used for the GWAS was 61,729

Discussion

This is the first study to identify genomic regions (other than WUR) associated with host response to PRRS MLV vaccination and co-infection with PRRSV and PCV2b. Significant regions were detected for PCV2b VL, ADG Post Vx, and ADG Post Co-X and results from functional annotation analyses provided biological evidence that supported these statistically associated regions. Results of the functional annotation analyses also supported the many regions with small effects that were detected. Multiple SNPs on SSC7 were significantly associated with PCV2b VL, regardless of prior vaccination for PRRS. However, regions with a significantly different effect on ADG, depending on prior vaccination for PRRS, were detected for ADG Post Vx and Post Co-X. These findings indicate that multiple genomic regions have the potential to be used to select pigs for decreased PCV2b VL following PRRSV/PCV2b co-infection, regardless of prior vaccination for PRRS, but the same is not true for ADG.

This study also introduced a novel approach of using previously-reported QTL to provide evidence for statistically-associated regions from GWAS. Other studies have assessed clustering of health QTL throughout the rice genome [18] and the gene density of QTL regions compared to the rest of the genome in cattle [19], but this is the first study to use a catalog of previously-reported QTL to assess overrepresentation of a QTL category as a means of providing evidence for regions identified from GWAS.

The WUR SNP, a genetic marker for a major QTL for PRRS on SSC4, was identified in a previous study. Results showed that WUR was associated with 15.7% and 11.2% of genetic variation in PRRS VL and weight gain (**WG**) under PRRSV-only infection, respectively [1]. Follow-up studies identified the putative causative mutation in *GBP5* [2], which has been shown to play a role in the innate immune response to infection in mice [20]. In recent years, additional studies have validated the effect of WUR on host response to PRRS using multiple breeds, populations, and following infection with two different PRRSV isolates [3–6]. Results from these studies showed that WUR had a significant effect on PRRS VL following infection with both the NVSL 97-7985 (**NVSL**) and KS2006-72109 (**KS06**) PRRSV isolates, but not WG following infection with KS06 [5]. Authors suggested that this non-significant effect was related to reduced virulence of the KS06 versus the NVSL strain. Taken together, results from these studies showed that WUR had a significant effect on VL with different isolates of PRRSV and for different genetic backgrounds.

Table 9 SNP Test results for the main/interaction effects of SNPs associated with PRRS and PCV2b VL

Trait	Effect[a]	SNP List[b]	# SNPs mapping to health QTL in SNP list[c]	# SNPs in SNP list[d]	P-value
PRRS VL	Main	2	120	528	0.53
		2.5	27	168	0.99
		3	6	37	0.88
	Interaction	2	155	516	8.4E-5
		2.5	45	159	0.06
		3	3	42	0.99
PCV2b VL	Main	2	221	823	3.0E-3
		2.5	105	350	1.0E-3
		3	52	160	3.0E-3
	Interaction	2	110	520	0.83
		2.5	33	156	0.72
		3	8	44	0.82

[a]Effect: The effect of SNP across groups vaccinated, or not, for PRRS (main) or the effect of SNP interacting with PRRS vaccination status (interaction)
[b]SNP List: Lists of SNPs from the GWAS with a $-\log_{10}$ p-value above 2, 2.5 or 3
[c]The number of unique SNPs mapping to health QTL in the genome is 14,063
[d]The total number of SNPs used for the GWAS was 61,729

Table 10 SNP Test results for the main/interaction effects of SNPs associated with ADG

Infection Period	Effect[a]	SNP List[b]	# SNPs mapping to growth QTL in SNP list[c]	# SNPs in SNP list[d]	P-value
Post Vaccination	Main	2	161	902	0.01
		2.5	87	365	7.1E-6
		3	26	141	0.16
	Interaction	2	103	687	0.53
		2.5	48	251	0.05
		3	36	110	2.8E-6
Post Co-Infection	Main	2	126	870	0.70
		2.5	62	364	0.16
		3	33	176	0.11
	Interaction	2	87	682	0.96
		2.5	33	291	0.97
		3	19	120	0.44

[a]Effect: The effect of SNP across groups vaccinated, or not, for PRRS (main) or the effect of SNP interacting with PRRS vaccination status (interaction)
[b]SNP List: Lists of SNPs from the GWAS with a $-\log_{10}$ p-value above 2, 2.5 or 3
[c]The number of unique SNPs mapping to growth QTL in the genome is 9294
[d]The total number of SNPs used for the GWAS was 61,729

A natural follow-up question to these studies was whether WUR also has a significant effect on host response to PRRSV-infection following co-infection with another pathogen. This is a practical question since PRRSV is known to suppress the host immune response, making pigs more susceptible to secondary infections [21]. Co-infection with PRRSV and PCV2b was used as a co-infection model to address this question, given the ubiquitous nature of PCV2b [22, 23], prevalence of PRRSV/PCV2b co-infections in the field (when pigs are not vaccinated for PCV2), and previous experience working with and conducting PCV2b-experimental infection trials. An additional objective was to estimate the effect of WUR on host response to PRRS MLV vaccination since commercial PRRS MLV vaccines are widely used [24, 25]. Although PRRS MLV vaccines are attenuated, modified live virus replicates in pigs post-vaccination.

Results from Dunkelberger et al. [7] showed that WUR had a significant effect on vaccination VL as well as PRRS VL and PCV2b VL of Vx pigs, but not PCV2b VL of Non-Vx pigs. Results also showed that, numerically, the effect of WUR on PRRS VL was greater following primary versus secondary PRRSV exposure, which is consistent with the biological role of GBP5. Using these same data, the first objective of the current study was to identify genomic regions other than WUR associated with host response to PRRS MLV vaccination and PRRSV/PCV2b co-infection.

Genome-wide association studies
Viral load
When Vx and Non-Vx pigs were analyzed separately, H3GA0020199, located at SSC7_20, was the only SNP

significantly associated with PCV2b VL, and only in Non-Vx pigs. This SNP is located within the gene *KIAA0319*, which has a known role in neuronal growth and migration in humans [26]. This SNP, in addition to several other SNPs located on SSC7 (at 7_19, 7_32, and 7_41) were significantly associated with the main effect of SNP genotype on PCV2b VL. Interestingly, these SNPs are in the vicinity of the swine leukocyte antigens (**SLA**) complex (7_23 to 7_31), one of the most gene-dense regions of the genome which is known to harbor many genes associated with the immune response [27, 28].

A SNP within the SLA was also associated with host response to PCV2b infection in a previous study. This SNP, (referred to as SNP1; name not published) located at 7_28 (SLA Class III) was associated with host response to experimental infection with PCV2b in commercial crossbred pigs [8]. Although the effect of SNP1 did not reach genome-wise significance for the current study, when the effect of SNP1 was fitted as a fixed effect with the effect of other candidates SNPs for PRRS and PCV2b simultaneously, the AA genotype for SNP1 was associated with significantly greater ADG and numerically lower PCV2b VL for Non-Vx pigs [7]. The direction of these effects are consistent with previously reported results, except that Engle et al. [8] did not detect a significant association of SNP1 with ADG post-infection.

There is also evidence of associations of SNPs within the SLA region with host response to PRRSV-infection from other studies, including experimental PRRSV-infection [29] and following a natural PRRSV outbreak [30]. Results from the study conducted by Hess [29] showed that SNPs at 7_26 (SLA Class I) and 7_29 (SLA Class II) were associated with 10-45% of the total genetic variation for PRRS antibody response, depending on infection with the NVSL versus KS06 PRRSV isolate. For the study conducted by Serão et al. [30], SNPs spanning 24 to 30 Mb on SSC7 (harboring SLA Classes I, II, and III) jointly explained 25% of the total genetic variation in antibody response following a natural outbreak in a commercial herd of gestating females.

Despite these sizeable associations of the SLA region with PRRS antibody response, no SNPs within this region, or any other region of the genome, were significantly associated with vaccination VL or PRRS VL for the current study. In general, the lack of significant associations for vaccination VL and PRRS VL is consistent with the conclusion reported by Waide et al. [6] that genomic regions other than WUR explained little to no genetic variation in PRRS VL following experimental infection with the NVSL or KS06 PRRSV isolate. One exception was the 7_30 Mb window, located within the SLA Class II region, which was associated with a small percentage (0.32%) of the total genetic variation in PRRS VL following infection with the KS06 PRRSV isolate [6]. Although not genome-wise significant, a SNP within this same window (ASGA0032151) was the second-most significant SNP $(-log_{10}P = 4.25)$ associated with PRRS VL in Vx pigs for the current study.

Average daily gain post vaccination

Contrary to analysis of vaccination VL and PRRS VL, we identified multiple SNPs that were significantly associated with ADG Post Vx and Post Co-X. This finding conflicts with results reported by Waide et al. [6] who reported no significant associations with WG post PRRSV-only infection, other than the WUR region.

Most genomic regions associated with ADG Post Vx in the current study were also associated with ADG in previous studies. However, for all but one of the identified regions, associations with ADG in other studies were in a disease-free environment. For example, the SSC15_129 region associated with ADG of Non-Vx pigs Post Vx, was also associated with ADG in non-challenged pigs by Rückert and Bennewitz [31]. Regions associated with ADG Post Vx of Vx pigs at SSC9_52 (ALGA0052956) and SSC11_53 (WU_10.2_11_53143619 and ALGA0062289) were also associated with ADG in other studies [31, 32]. A region associated with ADG of Non-Vx pigs (WU_10.2_5_5635354) and ADG of Vx pigs Post Vx (WU_10.2_5_5693454), located at 5_5, was the only region associated with ADG Post Vx that was not associated with ADG in a previous study. Therefore, this region may represent a novel QTL for ADG. Interestingly, this SNP (WU_10.2_5_5693454) is located within the growth factor receptor-bound protein 2 (**GRB2**)-reltated adaptor protein 2 gene in pigs, also known as GRB2-related adaptor downstream of Shc

(**GADS**). In humans, GADS is involved in the formation of the T cell receptor complex [33] and has a known role in T cell development and signaling [34]. However, there was no evidence that this SNP was significantly associated with any of the VL traits.

None of the SNPs that were significantly associated with ADG of Non-Vx pigs were also significantly associated with ADG of Vx pigs. However, several other SNPs had a significant effect on ADG Post Vx, regardless of vaccination status, which agrees with the high, positive (0.92 ± 0.92) genetic correlation identified between ADG of Vx and Non-Vx pigs Post Vx reported in our previous study using these same data [7]. These SNPs, with significant main effects, were located at 6_108, 7_10, 7_81, 7_82, and 18_4, all of which were associated with ADG in previous studies [35–38]. Given the significant main effect detected for these SNPs, these five SNPs may be used to select pigs for improved host response to PRRSV/PCV2b co-infection, regardless of prior vaccination for PRRS. Interestingly, the SNP located on SSC6 (ALGA0036437) is located within the Ring Finger Protein 125 (**RNF125**) gene, which is a negative regulator of the RIG-1 like receptor signaling pathway [39]. The RIG-1 like receptor is a known recognition receptor of RNA viruses. Activated RIG-1 like protein signals for the production of cytokines, including type I IFN in the innate immune response pathway [39]. Suggestive evidence of a significant $(-log_{10}P = 3.16)$ association of this SNP with the main effect of SNP genotype on PRRS VL was also detected.

The effect of ASGA0077518 (17_57) on ADG Post Vx depended on prior vaccination for PRRS, where the effect of this SNP on ADG was near genome-wise significance for Vx pigs, but not Non-Vx pigs. The nearest QTL for ADG reported in a previous study spans 17_64 to 17_66 [40]. During the vaccination period, ADG of Vx pigs represented growth under PRRSV-infection (albeit a modified PRRSV infection), while ADG of Non-Vx pigs represented growth under non-challenged conditions. Therefore, the significant interaction identified for ASGA0077518 indicates that this SNP had a significantly different effect on growth rate Post Vx, depending on PRRS vaccination status.

Average daily gain post co-infection

Similar to analysis of ADG Post Vx, several SNPs were also significantly associated with ADG Post Co-X, including the only SNP that was previously associated with growth under disease challenged conditions in a separate study [3]. This SNP, H3GA0020408, is located in the SLA region (7_27) and was associated with ADG of Non-Vx pigs Post Co-X. This same 1-Mb window was also associated with WG following PRRSV-only infection for analysis of PRRS Host Genetics Consortium trials 1-8 [3].

Other significant SNPs associated with ADG Post Co-X included a SNP (WU_10.2_15_140171163) located at 15_140. This same region was associated with ADG of non-challenged pigs in a previous study [41]. MARC0021766, located at 1_47, was associated with ADG of Vx pigs and this same region was also associated with ADG in a previous study [31]. Associations of SNPs located on SSC14 (H3GA0040428 and ALGA0077929) at 61 and 62 Mb, respectively, with ADG of Vx pigs have not been previously reported and may represent novel QTL for ADG under challenged conditions. No candidate genes were identified for H3GA0040428, but ALGA0077929 is located within the Pecanex Homolog 2 (*PCNXL2)* gene. In humans, *PCNXL2* has been associated with susceptibility to colorectal cancer [42].

Similar to results for ADG Post Vx, none of the SNPs that were significantly associated with ADG of Non-Vx pigs were also significantly associated with ADG of Vx pigs Post Co-X. However, several other SNPs had a significant effect on ADG regardless of prior vaccination for PRRS, which agrees with our previous finding of a moderate to high (0.75 ± 0.37) genetic correlation between ADG of Vx and Non-Vx pigs [7]. These SNPs included MARC0021766 on SSC1 (significant for Vx pigs) and WU_10.2_15_140171163 on SSC15 (significant for Non-Vx pigs). Therefore, MARC0021766 and WU_10.2_15_140171163 can be used to select pigs for improved host response to PRRSV/PCV2b co-infection, regardless of prior vaccination for PRRS. WU_10.2_4_6084304, located at 4_6, was also associated with ADG in a previous study [38].

Protein pathway analyses
In addition to identifying genomic regions associated with host response to PRRS vaccination and PRRSV and PCV2b co-infection, another objective of this study was to provide biological evidence for regions identified from GWAS. PANTHER software was used to test for enrichment of genes near SNPs associated with each trait.

Viral load
Pathways enriched in genes near SNPs associated with vaccination VL included cell proliferation, cell movement, cell signaling, and cytokine signaling. Enrichment of these pathways is consistent with the literature regarding PRRS MLV vaccination response. Compared to infection with a field isolate of PRRSV, PRRS MLV vaccination is known to result in a delayed humoral and cell-mediated immune response [43]. Cell-mediated immunity is often characterized by lymphocyte proliferation and increased cytokine production, mainly

production of IFN-γ [44, 45]. The pathways "cell proliferation" and "cell signaling" reflect these processes.

Consistent with results reported by Waide et al. [6], the G-protein coupled receptor signaling pathway was enriched for genes near SNPs associated with PRRS VL of Vx pigs. Enrichment of this pathway is an interesting result since this protein receptor class plays a role in T cell immunity, including T cell migration by regulating chemotaxis [46]. Results from a previous study also showed that AA and AB pigs differ regarding utility of the G-protein coupled receptor pathway. Pigs with the AA genotype showed extended expression of this pathway (which is essential for the phosphoinositide 3 kinase pathway which is related to PRRS virus entry) compared to AB pigs, thereby providing greater opportunity for PRRSV entry into host cells [47]. No pathways were enriched in genes near SNPs associated with the main effect of SNP genotype for PRRS VL, which is consistent with GWAS results for this trait.

Immune-related pathways were also enriched for genes near SNPs associated with PCV2b VL, including but not limited to: cellular defense response, cell-to-cell adhesion, and nucleobase metabolic processes for Vx pigs only, and the chromatin organization pathway for both Vx and Non-Vx pigs. This result agrees with the finding that several SNPs near the SLA region were significant for the main effects of SNP on PCV2b VL. It is also consistent with the high, positive genetic correlation between PCV2b VL of Vx and Non-Vx pigs (0.99 ± 0.94) reported for our previous analyses of these same data [7]. Based on this estimate, it is expected that the same genomic regions, and therefore the same protein pathways, are associated with PCV2b VL Post Co-X, regardless of previous vaccination for PRRS.

For PCV2b VL, genes near SNPs associated with the main effect of SNP were enriched for primary metabolic process and response to IFN-γ. The "primary metabolic process" pathway was previously found to be associated with PRRS VL following infection with the KS06 and NVSL PRRSV isolates [6]. This pathway likely reflects the need for metabolizable energy to mount and maintain an immune response [48]. Enrichment of genes in the "response to IFN-γ" pathway is particularly interesting, given the well-known role of IFN-γ in response to viral infection. There is evidence that IFN-γ production increases with increasing replication of PCV2b [49] and that PCV2b replication increases in pigs that are already infected with PRRSV [50, 51].

Average daily gain
Results of protein pathways analyses showed enrichment of metabolism-related pathways for genes near SNPs associated with ADG of Vx and Non-Vx pigs Post Co-X. Lipid transport was also enriched for genes near SNPs associated with the interaction effect of SNP genotype

by VxStatus on ADG Post Co-X. Metabolic processes are clearly associated with growth rate and such pathways were also enriched for WG following PRRSV-only infection in a previous study [6].

For the protein pathway analyses, it is important to note that all analyses were performed using the 10.2 build of the swine genome. Therefore, repeating these analyses using the newer (11.1) build could change results for any regions previously misassembled and/or regions for which additional annotation information is now available.

QTL and SNP tests

The QTL and SNP Tests proposed in this study were then used to provide another piece of evidence for regions identified from GWAS. A significant result for the QTL Test indicated that SNPs identified from GWAS mapped to significantly more unique trait QTL (i.e. health or growth QTL) reported in previous studies than expected by chance. The SNP Test was designed to answer a similar question, where the objective was to assess whether the proportion of SNPs within a SNP list that mapped to health or growth QTL was greater than expected by chance. In general, similar results were obtained for the QTL and SNP Tests for analyses of each trait.

Viral load

Results of the QTL Test for the main and interaction effects of SNP for PRRS and PCV2b VL were consistent with results of the SNP Test, except that for T3, the main effect of SNP was not significant for SNPs associated with PRRS VL for the SNP Test. When PRRS and PCV2b VL were analyzed separately by vaccination status, every SNP list showed significant overrepresentation of health QTL, except for T3 for PRRS VL of Non-Vx pigs. These results were consistent with those obtained for the SNP Test, except that none of the SNP lists were significant for analysis of PRRS VL of Non-Vx pigs. This was driven by the fact that, compared to PRRS VL of Vx pigs, similar numbers of SNPs were identified for each SNP list, but fewer SNPs mapped to health QTL for Non-Vx versus Vx pigs.

Collectively, results of these tests indicate that regions associated with PRRSV and PCV2b VL showed significant overrepresentation for health QTL identified from previous studies, especially when Vx and Non-Vx pigs were analyzed separately. Results of the SNP Test show that mapping of SNPs to health QTL was non-random. Non-significant results obtained for the main/interaction effects of SNP likely reflect noise since relaxed $-\log_{10}p$-value thresholds were used to construct the SNP lists, thereby including SNPs with smaller effects on the trait of interest. Non-significant results for the main/

interaction effects of SNP on PRRS VL and for the interaction effect for PCV2b VL indeed appear to be consistent with GWAS results for these traits, for which no SNPs reached genome-wise significance for these effects.

Average daily gain

Few significant results were obtained for the QTL Test for ADG. Only two SNP lists, both for the main effect of SNP on ADG Post Vx, showed a significant result. Overrepresentation of growth QTL was not detected for any of the SNP lists for SNPs associated with ADG Post Co-X, except for T3 for ADG of Non-Vx pigs. An additional SNP list (T2) showed significant overrepresentation of growth QTL for ADG of Non-Vx pigs Post Vx for the SNP Test. Possibly, significant results for the QTL and SNP Tests were obtained for Non-Vx pigs Post Vx only because this trait represents growth in a healthy environment, which is the same condition under which growth traits were measured for the majority of previously-identified growth QTL. Therefore, non-significant results for analysis of ADG of Vx pigs Post Vx or ADG Post Co-X might indicate the identification of novel QTL for ADG, and/or QTL specific to growth under challenge.

Collectively, results of the QTL and SNP Tests for SNPs associated with ADG indicate significant overrepresentation of growth QTL for growth rate under non-challenged conditions, but not under challenged conditions. For growth under challenged conditions, significant results for the SNP Test also indicate non-random associations of SNPs mapped to growth QTL.

The QTL Test and SNP Test used in this study are novel approaches used to provide evidence for statistically associated regions from GWAS. Although these tests have shown to provide valuable evidence for regions identified in this study, it is also important to note that results obtained from these tests are contingent on several factors. For example, several necessary steps for preparing the filtered QTL list are subject to modification, including the criteria used to determine a "unique" QTL entry and the length of the QTL interval used to retain unique entries. In addition, a limiting factor of the QTL and SNP Tests is that it is only possible to test for significant overrepresentation of QTL using QTL that have already been identified. Therefore, a non-significant result merely implies that QTL of a particular category are not significantly overrepresented for a subset of SNPs based on the current catalog of QTL for the pig genome. Consequently, non-significant results are obtained for SNPs mapping to undiscovered QTL.

Other possible limitations are that results are also subject to the window size used to identify genes mapping to significant SNPs, as well as the significance thresholds used to construct the SNP lists. For the current study, 1-

Mb SNP windows were used to allow for the possibility of trans-acting QTL and SNP list thresholds were selected to be consistent with the procedure described by Waide et al. [6], as previously mentioned. These thresholds were purposefully selected to include SNPs that did not reach genome-wise significance in order to capture the effects of all SNPs affecting the trait of interest, including those with small effects. This was important because disease-related traits are considered complex traits, and are therefore assumed to be controlled by many loci with small effects [52].

Conclusions

In conclusion, other than WUR, several other genomic regions were associated with host response to PRRS MLV vaccination and co-infection with PRRSV and PCV2b, but in general, host response was highly polygenic. Multiple SNPs near the SLA region were associated with PCV2b VL, regardless of previous vaccination for PRRS. Several regions associated with ADG Post Vx and Post Co-X were also identified, but SNPs with a significantly different effect on ADG, depending on vaccination status, were identified for ADG during both periods. Taken together, results indicate that multiple SNPs near the SLA region have the potential to be used to select pigs for decreased PCV2b VL following PRRSV/PCV2b co-infection, but that different SNPs were associated with ADG following PRRS vaccination and PRRSV/PCV2b co-infection, depending on previous vaccination for PRRS.

Results from the protein pathway enrichment analyses supported GWAS results, showing that immune-related pathways were enriched for genes near SNPs associated with vaccination VL, PRRS VL, and PCV2b VL and that metabolic pathways were enriched for genes near SNPs associated with ADG. Results of the QTL and SNP Tests provided additional evidence for the identified regions and similar results were obtained for both tests. Results showed that SNPs associated with vaccination VL, PRRS VL, and PCV2b VL were significantly overrepresented for health QTL when Vx and Non-Vx pigs were analyzed separately and results of the SNP Test showed that mapping of SNPs to health QTL was non-random. Results for ADG showed that, for Non-Vx pigs prior to co-infection, SNPs associated with ADG were significantly overrepresented for growth QTL and results of the SNP Test showed that mapping of SNPs to growth QTL was non-random. These findings likely reflect the fact that most QTL used for these tests were associated with growth under non-challenged conditions.

Collectively, results of functional annotation analyses provide valuable insight regarding the biological pathways underlying host response to PRRS MLV vaccination and PRRSV/PCV2b co-infection, biological evidence for regions statistically associated with the traits of interest, and a means of summarizing GWAS results for complex traits, including host response to disease challenge.

Additional file

Additional file 1: Table S1. Number of independent principal components per chromosome required to capture 99.5% of the genetic variation. **Table S2.** Number of GO terms corresponding to lists of SNPs associated with each trait. **Table S3.** Number of GO terms corresponding to SNP lists associated with the main/interaction effect of SNP. **Table S4.** Significantly enriched GO terms for genes near SNPs associated with ADG following PRRS vaccination. **Table S5.** Significantly enriched GO terms for genes near SNPs associated with PRRS VL following PRRSV/PCV2b co-infection. **Table S6.** Significantly enriched GO terms for genes near SNPs associated with PCV2b VL following PRRSV/PCV2b co-infection. **Table S7.** Significantly enriched GO terms for genes near SNPs associated with ADG following PRRSV/PCV2b co-infection. **Table S8.** QTL Test results for SNPs associated with PRRS and PCV2b VL. **Table S9.** QTL Test results for SNPs associated with ADG following PRRS vaccination and PRRSV/PCV2b co-infection. **Table S10.** SNP Test results for SNPs associated with PRRS and PCV2b VL. **Table S11.** SNP Test results for SNPs associated with ADG following PRRS vaccination and PRRSV/PCV2b co-infection. (DOCX 74 kb)

Abbreviations

ADG: Average daily gain; Dpi: Days post-infection; GADS: GRB2-related adaptor downstream of Shc; *GBP5*: Guanylate binding protein 5; GO: Gene ontology; GRB2: growth factor receptor-bound protein 2; GRM: Genomic relationship matrix; IFN: Interferon; Interaction effect: The interaction of SNP genotype with vaccination status; KS06: KS2006-72109; Main effect: The effect of each SNP averaged across vaccination status; Mb: Megabase; Non-Vx: Non-vaccinated; NVSL: NVSL 97-7985.; *PCNXL2*: Pecanex Homolog 2; PCV2b: Porcine circovirus type 2b; Post Co-X: Post co-infection; Post Vx: Post vaccination; PRRS: Porcine reproductive and respiratory syndrome; PRRSV: PRRS virus; QTL: Quantitative trait locus; *RNF125*: Ring Finger Protein 125; SLA: Swine leukocyte antigens; SNP: Single nucleotide polymorphism; SSC: *Sus scrofa* chromosome; T2: Lists of SNPs with a $-\log_{10}$p-value ≥ 2; T2.5: Lists of SNPs with a $-\log_{10}$ p-value ≥ 2.5; T3: Lists of SNPs with a $-\log_{10}$ p-value ≥ 3; VL: Viral load; Vx: Vaccinated; WG: Weight gain; WUR: WUR1000125

Acknowledgments

The authors would also like to acknowledge PIC, Choice Genetics, and the members of Dr. Bob Rowland's lab for their assistance with conducting and collecting data for the co-infection trials.

Funding

This project was funded by two USDA NIFA grants: 1.) 2012-38,420-19,286 – a National Needs Training Grant that supported data analysis/interpretation and manuscript preparation; and 2.) 2013-68,004-20,362 that supported the study design and data collection.

Authors' contributions

JRD wrote the manuscript, conducted analyses, and interpreted results; NVLS assisted with analyses and interpretation of results; ZW assisted with analyses; EHW assisted with analyses; MCN collected samples for the experimental co-infection trials; MAK collected samples for the experimental co-infection trials and helped with the design of the trials; JKL helped conceive the study; RRR led the co-infection trials and helped conceive the study; and JCMD assisted with preparation of the manuscript, oversaw analyses, aided with interpretation of results, and helped conceive the study. All authors read and approved the final manuscript.

Ethics approval

Experiments involving animals and virus were performed in accordance with the Federation of Animal Science Societies Guide for the Care and Use of Agricultural Animals in Research and Teaching, the USDA Animal Welfare Act and Animal Welfare Regulations, or according to the National Institutes of Health's Guide for the Care and Use of Laboratory Animals, and were approved by the Kansas State University and Iowa State University institutional animal care and use committees and institutional biosafety committees. Animals were humanely euthanized by pentobarbital overdose following the American Veterinary Medical Association (AVMA) guidelines for the euthanasia of animals, and all efforts were made to minimize suffering. Animals used for this study originated from commercial sources including PIC and Choice Genetics

Competing interests

The authors declare that they have no competing interests.

Author details

[1]Department of Animal Science, Iowa State University, Ames, IA 50011, USA. [2]Topigs Norsvin, USA, Burnsville, MN 55337, USA. [3]ABS Global Inc., DeForest, WI 53532, USA. [4]The Seeing Eye Inc., Morristown, NJ 07960, USA. [5]Department of Diagnostic Medicine/Pathobiology, College of Veterinary Medicine, Kansas State University, Manhattan, KS 66506, USA. [6]USDA, ARS, BARC, APDL, Beltsville, MD 20705, USA.

References

1. Boddicker N, Waide EH, Rowland RRR, Lunney JK, Garrick DJ, Reecy JM, et al. Evidence for a major QTL associated with host response to porcine reproductive and respiratory syndrome virus challenge. J Anim Sci. 2012;90:1733–46.
2. Koltes JE, Fritz-Waters E, Eisley CJ, Choi I, Bao H, Kommadath A, et al. Identification of a putative quantitative trait nucleotide in guanylate binding protein 5 for host response to PRRS virus infection. BMC Genomics. 2015;16:1–13.
3. Boddicker NJ, Bjorkquist A, Rowland RRR, Lunney JK, Reecy JM, Dekkers JCM. Genome-wide association and genomic prediction for host response to porcine reproductive and respiratory syndrome virus infection. Genet Sel Evol. 2014;46:1–14.
4. Boddicker NJ, Garrick DJ, Rowland RRR, Lunney JK, Reecy JM, Dekkers JCM. Validation and further characterization of a major quantitative trait locus associated with host response to experimental infection with porcine reproductive and respiratory syndrome virus. Anim Genet. 2014;45:48–58.
5. Hess AS, Islam Z, Hess MK, Rowland RRR, Lunney JK, Wilson AD, et al. Comparison of host genetic factors influencing pig response to infection with two north American isolates of porcine reproductive and respiratory syndrome virus. Genet Sel Evol. 2016;48:1–20.
6. Waide EH, Tuggle CK, Serão NVL, Schroyen M, Hess A, Rowland RRR, et al. Genomewide association of piglet responses to infection with one of two porcine reproductive and respiratory syndrome virus isolates. J Anim Sci. 2017;95:16–38.
7. Dunkelberger JR, Serão NVL, Niederwerder MC, Kerrigan MA, Lunney JK, Rowland RRR, et al. Effect of a major quantitative trait locus for porcine reproductive and respiratory syndrome (PRRS) resistance on response to coinfection with PRRS virus and porcine circovirus type 2b (PCV2b) in commercial pigs, with or without prior vaccination for PRRS. J Anim Sci. 2017;95:584–98.
8. Engle T, Jobman E, Moural T, McKnite A, Barnes S, Davis E, et al. Genome-wide analysis of the differential response to experimental challenge with porcine circovirus 2b. Proc 10th World Congr Genet Appl Livest Prod. Vancouver, Canada. 2014. p. 1–4.
9. Niederwerder MC, Bawa B, Serão NVL, Trible BR, Kerrigan MA, Lunney JK, et al. Vaccination with a porcine reproductive and respiratory syndrome (PRRS) modified live virus vaccine followed by challenge with PRRSV and porcine circovirus type 2 (PCV2) protects against PRRS but enhances PCV2 replication and pathogenesis when compared to results for nonvaccinated cochallenged controls. Clin Vaccine Immunol. 2015;22:1244–54.
10. VanRaden PM. Efficient methods to compute genomic predictions. J Dairy Sci. 2008;91:4414–23.
11. Gilmour AR, Gogel BJ, Cullis BR, Welham SJ, Thompson R. ASReml User Guide Release 4.1 Structural Specification. Hemel Hempstead: VSN International Ltd; 2015.
12. Gao X, Starmer J, Martin ER. A multiple testing correction method for genetic association studies using correlated single nucleotide polymorphisms. Genet Epidemiol. 2008;32:361–9.
13. Team RC. A language and environment for statistical computing. Vienna: R Foundation for Statistical Computing; 2015.
14. Ensembl. European Bioinformatics Institute and Wellcome Trust Sanger Institute, Hinxton, UK. http://www.ensembl.org. Accessed 27 Jan 2017.
15. Mi H, Muruganujan A, Thomas PD. PANTHER in 2013: modeling the evolution of gene function, and other gene attributes, in the context of phylogenetic trees. Nucleic Acids Res. 2013;41:377–86.
16. PANTHER. Thomas Lab at the University of Southern California, Los Angeles. http://pantherdb.org. Accessed 29 Jan 2017.
17. PigQTLdb. NAGRP Bioinformatics Coordination, Ames. http://www.animalgenome.org. Accessed 26 Sept 2016.
18. Wisser RJ, Sun Q, Hulbert SH, Kresovich S, Nelson RJ. Identification and characterization of regions of the rice genome associated with broad-spectrum, quantitative disease resistance. Genetics. 2005;169:2277–93.
19. Salih H, Adelson DL. QTL global meta-analysis: are trait determining genes clustered? BMC Genomics. 2009;10:1–8.
20. Shenoy AR, Wellington DA, Kumar P, Kassa H, Booth CJ, Cresswell P, et al. GBP5 promotes NLRP3 inflammasome assembly and immunity in mammals. Science. 2012;336:481–5.
21. Yin S-H, Xiao C-T, Gerber PF, Beach NM, Meng X-J, Halbur PG, et al. Concurrent porcine circovirus type 2a (PCV2a) or PCV2b infection increases the rate of amino acid mutations of porcine reproductive and respiratory syndrome virus (PRRSV) during serial passages in pigs. Virus Res. 2013;178:445–51.
22. Gillespie J, Opriessnig T, Meng XJ, Pelzer K, Buechner-Maxwell V. Porcine circovirus type 2 and porcine circovirus-associated disease. J Vet Intern Med. 2009;23:1151–63.
23. Segalés J. Best practice and future challenges for vaccination against porcine circovirus type 2. Expert Rev Vaccines. 2015;14:473–87.
24. Meng X. Heterogeneity of porcine reproductive and respiratory syndrome virus: implications for current vaccine efficacy and future vaccine development. Vet Microbiol. 2000;74:309–29.
25. Hu J, Zhang C. Porcine reproductive and respiratory syndrome virus vaccines: current status and strategies to a universal vaccine. Transbound Emerg Dis. 2014;61:109–20.
26. Eicher JD, Montgomery AM, Akshoomoff N, Amaral DG, Bloss CS, Libiger O, et al. Dyslexia and language impairment associated genetic markers influence cortical thickness and white matter in typically developing children. Brain Imaging Behav. 2016;10:272–82.
27. Lunney JK, Ho C-S, Wysocki M, Smith DM. Molecular genetics of the swine major histocompatibility complex, the SLA complex. Dev Comp Immunol. 2009;33:362–74.
28. Geffrotin C, Popescu CP, Cribiu EP, Boscher J, Renard C, Chardon P, et al. Assignment of MHC in swine to chromosome 7 by in situ hybridization and serological typing. Ann Genet. 1984;27:213–9.
29. Hess A. Genetic and biological factors influencing host response to porcine reproductive and respiratory syndrome virus in growing pigs. PhD Diss. Ames: Iowa State Univ; 2016.
30. Serão NVL, Matika O, Kemp RA, Harding JCS, Bishop SC, Plastow GS, et al. Genetic analysis of reproductive traits and antibody response in a PRRS outbreak herd. J Anim Sci. 2014;92:2905–21.
31. Rückert C, Bennewitz J. Joint QTL analysis of three connected F2-crosses in pigs. Genet Sel Evol. 2010;42:1–12.
32. van Wijk HJ, Buschbell H, Dibbits B, Liefers SC, Harlizius B, Heuven HCM, et al. Variance component analysis of quantitative trait loci for pork carcass composition and meat quality on SSC4 and SSC11. J Anim Sci. 2007;85:22–30.
33. Liu SK, Smith CA, Arnold R, Kiefer F, Mcglade CJ. The adaptor protein gads (Grb2-related adaptor downstream of Shc) is implicated in coupling

hemopoietic progenitor kinase-1 to the activated TCR. J Immunol. 2000;165:1417–26.

34. Witsenburg JJ, Sinzinger MD, Stoevesandt O, Ruttekolk IR, Roth G, Adjobo-hermans MJW, et al. A peptide-functionalized polymer as a minimal scaffold protein to enhance cluster formation in early T cell signal transduction. Chembiochem. 2015;16:602–10.

35. de Koning DJ, Varona L, Evans GJ, Giuffra E, Sanchez A, Plastow G, et al. Full pedigree quantitative trait locus analysis in commercial pigs using variance components. J Anim Sci. 2003;81:2155–63.

36. Nezer C, Moreau L, Wagenaar D, Georges M. Results of a whole genome scan targeting QTL for growth and carcass traits in a Piétrain × Large White intercross. Genet Sel Evol. 2002;34:371–87.

37. Nagamine Y, Visscher PM, Haley CS. QTL detection and allelic effects for growth and fat traits in outbred pig populations. Genet Sel Evol. 2004;36:83–96.

38. Liu G, Kim JJ, Jonas E, Wimmers K, Ponsuksili S, Murani E, et al. Combined line-cross and half-sib QTL analysis in Duroc-Pietrain population. Mamm Genome. 2008;19:429–38.

39. Hao Q, Jiao S, Shi Z, Li C, Meng X, Zhang Z, et al. A non-canonical role of the p 97 complex in RIG-I antiviral signaling. EMBO J. 2015;34:2903–20.

40. Thomsen HK, Lee HK, Rothschild MF, Malek M, Dekkers JCM. Characterization of quantitative trait loci for growth and meat quality in a cross between commercial breeds of swine. J Anim Sci. 2004;82:2213–28.

41. Soma Y, Uemoto Y, Sato S, Shibata T, Kadowaki H, Kobayashi E, et al. Genome-wide mapping and identification of new quantitative trait loci affecting meat production, meat quality, and carcass traits within a Duroc purebred population. J Anim Sci. 2011;89:601–8.

42. Poulogiannis G, Frayling IM, Arends MJ. DNA mismatch repair deficiency in sporadic colorectal cancer and Lynch syndrome. Histopathology. 2010;56:167–79.

43. Charerntantanakul W. Porcine reproductive and respiratory syndrome virus vaccines: immunogenicity, efficacy and safety aspects. World J Virol. 2012;1:23–30.

44. Zuckermann FA, Alvarez Garcia E, Diaz Luque I, Christopher-Hennings J, Doster A, Brito M, et al. Assessment of the efficacy of commercial porcine reproductive and respiratory syndrome virus (PRRSV) vaccines based on measurement of serologic response, frequency of gamma-IFN-producing cells and virological parameters of protection upon challenge. Vet Microbiol. 2007;123:69–85.

45. Martelli P, Gozio S, Ferrari L, Rosina S, De Angelis E, Quintavalla C, et al. Efficacy of a modified live porcine reproductive and respiratory syndrome virus (PRRSV) vaccine in pigs naturally exposed to a heterologous European (Italian cluster) field strain: clinical protection and cell-mediated immunity. Vaccine. 2009;27:3788–99.

46. Cinalli RM, Herman CE, Lew BO, Wieman HL, Craig B, Rathmell JC. T cell homeostasis requires G protein-coupled receptor-mediated access to trophic signals that promote growth and inhibit chemotaxis. Eur J Immunol. 2005;35:786–95.

47. Schroyen M, Eisley C, Koltes JE, Fritz-Waters E, Choi I, Plastow GS, et al. Bioinformatic analyses in early host response to porcine reproductive and respiratory syndrome virus (PRRSV) reveals pathway differences between pigs with alternate genotypes for a major host response QTL. BMC Genomics. 2016;17:1–16.

48. Rauw WM. Immune response from a resource allocation perspective. Front Genet. 2012;3:1–14.

49. Meerts P, Misinzo G, Nauwynck HJ. Enhancement of porcine circovirus 2 replication in porcine cell lines by IFN-gamma before and after treatment and by IFN-alpha after treatment. J Interf Cytokine Res. 2005;25:684–93.

50. Allan GM, McNeilly F, Ellis J, Krakowka S, Meehan B, McNair I, et al. Experimental infection of colostrum deprived piglets with porcine circovirus 2 (PCV2) and porcine reproductive and respiratory syndrome virus (PRRSV) potentiates PCV2 replication. Arch Virol. 2000;145:2421–9.

51. Harms PA, Sorden SD, Halbur PG, Bolin SR, Lager KM, Morozov I, et al. Experimental reproduction of severe disease in CD/CD pigs concurrently infected with type 2 porcine circovirus and porcine reproductive and respiratory syndrome virus. Vet Pathol. 2001;38:528–39.

52. Goddard ME, Hayes BJ. Mapping genes for complex traits in domestic animals and their use in breeding programmes. Nat Rev Genet. 2009;10:381–91.

2

Proteomics in non-human primates: utilizing RNA-Seq data to improve protein identification by mass spectrometry in vervet monkeys

J. Michael Proffitt[1], Jeremy Glenn[1], Anthony J. Cesnik[2], Avinash Jadhav[1,6], Michael R. Shortreed[2], Lloyd M. Smith[2,3], Kylie Kavanagh[4], Laura A. Cox[1,5] and Michael Olivier[1,5,6*]

Abstract

Background: Shotgun proteomics utilizes a database search strategy to compare detected mass spectra to a library of theoretical spectra derived from reference genome information. As such, the robustness of proteomics results is contingent upon the completeness and accuracy of the gene annotation in the reference genome. For animal models of disease where genomic annotation is incomplete, such as non-human primates, proteogenomic methods can improve the detection of proteins by incorporating transcriptional data from RNA-Seq to improve proteomics search databases used for peptide spectral matching. Customized search databases derived from RNA-Seq data are capable of identifying unannotated genetic and splice variants while simultaneously reducing the number of comparisons to only those transcripts actively expressed in the tissue.

Results: We collected RNA-Seq and proteomic data from 10 vervet monkey liver samples and used the RNA-Seq data to curate sample-specific search databases which were analyzed in the program Morpheus. We compared these results against those from a search database generated from the reference vervet genome. A total of 284 previously unannotated splice junctions were predicted by the RNA-Seq data, 92 of which were confirmed by peptide spectral matches. More than half (53/92) of these unannotated splice variants had orthologs in other non-human primates, suggesting that failure to match these peptides in the reference analyses likely arose from incomplete gene model information. The sample-specific databases also identified 101 unique peptides containing single amino acid substitutions which were missed by the reference database. Because the sample-specific searches were restricted to actively expressed transcripts, the search databases were smaller, more computationally efficient, and identified more peptides at the empirically derived 1 % false discovery rate.

Conclusion: Proteogenomic approaches are ideally suited to facilitate the discovery and annotation of proteins in less widely studies animal models such as non-human primates. We expect that these approaches will help to improve existing genome annotations of non-human primate species such as vervet.

Keywords: Proteogenomics, Proteomics, Liver, Vervet, RNA-Seq, Morpheus, Non-human primate, Galaxy-P

* Correspondence: molivier@wakehealth.edu
[1]Department of Genetics, Texas Biomedical Research Institute, San Antonio, Texas, USA
[5]Southwest National Primate Research Center, Texas Biomedical Research Institute, San Antonio, Texas, USA
Full list of author information is available at the end of the article

Background

Shotgun proteomic approaches employ a database search strategy to compare experimentally observed mass spectra to an in silico-generated library of theoretical spectra derived from gene annotation information of the organism(s) being studied. The successful matching of peptides is thus predicated upon the accuracy of the search database being utilized to make these comparisons. The outcome of proteomics experiments is therefore driven by the quality and completeness of the genomic information of the organism being studied. Proteomic studies of genetically well-characterized species such as mice and humans benefit from robust proteomic search databases and extensive genome annotations which can account for known genetic variability such as splice variants and sequence variation altering the amino acid sequence of encoded proteins. However, protein identification of other research model organisms is limited by the quality of reference genome annotations.

Proteogenomic methods attempt to improve the search library limitations by leveraging information about gene transcription to guide the curation of search databases customized to the tissues of individual organisms. Several groups have demonstrated that transcriptional profiling using massively parallel sequencing approaches (RNA-Seq) improves the detection of peptides in proteomics experiments in a wide range of different species, ranging from microorganisms [1–3] and plants [4–7] to crustaceans [8], squids [9, 10], honey bees [11], chicken [12], ground squirrels [13], pig [14], and sheep [15]. The approach improves peptide assignment primarily in three important ways [16–18].

First, RNA-Seq data reveal sample-specific genetic sequences which may differ from the reference genome including nucleotide insertions, deletions, or substitutions. Single nucleotide polymorphisms (SNPs) comprise the majority of genomic variation within coding exons of genes, and these genetic variants are divided into two broad categories; SNPs which change a coding triplet but do not result in an amino acid substitution are referred to as synonymous SNPs, while variants which result in amino acid substitutions are called nonsynonymous SNPs (nsSNPs). The identification and inclusion of nsSNPs has the potential to improve the search database annotation because these changes can alter the chemical properties of the fragmented peptides. Failure to account for the resultant mass and/or charge change arising from the amino acid substitutions introduces ambiguity in peptide matching because, unlike nucleotide sequencing where the base order and fragment size can be directly inferred from the raw data, proteomic matching of peptides is based solely on the atomic mass and charge expected to be derived from enzymatically fragmented proteins.

Second, RNA-Seq reads can be used to identify splice junctions (SJs) which characterize mRNA isoforms absent from the reference gene model. SJs can result from genetic variation that alters how the spliceosome interacts with mRNAs, or SJs might arise as a result of alternative exon usage in tissue- or condition-specific contexts. RNA-Seq can also detect chimeric RNAs which arise from gene fusion events. Search databases that inadequately account for this isoform variability will fail to accurately identify their translated peptide products, and organisms with incomplete gene model information are therefore more susceptible to peptide misidentification.

Finally, RNA-Seq reads can be used to estimate transcript abundance. Knowledge of which mRNAs are expressed can inform how search databases can be trimmed to minimize multiple testing that occurs within peptide spectral matching algorithms. As with any iterative comparison process, the likelihood of misidentifying peptides increases with the total number of comparisons made. Experiments which incorporate proteomics results are vulnerable to Type I error inflation because multiple comparisons are made first at the peptide spectral matching identification stage, and then again in the context of the experimental condition (e.g. identifying protein abundance differences between case and control groups).

Based on this summary, it is evident that sample-specific search databases derived from RNA-Seq analyses from the same tissue sample as the proteomics data provide an improved search database, as it will include those peptides mostly likely to be found within the tissue being sampled, while excluding records derived from extraneous genomic information. This approach becomes even more valuable for analyses in samples from species with poorly annotated genomes [19], for reasons we discuss below.

Several bioinformatics pipelines have been described which facilitate the conversion of RNA-Seq reads into customized peptide databases which can be used to search mass spectrometry (MS) data. A well-established approach, developed and described by Sheynkman et al. [16], leverages the web interface of the Galaxy bioinformatics project [20] to facilitate the coordination of independent bioinformatics tools into a functional, user-generated proteogenomic workflow. As part of our analysis, we specifically implemented the proteogenomics approach using the analysis program Morpheus [21] which is computationally less demanding than other programs, and effectively calculates empirical false discovery rates (FDR) for peptide matches. We utilized this approach to determine whether RNA-Seq data from vervet monkey liver samples, a non-human primate without a well-characterized and annotated genome, improves the detection of peptides in vervet liver proteomic data.

While several non-human primate animal models for disease have been extensively used for decades, only recently have the genomes of these organisms begun to be characterized. The African green monkey, or vervet monkey (*Chlorocebus aethiops sabeus*), is one such example. The vervet monkey has long been an important model in AIDS research, as vervets are natural carriers of SIV yet display no symptoms of illness upon infection [22, 23]. More recently, vervets have provided insight into neurologic [24, 25] and metabolic diseases [26–29]. With the recent release of the first vervet genome [30], this animal model is ideally suited to benefit from the expansion of its genome annotation that proteogenomic approaches can provide. In this paper, we utilize matched RNA-Seq and MS data from vervet liver samples to characterize the vervet liver proteome and demonstrate that proteogenomic methods improve the detection of peptides otherwise missed by search databases constructed from the current reference vervet genome annotation.

Methods
Sample collection
All experimental procedures involving animals were approved and complied with the guidelines of the Institutional Animal Care and Use Committee of Wake Forest University Health Sciences and conducted in AAALAC approved facilities. All animals included in this study were female vervet/African green monkeys (*Chlorocebus aethiops sabaeus*) from the Vervet Research Colony (VRC) at Wake Forest School of Medicine. All monkeys were US-colony born within the VRC, which is a multi-generational, pedigreed, and genotyped colony originally founded in 1975 by the University of California Los Angeles, with 57 animals imported from St. Kitts and Nevis. In early 2008, the VRC was transferred to Wake Forest School of Medicine and remains a continuously NIH-supported national research resource. To obtain the samples reported here, 10 vervet monkeys were sedated with ketamine (15 mg/kg intramuscularly), intubated, and anesthetized using isoflurane to facilitate the surgical retrieval of liver tissue via laparotomy. Liver tissue was immediately frozen in liquid nitrogen and stored at −80C until analysis.

RNA-Seq
Total RNA was extracted from vervet monkey livers using the Zymo Direct-zol™ kit (Zymo Research, R2070) and each sample was subsequently quantified by Qubit assay (Thermo Fisher, Q32852). RNA-Seq libraries were prepared from 500 ng of total RNA according to the Illumina TruSeq stranded mRNA protocol (Illumina,

RS-122-2101), which specifically retains polyadenylated mRNAs through the use of oligo dT coated magnetic beads. Sequencing library concentrations were quantified using the KAPA library quantification kit (Kapa Biosystems, KK4824). Clusters were generated by cBot (Illumina), and 2×100 base paired-end sequencing libraries were sequenced using the Illumina HiSeq 2500 with v3 sequencing reagents (Illumina, FC-401-3001).

Conversion of RNA-Seq data to customized peptide databases in galaxy-P
Methods for converting the RNA-Seq reads into searchable protein databases have been extensively described previously [16, 31, 32]. We adapted these approaches within Galaxy-P to create sample-specific search databases for each of the 10 vervet monkey liver samples, using the reference vervet monkey genome (ChlSab1.1) as the basis for the sequence alignments. General overviews of each component of the database construction, along with URLs pointing to the specific workflows with the Galaxy toolshed, are outlined below. Upon completion of the three workflows for each RNA-Seq sample, the records from the three pipelines were concatenated to create a completed sample-specific search database for each of the 10 animals in the study.

Single amino acid variant (SAV) database construction and workflow
Within the SAV workflow, RNA-Seq reads from one sample are aligned to the vervet reference genome using Tophat [33], single nucleotide variant calls are made using SAMtools [34], and the subset of identified SNPs which reside within exons are subsequently annotated using SnpEff [35]. A tool developed within Galaxy-P called "SNPeff to Peptide Fasta" is used to convert the nucleotide sequences into the expected corresponding amino acid sequences. The complete workflow can be found here: http://toolshed.g2.bx.psu.edu/view/galaxyp/proteomics_rnaseq_sap_db_workflow.

Splice junction (SJ) database construction and workflow
The SJ workflow begins by aligning the RNA-Seq reads to the reference vervet genome as well as the Ensembl gene models for the species. The coordinates of all the detected junctions are compared between the two, and only those junctions mapping to the reference genome but not the Ensembl gene model are retained for the SJ annotation. The Galaxy-P program "Translate BED sequences" is used to convert the SJs identified by the RNA-Seq reads into the corresponding polypeptide sequences. Full details are available here: http://toolshed.g2.bx.psu.edu/view/galaxyp/proteomics_rnaseq_splice_db_workflow.

Transcript abundance-based database reduction workflow

In order to reduce the records of proteins based on transcript abundance, RNA-Seq data is quantified by RSEM [36] within the Galaxy-P framework. Quantitative values are normalized and output in transcripts per million (TPM). Text manipulation tools in Galaxy concatenate the protein FASTA data with the transcript identifiers and TPM values, and all records where the values are less than one TPM are excluded from the search database, in accordance with our standard RNA-Seq quality control procedures. Including transcripts with lower abundance increases false-positive alignments, and would require validation through deep sequencing to confirm the presence of the transcript. The workflow repository with the Galaxy toolshed is listed here: http://toolshed.g2.bx.psu.edu/view/galaxyp/proteomics_rnaseq_reduced_db_workflow.

MS-based proteomics

Proteins were extracted from liver tissue using RIPA lysis buffer, and separated on 4–12% gradient Bis-Tris gel. Three gel slices were excised and each was reduced with 10 mM DTT for 30 min at room temperature and alkylated with 55 mM iodoacetamide in 100 mM ammonium bicarbonate for 30 min at room temperature. The gel pieces were subsequently washed with ultrapure 100 mM ammonium bicarbonate, dehydrated with 100% acetonitrile, and dried by Speedvac for 2–3 min.

Samples were then digested with trypsin (Promega, V5280) at 37 °C overnight. Formic acid (1%) was added to the trypsinized samples to quench the proteolysis, and the peptides were desalted and concentrated using C_{18} ZipTips (Millipore, Z720046-960EA). HPLC separation was performed on a 15 cm column of 3 μm diameter which was packed in house with C_{18} beads. Peptides were loaded onto the column at a flow rate of 400 nl/min for 3 h and MS data were acquired by a data dependent scanning on the Thermo Scientific Orbitrap Elite mass spectrometer utilizing a default top 15 method.

Raw mass spectrometry (MS) files were subsequently analyzed in the program Morpheus [21]. The following settings were used in all searches: Assumed Precursor Charge States, Minimum = 2; Assumed Precursor Charge States, Maximum = 4; MS/MS Peak Filtering, Maximum Number of Peaks = 400; MS/MS Analysis, Assign Charge States = enabled; Protease = trypsin (no proline rule); Maximum Missed Cleavages = 2; Initiator Methionine Behavior = variable; Fixed Modifications = carbamidomethylation of C; Variable Modifications = oxidation of M; Maximum Variable Modification Isoforms Per Peptide = 1024; Precursor Mass Tolerance = ± 2.1 Da (monoisotopic); Precursor Monoisotopic Peak Correction = disabled; Product Mass Tolerance = ± 0.025 Da (monoisotopic); Maximum False Discovery Rate = 1%.

For each liver sample, two sets of Morpheus output files were created; the first analysis was searched using the reference vervet monkey database and the second analysis was searched utilizing the sample-specific database created by the Galaxy-P pipelines described above.

Comparative proteomic analyses

Prior to comparison of the proteomic results, the six sets of output files from the Morpheus program for each of the liver samples (3 fractions per sample, run against 2 search databases = 6 files/sample) were combined and transformed to create unique identifiers for all of the peptide spectral match records. This permits the direct comparison of spectra matched from the raw MS files. The search database file size comparisons and wall clock times were extracted from the Morpheus summary files. The "VennDiagram" package in R (https://CRAN.R-project.org/package=VennDiagram) was used to create the lists of unique peptides and protein groups, as well as Venn diagram image files. An R markdown document outlining the tidying and concatenation of the Morpheus output files, along with the creation of the Venn diagrams, can be found in Additional file 1. Gene set enrichment analyses were conducted to identify classes of proteins overrepresented within the list of proteins identified by the reference database but not the sample-specific databases [37, 38].

Results

Search databases curated from RNA-Seq data are smaller and computationally more efficient than reference genome databases

To demonstrate the utility of RNA-Seq derived proteomics search databases, we created sample-specific databases (SSdb) for each liver sample from 10 different vervet monkeys based on sequenced mRNA extracted from the same tissue sample as the protein being analyzed by MS. As outlined above, this procedure creates a unique optimized search database for each sample from the RNA-Seq data, and each MS dataset for a given sample is searched against just the SSdb. For each of the 10 samples, the peptide spectral matching performance was compared between the SSdb and a search database created from the reference vervet genome (REFdb). The descriptive statistics for the RNA-Seq alignments and MS/MS raw data are outlined in Table 1. The RNA-Seq read depth ranged from 6.5 million to 10.5 million mapped reads for the 10 samples. Despite this variability, we found no relationship between RNA-Seq read depth and peptide spectral matches (PSMs) or unique peptides identified in the samples when searched by the SSdb.

Restricting the size of SSdb to transcripts with an abundance of 1 TPM or more condensed the search database size and Morpheus compute time when

Table 1 Descriptive statistics for the RNA-Seq and mass spectrometry analyses utilizing the Vervet reference search database (REFdb, 19,255 gene entries) and the sample-specific databases (SSdb)

Sample	RNA-Seq		SSdb Entries		Mass Spectra	PSMs		Peptide IDs	
	RNA-Seq reads	% reads aligned	Genes	Novel SJs		REFdb	SSdb	REFdb	SSdb
1030	7,040,525	55.5	13,804	4069	80,003	26,525	26,680	9765	9702
1211	6,585,341	68.8	15,782	7171	79,381	27,288	27,673	10,532	10,527
1238	6,594,936	67.1	15,659	6595	78,444	19,600	19,898	9349	9354
1245	6,730,432	64.0	13,901	4089	80,281	29,143	29,193	10,503	10,463
1248	10,504,974	69.4	15,513	7429	80,221	22,205	22,479	9120	9162
1254	9,127,588	62.5	15,936	7641	79,675	23,655	23,738	9334	9385
1291	6,575,182	67.9	13,354	3653	79,960	30,623	30,722	11,478	11,652
1347	6,637,842	56.6	13,284	3147	78,791	17,284	17,037	8633	8575
1448	8,019,158	65.0	15,668	6419	71,853	15,612	15,561	7582	7593
1467	9,983,615	66.2	16,176	7305	78,781	20,101	20,162	9177	9223

compared to the REFdb. On average, the SSdbs were 77% of the size of the REFdb, and compute times in Morpheus averaged 53% faster for SSdbs compared to the REFdb (defined by ([REFdb time/SSdb time]-1)).

Interestingly, for two samples (samples 1347 and 1448), the search against the REFdb resulted in more peptide spectral matches (aon average 0.6%) when compared to the search using the SSdb. Similarly, for three samples (samples 1030, 1245, and 1347) the analysis against the REFdb identified slightly more peptides compared to the search using the SSdb (on average 0.5%). Given the lower number of RNA-Seq reads or mass spectra obtained for some of these samples, it is conceivable that this difference is due to variation in sample preparation or sample quality. Partial degradation of tissue samples would affect both RNA and protein recovery, and may have impacted the analyses presented here.

RNA-Seq derived search databases identify peptides not annotated in the vervet reference genome

Next, we combined the search results across the 10 samples to compare the unique peptides and protein groups identified in vervet monkey liver samples by the REFdb versus the SSdb. These results are shown in the Venn diagram of Fig. 1. We identified 601 peptides in analyses using the SSdb that were not identified using the REFdb.

The first set of these peptides represents peptides that match newly identified SJs not annotated in the gene models of the reference genome. Of the 284 SJ peptide search records identified and annotated from the RNA-Seq data, we identified 47 peptides by MS in more than one sample, which suggests these matches represent incompletely annotated genes of the vervet genome. Consequently, the results of these proteomics analyses could aid in the improved annotation of the gene models. Another 45 peptides mapped to a single sample, bringing the total number of distinctly SJ-mapped peptides

identified in these samples to 92. A comparative analysis of these peptide sequences with other primates utilizing BLASTP revealed that the majority of the identified peptides (53/92) could be matched with an orthologous protein [39]. The complete catalogue of SJs with their corresponding peptides can be found in Additional file 2.

A second set of peptides uniquely identified in searches using the SSdb are peptides that map to an SAV record where the amino acid variant resides within the predicted tryptic peptide that matched in the search. In total, 192 peptide matches representing 101 distinct peptides were found in the 10 samples by the sample-specific analyses using the SSdb. Of these 101 distinct peptides that conformed SAVs identified in the RNA-Seq data, 37 peptides were identified in more than one sample. A list of all peptides matching SAV records are included in Additional file 2.

These first two categories of peptides identified within the SSdb analyses represent search results we expected to obtain from the RNA-Seq-based proteogenomic approach because the mRNA read data create search records which accurately predict the respective peptide fragments seen in the MS dataset.

Reduction of the search database size recovers peptide identifications

While we might expect the same number of matches to reference protein entries contained in both the SSdb and REFdb searches, a total of 313 peptides were identified as matches to reference proteins in the search against the SSdb but not the REFdb. Morpheus, like many other spectral matching algorithms, uses a decoy-based searching approach to empirically estimate and maintain a 1% false discovery rate (FDR). The inclusion of these peptides may result from the adjustment of the absolute FDR that arises from the reduction in the absolute size of the SSdb compared to the REFdb. This is supported

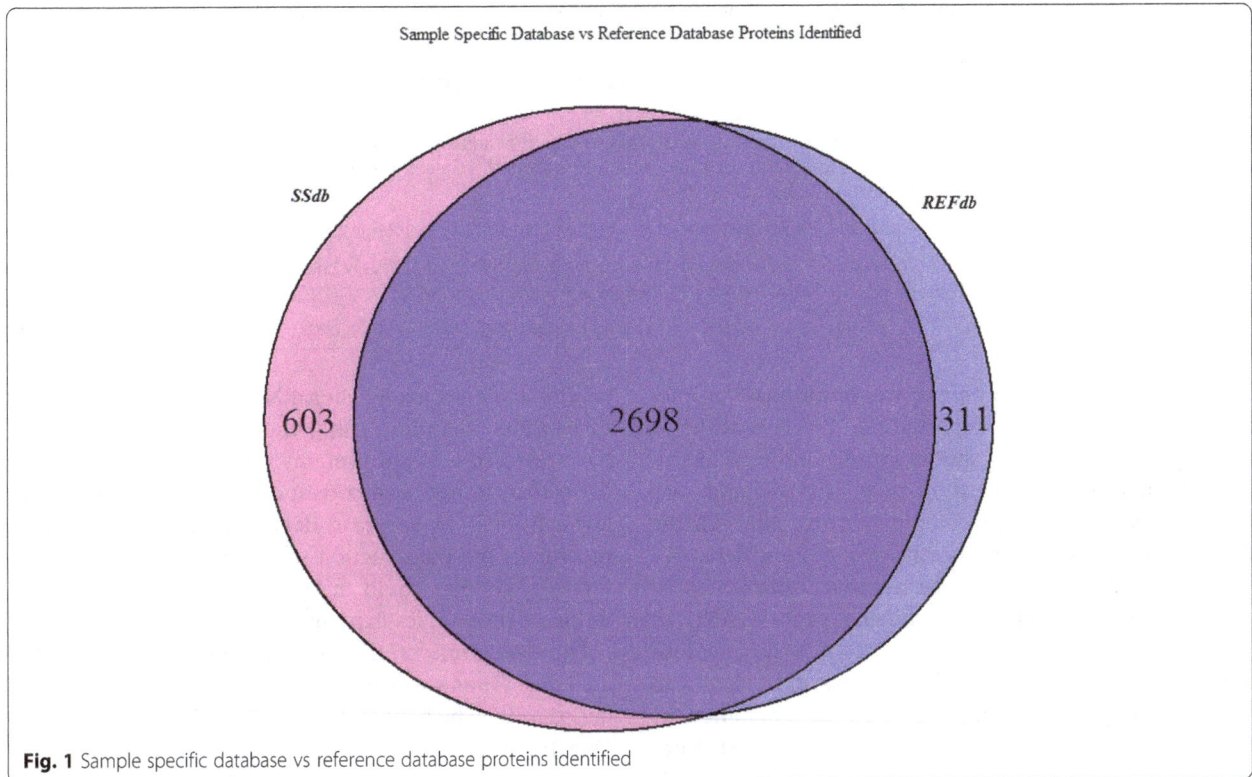

Sample Specific Database vs Reference Database Proteins Identified

SSdb

REFdb

603   2698   311

Fig. 1 Sample specific database vs reference database proteins identified

by the fact that the q-values of this subset of SSdb-identified peptides is higher, on average, compared to the overall average q-value of all peptides identified using the SSdb search. Of the "reference" peptides identified using the SSdb, only 16% (151 of 948) were seen in two or more samples. The peptides identified exclusively in the SSdb analyses are listed in Additional file 3.

Peptide identifications missed by RNA-Seq-derived databases are predominantly structural proteins

While the minimization of the search database helps recover some true positive matches, as described above, concerns about restricting proteome search databases based upon RNA-Seq data could arise when a protein's abundance is poorly correlated with its corresponding transcript abundance, or when proteins might be derived from a tissue of origin different from which they currently reside. Extracellular structural proteins or chromatin-associated proteins with long-half-lives would be likely candidates for the former category, while growth factors, cytokines or contaminating proteins might comprise the latter. We surveyed the list of peptides identified by REFdb but not SSdb analyses and identified 1950 total matches mapping to 891 unique peptides, of which 506 (57%) were identified in 2 or more liver samples. Those 506 peptides correspond to 238 unique protein entries in the vervet ENSEMBL database. Gene set enrichment analysis revealed that this list

of proteins was indeed overrepresented by structural, cytoskeletal, and ribonucleotide binding proteins. Table 2 lists the significantly enriched categories identified in this subset of proteins. The peptides identified by the REFdb but not the SSdbs searches can be found in Additional file 4.

Though distinct differences in the peptides identified by the SSdb versus REFdb have been revealed by this comparative exercise, it is perhaps equally important to highlight that SSdb analyses are capable of matching nearly 98% of the same spectra as the REFdb. The customized databases created from RNA-Seq robustly identify proteins from tissue samples, identify peptide variation that would otherwise be missed by searching against the reference annotation, and perform these functions using smaller file sizes and faster compute times.

Discussion

This work represents the implementation and proof-of-concept application of an established RNA-Seq proteogenomic approach to improve the identification of proteins in non-human primate proteomics experiments, such as the vervet monkey liver sample analyses reported here. Several non-human primate genomes have been drafted but incompletely annotated because a limited number of animals have been sequenced. Less common sequence and splice variants will continue to be incorporated as DNA sequencing sample sizes

Table 2 Gene Set Enrichment Analysis for proteins identified by reference but not sample-specific databases

GO Annotation	Description	p-value	FDR q-value
GO: STRUCTURAL MOLECULE ACTIVITY	The action of a molecule that contributes to the structural integrity of a complex or assembly within or outside a cell.	3.03×10^{-15}	3.42×10^{-11}
GO: OXIDATION REDUCTION PROCESS	A metabolic process that results in the removal or addition of one or more electrons to or from a substance, with or without the concomitant removal or addition of a proton or protons.	9.8×10^{-14}	5.52×10^{-10}
GO: DNA PACKAGING COMPLEX	A protein complex that plays a role in the process of DNA packaging.	7.61×10^{-13}	2.86×10^{-9}
GO: EXTRACELLULAR SPACE	That part of a multicellular organism outside the cells proper, usually taken to be outside the plasma membranes, and occupied by fluid.	1.25×10^{-11}	2.81×10^{-8}
GO: PROTEIN DNA COMPLEX	A macromolecular complex containing both protein and DNA molecules.	6.04×10^{-11}	1.13×10^{-7}

increase and more complete transcriptional profiles across tissue types are reported. In the meantime, a proteogenomic approach should improve protein identification in MS experiments in these species, and may offer robust and reliable data to improve the existing annotation using both transcript and protein data. We demonstrated that this analysis approach, implemented in Galaxy-P, is capable of identifying peptides derived from unannotated splice junctions and non-synonymous coding substitutions revealed from the RNA-Seq read data. These peptides would have gone unmatched by searching the MS data using the vervet reference genome annotation data. These novel peptides likely represent a combination of common, yet previously unannotated gene isoforms as well as isoforms and variants private to individual animals studied.

As with other "omics" scale analyses, the iterative search process of shotgun proteomics presents challenges in balancing type I and type II error rates. The optimal search database would incorporate only those potential proteins which are likely to be found within a given sample; however, if the repertoire of proteins were already known, it would obviate the need for conducting proteomics experiments in the first place. Instead, RNA-Seq data can facilitate a compromise in the database curation process by predicting non-reference protein isoforms for inclusion while also utilizing information about transcript abundance to exclude individual gene sequences from the search records as the corresponding mRNA are not expressed and therefore the translated protein (or a peptide thereof) is unlikely to be identified by MS. While RNA-Seq data can be a useful benchmark for restricting the size of the MS search database, it should be noted that certain exceptions to the database exclusion process should be considered. Examples include proteins whose tissue of origin is different from the sampled tissue, such as blood-derived albumin, immunoglobulins, and complement proteins, or long-lived structural proteins such as collagen or ribosomal proteins, whose protein abundance is uncoupled from their corresponding mRNA expression in the tissue of interest at the time of sample collection. By comparing

peptide matches made using RNA-Seq derived databases to peptides identified using the reference database, we have revealed a list of proteins routinely found in vervet livers that do not have corresponding mRNA abundance levels from RNA-Seq read data. Including these protein records in the construction of sample-specific databases for liver samples could mitigate the loss of information in future proteomics experiments.

The reduction of the search database size and the included protein records significantly impacts the confidence with which peptides and proteins are identified. Due to the smaller number of records to compare an experimental spectrum to, the confidence with which individual peptides are assigned to the correct record is higher. However, these "improved" matches have another consequence: some experimental peptide spectra may not be confidently assigned to a specific sequence in the REFdb analysis since the match may not fall under the stringent 1% FDR commonly required to confirm a match. However, using the smaller SSdb, the same match is now made with a FDR of less than 1% (simply based on the smaller number of searches), resulting in some additional "reference" proteins being identified in the SSdb searches at a 1% FDR but not in the REFdb searches. It is possible that these matches may include some low confidence peptide matches, but overall these additional peptide matches emphasize the additional power that is gained from reducing the search database to only relevant records of expressed proteins and peptide sequences.

Our analysis only used one standard established analysis approach for these comparisons. The analysis pipeline implemented in Galaxy includes the Morpheus search algorithm which empirically calculates the FDR for peptide matches obtained in the search. Numerous other approaches have been proposed for the analysis of proteogenomic data, and the assessment of FDR in peptide and protein identifications. It is likely that some of these approaches, such as analyzing FDR separately for REFdb matches and matches to novel SSdb records derived from RNA-Seq [19] or alternative programs to calculate FDR in these datasets [40–42], would improve

Proteomics in non-human primates: utilizing RNA-Seq data to improve protein identification by mass...

27

the results presented here, and further enhance the utility of this proteogenomic approach for non-human primate proteomics. However, a detailed comparison of these different analysis approaches was not the goal of the current study, and future studies will help define the optimal approach for a proteogenomic analysis in these species, including the optimal RNA-Seq coverage and the depth of proteomic analysis. Prior analyses have generated far more detailed mass spectral analysis data (500,000 mass spectra compared to 80,000 used in our study), and it remains to be seen what the optimal approach will be [16]. As proteogenomic approaches continue to gain momentum in shotgun proteomics experiments, we anticipate further refinement of search databases to account for biochemical variability in peptides which arise from post-translational modifications (PTMs). A recent publication has outlined an approach to parsimoniously account for peptide mass shifts caused by PTMs through incorporating Uniprot annotation data ([43, 44]). Similarly, proteogenomics can incorporate findings from complimentary NGS approaches, such as ribosomal profiling, to expand the prediction of the protein-coding products from novel coding sequences [45] and lncRNA molecules previously presumed to be untranslated [44, 46]. Continued refinements to search databases and proteomics search algorithms will accelerate the accurate identification and quantification of peptides in MS analyses, and it will complement and improve the genome annotation of animal research model organisms and help researchers utilize shotgun proteomics to implicate protein changes associated with pathophysiologic processes. Ultimately, the proteomic validation of novel splice variants and non-synonymous sequence variants will greatly enhance the ongoing efforts of genome annotation, especially in model species with poorly annotated genomes, such as many non-human primates.

Conclusions

A proteogenomic approach to the analysis of liver shotgun proteomic data from a nonhuman primate species, the vervet/African green monkey (*Chlorocebus aethiops sabaeus*), demonstrates that the use of sample-derived RNA-Seq data, as anticipated, improves peptide identification and the accuracy and confidence of protein identification, while simultaneously reducing the search database space and the resulting computing effort required for the data analysis. Novel peptides including sequence variants identified by RNA-Seq, as well as new splice variants uncovered in the transcriptional analysis account for the majority of the novel peptides identified, highlighting the importance of proteogenomic approaches in species with limited available genome sequence data and gene annotation, such as non-human primates.

Additional files

Additional file 1: Outlining the data transformations and analyses of the Morpheus file outputs. (HTML 641 kb)

Additional file 2: Listing the SAV and SJ derived peptides identified exclusively by the sample-specific databases. (XLSX 46 kb)

Additional file 3: Listing the reference genome peptides identified exclusively by the sample-specific databases. (XLS 343 kb)

Additional file 4: Listing the peptides identified by the REFdb but not the SSdb searches. (XLS 541 kb)

Abbreviations
AIDS: Acquired immune deficiency syndrome; DTT: Dithiothreitol; FDR: False discovery rate; GO: Gene ontology; GSEA: Gene set enrichment analysis; HPLC: High performance liquid chromatography; lncRNA: Long, non-coding ribonucleic acid; mRNA: Messenger ribonucleic acid; MS: Mass spectrometry; PSM: Peptide spectral match; REFdb: Reference database; SAV: Single amino variant; SIV: Simian immunodeficiency virus; SJ: Splice junction; SNP: Single nucleotide polymorphism; SSDb: Sample-specific database; TPM: Transcripts per million

Acknowledgements
Not applicable

Funding
Animal sample collection supported by P40 OD010965 and UL1 TR004120 (KK) Proteogenomic analyses and interpretation supported by NIGMS R01 GM109099 (MO and LMS). AJC supported by the CIBM Training Program 5T15LM007359. The authors also acknowledge grant support from NIH Grant P51 OD011133 and facilities support from NIH Research Facilities Improvement Program Grants C06 RR 1 C06 RR013556 and 1 C06 RR017515. The funding agencies did not have any role in the design of the study, collection, analysis, and interpretation of data, or in writing the manuscript.

Authors' contributions
JMP, MO, LAC, and KK conceived the study. KK oversaw the selection and sampling of the vervet liver biopies. LAC and JG generated the RNA-Seq data. MO and AJ generated the proteomics data. LMS, MRS, AJC, and JG contributed to the construction of the sample-specific databases. JMP analyzed the mass spectrometry data and conducted the comparative analyses. JMP and MO drafted the manuscript, and LMS, AJC, and LAC provided significant editorial revisions. All authors have read and approved of the final version of the manuscript.

Competing interests
The authors declare that they have no competing interests.

Author details

[1]Department of Genetics, Texas Biomedical Research Institute, San Antonio, Texas, USA. [2]Department of Chemistry, University of Wisconsin, Madison, Wisconsin, USA. [3]Genome Center of Wisconsin, University of Wisconsin, Madison, Wisconsin, USA. [4]Department of Pathology and Comparative Medicine, Wake Forest School of Medicine, Winston-Salem, North Carolina, USA. [5]Southwest National Primate Research Center, Texas Biomedical Research Institute, San Antonio, Texas, USA. [6]Current address: Department of Internal Medicine, Section of Molecular Medicine, Wake Forest School of Medicine, NRC Building, G-55, Winston-Salem, North Carolina 27157, USA.

References

1. Heunis T, Dippenaar A, Warren RM, van Helden PD, van der Merwe RG, Gey van Pittius NC, Pain A, Sampson SL, Tabb DL. Proteogenomic investigation of strain variation in clinical mycobacterium tuberculosis isolates. J Proteome Res. 2017; doi: 10.1021/acs.jproteome.7b00483.

2. Swearingen KE, Lindner SE, Flannery EL, Vaughan AM, Morrison RD, Patrapuvich R, Koepfli C, Muller I, Jex A, Moritz RL, Kappe SHI, Sattabongkot J, Mikolajczak SA. Proteogenomic analysis of the total and surface-exposed proteomes of plasmodium vivax salivary gland sporozoites. PLoS Negl Trop Dis. 2017;11(7):e0005791.

3. Krishna R, Xia D, Sanderson S, Shanmugasundram A, Vermont S, Bernal A, Daniel-Naguib G, Ghali F, Brunk BP, Roos DS, Wastling JM, Jones AR. A large-scale proteogenomics study of apicomplexan pathogens-Toxoplasma gondii and Neospora caninum. Proteomics. 2015;15(15): 2618–28.

4. Zhu FY, Chen MX, Ye NH, Shi L, Ma KL, Yang JF, Cao YY, Zhang Y, Yoshida T, Fernie AR, Fan GY, Wen B, Zhou R, Liu TY, Fan T, Gao B, Zhang D, Hao GF, Xiao S, Liu YG, Zhang J. Proteogenomic analysis reveals alternative splicing and translation as part of the abscisic acid response in Arabidopsis seedlings. Plant J. 2017;91(3):518–33.

5. Grossmann J, Fernández H, Chaubey PM, Valdés AE, Gagliardini V, Cañal MJ, Russo G, Grossniklaus U. Proteogenomic analysis greatly expands the identification of proteins related to reproduction in the Apogamous Fern Dryopteris Affinis Ssp. Affinis. Front Plant Sci. 2017;8:336.

6. Zargar SM, Mahajan R, Nazir M, Nagar P, Kim ST, Rai V, Masi A, Ahmad SM, Shah RA, Ganai NA, Agrawal GK, Rakwal R. Common bean proteomics: present status and future strategies. J Proteome. 2017;169:233–38. doi:10.1016/j.jprot.2017.04.010.

7. Bryant L, Patole C, Cramer R. Proteomic analysis of the medicinal plant Artemisia Annua: data from leaf and trichome extracts. Data Brief. 2016;7:325–31.

8. Trapp J, Gaillard JC, Chaumot A, Geffard O, Pible O, Armengaud J. Ovary and embryo proteogenomic dataset revealing diversity of vitellogenins in the crustacean Gammarus fossarum. Data Brief. 2016;8:1259–62.

9. Caruana NJ, Cooke IR, Faou P, Finn J, Hall NE, Norman M, Pineda SS, Strugnell JM. A combined proteomic and transcriptomic analysis of slime secreted by the southern bottletail squid, Sepiadarium Austrinum (Cephalopoda). J Proteome. 2016;148:170–82.

10. Whitelaw BL, Strugnell JM, Faou P, da Fonseca RR, Hall NE, Norman M, Finn J, Cooke IR. Combined Transcriptomic and proteomic analysis of the posterior salivary gland from the southern blue-ringed octopus and the southern sand octopus. J Proteome Res. 2016;15(9):3284–97.

11. McAfee A, Harpur BA, Michaud S, Beavis RC, Kent CF, Zayed A, Foster LJ. Toward an upgraded honey bee (Apis Mellifera L.) genome annotation using Proteogenomics. J Proteome Res. 2016;15(2):411–21.

12. Bottje WG, Lassiter K, Piekarski-Welsher A, Dridi S, Reverter A, Hudson NJ, Kong BW. Proteogenomics reveals enriched ribosome assembly and protein translation in Pectoralis major of high feed efficiency pedigree broiler males. Front Physiol. 2017;8:306.

13. Anderson KJ, Vermillion KL, Jagtap P, Johnson JE, Griffin TJ, Andrews MT. Proteogenomic analysis of a hibernating mammal indicates contribution of skeletal muscle physiology to the hibernation phenotype. J Proteome Res. 2016;15(4):1253–61.

14. Marx H, Hahne H, Ulbrich SE, Schnieke A, Rottmann O, Frishman D, Kuster B. Annotation of the domestic pig genome by quantitative Proteogenomics. J Proteome Res. 2017;16(8):2887–98.

15. Chemonges S, Gupta R, Mills PC, Kopp SR, Sadowski P. Characterisation of the circulating acellular proteome of healthy sheep using LC-MS/MS-based proteomics analysis of serum. Proteome Sci. 2017;15:11.

16. Sheynkman GM, Johnson JE, Jagtap PD, Shortreed MR, Onsongo G, Frey BL, et al. Using galaxy-P to leverage RNA-Seq for the discovery of novel protein variations. BMC Genomics. 2014;15:703.

17. Wen B, Xu S, Sheynkman GM, Feng Q, Lin L, Wang Q, et al. sapFinder: an R/bioconductor package for detection of variant peptides in shotgun proteomics experiments. Bioinformatics. 2014;30(21):3136–8.

18. Wen B, Xu S, Zhou R, Zhang B, Wang X, Liu X, et al. PGA: an R/bioconductor package for identification of novel peptides using a customized database derived from RNA-Seq. BMC Bioinformatics. 2016;17(1):244.

19. Nesvizhskii AI. Proteogenomics: concepts, applications and computational strategies. Nat Methods. 2014;11(11):1114–25.

20. Afgan E, Baker D, van den Beek M, Blankenberg D, Bouvier D, Čech M, et al. The galaxy platform for accessible, reproducible and collaborative biomedical analyses: 2016 update. Nucleic Acids Res. 2016;44(W1):W3–W10.

21. Wenger CD, Coon JJ. A proteomics search algorithm specifically designed for high-resolution tandem mass spectra. J Proteome Res. 2013;12(3):1377–86.

22. Schmitz JE, Zahn RC, Brown CR, Rett MD, Li M, Tang H, et al. Inhibition of adaptive immune responses leads to a fatal clinical outcome in SIV-infected pigtailed macaques but not vervet African green monkeys. PLoS Pathog. 2009;5(12):e1000691.

23. Zahn RC, Rett MD, Korioth-Schmitz B, Sun Y, Buzby AP, Goldstein S, et al. Simian immunodeficiency virus (SIV)-specific CD8+ T-cell responses in vervet African green monkeys chronically infected with SIVagm. J Virol. 2008;82(23):11577–88.

24. Burke MW, Ptito M, Ervin FR, Palmour RM. Hippocampal neuron populations are reduced in vervet monkeys with fetal alcohol exposure. Dev Psychobiol. 2015;57(4):470–85.

25. Bouskila J, Harrar V, Javadi P, Beierschmitt A, Palmour R, Casanova C, et al. Cannabinoid receptors CB1 and CB2 modulate the Electroretinographic waves in Vervet monkeys. Neural Plast. 2016;2016:1253245.

26. Jasinska AJ, Schmitt CA, Service SK, Cantor RM, Dewar K, Jentsch JD, et al. Systems biology of the vervet monkey. ILAR J. 2013;54(2):122–43.

27. Voruganti VS, Jorgensen MJ, Kaplan JR, Kavanagh K, Rudel LL, Temel R, et al. Significant genotype by diet (G × D) interaction effects on cardiometabolic responses to a pedigree-wide, dietary challenge in vervet monkeys (Chlorocebus Aethiops Sabaeus). Am J Primatol. 2013;75(5):491–9.

28. Kavanagh K, Flynn DM, Nelson C, Zhang L, Wagner JD. Characterization and validation of a streptozotocin-induced diabetes model in the vervet monkey. J Pharmacol Toxicol Methods. 2011;63(3):296–303.

29. Kavanagh K, Fairbanks LA, Bailey JN, Jorgensen MJ, Wilson M, Zhang L, et al. Characterization and heritability of obesity and associated risk factors in vervet monkeys. Obesity (Silver Spring). 2007;15(7):1666–74.

30. Warren WC, Jasinska AJ, García-Pérez R, Svardal H, Tomlinson C, Rocchi M, et al. The genome of the vervet (Chlorocebus Aethiops Sabaeus). Genome Res. 2015;25(12):1921–33.

31. Jagtap PD, Johnson JE, Onsongo G, Sadler FW, Murray K, Wang Y, et al. Flexible and accessible workflows for improved proteogenomic analysis using the galaxy framework. J Proteome Res. 2014;13(12):5898–908.

32. Sheynkman GM, Shortreed MR, Frey BL, Smith LM. Discovery and mass spectrometric analysis of novel splice-junction peptides using RNA-Seq. Mol Cell Proteomics. 2013;12(8):2341–53.

33. Trapnell C, Pachter L, Salzberg SL. TopHat: discovering splice junctions with RNA-Seq. Bioinformatics. 2009;25(9):1105–11.

34. Li H. A statistical framework for SNP calling, mutation discovery, association mapping and population genetical parameter estimation from sequencing data. Bioinformatics. 2011;27(21):2987–93.

35. Cingolani P, Platts A, Wang LEL, Coon M, Nguyen T, Wang L, et al. A program for annotating and predicting the effects of single nucleotide polymorphisms, SnpEff: SNPs in the genome of Drosophila Melanogaster strain w1118; iso-2; iso-3. Fly (Austin). 2012;6(2):80–92.

36. Li B, Dewey CN. RSEM: accurate transcript quantification from RNA-Seq data with or without a reference genome. BMC Bioinformatics. 2011;12:323.

37. Subramanian A, Tamayo P, Mootha VK, Mukherjee S, Ebert BL, Gillette MA, et al. Gene set enrichment analysis: a knowledge-based approach for interpreting genome-wide expression profiles. Proc Natl Acad Sci U S A. 2005;102(43):15545–50.

38. Gene Ontology Consortium. Gene ontology consortium: going forward. Nucleic Acids Res. 2015;43(Database issue):D1049–56.

39. Altschul SF, Madden TL, Schäffer AA, Zhang J, Zhang Z, Miller W, et al. Gapped BLAST and PSI-BLAST: a new generation of protein database search programs. Nucleic Acids Res. 1997;25(17):3389–402.

Proteomics in non-human primates: utilizing RNA-Seq data to improve protein identification by mass...

29

40. Wang X, Slebos RJ, Wang D, Halvey PJ, Tabb DL, Liebler DC, Zhang B. Protein identification using customized protein sequence databases derived from RNA-Seq data. J Proteome Res. 2012;11(2):1009–17.

41. Li H, Park J, Kim H, Hwang KB, Paek E. Systematic comparison of false-discovery-rate-controlling strategies for Proteogenomic search using spike-in experiments. J Proteome Res. 2017;16(6):2231–9.

42. Ma C, Xu S, Liu G, Liu X, Xu X, Wen B, Liu S. Improvement of peptide identification with considering the abundance of mRNA and peptide. BMC Bioinformatics. 2017;18(1):109.

43. Cesnik AJ, Shortreed MR, Sheynkman GM, Frey BL, Smith LM. Human proteomic variation revealed by combining RNA-Seq Proteogenomics and global post-translational modification (G-PTM) search strategy. J Proteome Res. 2016;15(3):800–8.

44. Ingolia NT, Lareau LF, Weissman JS. Ribosome profiling of mouse embryonic stem cells reveals the complexity and dynamics of mammalian proteomes. Cell. 2011;147(4):789–802.

45. Raj A, Wang SH, Shim H, Harpak A, Li YI, Engelmann B, et al. Thousands of novel translated open reading frames in humans inferred by ribosome footprint profiling. elife. 2016;27:5.

46. Ji Z, Song R, Regev A, Struhl K. Many lncRNAs, 5'UTRs, and pseudogenes are translated and some are likely to express functional proteins. elife. 2015;4:e08890.

3

Transcriptome profiling of aging *Drosophila* photoreceptors reveals gene expression trends that correlate with visual senescence

Hana Hall[1], Patrick Medina[2], Daphne A. Cooper[3], Spencer E. Escobedo[1], Jeremiah Rounds[2], Kaelan J. Brennan[1], Christopher Vincent[2], Pedro Miura[3], Rebecca Doerge[4] and Vikki M. Weake[1,5*]

Abstract

Background: Aging is associated with functional decline of neurons and increased incidence of both neurodegenerative and ocular disease. Photoreceptor neurons in *Drosophila melanogaster* provide a powerful model for studying the molecular changes involved in functional senescence of neurons since decreased visual behavior precedes retinal degeneration. Here, we sought to identify gene expression changes and the genomic features of differentially regulated genes in photoreceptors that contribute to visual senescence.

Results: To identify gene expression changes that could lead to visual senescence, we characterized the aging transcriptome of *Drosophila* sensory neurons highly enriched for photoreceptors. We profiled the nuclear transcriptome of genetically-labeled photoreceptors over a 40 day time course and identified increased expression of genes involved in stress and DNA damage response, and decreased expression of genes required for neuronal function. We further show that combinations of promoter motifs robustly identify age-regulated genes, suggesting that transcription factors are important in driving expression changes in aging photoreceptors. However, long, highly expressed and heavily spliced genes are also more likely to be downregulated with age, indicating that other mechanisms could contribute to expression changes at these genes. Lastly, we identify that circular RNAs (circRNAs) strongly increase during aging in photoreceptors.

Conclusions: Overall, we identified changes in gene expression in aging *Drosophila* photoreceptors that could account for visual senescence. Further, we show that genomic features predict these age-related changes, suggesting potential mechanisms that could be targeted to slow the rate of age-associated visual decline.

Keywords: Aging, Transcriptome, Drosophila, Neurons, Photoreceptors

Background

The incidence of ocular disease increases with age leading to an increase in reported visual impairment from 5.7% in 18 – 44 year old people to 21% in people older than 75 years [1]. Whereas theories of aging in the eye have traditionally focused on the role of oxidative damage to the genome and mitochondrial dysfunction [2], changes in expression of genes in the aging retina could also contribute to the age-associated increase in disease susceptibility [3]. Photoreceptor neurons, and in particular rod photoreceptors, which comprise the major retinal cell type in humans, show age-associated decreases in both visual function and in number [4–14]. Loss of rod photoreceptors is the major factor leading to ocular disease-associated blindness [15]. Microarray analysis of aging mouse rod photoreceptors indicates that gene expression changes begin as early as five months of age in rodents, preceding pathological changes by two years [12, 13]. Thus, gene expression changes precede the onset of visual dysfunction and disease. Identifying these signature early gene expression changes could therefore provide the opportunity to prevent or delay the onset of ocular disease.

* Correspondence: vweake@purdue.edu
[1]Department of Biochemistry, Purdue University, West Lafayette, IN 47907, USA
[5]Purdue University Center for Cancer Research, Purdue University, West Lafayette 47907, USA
Full list of author information is available at the end of the article

As in humans, the fruitfly *Drosophila melanogaster* shows visual senescence, defined as a progressive decline in visual function with age. While young flies are attracted towards light and show positive phototaxis, this phototactic behavior decreases with age [16–18]. Moreover, the rate of visual senescence is influenced by genetic variation [16], indicating that genetic factors regulate the age-related loss of visual function. Here, we sought to identify gene expression changes and the genomic features of differentially regulated genes in aging photoreceptor neurons that contribute to visual senescence.

Since the retina is comprised of multiple cell types, we focused our analysis on the outer photoreceptor neurons in *Drosophila*, R1 – R6 cells. These six outer photoreceptor neurons share several key functional similarities with human rods. First, both cell types represent the majority of photoreceptors in the retina, function in dim light, and express a single rhodopsin protein, Rhodopsin 1 (Rh1) [19]. Second, phototransduction in both human rods and in R1 – R6 photoreceptor cells initiates with the light-induced isomerization of photosensitive rhodopsin [20]. The rapid life cycle of *Drosophila*, coupled with our ability to genetically label and isolate photoreceptors in an intact organism, permitted us to examine the photoreceptor transcriptome at multiple time points during aging, prior to the first signs of retinal degeneration. Here, we show that subsets of genes in photoreceptor neurons are age-regulated. We find that combinations of sequence motifs and gene characteristics such as gene length and exon content identify age-regulated genes. Further, we show that circular RNAs (circRNAs) accumulate in aged photoreceptors. Together, these data indicate that targeting gene expression mechanisms might provide a way to slow the rate of visual decline associated with aging and thereby delay the onset of ocular disease.

Results

Visual function declines with age independent of retinal degeneration

In this study, we sought to identify gene expression changes in aging photoreceptor neurons that could contribute to visual senescence. While decreased visual behavior is observed by 3 – 4 weeks of age, little retinal degeneration is observed in wild-type flies at these ages [16, 17, 21, 22]. We directly compared retinal degeneration, phototaxis and lifespan in *Rh1-Gal4 > KASH-GFP* flies to identify an age at which flies show decreased visual behavior in the absence of either retinal degeneration or significant morbidity. Male flies were used for all experiments because sex-specific differences have been reported for both phototaxis and visual senescence [16, 17]. Similar to observations from other groups [22],

we found that 99.6% and 99.2% of rhabdomeres in the outer photoreceptors (R1 - R6) are intact in male flies at day 10 and day 40 post-eclosion (emergence from the pupal case), respectively (n = 5; Fig. 1a). Rhabdomere loss is observed during retinal degeneration in flies, and provides a stringent measure of photoreceptor health; thus, we conclude that there is no significant retinal degeneration by 40 days post-eclosion. In addition, 93% of male flies survived until day 40 under our standard laboratory conditions with 12:12 h light/dark cycles (Fig. 1b). While we did not observe significant morbidity or retinal degeneration by day 40, we did observe a significant decrease in positive phototaxis in male day 40 flies compared to day 10 flies using a two-choice T-maze assay (Fig. 1c). Significantly decreased phototaxis was also observed in day 25 flies relative to day 10; however, day 25 flies also showed more variability in phototaxis as compared with either the day 10 or day 40 flies, suggesting that these flies are more heterogeneous with respect to visual behavior. Although older flies are known to have decreased locomotion [17], the T-maze assay minimizes the effect of locomotive behavior on phototaxis since flies are presented with a single choice between light and dark [23]. In addition, published reports show that the age-related increase in visual senescence reflects visual behavior rather than locomotion [16]. Thus, day 40 flies show decreased visual behavior in the absence of retinal degeneration, indicating that cellular function is compromised in the aging eye.

Transcriptome profiling of photoreceptor neurons

The *Drosophila* eye consists of repeating units termed ommatidia that each contain about 20 different cells including eight photoreceptor neurons [24, 25]. We sought to profile the transcriptome of aging photoreceptors to identify genes that show age-dependent changes in expression in this visual cell type. Photoreceptors are highly polarized epithelial cells that extend axons to the neuropils in the brain [24]. To profile the photoreceptor transcriptome, we examined nuclear RNA, which has been shown to correlate well with levels of active transcription [26]. To do this, we labeled photoreceptor nuclei with nuclear membrane-localized GFP (KASH-GFP). Whereas the outer photoreceptors (R1 – R6) express Rh1, the inner photoreceptors (R7 and R8) express Rh3/4 and Rh5/6 respectively [19, 27]. We labeled R1 – R6 photoreceptors using *Rh1-Gal4* [28] driven *UAS-KASH-GFP*. The Klarsicht, Anc-1, Syn3-1 homology (KASH) domain of Msp300 localizes GFP to the cytoplasmic face of the nuclear membrane, allowing subsequent affinity-enrichment of labeled nuclei with GFP antibodies coupled to magnetic beads (Fig. 1d) [29, 30].

To determine the enrichment of our target nuclei versus nonspecific background levels, we mixed equal

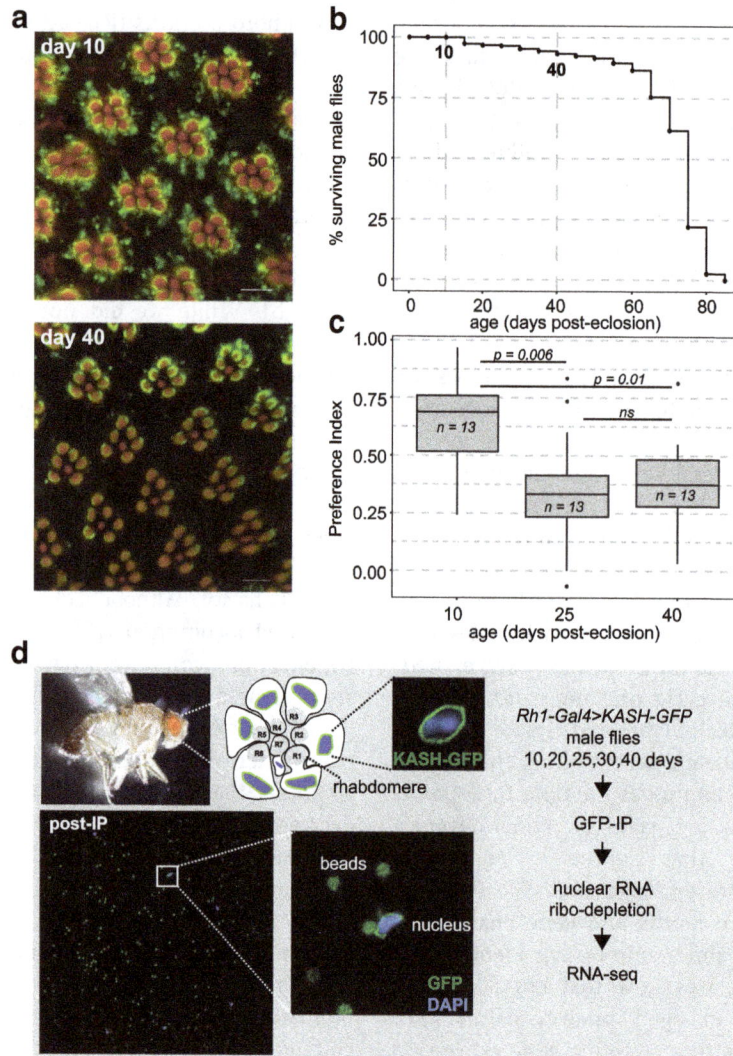

Fig. 1 Visual function declines with age independent of retinal degeneration. **a** Representative confocal images of adult retinas stained with phalloidin (red) and 4C5 (Rh1, green) from male *Rh1-Gal4 > KASH-GFP* flies 10 and 40 days post-eclosion (*n* = 5). Scale bars: 5 μm. **b** Survival curve showing the percentage of viable *Rh1-Gal4 > KASH-GFP* male flies at each age (*n* = 345). **c** Box plots showing the light preference indices (positive phototaxis) for *Rh1-Gal4 > KASH-GFP* flies at day 10, 25 and 40 (*n* = 13 experiments; 27 - 33 male flies/experiment). *p* value, normally-distributed data were analyzed using ANOVA followed by Tukey's honest significant different (HSD) post hoc test. **d** Photoreceptor R1 – R6 nuclei in each ommatidium were labeled with nuclear membrane-localized GFP in *Rh1-Gal4 > KASH-GFP* flies. Affinity-enriched GFP-labeled nuclei bound to antibody-coated magnetic beads are shown in the two lower panels. DAPI, blue; GFP, green. Schematic of the RNA-seq experimental design is shown in the right panel. Graphic generated by authors

numbers of flies that expressed either *KASH-GFP* or *KASH-mCherry* in photoreceptors under *Rh1-Gal4* control, and generated head homogenates in which an equal number of GFP- and mCherry-labeled nuclei were present. We then performed GFP affinity-enrichment, and measured GFP and mCherry transcript levels in the pre-isolation (head homogenate) and post-isolation samples by qPCR. We observed 82 ± 22 fold enrichment of GFP transcripts in the post-isolation samples, with no corresponding increase in mCherry levels, demonstrating that the affinity-enrichment of GFP-labeled nuclei is highly specific (Additional file 1: Figure S1A). Next, we

profiled the transcriptome of affinity-enriched GFP-labeled nuclei from 10 day old male flies using Illumina sequencing (RNA-seq) and compared this to the corresponding pre-isolation samples. The post-isolation samples for each of the three biological replicates grouped together by principal component analysis and were distinct from each of their respective pre-isolation samples (Additional file 1: Figure S1B). Using edgeR, we identified 447 post-enriched and 444 post-reduced genes (False Discovery Rate, FDR < 0.05, fold change, FC > 2) (Additional file 1: Figure S1C, Additional file 2: Table S1). Supporting an enrichment of photoreceptors, gene ontology (GO) term analysis of the 447 photoreceptor post-enriched genes identified terms including photo-transduction, calcium signaling and retina homeostasis (Additional file 1: Figure S1D, Additional file 3: Table S2). In contrast, the post-reduced genes were enriched for a variety of metabolic pathways. Consistent with a depletion of cytoplasmic and mitochondrial RNAs in affinity-enriched nuclear RNA, 11 of 13 detected mitochondrial-encoded genes were significantly reduced in the post-isolation samples. These data demonstrate that RNA isolated using our approach is highly enriched for nuclear RNA transcribed in the target cell population.

Surprisingly, we identified the GO term *Sensory perception of sound* (GO:0007605) as being significantly overrepresented in our post-enriched gene group. This GO term shares a number of common gene members with the GO terms describing phototransduction and light response including the R7 – R8 rhodopsins *Rh5* and *Rh6* (Additional file 1: Figure S1D). The auditory organ (Johnston's organ: JO) in *Drosophila* is composed of a sound receiver and an auditory sensory organ that contains mechanosensory neurons [31]. Intriguingly, several genes that function in phototransduction are expressed in the auditory organ including *Arr2, Rh3, Rh5, Rh6, inaD, trp* and *trpl* [32]. Since, *Rh1* transcription is restricted to R1 – R6 photoreceptors, which do not express other rhodopsins [27, 33–37], we hypothesized that expression of *Rh1-Gal4* in mechanosensory neurons could account for the observed enrichment of R7 – R8 rhodopsins in our study. To determine if *Rh1-Gal4* was expressed in antennae, we purified total RNA from dissected heads, eyes, antennae and bodies of *Rh1-Gal4 > KASH-GFP* flies and examined *Rh1* and *GFP* transcript levels by qPCR. In line with our hypothesis, we found that both *Rh1* and *GFP* genes are expressed in the antennae at ~10 - 20% levels found in the eye (Additional file 1: Figure S2). This therefore accounts for enrichment of R7 – R8 photoreceptor-specific markers such as *Rh3* and *Rh5* in our affinity-enriched photoreceptor nuclei. We note that similar enrichment of R7 – R8 *rhodopsins* was previously reported by Yang et al. who profiled R1 – R6 mRNAs by expressing polyA-binding protein under *Rh1-Gal4* control and purifying bound-mRNAs from whole heads [38]. Since there are approximately 9600 outer photoreceptor neurons and 1000 mechanosensory neurons per head [25, 39], we conclude that using *Rh1-Gal4 > KASH-GFP* flies, our approach predominantly enriches photoreceptor neurons, but that ~10% of our enriched nuclei are most likely contributed by mechanosensory neurons.

Age-related changes in the photoreceptor transcriptome

To identify genes that show age-regulated expression in photoreceptors, we affinity-enriched *Rh1-Gal4 > KASH-*

GFP labeled nuclei from adult male flies. To avoid changes in gene expression associated with the transition from development to adulthood, and to identify changes in gene expression that contribute to decreased photo-taxis between day 10 and 40, we profiled the photoreceptor nuclear transcriptome at 10, 20, 25, 30 and 40 days post-eclosion (Fig. 1d). We obtained similar RNA yields across each time point (Additional file 1: Figure S3A) that yielded an average of 32 million high-quality paired-reads for each biological replicate ($n = 3$). We discarded one sample (day 30 replicate 3) due to poor alignment. We then analyzed the RNA-seq time series data using maSigPro, which is a generalized linear model-based approach [40]. Utilizing maSigPro and multiple time points enabled us to identify genes with robust expression changes that correlate strongly with chronological age. Notably, maSigPro has a much lower false positive rate for time series data when compared with pair-wise differential expression methods such as edgeR [41]. Using maSigPro, we identified 604 age-regulated genes (FDR < 0.05). To limit the age-regulated genes to those that were expressed specifically in photoreceptors, we excluded 49 age-regulated genes that were significantly reduced in the post-isolation samples from day 10 flies (Additional file 2: Table S1). Thus, 555 genes were differentially expressed with age in *Drosophila* photoreceptors (Additional file 4: Table S3). This differential expression did not reflect differences in the relative GFP-labeling of photoreceptors and mechanosensory neurons because GFP mRNA and protein levels in the eye did not change with age (Additional file 1: Figure S3B,C). Moreover, we did not observe consistent patterns of change in expression of neuronal cell-type specific genes during aging (Additional file 1: Figure S3D). Further, 5 of 7 selected age-regulated genes showed significant differences in expression between day 10 and 40 in dissected eyes from male flies and in independent affinity-enriched samples by qPCR (Additional file 1: Figure S4). Thus, the majority of age-regulated genes identified are differentially expressed in photoreceptors. Phototaxis differs between male and female flies with one study reporting 20% lower phototaxis in females at 4 weeks, while another showed 10% higher phototaxis at the same age [16, 21]. Further, a recent study has shown that age-related changes in gene expression in the retina differ between male and female mice [42]. To test if female flies showed similar patterns of gene expression changes in aging photoreceptors, we examined a subset of the age-regulated genes by qPCR in female eyes. We observed the same trends in gene expression for the age-regulated genes examined between day 10 and 40 in dissected eyes from male and female flies (Additional file 1: Figure S4). However, these gene expression changes were not significant in the female flies due both to high variability in the young samples, and to smaller magnitude of changes between young and old eyes. Thus, we conclude that gene expression changes in aging photoreceptors are likely similar between male and female flies, but that female flies might show delayed onset of these gene expression changes compared with males.

We next characterized the direction and temporal pattern of the changes in gene expression for the age-regulated genes. To do this, we used k-means clustering to group the 555 age-regulated genes based on their temporal pattern of gene expression. We found that the age-regulated genes clustered into 11 expression clusters, whose changes in relative gene expression over age were best described by second degree polynomial equations (Fig. 2a). Only moderate improvement was obtained from using higher order polynomials or by increasing the number of clusters used for k-means clustering (Additional file 1: Figure S5A). We next determined the overall direction of the change in gene expression for age-regulated genes in each clusters based on the slope of the fitted curve: using these criteria, 288 age-regulated genes were upregulated, and 267 age-regulated genes were downregulated by day 40 (Additional file 1: Figure S5B). We then determined when the change in gene expression occurred: early clusters showed maximal changes in expression between days 10 and 20, late clusters between days 30 and 40, while the middle clusters showed little to no change in the rate of expression (slope) throughout the time course. Most of the downregulated genes were found in the early clusters, with only 39 genes (cluster 11) being downregulated late. In contrast, only 44 of the upregulated genes fell into an early cluster, while 154 upregulated genes were in late clusters. Thus, the temporal expression clustering suggests that most age-related changes in gene expression in photoreceptors do not occur gradually or linearly. Instead, most downregulated genes showed the highest rates of changes in gene expression at the earliest stages of the aging process, while more than half of upregulated genes increased later during aging. These data suggest that distinct mechanisms underlie the changes in gene expression observed in these subsets of age-regulated genes.

Aging is associated with upregulation of stress-inducible genes and downregulation of genes required for neuronal function

Next, we asked if the gene expression changes observed in aging photoreceptors could contribute to the observed visual senescence between day 10 and days 25 and 40. GO term analysis of the upregulated genes revealed an enrichment for genes indicative of an induced stress response such as DNA repair and the unfolded protein response (Fig. 2b, Additional file 5: Table S4). In contrast to the upregulated genes, the downregulated genes were enriched for GO terms including ion transport, synaptic transmission and behavior (Fig. 2b, Additional file 5: Table S4). Further analysis using DAVID [43] identified additional functional categories associated with the age-regulated genes; we then examined published reports to identify specific age-regulated genes involved in these processes which could potentially impact visual function (Fig. 2c).

In-depth analysis of the age-regulated genes revealed that multiple genes in the DNA damage response pathway were upregulated with age including those that function in non-homologous end-joining repair (*mre11*, *rad50*, *Ku80* and *mus308*) and in translesion DNA synthesis (*mus205* and *DNApol-eta*) [44–46]. Genes that encoded enzymes with antioxidant properties, such as the thioredoxin reductase *Trxr-1*, and antioxidant genes involved in glutamate metabolism, such as *GlnRS*, *isoQC* and *QC*, were also upregulated with age [47–50]. We also observed increased age-associated expression of chaperone genes (*Cct1*, *Cct4*, *Cct5*, *Cct6*, *Hsc70-4*) and the unfolded protein response transcription factor *Xbp1*, consistent with an induction of the unfolded protein response [51–53]. Under stress conditions, there is a translational switch that favors production of stress-related proteins while decreasing translation of other proteins [54]. Paralogs of canonical translation factors such as *NAT1* and *Rack1*, which were both upregulated, promote this switch to cap-independent translation [55, 56]. Notably, *Rheb*, which is downregulated with age, positively regulates ribosome production and cap-dependent translation by activating the mechanistic target of rapamycin (mTOR) kinase pathway [57]. Thus, decreased *Rheb* levels during aging could decrease mTOR pathway activity, which extends lifespan and is protective against age-related pathology [58]. Together, these data suggest that multiple genes are induced in aging photoreceptors to mitigate the effects of oxidative stress, protein misfolding and DNA damage.

In contrast to the upregulated genes, many of the genes that were downregulated with age are required for the proper response to light in photoreceptors. For example, sodium and potassium ion channels, such as those encoded by *Atpα*, *Sh*, *shakB* and *Shal*, are required for the sensitivity and dynamic range of photoreceptors in response to bright light [59–62]. In addition, ion channels, such as Sh, and genes such as *Csp*, *Hdc* and *Sap47* are necessary for proper synaptic function, which is required to transmit the light signal from the retina to the brain [63–66]. Further, calcium-binding proteins such as Cpn function in photoreceptors to buffer potentially toxic levels of intracellular calcium induced by

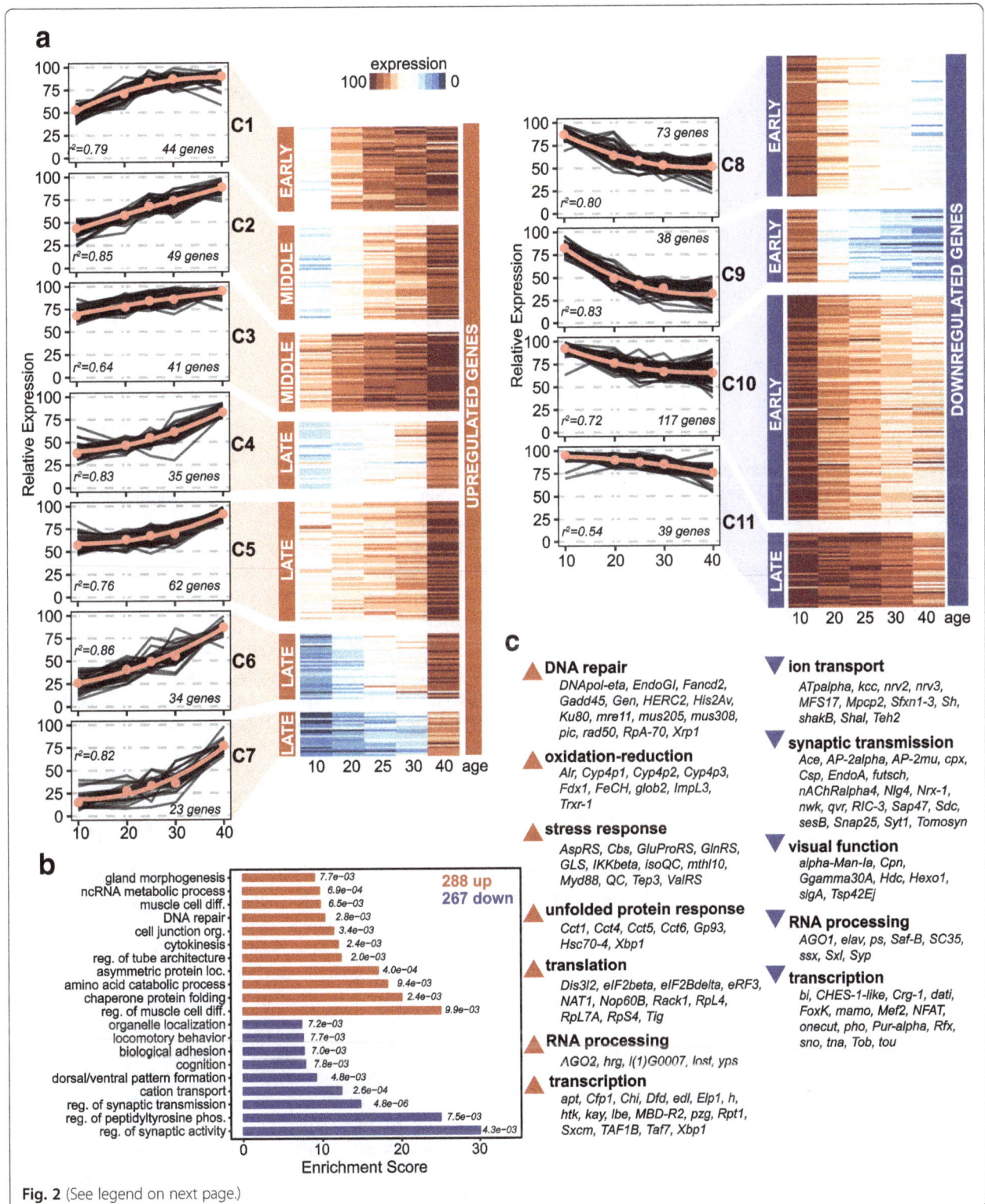

Fig. 2 (See legend on next page.)

(See figure on previous page.)
Fig. 2 Age-related changes in gene expression in adult photoreceptors. **a** Age-regulated genes identified by time-series analysis using maSigPro (555 genes, FDR < 0.05) were clustered using k-means into 11 clusters based on temporal expression pattern (relative expression). The median expression values (circles) and fitted curves with indicated r^2 values are shown in red on the line graphs. Age-regulated genes were designated as upregulated or downregulated, and early, middle or late based on the fitted curves for their respective cluster (see Additional file 1: Fig. S4). **b** Over-represented GO terms ($p < 0.01$, Fisher's exact test) were identified for 288 upregulated or 267 downregulated genes relative to all 7579 expressed genes using TopGO (Additional file 5: Table S4). Similar GO terms were grouped based on intersecting gene members, and a single representative GO term is shown from each group in the bar plot. Enrichment score indicates the number of genes with the GO term in the target gene set versus the number of expected genes, with p-values shown to the right of each bar. **c** Representative functional categories identified using GO term analysis and DAVID for upregulated and downregulated genes. Selected age-regulated genes involved in the indicated functions are shown below each term based on published reports

prolonged phototransduction [67–70]. Several other genes including *Hexo1*, *α-Man-Ia* and *Tsp42Ej* encode proteins required for post-translational modification or degradation of Rh1 [71, 72]. Notably, mutations that prevent either processing of Rh1, or degradation of activated endocytosed Rh1, cause retinal degeneration [72–75]. In addition to genes that are necessary for photoreceptor function, some of the age downregulated genes have been directly shown to impact phototaxis including *Ace*, *slgA* and *Rheb* [76–78]. Together, these data show that the cumulative downregulation of genes involved in processes required for photoreceptor function could indeed account for the decrease in vision we observed by days 25 and 40.

Intriguingly, functional analysis using DAVID showed an enrichment for RNA processing and transcription in both the up- and downregulated genes. Several splicing factor genes including *ps*, *Saf-B* and *SC35* were downregulated with age [79, 80]. In contrast, *hrg*, which encodes the single poly(A) polymerase enzyme required for mRNA polyadenylation in flies [81], is upregulated with age. Further, whereas *AGO1*, which regulates microRNA-induced silencing is downregulated with age, *AGO2*, which regulates small interfering-RNA silencing is upregulated [82]. *AGO2* has been linked to repair of double-stranded DNA breaks [83], suggesting that changes in expression of some RNA processing factors might occur as part of the stress response. In addition to genes involved in RNA processing, a large number of genes that encode transcription factors or transcriptional regulatory proteins were age regulated. Some of these transcription factors could control expression of other age-regulated genes. For example, the calcium-regulated transcription factor NFAT, which is downregulated with age, is required for neural development, including pre--synaptic growth, and plasticity [84]. Similarly, the transcription factor onecut that is required to maintain neuronal identity, is also downregulated with age [85]. Thus, decreases in levels of these transcription factors could contribute to downregulation of neuronal-specific genes, such as those involved in synaptic transmission. Similarly, upregulation of *Xbp1* could contribute to upregulation of genes involved in the UPR, although Xbp1

activity is primarily regulated through alternative splicing [86]. In addition to transcription factors, epigenetic regulators such as the TFIID subunit Taf7 or the Set1/COMPASS histone methyltransferase subunit Cfp1 are also age regulated [87, 88]. These data suggest that multiple factors converge to drive changes in the transcriptional landscape of aging photoreceptors.

Combinations of promoter sequence motifs identify age-regulated genes

We sought to identify factors involved in the regulation of gene expression that drive changes in the transcriptional landscape of aging photoreceptors. Since several transcription factors are themselves age-regulated, and because alterations in age-related signaling pathways converge on transcription factors, we first asked whether the age-regulated genes were targeted by common transcription factors. To do this, we examined the upregulated genes (clusters 1 – 7) or the downregulated genes (clusters 8 – 11) to identify shared promoter sequence motifs. We used HOMER (Hypergeometric Optimization of Motif EnRichment) [89] to identify significantly enriched ($p < 0.001$) sequence motifs in the promoters of age up- or downregulated genes. Using this approach, 40 significantly enriched sequence motifs were identified for the upregulated genes and 41 significantly enriched motifs were identified for the downregulated genes (Additional file 6: Table S5).

We then asked if the presence of any combination of the individual enriched sequence motifs were associated with a gene that is up or downregulated with age. To do this, we generated ROC (Receiver Operating Characteristic) curves to assess the ability of each individual sequence motif to identify whether a gene would be up or downregulated with age. We then compared the AUC (area under the curve) for each ROC curve; higher AUC values indicate an improved ability to identify genes that were age-regulated. Not surprisingly, no single motif provided a strong ability to identify the direction of regulation with age. However, when we analyzed combinations of promoter motifs generated by iteratively combining the motif with the highest AUC score with other motifs, we found that increasing combinations of sequence motifs could strongly identify whether a gene would be up or downregulated with age (Fig. 3a). A combination of all 40 sequence motifs resulted in AUC values of 0.88 or 0.85 for the up and downregulated genes respectively, versus 0.58 for the control (see methods); the AUC value for the motif combination was significantly improved by the addition of a single motif for the first 14 motifs, which we termed *top motifs* (Fig. 3b, Additional file 1: Figure S6). Network analysis indicated that these top motifs co-occurred frequently with other motifs at gene promoters (Additional file 1: Figure S7A). Further, the top motifs showed lower clustering coefficients in the network, indicating that they co-occurred with a wider variety of the other enriched-motifs (Additional file 1: Figure S7B). Together, these data are consistent with combinatorial activity of transcription factors at age-regulated genes, and suggest that the top motifs identified represent binding sites for transcription

factors that integrate multiple types of signaling pathways during aging.

Next, we asked which transcription factors were most likely to bind the top sequence motifs. To do this, we used HOMER to compare the top motifs with known insect transcription factor binding sites including those recently identified by Nitta et al. [90]. We then asked whether the putative transcription factor matches were expressed in photoreceptors based on our RNA-seq data. A complete list of matching, expressed transcription factors is provided for all top motifs in Additional file 7: Table S6, and the best matches for the top motifs based on score are described in Additional file 1: Figure S6. The age-regulated genes with promoter motifs putatively bound by these transcription factors are shown in Additional file 1: Figure S8 and Figure S9. We observed differences in the occurrence of specific motifs in each expression cluster, suggesting that different groups of transcription factors regulate early and late genes (Additional file 1: Figure S10). Further, several of the transcription factors that match the top motifs were themselves age regulated at the gene expression level. For example, the transcription repressor hairy (*h*) [91] was upregulated early during aging, and matched one of the top motifs (motif 15) identified for the downregulated genes. In contrast, the transcription activator one-cut [85], which was downregulated early during aging, matched one of the top motifs (motif 9) for the downregulated genes. The transcription factor Deformed (*Dfd*) was present in one of the middle upregulated gene clusters (cluster 2), and matched one of the top motifs for the upregulated genes (motif 17). Another early upregulated gene, the FOS homolog *kayak* (kay) that together with Jun-related antigen (*Jra*) forms the AP-1 transcription factor [92], corresponded to another of the top motifs for the upregulated genes (motif 18). AP-1 acts downstream of Jun-N-terminal kinase signaling in response to UV-induced DNA damage in the *Drosophila* retina [93], consistent with a putative role for AP-1 in upregulating stress-responsive aging genes such as the DNA repair gene *Xrp1* (Additional file 1: Figure S9). Together, these data suggest that transcription factors play a key role in driving the gene expression changes observed in aging photoreceptors.

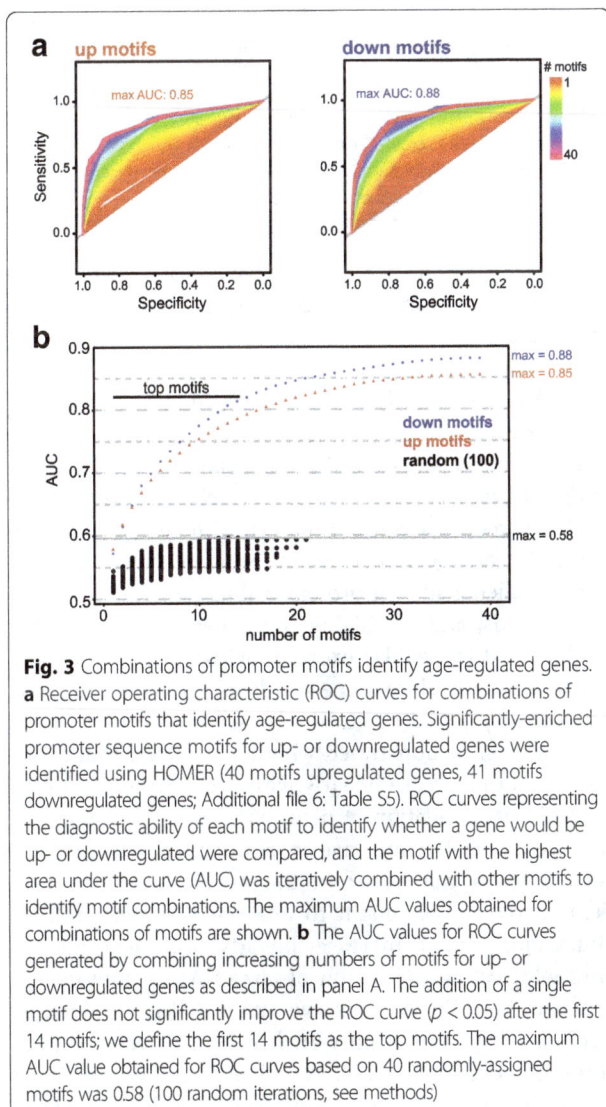

Fig. 3 Combinations of promoter motifs identify age-regulated genes. **a** Receiver operating characteristic (ROC) curves for combinations of promoter motifs that identify age-regulated genes. Significantly-enriched promoter sequence motifs for up- or downregulated genes were identified using HOMER (40 motifs upregulated genes, 41 motifs downregulated genes; Additional file 6: Table S5). ROC curves representing the diagnostic ability of each motif to identify whether a gene would be up- or downregulated were compared, and the motif with the highest area under the curve (AUC) was iteratively combined with other motifs to identify motif combinations. The maximum AUC values obtained for combinations of motifs are shown. **b** The AUC values for ROC curves generated by combining increasing numbers of motifs for up- or downregulated genes as described in panel A. The addition of a single motif does not significantly improve the ROC curve (p < 0.05) after the first 14 motifs; we define the first 14 motifs as the top motifs. The maximum AUC value obtained for ROC curves based on 40 randomly-assigned motifs was 0.58 (100 random iterations, see methods)

Gene length, expression and splicing correlate with age-downregulation

Several data suggest that the regulation of transcription elongation and RNA processing events could also contribute to age-related gene expression changes. Neuronal genes tend to be longer than average [94], and are often heavily alternatively spliced [95], implying that transcription elongation or splicing might be critical for proper expression of these genes. Expression of long genes is more dependent on topoisomerases, which relieve transcription-induced torsional stress [96], and on proper DNA repair because long genes stochastically accumulate more DNA damage [97]. Our data indicate that genes involved in splicing are age-regulated in photoreceptors. Notably, age-related changes in splicing, which could contribute to alterations in mRNA levels, have been observed in several studies [98]. Since the genes that are downregulated during aging include a large number of neuron-specific genes, we wondered whether a bias in their genomic features, such as gene length or number of exons, could contribute to their age-associated decline.

To examine this question, we analyzed gene length, overall expression level across all time points (RPKM), and the number of expressed exons and transcript isoforms for genes in each expression cluster C1 – C11 (Additional file 1: Figure S11). We used Wilcoxon Rank-Sum test to compare pair-wise differences in the distribution of gene length or other characteristics between each expression cluster and the genes that were not age-regulated (non-significant genes). Three of the 4 downregulated clusters showed significantly different distributions of gene length, expression or transcription isoform number compared with the nonsignificant genes. In each of these downregulated clusters, genes showed longer median gene lengths, higher median expression, and higher median numbers of expressed transcript isoforms. Further, 2 of the 4 downregulated clusters had significantly different distributions of exon numbers, with higher median numbers of expressed exons. In contrast, only one of the upregulated clusters showed significantly different distribution of expression (higher median expression) or transcript isoforms (smaller median number). Although increased expression level and gene length could contribute to enhanced statistical power in differential gene expression analysis [99], we would expect this statistical power to apply equally to genes in all of the expression clusters whether they were up or downregulated. Thus, these data demonstrate that the age down-regulated genes show a bias towards long, highly expressed and heavily spliced genes.

Next, we asked if any of the characteristics that showed a bias in the downregulated gene set could identify whether a given gene would be downregulated with age. To do this, we generated ROC curves based on the ability of each characteristic, such as gene length, to identify whether a gene would be up- or downregulated (Fig. 4). Whereas the ROC curves for gene length, expression, exon number or transcript isoforms showed no ability to identify upregulated genes, all four characteristics showed modest ability to identify downregulated genes with AUCs between 0.60 – 0.74. These data suggest that long, highly transcribed genes are susceptible

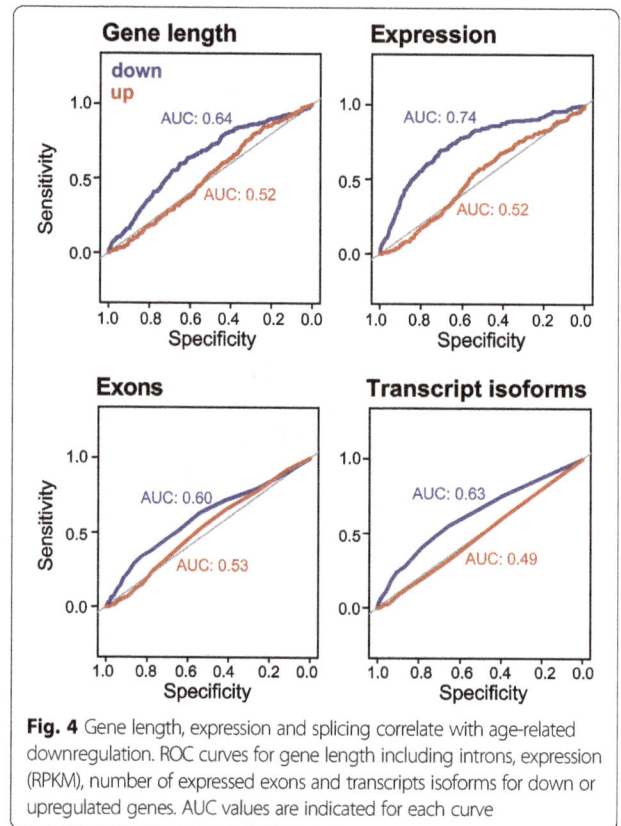

Fig. 4 Gene length, expression and splicing correlate with age-related downregulation. ROC curves for gene length including introns, expression (RPKM), number of expressed exons and transcripts isoforms for down or upregulated genes. AUC values are indicated for each curve

to downregulation with age, implying that in addition to transcription factors, other gene regulatory mechanisms might become altered with age.

circRNA levels strongly correlate with age in photoreceptors

In addition to changes in the expression of specific genes during aging, specific classes of RNA known as circRNAs show increased abundance with age [100]. These circRNAs are generated from back-splicing events at known splicing sites of protein coding genes, lack free 5′ and 3′ ends, and are highly stable because they cannot be degraded by cellular exoribonucleases [101] (Fig. 5a). While circRNA abundance has been shown to increase between day 1 and day 20 in *Drosophila* heads [100], it has not been demonstrated whether circRNA abundance continues to increase linearly with age. Further, although circRNA accumulation in neurons is thought to underlie the age-associated increase in abundance, whether this accumulation can occur in the nucleus has not been demonstrated. We sought to examine whether circRNA abundance would increase linearly with age due to chronological accumulation of circRNAs in photoreceptor neurons. To do this, we used CIRI2 to identify circRNAs from our affinity-enriched photoreceptor transcriptome data [102, 103]. We identified 1209 circRNAs in the sensory neuron data with at least two counts from

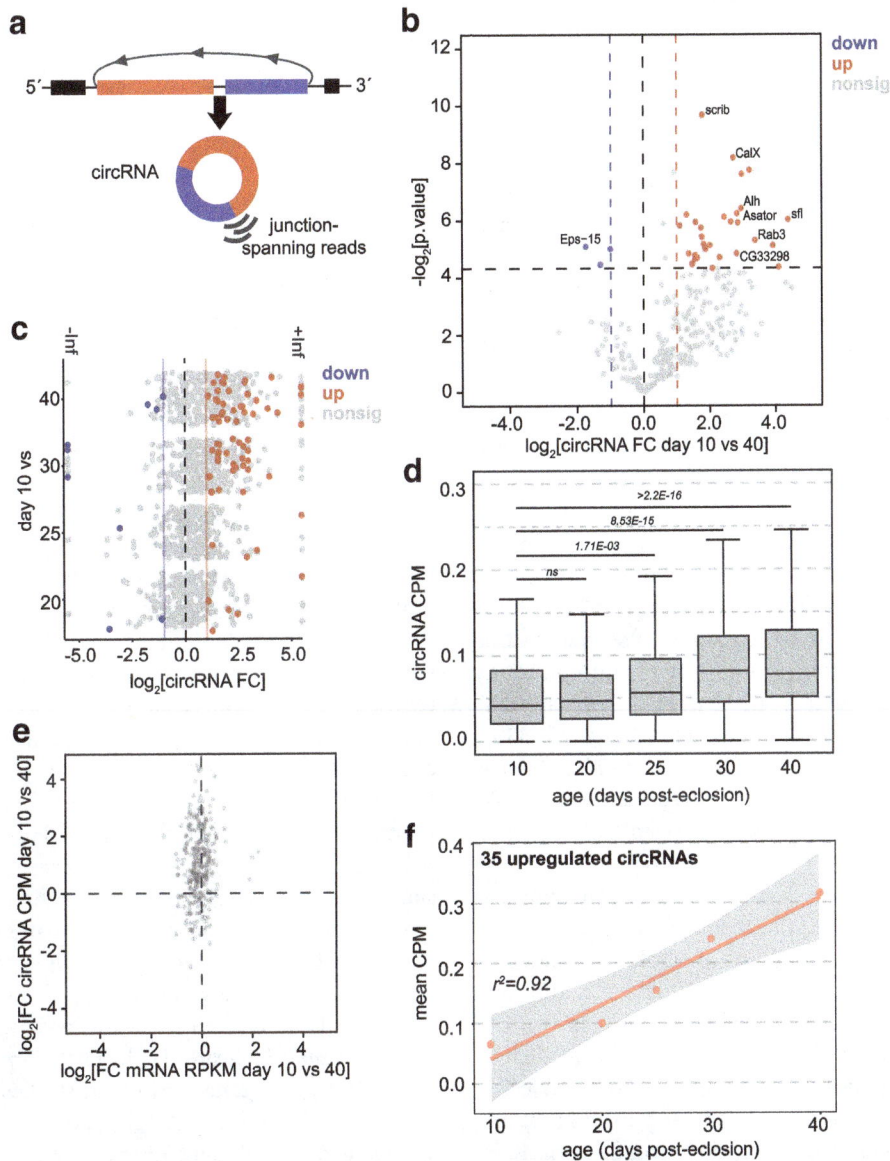

Fig. 5 circRNA levels increase in aging photoreceptors. **a** Schematic showing how junction-spanning reads were used to detect circRNAs resulting from back-splicing events. **b** Volcano plot showing the fold change in circRNA abundance plotted as $\log_2$(fold change in counts per million reads, CPM) for each circRNA relative to its p value ($-\log_2$[p.value]). CircRNAs with significantly differential expression ($p \leq 0.05$ and FC ≥ 2, dotted lines) are highlighted. Labels indicate the corresponding host gene for selected circRNAs. **c** Fold changes in circRNA abundance for pairwise comparisons between the indicated aging time-point. CircRNAs with significantly differential expression ($p \leq 0.05$ and FC ≥ 2, blue/red) are highlighted. **d** Total abundance of circRNAs (CPM) identified at each age. p values, non-parametrical Kruskal-Wallis with Nemenyi post-hoc test for multiple comparisons. **e** Density plots comparing the $\log_2$ fold changes in circRNA CPM with fold change in linear RNA RPKM from the corresponding gene for 10 versus 40 day sensory neurons. **f** Linear regression analysis of the mean CPM of the 35 significantly upregulated circRNAs versus age

the 14 libraries (Additional file 8: Table S7). For these circRNAs, 1095 were previously annotated [100], and 114 were novel annotations. We quantified circRNA abundance as counts per million reads (CPM), and calculated pairwise differential expression statistics across all pair-wise comparisons. Using this approach, we identified 38 out of 315 abundant circRNAs (>6 total counts per circRNA) that were differentially expressed

between day 10 and 40 ($p < 0.05$) (Fig. 5b, Additional file 8: Table S7). Thirty-five out of these 38 circRNAs increased with age, supporting an overall increase in circRNA abundance with age. Notably, there was a strong trend for circRNA accumulation that was not statistically significant most likely due to low read counts (Fig. 5b, bottom right quadrant). When we examined levels of a subset of the identified age-regulated circRNAs in independent samples from day 10 and day 40 flies, we validated that 6/6 circRNAs that were significantly increased in the RNA-seq data also showed significant increases in expression by RT-qPCR (Additional file 1: Figure S12). The single downregulated circRNA examined, *Eps-15*, did not show significantly decreased expression by RT-qPCR. We conclude that the downregulated circRNAs identified are likely false positives, and that our analysis probably underestimates the number of circRNAs that are upregulated with age (also see [104]).

Next, we performed pairwise comparisons to identify significantly age-regulated circRNAs between the different ages. Notably, a comparison between day 10 and day 30 or 40 identified many more significantly upregulated circRNAs than day 10 compared with day 20 or 25 (Fig. 5c). Next, we plotted expression of all detected circRNAs that met a minimum read cutoff of one unique read per library (14 read count minimum per circRNA) for all time points, and compared circRNA abundance between day 10 and each subsequent time point (Fig. 5d). Wilcoxon rank sum test with continuity correction revealed significant increases in circRNA abundance during aging between days 10 and 25, 30 or 40, but not between day 10 and 20. Importantly, changes in levels of individual circRNAs between days 10 and 40 occurred independent of the host gene mRNA (Fig. 5e), indicating that changes in expression of the host gene do not influence circRNA levels. Thus, we conclude that either enhanced biogenesis through increased backsplicing, or exceptional stability of circRNAs underlies their age-dependent accumulation. While we cannot distinguish between these possibilities, the genes from which abundant circRNAs are generated were significantly longer and more heavily spliced than all other genes (Additional file 1: Figure S13), suggesting that any alterations in splicing would be likely to affect circRNA biogenesis. Although circRNA host genes shared these characteristics (gene length, splicing) with the age downregulated genes, only 27 of the 218 genes with highly-expressed circRNA were significantly downregulated with age, whereas 5 circRNA host genes were upregulated. Together, these data suggest that circRNA biogenesis does not substantially influence expression of most host genes.

If circRNA abundance progressively increases with age, then we would expect that overall circRNA levels would correlate with age in photoreceptor neurons. Indeed, the mean circRNA expression level (CPM) for the 35 age upregulated circRNAs correlated highly with age (Fig. 5f, $r^2 = 0.92$). Because the photoreceptor transcriptome data is based on nuclear RNA, and showed lower levels of cytoplasmic transcripts such as mitochondrial genes, the increase in circRNA abundance with age is likely to reflect circRNAs that are retained in the nucleus. Together, these results suggest that the increased circRNA levels observed in aging heads may be driven by accumulation of circRNAs in neurons, including photoreceptors.

Discussion

In this study, we describe gene expression changes in aging photoreceptor neurons that correlate with visual senescence. Many genes required for neuronal function, such as those involved in ion transport or synaptic transmission, are downregulated in photoreceptors during aging. Moreover, the decreased expression of these genes begins at the earliest stages of the aging process when decreased phototaxis is first observed (day 25), and precedes the upregulation of most stress-response genes. The age-related decrease in expression of genes required for neuron function is evolutionarily conserved across a variety of organisms including flies, worms and vertebrates [105]; the mouse retina shows decreased expression of phototransduction genes with age [106, 107], and there is reduced expression of genes involved in synaptic plasticity in the brain of elderly humans [108–110]. Since 40 day old flies show no significant retinal degeneration, the decreased expression of genes required for neuronal function could account for the decreased phototaxis observed in these flies. Humans and rhesus macaques show much stronger age-dependent repression of neuronal genes in the brain than mice, leading to the conclusion that repression mechanisms have evolved recently [109]. However, genes involved in synaptic transmission are downregulated in our aging photoreceptor data, and in two independent studies from aging *Drosophila* heads [111, 112]. Similar to observations from aging gene expression studies in other tissues [98], we observed a general upregulation of stress response genes, including unfolded protein response and DNA damage response genes in aging photoreceptors. Notably, these changes occurred mostly mid- to late in aging photoreceptors, subsequent to the downregulation of many of the neuronal-specific genes discussed above.

Interestingly, 380 of the 555 age-regulated genes in *Drosophila* photoreceptors have identifiable human homologs, six of which are associated with retinal disease (RetNet: http://www.sph.uth.tmc.edu/RetNet/): *Cyp4c3* (*CYP4V2*), *l(1)G0007* (*DHX38*), *CG5291* (*PDZD7*), *CG3662* (*ITM2B*), *krz* (*SAG*), and *Cct1* (*PCYT1A*). For example, the human homolog of the age down-regulated gene *Cyp4c3*, *CYP4V2*, is associated with a recessive inherited retinal disorder, Bietti crystalline corneoretinal dystrophy that involves progressive age-associated retinal dystrophy. In addition, heterozygous mutation in *ITM2B*, the human homolog of the age down-regulated gene *CG3662*, is associated with another late-onset retinal dystrophy. The human homolog of *l(1)G0007*, *DHX38* (*PRP16*), which encodes an ATP-dependent RNA helicase involved in splicing that is upregulated with age, is associated with a recessive early-onset form of retinitis pigmentosa. The age-regulated genes with human homologs associated with retinal disease provide additional candidate genes that could impact either visual function or photoreceptor survival at older ages.

Here, we show that combinations of promoter motifs strongly predict whether a gene will be up- or downregulated with age, indicating an important role of transcription factors in driving changes in the transcriptional landscape of aging photoreceptors. We find that three of the transcription factors that provide best matches to the top motifs for the downregulated genes are annotated as being negative regulators of transcription (*Negative regulation of transcription from RNA polymerase II promoter, GO:0000122; rn, h, ovo*) whereas seven transcription factors are described as positive regulators of transcription (*Positive regulation of transcription from RNA polymerase II promoter, GO:0045944; sd, Cf2, br, ovo, onecut, SoxN, Adf1*). Similarly, three transcription factors that match the top motifs for the upregulated genes are negative regulators (*Dfd, dsx, Blimp-1*), whereas eight are positive regulators (*Dfd, Mad, kay, br, dsx, Trl, vvl, Mef2*). These data suggest that alterations in both transcription activation and repression are involved in the gene expression changes observed in aging photoreceptors. We note that caution should be used in definitively assigning specific transcription factors to each of the top motifs, since some of the lower scoring transcription factor matches for particular motifs might be more biologically relevant to aging; matches assigned in this study represent highest scoring matches for expressed transcription factors. For example, the CHES-1-like transcription factor is a close match to top motifs for both the upregulated (motif 8) and downregulated (motif 19) genes, but is outscored by broad (*br*) in both instances (Additional file 7: Table S6). CHES-1-like is required for hypoxia-induced inhibition of protein translation in cultured *Drosophila* cells [113], whereas broad is expressed in neurons and is required for some behaviors in *Drosophila* [114, 115]. Thus, both CHES-1-like and broad represent candidate transcription factors that could bind top motifs in the up or downregulated genes, and functional studies would be required to determine whether either of these transcription factors bound the target genes that we identified via motif analysis in photoreceptors. An additional limitation of the motif analysis used in this study was that we restricted our search to sequence motifs within 500 bp of the transcription start site. Many transcription factors are known to have longer range effects, and our sequence motif analysis would not identify these factors. Gene network analysis

of the age-regulated genes using GeneMania reveals some additional transcription factors and signaling kinases that could be involved in the mechanisms regulating these genes (Additional file 9: Table S8). For example, the transcription factor *crooked legs* (*crol*) and the MAP kinase *misshapen* (*msn*) are not age-regulated at the transcript level in photoreceptors, but are highly co-regulated with the age-regulated genes in terms of physical and genetic interactions, and co-expression analysis.

One of the challenges in identifying mechanisms that could drive age-related changes in gene expression is the cellular heterogeneity present in many aging gene expression studies [98]. Previous studies of gene expression in *Drosophila* heads using microarrays compared ages ranging from day 1 to 80 [111, 116, 117]; recently, age-regulated genes were also identified in heads using RNA-seq [112, 118]. A comparison of our data with the 2914 age-regulated genes identified in the most recent RNA-seq study showed an overlap of only 348 of our 555 age-regulated genes [112]. These data suggest that our cell-type specific approach identifies a subset of age-regulated genes that are masked by the cellular heterogeneity present in whole heads. However, the differences in the age-regulated genes identified might also reflect the inherent difficulty in comparing the nuclear and cytoplasmic transcriptomes. Since the nuclear enrichment strategy used in our approach biases the data towards genes that are actively transcribed, age-associated changes in the storage pool of cytoplasmic mRNAs available for translation are not reflected in our current photoreceptor data. However, we note that our qPCR analysis of individual age-regulated genes in dissected eyes largely mirrored the results from photoreceptor nuclei. Complementary cell type-specific data on the ribosome-bound pool of mRNAs destined for translation could be obtained using the translating ribosome affinity purification (TRAP) technique, based on cell-specific expression of an EGFP-tagged ribosomal subunit [119, 120].

Conclusion

The long-lived, post-mitotic nature of neurons makes them uniquely vulnerable to the long-term accumulation of genomic damage and oxidative stress. Here, we show that gene expression changes in photoreceptors precede retinal degeneration and correlate with observed decreases in visual function. Further, we demonstrate that the transcriptional landscape of aging photoreceptors is driven in large part by transcription factors. Notably, highly expressed, long and heavily spliced genes show a bias towards downregulation while circRNA levels increase strongly with age, suggesting that other gene expression mechanisms become altered with age.

The identification of potential regulatory mechanisms that drive changes in photoreceptors provides targets to delay gene expression changes in aging neurons, and thereby postpone the onset of ocular disease.

Methods

Fly strains, aging and phototaxis assays

Flies homozygous for KASH-GFP, $P\{w^{+mC} = UAS\text{-}GFP\text{-}Msp300KASH\}attP2$, under the control of Rh1-Gal4 $(P\{ry^{+t7.2} = rh1\text{-}GAL4\}3, ry^{506}$, BL8691] were raised in 12:12 h light:dark cycle at 25 °C on standard fly food [121]. For aging studies, flies were collected on the day of eclosion (day 1) and transferred to fresh vials every two days. For RNA-seq studies, 400 male flies were harvested between 10 am and 12 pm on the indicated day for each time point. For the survival curve, male flies were counted every 5 days until death. For phototaxis assays, 27-33 male flies were tested per assay ($n = 13$ assays) using a custom-built T-maze apparatus [23, 122] with dark equilibration time of 10 min and choice time of 2 min.

Immunostaining and western blotting

Retinal degeneration was assessed by confocal microscopy of adult retinas immunostained with phalloidin (Thermo Fisher; cat# A22287) and anti-Rhodopsin 1 (1:50, Developmental Studies Hybridoma Bank, cat# 4C5). Detailed protocols are provided via PURR. Western blotting analysis was performed using 40 µg of protein extracted from dissected eyes using the following antibodies: anti-GFP (rabbit; BioVision; 1:1000; cat# 3992).

Nuclei immuno-enrichment

For each sample, 400 male adult flies were anesthetized, frozen in liquid nitrogen and stored at −80 °C. For nuclei isolation, frozen flies were submitted to five rounds of vortexing and cooling in liquid nitrogen and heads were separated from thoracicoabdominal segments, wings and legs using two different-sized pre-chilled sieves (Hogentogler, 710 µm and 425 µm pore sizes). Separated heads were transferred into 1 mL of Nuclei Extraction Buffer (15 mM Hepes [Na+], pH 7.5, 10 mM KCl, 5 mM MgCl$_2$) in a Dounce homogenizer and incubated on ice for 5 min. Nuclei were extracted by using five strokes with a loose pestle, followed by an incubation on ice for 5 min and subsequent 5 strokes with a loose pestle. Head homogenate was filtered through a 40 µm Falcon cell strainer (VWR, cat # 21008-949) and immunoprecipitated with 10 µg of GFP antibody (Roche, cat # 11814460001) as previously described [29] with the following modifications: The salt concentration in the PBS wash buffer was supplemented to a final concentration of 300 mM NaCl, and five washes were conducted. Detailed protocols are provided via PURR.

RNA isolation and qPCR analysis

RNA for RNA-seq experiments was isolated using Trizol (Invitrogen). Quantitative real time PCR (qPCR) analysis for mRNA or circRNAs were performed on independent post-IP samples for day 10 and 40 relative to a standard curve of serially diluted cDNA generated using random hexamers as previously described [29]. qPCR analysis of heads, eyes, antennae and bodies from mixed flies was performed on total RNA isolated using Direct-zol RNA Micro-prep kit (Zymo Research, Cat. # R2062). Relative expression for each gene was normalized to the geometric mean of two reference genes based on MIQE guidelines [123]. Primers are listed in Additional file 10: Table S9.

Transcriptome library construction and high throughput sequencing (RNA-seq)

The cDNA libraries were generated from 10 ng of total nuclear RNA using the NuGEN Ovation RNA seq Systems 1-16 for Model Organism (NuGEN, cat #0350). RNA was DNAse treated, and single-stranded DNA was generated using both random hexamer and oligo-dT primers, and then depleted for ribosomal DNA. The cDNA libraries were ligated to unique adaptors and multiplexed libraries were sequenced with Illumina HiSeq 2500 technology. Three biological replicates were analyzed for each sample. Day 30 sample #3 was subsequently discarded due to poor mapping. Single-end 50 bp reads were sequenced for the post and pre day 10 samples, and paired-end 100 bp reads were sequenced for all aging samples (days 10 – 40).

RNA-seq data analysis

Reads were trimmed using Trimmomatic (v0.36). Quality trimmed reads were mapped to the *D. melanogaster* genome (BDGP6.89) using bowtie-2 (v2.3.2) and Tophat (v2.1.1). Counts were identified for each gene or exon using Htseq-count (v0.7.1) with strand-specific conditions (fr-secondstrand) and default parameters. Differential expression analysis was performed on genes with CPM > 1 in at least three of the samples. Differentially expressed genes between post and pre samples were identified using the *glmTreat* function in edgeR (v3.18.1) [124] with a FDR < 0.05 and FC > 2. Age-regulated genes were identified using maSigPro [40] with a FDR < 0.05 and clustered using k-means clustering based on relative mean expression values (maximum normalized count value set to one) for each time point.

Functional annotation analysis

All functional enrichment analyses were performed relative to the background gene set of all expressed genes with

CPM > 1 in at least three of the samples. GO term enrichment analysis on post-enriched or reduced genes was performed using topGO (v2.28.0) [125]. Related significantly-enriched GO terms were grouped using hierarchical clustering based on shared gene members. The DAVID functional annotation tool [43] was used to identify enriched functional classes in the age-regulated genes using the default parameters. Human homologs of *Drosophila* age-regulated genes were identified using the RetNet Database (RetNet, http://www.sph.uth.tmc.edu/RetNet/).

Motif analysis

Significantly-enriched promoter motifs were identified separately for up and downregulated genes using HOMER (v4.9, Hypergeometric Optimization of Motif EnRichment) [89] with the following parameters: motif length of 8, 10 or 12 bp within 500 bp upstream or downstream from the transcription start site. Motif enrichment analyses were performed relative to the background gene set of all expressed genes with CPM > 1 in at least three of the samples (7580 genes). The presence (1) or absence (0) of each of the 40 (up) or 41 (down) significantly-enriched motifs identified was determined for each gene in the background gene set, and these data were used for subsequent ROC analysis using pROC (v1.10.0) [126]. As a control, the maximum AUC computed from 100 random matrices was used. Each of the matrices were composed of 7580 rows (genes) with either presence (1) or absence (0) of a motif assigned based on random probabilities between 0 and 0.145 (the maximum number of genes in the background gene set matching to any specific motif was 14.5%). Sequence motifs were plotted using seqLogo (v1.42.0). Network analysis for co-occurring motifs was performed using igraph (v1.0.1), and for age-regulated genes using GeneMania [127]. To identify candidate transcription factors that bind each sequence motif, we compared the top motifs to the HOMER insect transcription factor database using the default parameters. The position weight matrices for an additional recently characterized 242 *Drosophila* transcription factors were added to the HOMER insect database [90].

Gene characteristic analysis

The total gene length (including introns) was determined as the gene length of the most abundant expressed transcript isoform. Transcript abundance and exon expression (exons with count >1 in at least one sample) were determined using exon counts obtained using Htseq-count from the photoreceptors aging RNA-seq data. Expression levels (RPKM) represent mean RPKM value across all time points and samples for a given gene. Kruskal–Wallis tests were performed to identify significant differences in the distribution of gene characteristics between any group. Pairwise Wilcoxon Rank Sum Tests were then performed to identify which groups exhibited significant differences in the distribution of the examined characteristic and FDR values were determined using a Benjamini and Hochberg correction. ROC analysis was performed using pROC (v1.10.0) [126].

circRNA analysis

Trimmed reads were also used as input for CIRI2 [102, 103] to identify circRNAs that mapped to annotated splice sites. A circRNA junction scaffold of 170 nts in length (85 nts of the downstream exonic junction and 85 nts of the upstream exonic junction) was generated for each circRNA using Bedtools getfasta [128]. Reads were mapped to the circRNA junction scaffold with Bowtie2 using the following option *–score-min = C,-15,0*. PCR and sequencing duplicates were removed using Picard MarkDuplicates (http://broadinstitute.github.io/picard/). Custom scripts were used to ensure that reads overlapped the circRNA junction by a minimum of 15 nts. Reads were assigned to individual circRNA records using Featurecounts [129], and only circRNA records with a minimum average of 1 read per library were used for analysis (i.e. 6 read minimum across 6 libraries). Read counts were normalized to CPM to account for library size variation. For pairwise comparisons between time points, we required a minimum of one count per library (i.e. six read minimum cut-off between day 10 and 40).

Database accession numbers

RNA-seq expression data are available in the Gene Expression Omnibus (GEO) repository through GEO series accession numbers GSE93128 and GSE83431. Raw data, detailed protocols and R custom scripts used for analysis have been deposited in the Purdue University Research Repository (PURR) as a publically available, archived data set and can be accessed using https://doi.org/doi:10.4231/R7736P29.

Additional files

Additional file 1: Complete Supplemental Figures. **Figure S1.** Affinity-purified nuclear RNA is enriched for photoreceptor-expressed genes. **Figure S2.** Rh1-Gal4 drives GFP expression in antennal sensory neurons. **Figure S3.** Relative sensory neuron proportions and yields of affinity-purified nuclear RNA do not change with age. **Figure S4.** qPCR of selected age-regulated genes. **Figure S5.** K-means clustering of age-regulated genes based on temporal expression pattern. **Figure S6.** Top promoter motifs that predict age-related expression changes. **Figure S7.** Promoter motifs with the best predictive power co-occur frequently with a variety of other motifs. **Figure S8.** Distribution of the top motifs in age upregulated genes. **Figure S9.** Distribution of the top motifs in age upregulated genes. **Figure S10.** Distribution of the top motifs between expression clusters. **Figure S11.** Downregulated gene clusters are enriched for longer, more highly expressed and more heavily spliced genes. **Figure S12.** qPCR of selected age-regulated circRNAs. **Figure S13.** circRNA-containing host genes are enriched for longer and more heavily spliced genes. (PDF 4807 kb)

Additional file 2: Table S1. Significantly post-enriched or post-reduced genes in affinity-enriched photoreceptor nuclear RNA. (XLSX 89 kb)

Additional file 3: Table S2. GO term analysis of significantly post-reduced genes. (XLSX 16 kb)

Additional file 4: Table S3. Age-regulated genes in photoreceptor neurons. (XLSX 121 kb)

Additional file 5: Table S4. GO term analysis of age-regulated genes in photoreceptor neurons. (XLSX 14 kb)

Additional file 6: Table S5. Enriched sequence motifs for age-regulated genes. (XLSX 48 kb)

Additional file 7: Table S6. Transcription factors matches for top motifs identified for age-regulated genes. (XLSX 13 kb)

Additional file 8: Table S7. circRNA lists. (XLSX 96 kb)

Additional file 9: Table S8. GeneMania analysis of age-regulated genes in photoreceptors. (XLSX 174 kb)

Additional file 10: Table S9. Primers used in this study. (XLSX 13 kb)

Abbreviation

AUC: Area under the curve; circRNA: Circular RNA; CPM: Counts per million; GEO: Gene Expression Omnibus; GFP: Green fluorescent protein; GO: Gene ontology; HOMER: Hypergeometric Optimization of Motif EnRichment; HSD: Honest significant different; KASH: Klarsicht, Anc-1, Syn3-1 homology; mTOR: Mechanistic target of rapamycin; qPCR: Quantitative PCR; Rh1: Rhodopsin 1; ROC: Receiver Operating Characteristic; RPKM: Reads per kilobase per million reads; TRAP: Translating ribosome affinity purification

Acknowledgements

Fly stocks from the Bloomington Drosophila Stock Center (NIH P40OD018537) and information from FlyBase were used in this study.

Funding

Support from the American Cancer Society Institutional Research Grant (IRG #58-006-53) to the Purdue University Center for Cancer Research, and the Indiana Clinical and Translational Sciences Institute funded by UL1TR001108 are gratefully acknowledged. RNA-seq data were collected using the genomics core facility and images were obtained at Bindley Bioscience Imaging Facility, Purdue University, supported by NIH P30 CA023168 to the Purdue University Center for Cancer Research. Research reported in this publication was supported by the National Eye Institute of the NIH under Award Number R01EY024905 to VW, and the National Institutes on Aging of the NIH under Award Number R15AG052931 to PM. The content is solely the responsibility of the authors and does not necessarily represent the official views of the NIH.

Authors' contributions

HH, with support from KB, conducted all biological experiments except for phototaxis assays by SE. PM, JR, CV and VW analyzed the data with consultation from RD. DC and PM performed the circRNA analysis. HH and VW wrote the manuscript in consultation with the other authors. All authors read and approved the final manuscript.

Competing interests

The authors declare that they have no competing interests.

Author details

[1]Department of Biochemistry, Purdue University, West Lafayette, IN 47907, USA. [2]Department of Statistics, Purdue University, West Lafayette, IN 47907, USA. [3]Department of Biology, University of Nevada, Reno, NV 89557, USA. [4]Carnegie Mellon University, Pittsburgh, PA 15213, USA. [5]Purdue University Center for Cancer Research, Purdue University, West Lafayette 47907, USA.

References

1. Klein R, Klein BE. The prevalence of age-related eye diseases and visual impairment in aging: current estimates. Invest Ophthalmol Vis Sci. 2013; 54(14):ORSF5–ORSF13.
2. Alavi MV. Aging and vision. Adv Exp Med Biol. 2016;854:393–9.
3. Yang HJ, Ratnapriya R, Cogliati T, Kim JW, Swaroop A. Vision from next generation sequencing: multi-dimensional genome-wide analysis for producing gene regulatory networks underlying retinal development, aging and disease. Prog Retin Eye Res. 2015;46:1–30.
4. Curcio CA, Millican CL, Allen KA, Kalina RE. Aging of the human photoreceptor mosaic: evidence for selective vulnerability of rods in central retina. Invest Ophthalmol Vis Sci. 1993;34(12):3278–96.
5. Curcio CA. Photoreceptor topography in ageing and age-related maculopathy. Eye (Lond). 2001;15(Pt 3):376–83.
6. Gao H, Hollyfield JG. Aging of the human retina. Differential loss of neurons and retinal pigment epithelial cells. Invest Ophthalmol Vis Sci. 1992;33(1):1–17.
7. Birch DG, Anderson JL. Standardized full-field electroretinography. Normal values and their variation with age. Arch Ophthalmol. 1992;110(11):1571–6.
8. Bonnel S, Mohand-Said S, Sahel JA. The aging of the retina. Exp Gerontol. 2003;38(8):825–31.
9. Freund PR, Watson J, Gilmour GS, Gaillard F, Sauve Y. Differential changes in retina function with normal aging in humans. Doc Ophthalmol. 2011;122(3):177–90.
10. Shinomori K, Werner JS. Aging of human short-wave cone pathways. Proc Natl Acad Sci U S A. 2012;109(33):13422–7.
11. Gresh J, Goletz PW, Crouch RK, Rohrer B. Structure-function analysis of rods and cones in juvenile, adult, and aged C57bl/6 and Balb/c mice. Vis Neurosci. 2003;20(2):211–20.
12. Kolesnikov AV, Fan J, Crouch RK, Kefalov VJ. Age-related deterioration of rod vision in mice. J Neurosci. 2010;30(33):11222–31.
13. Parapuram SK, Cojocaru RI, Chang JR, Khanna R, Brooks M, Othman M, Zareparsi S, Khan NW, Gotoh N, Cogliati T, et al. Distinct signature of altered homeostasis in aging rod photoreceptors: implications for retinal diseases. PLoS One. 2010;5(11):e13885.
14. Samuel MA, Zhang Y, Meister M, Sanes JR. Age-related alterations in neurons of the mouse retina. J Neurosci. 2011;31(44):16033–44.
15. Wright AF, Chakarova CF, Abd El-Aziz MM, Bhattacharya SS. Photoreceptor degeneration: genetic and mechanistic dissection of a complex trait. Nat Rev Genet. 2010;11(4):273–84.
16. Carbone MA, Yamamoto A, Huang W, Lyman RA, Meadors TB, Yamamoto R, Anholt RR, Mackay TF. Genetic architecture of natural variation in visual senescence in drosophila. Proc Natl Acad Sci U S A. 2016;113(43):E6620–9.
17. Simon AF, Liang DT, Krantz DE. Differential decline in behavioral performance of Drosophila Melanogaster with age. Mech Ageing Dev. 2006;127(7):647–51.
18. Grotewiel MS, Martin I, Bhandari P, Cook-Wiens E. Functional senescence in Drosophila Melanogaster. Ageing Res Rev. 2005;4(3):372–97.
19. Rister J, Desplan C. The retinal mosaics of opsin expression in invertebrates and vertebrates. Dev Neurobiol. 2011;71(12):1212–26.
20. Montell C. Drosophila visual transduction. Trends Neurosci. 2012;35(6):356–63.
21. Lebourg E, Badia J. Decline in photopositive tendencies with age in drosophila-Melanogaster (Diptera, Drosophilidae). J Insect Behav. 1995;8(6):835–45.
22. Kurada P, O'Tousa JE. Retinal degeneration caused by dominant rhodopsin mutations in drosophila. Neuron. 1995;14(3):571–9.
23. Gorostiza EA, Colomb J, Brembs B. A decision underlies phototaxis in an insectE. Axel Gorostiza, Julien Colomb, Björn Brembs Open Biol. 2016;6: 160229. doi:10.1098/rsob.160229. Published 21 December 2016
24. Katz B, Minke B. Drosophila photoreceptors and signaling mechanisms. Front Cell Neurosci. 2009;3:2.
25. Ready DF, Hanson TE, Benzer S. Development of the Drosophila Retina, a neurocrystalline lattice. Dev Biol. 1976;53(2):217–40.
26. Steiner FA, Talbert PB, Kasinathan S, Deal RB, Henikoff S. Cell-type-specific nuclei purification from whole animals for genome-wide expression and chromatin profiling. Genome Res. 2012;22(4):766–77.
27. Chou WH, Hall KJ, Wilson DB, Wideman CL, Townson SM, Chadwell LV, Britt SG. Identification of a novel drosophila opsin reveals specific patterning of the R7 and R8 photoreceptor cells. Neuron. 1996;17(6):1101–15.
28. Yoshihara Y, Mizuno T, Nakahira M, Kawasaki M, Watanabe Y, Kagamiyama H, Jishage K, Ueda O, Suzuki H, Tabuchi K, et al. A genetic approach to visualization of multisynaptic neural pathways using plant lectin transgene. Neuron. 1999;22(1):33–41.

29. Ma J, Weake VM. Affinity-based isolation of tagged nuclei from drosophila tissues for gene expression analysis. J Vis Exp. 2014;85

30. Ma J, Brennan KJ, D'Aloia MR, Pascuzzi PE, Weake VM. Transcriptome profiling identifies Multiplexin as a target of SAGA Deubiquitinase activity in Glia required for precise axon guidance during drosophila visual development. G3 (Bethesda). 2016;6(8):2435–45.

31. Gopfert MC, Robert D. Biomechanics. Turning the key on drosophila audition. Nature. 2001;411(6840):908.

32. Senthilan PR, Piepenbrock D, Ovezmyradov G, Nadrowski B, Bechstedt S, Pauls S, Winkler M, Mobius W, Howard J, Gopfert MC. Drosophila auditory organ genes and genetic hearing defects. Cell. 2012;150(5):1042–54.

33. Chou WH, Huber A, Bentrop J, Schulz S, Schwab K, Chadwell LV, Paulsen R, Britt SG. Patterning of the R7 and R8 photoreceptor cells of drosophila: evidence for induced and default cell-fate specification. Development. 1999;126(4):607–16.

34. Fortini ME, Rubin GM. Analysis of cis-acting requirements of the Rh3 and Rh4 genes reveals a bipartite organization to rhodopsin promoters in Drosophila Melanogaster. Genes Dev. 1990;4(3):444–63.

35. Mismer D, Michael WM, Laverty TR, Rubin GM. Analysis of the promoter of the Rh2 opsin gene in Drosophila Melanogaster. Genetics. 1988;120(1):173–80.

36. Mismer D, Rubin GM. Analysis of the promoter of the ninaE opsin gene in Drosophila Melanogaster. Genetics. 1987;116(4):565–78.

37. Papatsenko D, Sheng G, Desplan C. A new rhodopsin in R8 photoreceptors of drosophila: evidence for coordinate expression with Rh3 in R7 cells. Development. 1997;124(9):1665–73.

38. Yang Z, Edenberg HJ, Davis RL. Isolation of mRNA from specific tissues of drosophila by mRNA tagging. Nucleic Acids Res. 2005;33(17):e148.

39. Ishikawa Y, Kamikouchi A. Auditory system of fruit flies. Hear Res. 2016;338:1–8.

40. Nueda MJ, Tarazona S, Conesa A. Next maSigPro: updating maSigPro bioconductor package for RNA-seq time series. Bioinformatics. 2014;30(18):2598–602.

41. Spies D, Ciaudo C. Dynamics in Transcriptomics: advancements in RNA-seq time course and downstream analysis. Comput Struct Biotechnol J. 2015;13:469–77.

42. Du M, Mangold CA, Bixler GV, Brucklacher RM, Masser DR, Stout MB, Elliott MH, Freeman WM. Retinal gene expression responses to aging are sexually divergent. Mol Vis. 2017;23:707–17.

43. Huang DW, Sherman BT, Lempicki RA. Systematic and integrative analysis of large gene lists using DAVID bioinformatics resources. Nat Protoc. 2009;4(1):44–57.

44. Chan SH, Yu AM, McVey M. Dual roles for DNA polymerase theta in alternative end-joining repair of double-strand breaks in drosophila. PLoS Genet. 2010;6(7):e1001005.

45. Kane DP, Shusterman M, Rong Y, McVey M. Competition between replicative and translesion polymerases during homologous recombination repair in drosophila. PLoS Genet. 2012;8(4):e1002659.

46. Eeken JC, Romeijn RJ, de Jong AW, Pastink A, Lohman PH. Isolation and genetic characterisation of the drosophila homologue of (SCE)REV3, encoding the catalytic subunit of DNA polymerase zeta. Mutat Res. 2001;485(3):237–53.

47. Missirlis F, Ulschmid JK, Hirosawa-Takamori M, Gronke S, Schafer U, Becker K, Phillips JP, Jackle H. Mitochondrial and cytoplasmic thioredoxin reductase variants encoded by a single drosophila gene are both essential for viability. J Biol Chem. 2002;277(13):11521–6.

48. Pathak C, Jaiswal YK, Vinayak M. Queuine promotes antioxidant defence system by activating cellular antioxidant enzyme activities in cancer. Biosci Rep. 2008;28(2):73–81.

49. Hofling C, Indrischek H, Hopcke T, Waniek A, Cynis H, Koch B, Schilling S, Morawski M, Demuth HU, Rossner S, et al. Mouse strain and brain region-specific expression of the glutaminyl cyclases QC and isoQC. Int J Dev Neurosci. 2014;36:64–73.

50. Ognjenovic J, Wu J, Matthies D, Baxa U, Subramaniam S, Ling J, Simonovic M. The crystal structure of human GlnRS provides basis for the development of neurological disorders. Nucleic Acids Res. 2016;44(7):3420–11.

51. Garcia-Huerta P, Bargsted L, Rivas A, Matus S, Vidal RL. ER chaperones in neurodegenerative disease: folding and beyond. Brain Res. 2016;1648(Part B):580–7.

52. Kubota H, Hynes G, Carne A, Ashworth A, Willison K. Identification of six Tcp-1-related genes encoding divergent subunits of the TCP-1-containing chaperonin. Curr Biol. 1994;4(2):89–99.

53. Ohtsuka K, Suzuki T. Roles of molecular chaperones in the nervous system. Brain Res Bull. 2000;53(2):141–6.

54. de Nadal E, Ammerer G, Posas F. Controlling gene expression in response to stress. Nat Rev Genet. 2011;12(12):833–45.

55. Marash L, Liberman N, Henis-Korenblit S, Sivan G, Reem E, Elroy-Stein O, Kimchi A. DAP5 promotes cap-independent translation of Bcl-2 and CDK1 to facilitate cell survival during mitosis. Mol Cell. 2008;30(4):447–59.

56. Majzoub K, Hafirassou ML, Meignin C, Goto A, Marzi S, Fedorova A, Verdier Y, Vinh J, Hoffmann JA, Martin F, et al. RACK1 controls IRES-mediated translation of viruses. Cell. 2014;159(5):1086–95.

57. Hall DJ, Grewal SS, de la Cruz AF, Edgar BA. Rheb-TOR signaling promotes protein synthesis, but not glucose or amino acid import, in drosophila. BMC Biol. 2007;5:10.

58. Johnson SC, Rabinovitch PS, Kaeberlein M. mTOR is a key modulator of ageing and age-related disease. Nature. 2013;493(7432):338–45.

59. Juusola M, Niven JE, French AS. Shaker K+ channels contribute early nonlinear amplification to the light response in drosophila photoreceptors. J Neurophysiol. 2003;90(3):2014–21.

60. Vahasoyrinki M, Niven JE, Hardie RC, Weckstrom M, Juusola M. Robustness of neural coding in drosophila photoreceptors in the absence of slow delayed rectifier K+ channels. J Neurosci. 2006;26(10):2652–60.

61. Palladino MJ, Bower JE, Kreber R, Ganetzky B. Neural dysfunction and neurodegeneration in drosophila Na+/K+ ATPase alpha subunit mutants. J Neurosci. 2003;23(4):1276–86.

62. Hardie RC. Voltage-sensitive potassium channels in drosophila photoreceptors. J Neurosci. 1991;11(10):3079–95.

63. Bronk P, Nie Z, Klose MK, Dawson-Scully K, Zhang J, Robertson RM, Atwood HL, Zinsmaier KE. The multiple functions of cysteine-string protein analyzed at drosophila nerve terminals. J Neurosci. 2005;25(9):2204–14.

64. Frolov RV, Bagati A, Casino B, Singh S. Potassium channels in drosophila: historical breakthroughs, significance, and perspectives. J Neurogenet. 2012;26(3-4):275–90.

65. Saumweber T, Weyhersmuller A, Hallermann S, Diegelmann S, Michels B, Bucher D, Funk N, Reisch D, Krohne G, Wegener S, et al. Behavioral and synaptic plasticity are impaired upon lack of the synaptic protein SAP47. J Neurosci. 2011;31(9):3508–18.

66. Burg MG, Sarthy PV, Koliantz G, Pak WL. Genetic and molecular identification of a drosophila histidine decarboxylase gene required in photoreceptor transmitter synthesis. EMBO J. 1993;12(3):911–9.

67. Weiss S, Kohn E, Dadon D, Katz B, Peters M, Lebendiker M, Kosloff M, Colley NJ, Minke B. Compartmentalization and Ca2+ buffering are essential for prevention of light-induced retinal degeneration. J Neurosci. 2012;32(42):14696–708.

68. Ballinger DG, Xue N, Harshman KD. A drosophila photoreceptor cell-specific protein, calphotin, binds calcium and contains a leucine zipper. Proc Natl Acad Sci U S A. 1993;90(4):1536–40.

69. Martin JH, Benzer S, Rudnicka M, Miller CA. Calphotin: a drosophila photoreceptor cell calcium-binding protein. Proc Natl Acad Sci U S A. 1993;90(4):1531–5.

70. Jeong K, Lee S, Seo H, Oh Y, Jang D, Choe J, Kim D, Lee JH, Jones WD. Ca-alpha1T, a fly T-type Ca2+ channel, negatively modulates sleep. Sci Rep. 2015;5:17893.

71. Rosenbaum EE, Vasiljevic E, Brehm KS, Colley NJ. Mutations in four glycosyl hydrolases reveal a highly coordinated pathway for rhodopsin biosynthesis and N-glycan trimming in Drosophila Melanogaster. PLoS Genet. 2014;10(5): e1004349.

72. Han J, Reddig K, Li HS. Prolonged G(q) activity triggers fly rhodopsin endocytosis and degradation, and reduces photoreceptor sensitivity. EMBO J. 2007;26(24):4966–73.

73. O'Tousa JE. Requirement of N-linked glycosylation site in drosophila rhodopsin. Vis Neurosci. 1992;8(5):385–90.

74. Brown G, Chen DM, Christianson JS, Lee R, Stark WS. Receptor demise from alteration of glycosylation site in drosophila opsin: electrophysiology, microspectrophotometry, and electron microscopy. Vis Neurosci. 1994;11(3):619–28.

75. Katanosaka K, Tokunaga F, Kawamura S, Ozaki K. N-linked glycosylation of drosophila rhodopsin occurs exclusively in the amino-terminal domain and functions in rhodopsin maturation. FEBS Lett. 1998;424(3):149–54.

76. Hall JC, Alahiotis SN, Strumpf DA, White K. Behavioral and biochemical defects in temperature-sensitive acetylcholinesterase mutants of Drosophila Melanogaster. Genetics. 1980;96(4):939–65.

77. Perrimon N, Smouse D, Miklos GLG. Developmental genetics of loci at the base of the X-chromosome of drosophila-Melanogaster. Genetics. 1989;121(2):313–31.

78. Dimitroff B, Howe K, Watson A, Campion B, Lee HG, Zhao N, O'Connor MB, Neufeld TP, Selleck SB. Diet and energy-sensing inputs affect TorC1-mediated axon misrouting but not TorC2-directed synapse growth in a drosophila model of tuberous sclerosis. PLoS One. 2012;7(2):e30722.

79. Mount SM, Salz HK. Pre-messenger RNA processing factors in the drosophila genome. J Cell Biol. 2000;150(2):F37–44.

80. Park JW, Parisky K, Celotto AM, Reenan RA, Graveley BR. Identification of alternative splicing regulators by RNA interference in drosophila. Proc Natl Acad Sci U S A. 2004;101(45):15974–9.

81. Juge F, Zaessinger S, Temme C, Wahle E, Simonelig M. Control of poly(a) polymerase level is essential to cytoplasmic polyadenylation and early development in drosophila. EMBO J. 2002;21(23):6603–13.

82. Okamura K, Ishizuka A, Siomi H, Siomi MC. Distinct roles for Argonaute proteins in small RNA-directed RNA cleavage pathways. Genes Dev. 2004;18(14):1655–66.

83. Wei W, Ba Z, Gao M, Wu Y, Ma Y, Amiard S, White CI, Rendtlew Danielsen JM, Yang YG, Qi Y. A role for small RNAs in DNA double-strand break repair. Cell. 2012;149(1):101–12.

84. Freeman A, Franciscovich A, Bowers M, Sandstrom DJ, Sanyal S. NFAT regulates pre-synaptic development and activity-dependent plasticity in drosophila. Mol Cell Neurosci. 2011;46(2):535–47.

85. Nguyen DNT, Rohrbaugh M, Lai ZC. The drosophila homolog of Onecut homeodomain proteins is a neural-specific transcriptional activator with a potential role in regulating neural differentiation. Mech Develop. 2000;97(1-2):57–72.

86. Ryoo HD, Domingos PM, Kang MJ, Steller H. Unfolded protein response in a drosophila model for retinal degeneration. EMBO J. 2007;26(1):242–52.

87. Mohan M, Herz HM, Smith ER, Zhang Y, Jackson J, Washburn MP, Florens L, Eissenberg JC, Shilatifard A. The COMPASS family of H3K4 methylases in drosophila. Mol Cell Biol. 2011;31(21):4310–8.

88. Aoyagi N, Wassarman DA. Genes encoding Drosophila Melanogaster RNA polymerase II general transcription factors: diversity in TFIIA and TFIID components contributes to gene-specific transcriptional regulation. J Cell Biol. 2000;150(2):F45–50.

89. Heinz S, Benner C, Spann N, Bertolino E, Lin YC, Laslo P, Cheng JX, Murre C, Singh H, Glass CK. Simple combinations of lineage-determining transcription factors prime cis-regulatory elements required for macrophage and B cell identities. Mol Cell. 2010;38(4):576–89.

90. Nitta KR, Jolma A, Yin Y, Morgunova E, Kivioja T, Akhtar J, Hens K, Toivonen J, Deplancke B, Furlong EE, et al. Conservation of transcription factor binding specificities across 600 million years of bilateria evolution. eLife 2015;4:e04837. doi:10.7554/eLife.04837.

91. Vandoren M, Bailey AM, Esnayra J, Ede K, Posakony JW. Negative regulation of proneural gene activity - hairy is a direct transcriptional repressor of Achaete. Genes Dev. 1994;8(22):2729–42.

92. Perkins KK, Admon A, Patel N, Tjian R. The drosophila Fos-related Ap-1 protein is a developmentally regulated transcription factor. Genes Dev. 1990;4(5):822–34.

93. Luo X, Puig O, Hyun J, Bohmann D, Jasper H. Foxo and Fos regulate the decision between cell death and survival in response to UV irradiation. EMBO J. 2007;26(2):380–90.

94. Gabel HW, Kinde B, Stroud H, Gilbert CS, Harmin DA, Kastan NR, Hemberg M, Ebert DH, Greenberg ME. Disruption of DNA-methylation-dependent long gene repression in Rett syndrome. Nature. 2015;522(7554):89–93.

95. Mohr C, Hartmann B. Alternative splicing in Drosophila neuronal development. J Neurogenet. 2014;28(3-4):199–215.

96. King IF, Yandava CN, Mabb AM, Hsiao JS, Huang HS, Pearson BL, Calabrese JM, Starmer J, Parker JS, Magnuson T, et al. Topoisomerases facilitate transcription of long genes linked to autism. Nature. 2013;501(7465):58–62.

97. Vermeij WP, Dollé MET, Reiling E, Jaarsma D, Payan-Gomez C, Bombardieri CR, Wu H, Roks AJM, Botter SM, van der Eerden BC, Youssef SA, Kuiper RV, Nagarajah B, van Oostrom CT, Brandt RMC, Barnhoorn S, Imholz S, Pennings JLA, de Bruin A, Gyenis Á, Pothof J, Vijg J, van Steeg H, Hoeijmakers JHJ. Restricted diet delays accelerated ageing and genomic stress in DNA-repair-deficient mice. Nature. 2016;537:427–31. doi:10.1038/nature19329.

98. Stegeman R, Weake VM. Transcriptional signatures of aging. J Mol Biol. 2017;

99. Conesa A, Madrigal P, Tarazona S, Gomez-Cabrero D, Cervera A, McPherson A, Szczesniak MW, Gaffney DJ, Elo LL, Zhang X, et al. A survey of best practices for RNA-seq data analysis. Genome Biol. 2016;17:13.

100. Westholm Jakub O, Miura P, Olson S, Shenker S, Joseph B, Sanfilippo P, Celniker Susan E, Graveley Brenton R, Lai Eric C. Genome-wide analysis of drosophila circular RNAs reveals their structural and sequence properties and age-dependent neural accumulation. Cell Rep. 2014;9(5):1966–80.

101. Ashwal-Fluss R, Meyer M, Pamudurti NR, Ivanov A, Bartok O, Hanan M, Evantal N, Memczak S, Rajewsky N. Kadener S: circRNA biogenesis competes with pre-mRNA splicing. Mol Cell. 2014;56(1):55–66.

102. Gao Y, Wang J, Zhao F. CIRI: an efficient and unbiased algorithm for de novo circular RNA identification. Genome Biol. 2015;16:4.

103. Gao Y, Zhang J, Zhao F. Circular RNA identification based on multiple seed matching. Brief Bioinform. 2017;bbx014. https://doi.org/10.1093/bib/bbx014.

104. Gruner H, Cortes-Lopez M, Cooper DA, Bauer M, Miura P. CircRNA accumulation in the aging mouse brain. Sci Rep. 2016;6:38907.

105. Bishop NA, Lu T, Yankner BA. Neural mechanisms of ageing and cognitive decline. Nature. 2010;464(7288):529–35.

106. Chowers I, Liu D, Farkas RH, Gunatilaka TL, Hackam AS, Bernstein SL, Campochiaro PA, Parmigiani G, Zack DJ. Gene expression variation in the adult human retina. Hum Mol Genet. 2003;12(22):2881–93.

107. Sharon D, Blackshaw S, Cepko CL, Dryja TP. Profile of the genes expressed in the human peripheral retina, macula, and retinal pigment epithelium determined through serial analysis of gene expression (SAGE). Proc Natl Acad Sci U S A. 2002;99(1):315–20.

108. Lu T, Pan Y, Kao SY, Li C, Kohane I, Chan J, Yankner BA. Gene regulation and DNA damage in the ageing human brain. Nature. 2004;429(6994):883–91.

109. Loerch PM, Lu T, Dakin KA, Vann JM, Isaacs A, Geula C, Wang J, Pan Y, Gabuzda DH, Li C, et al. Evolution of the aging brain transcriptome and synaptic regulation. PLoS One. 2008;3(10):e3329.

110. Berchtold NC, Cribbs DH, Coleman PD, Rogers J, Head E, Kim R, Beach T, Miller C, Troncoso J, Trojanowski JQ, et al. Gene expression changes in the course of normal brain aging are sexually dimorphic. Proc Natl Acad Sci U S A. 2008;105(40):15605–10.

111. Girardot F, Lasbleiz C, Monnier V, Tricoire H. Specific age-related signatures in drosophila body parts transcriptome. BMC Genomics. 2006;7:69.

112. Kuintzle RC, Chow ES, Westby TN, Gvakharia BO, Giebultowicz JM, Hendrix DA. Circadian deep sequencing reveals stress-response genes that adopt robust rhythmic expression during aging. Nat Commun. 2017;8:14529.

113. Lee SJ, Feldman R, O'Farrell PH. An RNA interference screen identifies a novel regulator of target of Rapamycin that mediates hypoxia suppression of translation in drosophila S2 cells. Mol Biol Cell. 2008;19(10):4051–61.

114. Zhou B, Williams DW, Altman J, Riddiford LM, Truman JW. Temporal patterns of broad isoform expression during the development of neuronal lineages in drosophila. Neural Dev. 2009;4:39.

115. Armstrong JD, Texada MJ, Munjaal R, Baker DA, Beckingham KM. Gravitaxis in Drosophila Melanogaster: a forward genetic screen. Genes Brain Behav. 2006;5(3):222–39.

116. Kim SN, Rhee JH, Song YH, Park DY, Hwang M, Lee SL, Kim JE, Gim BS, Yoon JH, Kim YJ, et al. Age-dependent changes of gene expression in the drosophila head. Neurobiol Aging. 2005;26(7):1083–91.

117. Neretti N, Wang P-Y, Brodsky AS, Nyguyen HH, White KP, Rogina B, Helfand SL. Long-lived Indy induces reduced mitochondrial reactive oxygen species production and oxidative damage. Proc Natl Acad Sci. 2009;106(7):2277–82.

118. Wood JG, Jones BC, Jiang N, Chang C, Hosier S, Wickremesinghe P, Garcia M, Hartnett DA, Burhenn L, Neretti N, et al. Chromatin-modifying genetic interventions suppress age-associated transposable element activation and extend life span in drosophila. Proc Natl Acad Sci U S A. 2016;113(40): 11277–82.

119. Doyle JP, Dougherty JD, Heiman M, Schmidt EF, Stevens TR, Ma G, Bupp S, Shrestha P, Shah RD, Doughty ML, et al. Application of a translational profiling approach for the comparative analysis of CNS cell types. Cell. 2008;135(4):749–62.

120. Heiman M, Kulicke R, Fenster RJ, Greengard P, Heintz N. Cell type-specific mRNA purification by translating ribosome affinity purification (TRAP). Nat Protoc. 2014;9(6):1282–91.

121. Lewis EB. A new standard food medium. Drosophila Information Service. 1960;34:117–8.

122. Tully T, Quinn WG. Classical conditioning and retention in normal and mutant Drosophila Melanogaster. J Comp Physiol A. 1985;157(2):263–77.

123. Bustin SA, Benes V, Garson JA, Hellemans J, Huggett J, Kubista M, Mueller R, Nolan T, Pfaffl MW, Shipley GL, et al. The MIQE guidelines: minimum information for publication of quantitative real-time PCR experiments. Clin Chem. 2009;55(4):611–22.

124. Robinson MD, McCarthy DJ, Smyth GK. edgeR: a Bioconductor package for differential expression analysis of digital gene expression data. Bioinformatics. 2010;26(1):139–40.

125. Alexa A, Rahnenfuhrer J. topGO: enrichment analysis for gene ontology. R Package Version. 2016;2260.https://bioconductor.org/packages/release/bioc/html/topGO.html.

126. Robin X, Turck N, Hainard A, Tiberti N, Lisacek F, Sanchez JC, Muller M. pROC: an open-source package for R and S+ to analyze and compare ROC curves. BMC Bioinformatics. 2011;12:77.

127. Montojo J, Zuberi K, Rodriguez H, Bader GD, Morris Q. GeneMANIA: fast gene network construction and function prediction for Cytoscape. F1000Res. 2014;3:153.

128. Quinlan AR, Hall IM. BEDTools: a flexible suite of utilities for comparing genomic features. Bioinformatics. 2010;26(6):841–2.

129. Liao Y, Smyth GK, Shi W. featureCounts: an efficient general purpose program for assigning sequence reads to genomic features. Bioinformatics. 2014;30(7):923–30.

4

Genome-wide association studies of fertility and calving traits in Brown Swiss cattle using imputed whole-genome sequences

Mirjam Frischknecht[1,2]* (iD), Beat Bapst[1], Franz R. Seefried[1], Heidi Signer-Hasler[2], Dorian Garrick[3], Christian Stricker[4], Intergenomics Consortium[5], Ruedi Fries[6], Ingolf Russ[7], Johann Sölkner[8], Anna Bieber[9], Maria G. Strillacci[10], Birgit Gredler-Grandl[1] and Christine Flury[2]

Abstract

Background: The detection of quantitative trait loci has accelerated with recent developments in genomics. The introduction of genomic selection in combination with sequencing efforts has made a large amount of genotypic data available. Functional traits such as fertility and calving traits have been included in routine genomic estimation of breeding values making large quantities of phenotypic data available for these traits. This data was used to investigate the genetics underlying fertility and calving traits and to identify potentially causative genomic regions and variants.

We performed genome-wide association studies for 13 functional traits related to female fertility as well as for direct and maternal calving ease based on imputed whole-genome sequences. Deregressed breeding values from ~1000–5000 bulls per trait were used to test for associations with approximately 10 million imputed sequence SNPs.

Results: We identified a QTL on BTA17 associated with non-return rate at 56 days and with interval from first to last insemination. We found two significantly associated non-synonymous SNPs within this QTL region. Two more QTL for fertility traits were identified on BTA25 and 29. A single QTL was identified for maternal calving traits on BTA13 whereas three QTL on BTA19, 21 and 25 were identified for direct calving traits. The QTL on BTA19 co-localizes with the reported BH2 haplotype. The QTL on BTA25 is concordant for fertility and calving traits and co-localizes with a QTL previously reported to influence stature and related traits in Brown Swiss dairy cattle.

Conclusion: The detection of QTL and their causative variants remains challenging. Combining comprehensive phenotypic data with imputed whole genome sequences seems promising. We present a QTL on BTA17 for female fertility in dairy cattle with two significantly associated non-synonymous SNPs, along with five additional QTL for fertility traits and calving traits. For all of these we fine mapped the regions and suggest candidate genes and candidate variants.

Keywords: Whole genome sequencing, Genome-wide association study, QTL discovery, Functional traits, Brown Swiss, Dairy cattle, Calving ease, Fertility

* Correspondence: Mirjam.frischknecht@qualitasag.ch
[1]Qualitas AG, Chamerstrasse 56a, 6300 Zug, Switzerland
[2]School of Agricultural, Forest and Food Sciences HAFL, Bern University of Applied Sciences, Länggasse 85, 3052 Zollikofen, Switzerland
Full list of author information is available at the end of the article

Background

Inadequate fertility and problems associated with calving have high economic importance because they collectively cause 40% of the involuntary culling that occurs in the Brown Swiss population [1]. Calving ease is also an important because it influences animal welfare. Fertility and calving ease traits are included in routine national evaluations of several countries [2] including the Swiss dairy breeding programs [3, 4]. Genome-wide association studies (GWAS) for fertility and calving ease traits have been performed for several cattle breeds. For female fertility a number of quantitative trait loci (QTL) have been reported in different populations [5–7]. In Danish Jersey cattle the use of imputed whole-genome sequence data allowed the identification of various QTL influencing their national fertility index [8], which includes traits such as number of inseminations per conception, interval from calving to first insemination, 56-day non-return rate and days from first to last insemination. Those QTL were located on bovine chromosome (BTA) 7, 9, 20, 23, 25. Most variants found to have the highest significance of association with that fertility index were intergenic, except one missense variant associated with non-return rate on BTA23. In Italian Holstein, QTL have been identified using 50 K SNP chip data on BTA5, 7, 8, 13, 16, 18 and 27 for days to first service and on BTA2, 17 and 19 for their aggregate fertility index [9].

A locus on BTA18 has been associated with calving ease in Holstein [10]. Other QTL for calving ease have been identified in German Fleckvieh on BTA14 and 21 [5]. The QTL on BTA14 has been associated with stature in German Fleckvieh [11]. In Nordic Red cattle a QTL has been identified to be associated with a sire calving index that includes calving ease, stillbirth rate and a body conformation index including stature [12]. Further genomic associations with calving ease have been found on BTA2 in Limousin and Charolais beef breeds [6]. Overall these studies revealed breed specific loci, often located in regions that are also associated with stature.

Genomic information on thousands of progeny tested bulls is available today as a result of the introduction and now widespread use of genomic selection [13]. With the 1000 Bull Genomes Project [14, 15], a reference panel for imputation to sequence level has become available to project partners. Imputed whole-genome sequence data used in GWAS enhances the discovery of causative variants [16].

The objective of this study was to identify QTL affecting fertility and calving ease traits using imputed whole-genome sequence genotypes of Brown Swiss bulls. Furthermore we aimed to fine-map those QTL to identify potentially causative genes and variants.

Methods

Animals

A total of ~23,000 Brown Swiss animals genotyped at various densities were available from routine genomic prediction including those involving data sharing among the InterGenomics consortium [17] and the LowInputBreeds project (FP7-project no. KBBE 222623).

Phenotypes

Estimated breeding values and corresponding reliabilities for five fertility traits were available: non-return rates for heifers (NRH) and cows (NRC) after 56 days; days to first service (DFS); interval between first and last insemination in heifers (IFLH) and in cows (IFLC) [4].

Deregressed breeding values for stillbirth (SB), calving ease (CE), gestation length (GL) and birth weight (BW) were available. For these four calving traits, a GWAS was carried out separately for maternal (m) and for direct (d) effects (Table 1). Additional analyses were undertaken for CEd (calving ease direct) with stature (s) as a covariate (CEd_sc) and for SBd and SBm with exclusion of some of the individuals (SBd corr; SBm corr). In total, breeding values for 13 traits were obtained from the Swiss Brown Swiss routine genomic evaluation [3, 4] and deregressed according to Garrick et al. [18]. We limited the analyses to breeding values of progeny tested bulls with reliabilities of estimated breeding values above 0.55 for fertility traits (except for DFS where the cutoff was 0.65) (Additional file 1: Figure S1) and above 0.20 for calving traits (Additional file 2: Figure S2). These thresholds were chosen to be the same as those used for choosing bulls to be included in the training set for routine genomic prediction. After these filters were applied, there were deregressed breeding values (deregBV) available for GWAS with 1136–4975 bulls depending upon trait (Table 1).

Imputation

We performed a two-step imputation as this has previously been shown to be more accurate than imputation directly to sequence density [19]. The first step included the imputation from 50 k single nucleotide polymorphism (SNP) chip data to high density (HD) SNP chip data using the software package FImpute with default parameter settings and pedigree information [20]. In the second step we imputed the HD SNPs to sequence density using Minimac with default settings [21] based on sequence variants from the 123 Brown Swiss (BSW) cattle that had been included in run 5 of the 1000 Bull Genomes Project [14, 15]. Only sequence SNPs and biallelic indels with a minor allele frequency (MAF) > 0.01 were imputed and only those with R-sq. > 0.1 in the .info file from Minimac were retained for GWAS (Table 1). R-sq.

Table 1 Minimum, maximum and mean deregressed breeding value (deregBV) and genomic inflation factor lambda and number of individuals included in GWAS per trait

Trait	Min deregBV	Max deregBV	Mean (sd) deregBV	Genomic inflation factor (Lambda)	Number of individuals
NRH	−75.08	64.28	0.90 (15.68)	0.983	2506
NRC	−58.99	46.41	1.22 (14.67)	0.994	3615
IFLH	−62.74	50.40	−0.17 (15.34)	0.976	1484
IFLC	−52.94	46.50	−1.39 (13.75)	0.990	4122
DFS	−61.88	41.91	−2.69 (12.52)	0.990	3619
CEd	−210.01	121.45	−3.37 (17.93)	0.974	4975
CEd_sc	−210.01	121.45	−4.32 (17.36)	0.992	4159
SBd	−114.71	291.12	−13.00 (27.54)	1.010	1610
GLd	−121.55	156.02	0.39 (18.26)	0.935	2753
BWd	−110.60	209.64	−0.59 (19.69)	0.948	2561
CEm	−122.15	107.57	1.09 (25.09)	0.990	2862
SBm	−187.92	180.70	−8.84 (28.73)	1.005	1136
GLm	−113.34	132.43	−1.78 (23.51)	0.981	2756
BWm	−92.46	97.23	0.82 (18.92)	0.978	2683
SBd corr	−114.71	291.12	−14.28 (27.80)	1.014	1496
SBm corr	−187.92	180.70	−8.58 (29.66)	1.003	1051

Trait: *NRH* non-return rate in heifers, *NRC* non-return rate in cows, *IFLH* interval from first to last insemination in heifers, *IFLC* interval from first to last insemination in cows, *DFS* Days to first service, *CEd* calving ease direct, *CEd_sc* calving ease direct with stature deregressed breeding value (deregBV) as covariate, *SBd* stillbirth direct, *GLd* gestation length direct, *BWd* birth weight direct, *CEm* calving ease maternal, *SBm* stillbirth maternal, *GLm* gestation length maternal, *BWd* birth weight maternal, *SBd corr* Stillbirth direct, excluding the smaller cluster, *SBm corr*, Stillbirth maternal, excluding the smaller cluster
SNPs: number of SNPs considered for GWAS after filtering
Min deregBV: Minimum deregressed breeding value for the trait
Max deregBV: Maximum deregressed breeding value for the trait
Mean (sd) deregBV: Mean and standard deviation of the deregressed breeding value for the trait

is an internally calculated value given by Minimac that reflects the imputation quality [21].

Genome-wide association studies

Genome-wide association studies were conducted using the mixed model approach implemented in EMMAX [22]. We used the −Z option along with dosage data from imputation as genetic information. Individuals having a pedigree-based gene proportion of Original Braunvieh (OB) [23] > 0.3 were excluded from analysis as animals with high OB-gene proportion would create sub-structures in the population that otherwise mostly comprises individuals with little or no OB-gene proportion. The individual OB-gene proportion and reliability of deregBV were used as covariates in the model. The genomic relationship matrix was calculated for animals included in the analysis for each trait from HD SNP chip genotypes using GCTA [24]. The genomic relationship matrix was used in the mixed model fitted by EMMAX. We performed a principal component analysis (PCA) in R [25] using the princomp function and plotted the first and the second PC to visualize relatedness captured by the genomic relationship matrix. Using the R package wasim [26] we colored the individuals according to their deregBV. We filtered alleles with MAF < 0.05 and those deviating from Hardy-

Weinberg equilibrium ($HWE < 1 \times 10^{-20}$). After filtering 9,748,130–9,999,287 variants were included in the GWAS. We used the Bonferroni corrected 0.05 significance threshold and for suggestive threshold we used $p < 1 \times 10^{-6}$.

Genomic inflation factor lambda was calculated in R applying the following formula:

$$\text{lambda} = \text{round}\left(\text{median}\left(\text{qnorm}(\mathbf{p}/2)\hat{\ }2\right)/0.454, 3\right)$$

where $\mathbf{p}$ is the vector of *p*-values from the EMMAX GWAS.

Variant annotation and description

All sequence variants were annotated using the Variant effect predictor (VEP) [27]. Frequency estimation of the sequence variants of interest were calculated within and across breeds using data from run 5 data of the 1000 Bull Genomes Project [14]. Linkage disequilibrium between variants was calculated using the −ld funcion in PLINK [28, 29]. We performed in silico prediction of the impact of missense variants using PolyPHEN2 [30]. Additionally the prediction of the impact of the variant from SIFT was available from VEP. Multiple sequence alignments were done using MAFFT (http://mafft.cbrc.jp/alignment/server/).

Results and discussion

We performed GWAS for five fertility and eight calving traits using filtered imputed whole-genome sequence SNPs assuming that the causative variants were included in the data set. We investigated the QTL we identified for significantly associated variants, in terms of variants with direct impacts on proteins (e.g. missense or frameshift mutations). Most traits we investigated have low heritability (exception GLd with 0.46) [3, 4]. We tried to keep our population relatively uniform by excluding animals with an OB-gene proportion > 0.3. The lambda values indicating genomic inflation (Table 1) reveal that for all traits except SBd and SBm, EMMAX tends to overcorrect for stratification (lambda < 1). In the PCA plot of the genomic relationship matrix for those individuals included for these two traits reveals two clusters (Additional file 3: Figure S3), while for the other traits the PCA plots show that the individuals are uniformly and continuously distributed with respect to the first two principal components (Additional file 3: Figure S3 and Additional file 4: Figure S4). Inspecting the relationships in the smaller cluster of the PCA plot from the individuals included in the GWAS for SB, we found one bull with about 90 male offspring as well as the sire of this bull. This heavily imbalanced relationship is likely to cause the substructure.

Fertility traits

We found significant associations for each of the five female fertility traits NRH, NRC, IFLH, IFLC and DFS (Fig. 1, Additional file 5: Table S1). The significant QTL were identified on BTA17 (NRH, IFLH and NRC), 25 (IFLC) and 29 (DFS). Beside the three significant associations of the QTL on BTA17, it was also suggestively associated with IFLC. Similarly, the QTL for IFLC on BTA25 was also suggestively associated with DFS. Additionally we found a suggestive QTL for DFS on BTA15.

The QTL on BTA17 is located at around 70–73 Mb. A zoom in this region reveals that the region with the most associated SNPs for IFLC, IFLH and NRH is not significantly associated with NRC (Fig. 2). For that trait, the peak is shifted about 3 Mb and is located around 73 Mb on the same autosome. We speculate that this is a second QTL, which is significant for NRC and NRH. For this QTL, we could identify two missense variants (Table 2). One of the missense variants is significantly associated with NRH and located in the *CABIN1* gene. That gene has been found to be significantly associated with fertility traits in Holstein [31] and also differentially expressed in cows from a high

and a low fertility group, based on estimated breeding value for calving interval [32]. Even more interesting is the nonsense variant in *ENSBTAG00000048030*, also located in the 73 Mb QTL. Nonsense variants introduce a premature stop codon and are therefore unlikely to produce a functional protein. Using BLAST we found that the sequence of the corresponding protein has an identity of 84% with XP_015322696.1, which is encoded by the *IGLL1* gene on BTA17. That protein is implicated in immune-response and has been shown to be differentially expressed in the endometrium of cows that either showed or did not show signs of estrus around artificial insemination [33]. This gene may play a role in fertilization and therefore we propose the variant in the *IGLL1* gene as a candidate variant for fertility traits in dairy cattle. We think that the 73 Mb QTL is present in multiple breeds, such as Italian Holstein [9], while the peak around 71 Mb seems to be specific to the Brown Swiss breed in relation to fertility traits.

There is additional evidence that the genomic regions of these QTL might affect quantitative traits in the Brown Swiss population because two complex copy number variant (CNV) regions have been found to be located at 72–73 Mb and 73–75 Mb [34] using the consensus map of CNV detected from PennCNV and SVS algorithms. A similar region for the Mexican Holstein breed was reported by Duran Aguilar et al. [35] comprising three CNV regions from 70.7 Mb to 70.9 Mb mapped using PennCNV. This latter study also reports a significant association with SCS, a functional trait under selection in most of the dairy cattle populations.

The peak region for IFLH is located between 70,462,351 bp and 71,559,004 bp on BTA17, which is the center of the 71 Mb QTL. The 70 most associated SNPs of NRH are within this same interval. In this interval two missense variants can be identified to be significantly associated in the GWAS and in the sequenced Brown Swiss animals they are in perfect LD with the top associated variant from the GWAS. The two variants are located in the *GAS2L1* gene (g.70,724,328C > T) and in the *ASCC2* gene (g.71,084,044G > A).

For both variants the alternative allele has a negative effect on the deregBV. Using in silico effect prediction on the protein, SIFT (provided by VEP) and PolyPHEN2 revealed that the variant in *ASCC2* (p.P42L) is likely to have deleterious (0.02 - SIFT) or probably damaging (0.992 - PolyPHEN2) impact. The effect of the variant in GAS2L1 (p.P655L) could not reliably be predicted

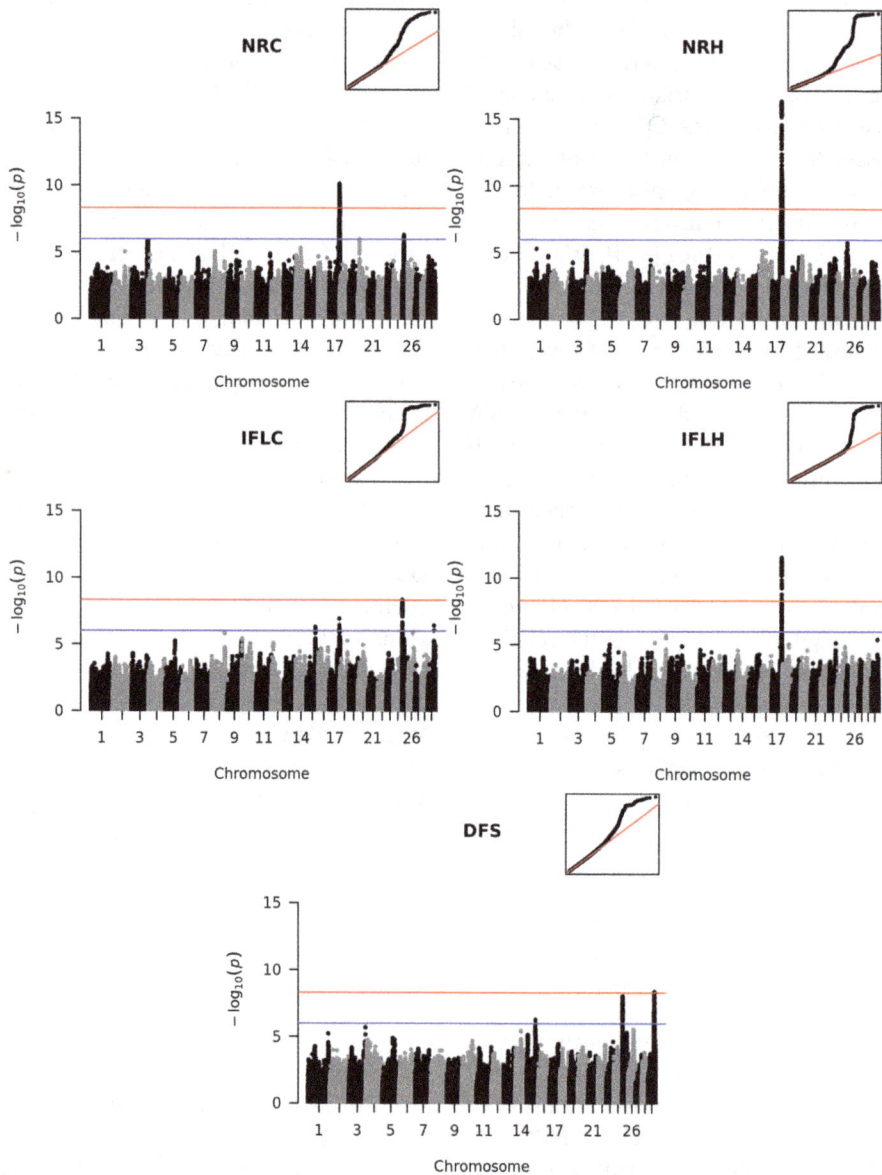

Fig. 1 Manhattan plots for genome-wide association studies for fertility traits. The red line marks the Bonferroni corrected significance threshold. The blue line shows the threshold for suggestive variants. Small figures in the upper right corner show the qqplot of the *p*-values. NRC: Non-return rate in cows; NRH: Non-return rate in heifers; IFLC: Interval from first to last insemination in cows; IFLH: Interval from first to last insemination in heifers; DFS: Days to first service

(deleterious low confidence (0.02); unknown). The variant in *ASCC2* is specific to Brown Swiss cattle. In the run 5 of the 1000 Bull Genomes Project, 22 Brown Swiss were found to be heterozygous for this variant whereas 1 animal was homozygous (Additional file 6: Table S2). Additionally we found the variant in two Danish Red cattle, however according to their pedigree, both are sired by animals with breed code BSW. The MAF of this variant is close to 10% in BSW, while below 1% across all the sequenced individuals of different breeds. The *GAS2L1* p.P655L variant was not only found in BSW but

also in Simmental, Angus, Jersey and Hereford. Due to the high LD between these two variants we speculate that the *GAS2L1* variant is older and the *ASCC2* variant occurred in Brown Swiss on the haplotype carrying the *GAS2L1* variant. Since we found homozygous individuals in the data set of sequenced animals for both variants, neither of the variants can be homozygous lethal. *ASCC2* is involved in gene activation and repression as part of the ASC1 complex. In order to have support of the hypothesis that the p.P42L variant in ASCC2 is more likely the causative mutation rather than *GAS2L1*

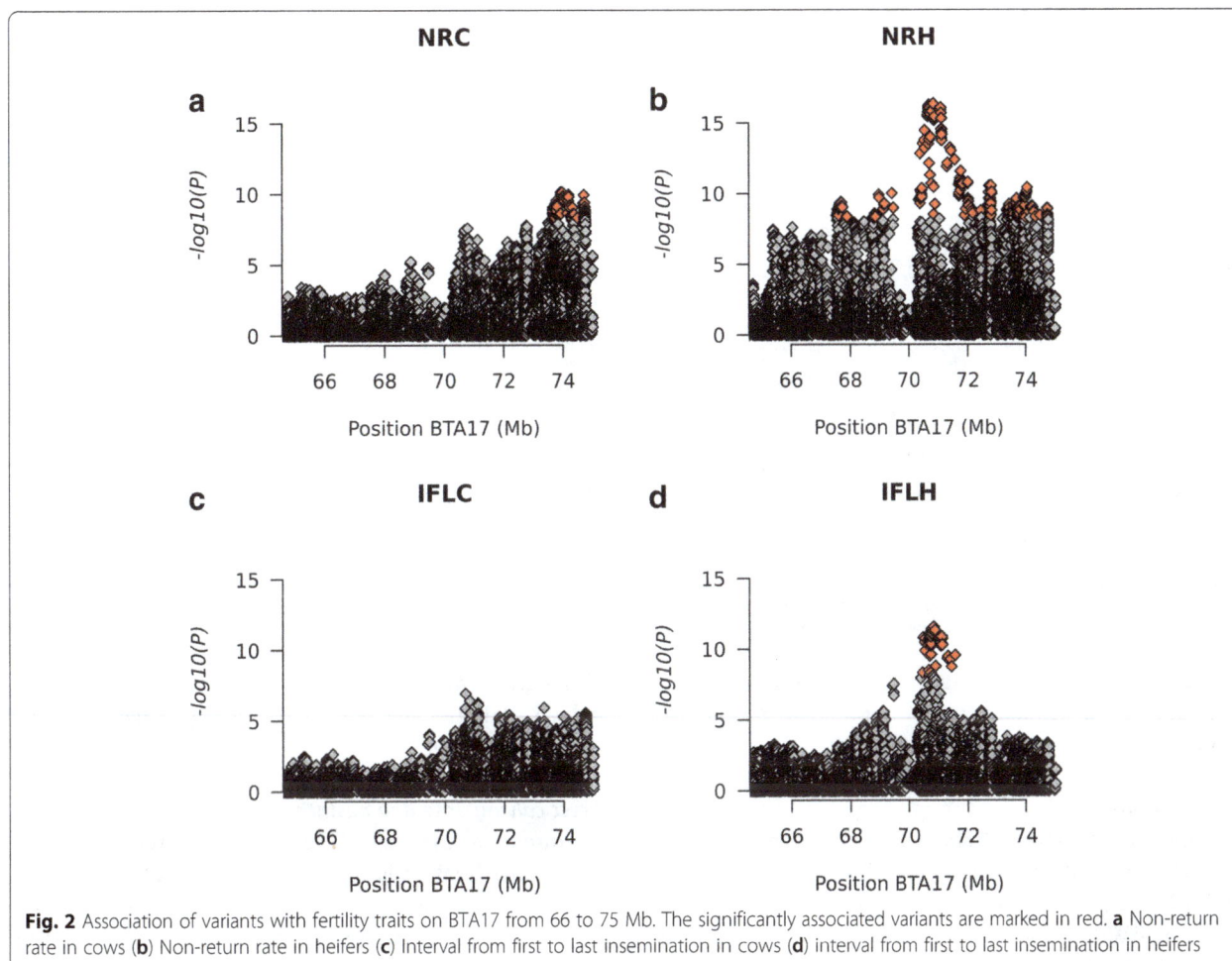

Fig. 2 Association of variants with fertility traits on BTA17 from 66 to 75 Mb. The significantly associated variants are marked in red. **a** Non-return rate in cows (**b**) Non-return rate in heifers (**c**) Interval from first to last insemination in cows (**d**) interval from first to last insemination in heifers

variant. We performed multiple sequence alignments for both p.P42L in *ASSC2* and p.P655L in *GAS2L1* (Fig. 3). In ASCC2 the amino acid at this position is conserved among all mammals in our comparison and in other species down to the zebra fish. The proline at position 655 in *GAS2L1* on the other hand is not conserved at all.

The locus on BTA25 is associated with IFLC. This region on BTA25 has been associated with stature in Brown Swiss cattle [36] and we found it to be associated with calving ease (see below). In Brahman cows it has been shown that larger cattle had lower pregnancy rates [37]. That author suggests the lower pregnancy rate is mainly due to negative energy balance in lactating cows, which cannot be compensated by increased feed intake [37]. This might explain why we do not find this locus to be associated with IFLH, the same fertility trait in heifers, which are not lactating when they become pregnant. For DFS, which is also a trait only measurable in lactating cows, we found a suggestive association for this locus. In a previous study in Brown Swiss this locus has

been shown to be associated with a fertility trait (cows ability to recycle after calving) and other traits [36].

The second QTL we identified for a single trait was on BTA29 at around 44 Mb and was associated with DFS. The only significantly associated variant is located in an intron of *PYGM*. *PYGM* encodes myophosphorylase, which is a glycogen phosphorylase expressed in muscle. Non-synonymous variants in this gene have been shown to cause protein alterations involved in glycogen storage disease type V [38]. It remains unclear how this gene could be related to fertility. In the QTL region on BTA29 a second plausible candidate gene is located, which is *PLCB3*. That gene has been found to be differentially expressed in low and high fertility Holstein cows [32]. Among the significantly associated variants, none were located in this gene. However there is a suggestively associated synonymous variant in *PLCB3*. Synonymous variants can influence translation efficiency if a rare codon is used [39], which would subsequently influence the level of gene expression and amount of protein available. Since there is differential expression in

Table 2 Missense and nonsense variants that were significantly ($p < 5*10^{-9}$) associated with fertility traits

Trait	BTA	Position	p-value	Gene	Variant	PolyPHEN2	SIFT
NRH	17	70,724,328	1.63×10^{-16}	GAS2L1	p.P655L	unknown	deleterious (low confidence)
IFLH			2.20×10^{-11}				
NRH	17	71,084,044	8.03×10^{-17}	ASCC2	p.P42L	probably damaging	deleterious
IFLH			3.52×10^{-11}				
NRH	17	72,747,746	8.78×10^{-10}	SLC5A4	p.G608S[a]/p.G235S[b]	probably damaging	deleterious
NRH	17	72,815,579	1.61×10^{-9}	ENSBTAG00000048030	p.Y108*	–	–
NRH	17	73,344,409	1.39×10^{-9}	CABIN1	p.R651Q	benign	tolerated
NRH	17	73,393,194	1.09×10^{-9}	CABIN1	p.A1721V	benign	tolerated (low confidence)
NRH	17	73,442,633	2.37×10^{-9}	ENSBTAG00000046900	p.T234A	benign	tolerated
NRC	17	74,739,013	3.25×10^{-9}	CDC45L	p.A263T	benign	tolerated
NRH			3.49×10^{-9}				

Trait: *NRC* non-return rate in heifers, *NRH* non-return rate in heifers, *IFLH* interval from first to last insemination in heifers
BTA: Chromosome
P-value: From GWAS with EMMAX
Variant: Amino acid change caused by the variant
PolyPhen2: Predicted effect of the variant on the protein function from PolyPhen2
SIFT: Predicted effect of the variant on the protein function from SIFT obtained via Variant effect predictor (VEP)
[a/b]The transcripts ENSBTAT00000010678[a] and ENSBTAT00000052915[b] for *SLC5A4*

low and high fertility Holstein cows a lower translation rate and a subsequent lower amount of protein for PLCB3 could potentially impact fertility.

Calving traits

For the direct calving traits we could identify significantly associated regions for the three traits CEd, SBd and GLd (Fig. 4, left side, Additional file 5: Table S1). Only for BWd no locus was found to be significantly associated with the trait. Unlike for the fertility traits, we found different QTL for each calving trait. Those QTL

with significant association were located on BTA19, 21 and 25. Additionally we identified suggestive QTL on BTA5, 22 and 29.

Direct calving ease and gestation length

For direct calving ease (CEd) two QTL were identified, on the proximal ends of BTA21 and BTA25. The locus on BTA25 has been associated with stature in Brown Swiss cattle [36]. Intuitively, it seems logical that large calves would determine more birth difficulties than small calves when born to cows of the same stature. However

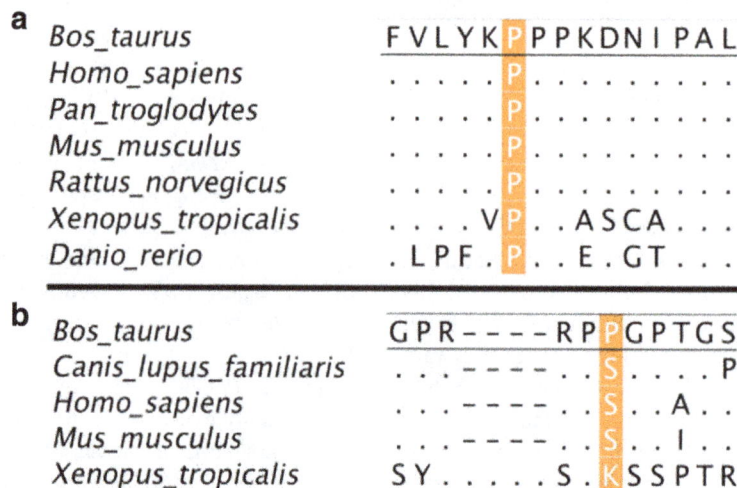

Fig. 3 Multiple sequence alignment of (**a**) ASCC2 (amino acids 37–51 (*Bos taurus*)) the orange shading marks the p.42 position. The sequences for ASCC2 were derived from the following accession numbers: *Bos taurus*, NP001015524.1; *Homo sapiens*, NP_115580.2; *Pan troglotydes*, XP_515064.3; *Mus musculus*, NP_083567.1; *Rattus norvegicus*, NP_001102561.1; *Xenopus tropicalis*, NP_001016871.1; *Danio rerio*, NP_956736.1 and (**b**) GAS2L1 (amino acids 650- the orange shading marks the p.655 position. The sequences for GAS2L1 were derived from the following accession numbers: *Bos taurus*, NP_001077167.1; *Canis lupus familiaris*, XP_543468.2; *Homo sapiens*, NP *Mus musculus*, NP_653146.1; Xenopus tropicalis, XP_002934334.1

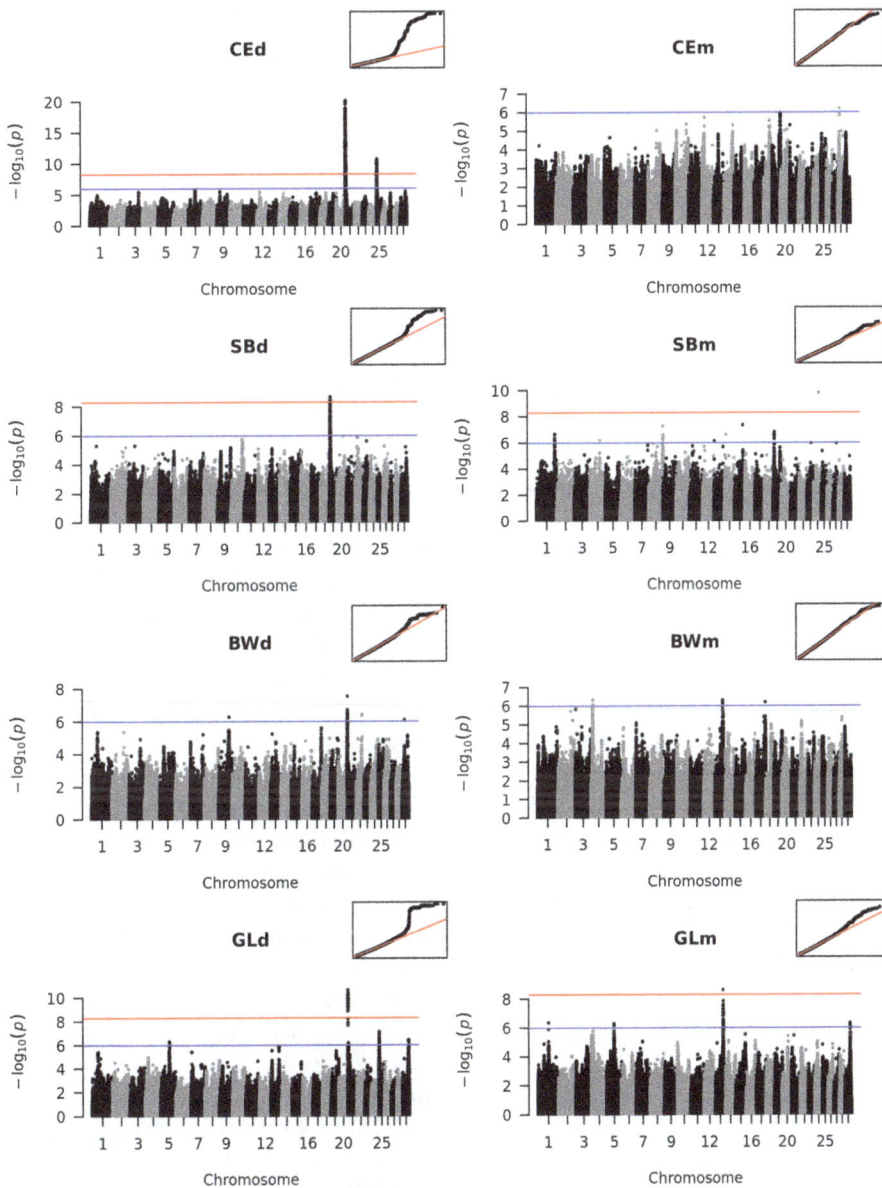

Fig. 4 Manhattan plots for the genome-wide association studies for calving traits. The red line marks the Bonferroni corrected significance threshold. The blue line shows the threshold for suggestive variants. Small figures in the upper right corner show the corresponding qqplot. CEd: Calving ease direct; CEm: Calving ease maternal; SBd: Stillbirth direct; SBm: Stillbirth maternal; BWd: Birth weight direct BWm: Birth weight maternal; GLd: Gestation length direct; GLm: Gestation length maternal

genetic correlations between stature and calving ease are non-significant in UK Holstein-Friesian cattle [40]. We also found only a low correlation between those two traits of 0.1 for the phenotypic correlation of the deregressed breeding values of the two traits. This might be explained by the fact that not all loci influencing adult size have an impact on the size of the calf. However, we additionally performed GWAS for calving ease using the deregressed breeding value for stature as a covariate. With stature as a covariate the signal on BTA25 disappears. (Additional file 7: Figure S5). We interpret,

that the locus on BTA25 is either pleiotropic or that the phenotypes observed are indeed due to differences in the size of the calves, with small calves being delivered more easily than tall calves. Even though this QTL has been described previously, no possible causative variant has been identified. For CEd, we found three missense variants among the significantly associated variants on BTA25, located in *CRAMP1L*, *PTX4* and *TELO2* genes (Table 3). All three variants were significantly associated with stature in BSW as well as CEd. The direction of the SNP effects was however opposite. According to

Table 3 Missense and nonsense variants that were significantly ($p < 5*10^{-9}$) associated with calving traits

Trait	BTA	Position	p-value	Gene	Variant	PolyPHEN2	SIFT
CEd	25	1,277,577	5.59×10^{-10}	CRAMP1L	p.R885Q	probably damaging	deleterious
CEd	25	1,177,995	1.93×10^{-9}	TELO2	p.R536W	probably damaging	deleterious
CEd	25	1,161,904	4.05×10^{-9}	PTX4	p.P207L	probably damaging	tolerated

Trait: *CEd* calving ease direct
BTA: Chromosome
P-value: From GWAS with EMMAX
Variant: Amino acid change caused by the variant
PolyPhen2: predicted effect of the variant on the protein function from PolyPhen2
SIFT: predicted effect of the variant on the protein function from SIFT obtained via Variant effect predictor (VEP)
* : Translation stop codon

PolyPHEN2 all three variants are probably damaging and therefore potentially causal. As for regions on BTA17, the CNVR_510 reported by Prinsen et al. [34] on BTA25 in BSW spans these significantly associated missense variants. The genomic variation CNVR_510 is detected in 86 individuals indicating that the CNV in this location may represent an important source of variation that may affect quantitative traits.

The locus on BTA21 remained significantly associated with calving ease when including stature as a covariate (Additional file 7: Figure S5). Therefore this locus is likely not involved in normal variation in adult stature. Interestingly the same locus had been identified earlier in German Fleckvieh cattle to be associated with the complex of calving traits [5]. However in that study an association of SB with the same QTL was reported, whereas we could not detect any association with SB in Brown Swiss. Moreover, we found this region to be associated with gestation length. Generally a longer gestation length is associated with higher birth weight and a subsequent increase in calving difficulties [3].

The direction of the SNP effects within the QTL on BTA 21 highlight this relationship between gestation length and calving ease. The mechanism by which this region could impact calving ease is unknown. It is known that the shared syntenic region in human is imprinted and associated with Prader-Willi syndrome for defects in the paternal allele and with Angelman syndrome for the maternal allele [41]. Human babies with Prader-Willi syndrome have an increased risk of being born by cesarean section [42], which would be in agreement with a decreased breeding value for calving ease. However on the other hand a further feature of Prader-Willi syndrome is that offspring are at increased risk of preterm birth, which does not match the expectation and known relationship between GL and CE.

We found 7 synonymous variants significantly associated with *ATP10A*. The *ATP10A* gene is maternally imprinted in human and associated with Angelman syndrome [43]. However, there is some evidence that this gene is not imprinted in mice [44]. We found suggestively associated non-synonymous variants in *ATP10A*

and *MAGEL2*. The one located in the *ATP10A* gene, causes p.M655 L, predicted by SIFT to be tolerated and by PolyPHEN2 to be benign. The other variant causes an amino acid exchange from Glutamic acid to Glycine at the 851 amino acid position. For this variant no prediction about its effect size could be done. *MAGEL2* is implicated in Prader-Willi syndrome [45], which phenotypically seems to have a larger influence on birth traits than Angelman syndrome. For *MAGEL2* in bovine parthenogenic and normal embryonic cells no difference in the expression pattern could be detected suggesting this gene is not imprinted in cattle [46]. For mice that are heterozygous for *MAGEL2* deficiency, an increased perinatal lethality has been described [47].

For GLd, the top associated variants are about 800 kb downstream compared to those for CEd. The SNPs in this region are also significantly associated with CEd. When analysing the SNP density in this region we found that there is a decreased SNP density at the beginning of BTA21 (less than 1000 SNPs per Mb) relative to other locations. This observation is displayed in the zoomed plot (Additional file 8: Figure S6). The SNP density is similar for both traits in this region. We speculate that the causative variant is indeed the same for both traits.

Stillbirth

We identified several SNPs on BTA19 with significant associations with SB at around 10 Mb. That locus co-localizes with the BH2 haplotype for which a causative mutation has been identified in the *TUBD1* gene [48]. Calves homozygous for the *TUBD1* variant show high mortality rates during or shortly after birth. In our data set the MAF of this variant was lower than 0.05 and therefore it was not included in the GWAS, and in any event the imputation R-sq. was rather low (0.12) for this variant. However, because this variant is an obvious candidate causative variant for this gene we ran an additional single SNP regression for this variant generating a non-significant p-value of 4.68×10^{-4}. Interestingly having a closer look at the p-values in the BH2 haplotype region, there were reduced $-\log_{10}$(p-values) observed compared to slightly upstream and

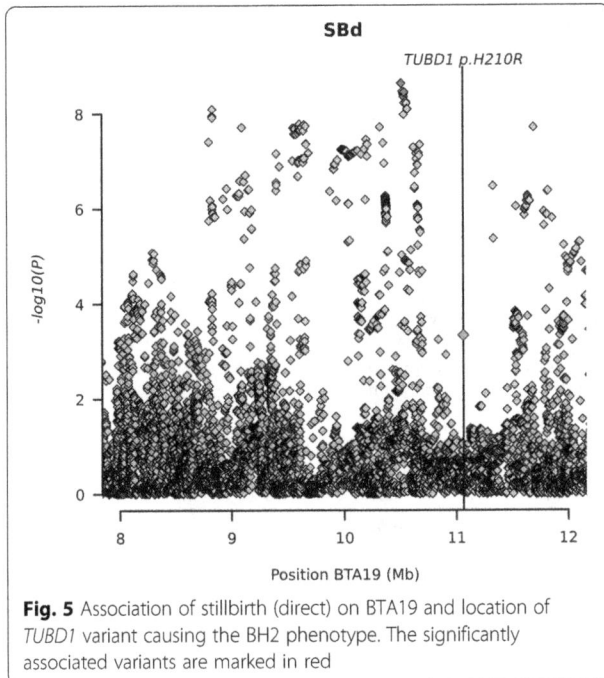

Fig. 5 Association of stillbirth (direct) on BTA19 and location of *TUBD1* variant causing the BH2 phenotype. The significantly associated variants are marked in red

downstream from the haplotype (Fig. 5). We have to remember that the model assumption for which association testing is performed is an additive mode of inheritance, but BH2 is never present in homozygous form. It is possible that another QTL is responsible for the signal. In order to verify that the association is truly caused by BH2, we removed BH2 carriers in an additional GWAS and in a second step we used BH2 status as a covariate. In both cases no significant SNP remained on BTA19. Therefore we conclude that the signal is indeed caused by BH2 (Additional file 9: Figure S7).

Maternal traits

We also performed GWAS for the maternal calving traits CEm, SBm, BWm and GLm (Fig. 4, right side, Additional file 5: Table S1). We found one SNP significantly associated with SBm. That SNP has an imputation R-sq. of 0.17 and MAF of 0.051. These values are only slightly above the quality control thresholds and therefore we believe that the significant effect of this SNP may represent a false positive. This is corroborated by the fact that no other SNP in that region shows either significant or suggestive association. There is a substructure in the individuals used for SBm and we believe this false-positive association may be due to that (Additional file 3: Figure S3 C and D). To support this hypothesis, we reran the GWAS excluding all individuals in the smaller cluster of the PCA plot (the sire with 90 offspring, its father and its siblings). The Manhattan plot clearly demonstrates that there is no longer any significantly associated SNP (Additional file 10: Figure S8).

Because a reduction in the number of individuals also leads to a reduction of power to detect true associations, we removed the same individuals for SBd, and there the peak on BTA19 remained significant. These results show how important a good correction and an eventual exclusion of outlier individuals is in order to avoid false-positive findings.

The other maternal trait for which we found a significant association was GL, the association for GLm is only due to a single significant SNP but a number of nearby SNPs are suggestively associated, just below the significance threshold. Additionally this SNP has a higher imputation R-sq. (0.781) and MAF (0.383). The QTL for GLm was located on BTA13. Zooming on this region reveals that there could actually be 2 QTL: at 65.5 Mb with the significant SNP and a suggestive QTL at 67.1 Mb (Fig. 6). The two loci are not in strong LD ($r^2 = 0.35$), supporting the hypothesis of two QTL in this region. The significantly associated variant at around 65.5 Mb, is located in an intron of *CPNE1*. *CPNE1* encodes a calcium-dependent phospholipid binding protein but little is known about the exact role of the protein. Recently it has been found that this gene is expressed in human placenta [49] but not found to be expressed above 50 fragments per kilobase million (FPKM) in the cattle transcriptome [49]. However the placentas used in that study came from term births and therefore differential expression over gestation cannot be ruled out. Totally 142 suggestively associated variants are distributed on the two QTL. Of these 15 are associated with the first QTL and include a missense variant in the *SPAG4* gene. The variant is p.Y289F and is predicted to be tolerated

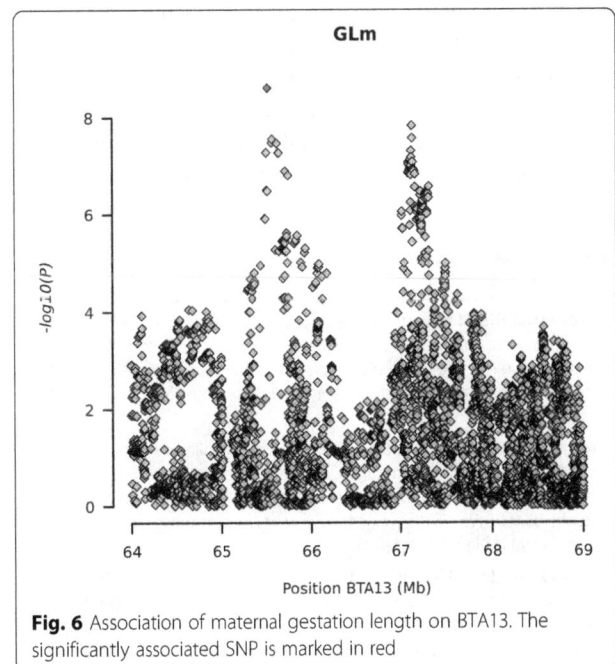

Fig. 6 Association of maternal gestation length on BTA13. The significantly associated SNP is marked in red

by SIFT, but PolyPHEN2 predicts that the variant is probably damaging. *SPAG4* encodes the sperm-associated antigen protein 4. As the name indicates, disruption of this gene leads to abnormal sperm development and decreased male fertility [50]. Additionally this gene has also been found to be differentially expressed in pre-eclamptic placenta in humans [51]. This gene is an interesting candidate gene for GLm.

At the second locus 127 variants are suggestively associated. Among these variants we find a single synonymous variant in the *CTNNBL1* gene. This gene has been found to be expressed in the placenta of *Ateles fusciceps* but not with FPKM > 50 in cattle. Deficiency of *CTNNBL1* in mice leads to embryonic lethality [52]. More interestingly for GLm are the genes *NNAT* and *BLCAP*. The peak of this second QTL is located between these two genes. Both genes are imprinted and expressed in the placenta [53]. Therefore these two genes provide good candidate genes for a maternal trait. We cannot suggest a possible causative variant. Variants with influence on gene expression but with no alteration of the protein are hard to detect. It is likely that the causative variant is located outside the coding region.

Conclusion

In this study we used readily available phenotypes that are collected for routine genomic prediction to identify QTL for traits related to fertility and calving ease. We detected a novel QTL for fertility in Brown Swiss on BTA17 and suggest a candidate that may represent the causal variant. We further identified regions associated with other birth and fertility traits. We detected a region with a known deleterious variant to be significantly associated with stillbirth and a region associated with maternal gestation length including genes with placental expression in that associated region. This study gives insight into the genetic architecture of the functional traits that characterize fertility and calving ease.

Additional files

Additional file 1: Figure S1. Distribution of reliabilities of estimated breeding values for fertility traits. NRC: Non-return rate in cows; NRH: Non-return rate in heifers; IFLC: Interval from first to last insemination in cows; IFLH: Interval from first to last insemination in heifers; DFS: Days to first service. (PNG 51 kb)

Additional file 2: Figure S2. Distribution of reliabilities of estimated breeding values for calving traits. CEd: Calving ease direct; CEm: Calving ease maternal; SBd: Stillbirth direct; SBm: Stillbirth maternal; BWd: Birth weight direct BWm: Birth weight maternal; GLd: Gestation length direct; GLm: Gestation length maternal. (PNG 57 kb)

Additional file 3: Figure S3. The first two principal coponents for the genomic relationship matrix comprising individuals used for genome-wide association of calving traits (A) Calving ease direct (B) Calving ease maternal (C) Stillbirth direct (D) Stillbirth maternal (E) Birth weight direct (F) Birth

weight maternal (G) Gestation length direct (H) Gestation length maternal. (PNG 816 kb)

Additional file 4: Figure S4. The first two principal coponents for the genomic relationship matrix comprising individuals used for genome-wide association of fertility traits (A) Non-return rate in cows (B) Non-return rate in heifers (C) Interval from first to last insemination in cows (D) Interval from first to last insemination in heifers (E) Days to first service. (PNG 677 kb)

Additional file 5: Table S1. GWAS results per trait. SNPs with p-values < 10^{-6} for each trait. (XLSX 415 kb)

Additional file 6: Table S2. Minor allele frequency per breed among sequenced individuals for *ASCC2* and *GAS2L1*. (XLSX 44 kb)

Additional file 7: Figure S5. GWAS for calving ease using stature as a covariate. The red line marks the Bonferroni corrected significance threshold. The blue line shows the threshold for suggestive variants. The figure in the upper right corner shows the corresponding qqplot. (PNG 74 kb)

Additional file 8: Figure S6. Association for calving ease direct (CEd) and gestation length direct (GLd) on BTA21 from 1 to 4 Mb. The red line indicates the Bonferroni corrected significance threshold. (PNG 582 kb)

Additional file 9: Figure S7. Association on BTA19 for stillbirth using only individuals not carrying BH2 (top) or using BH2 carrier status as a covariate (bottom). The red line indicates the Bonferroni corrected significance threshold. (PNG 563 kb)

Additional file 10: Figure S8. Manhattan plots for the GWAS for stillbirth direct and maternal excluding individuals from the smaller cluster in the PCA plot. The red lines mark the Bonferroni corrected significance threshold. The blue lines show the threshold for suggestive variants. Small figures in the upper right corner shows the corresponding qqplot. (PNG 506 kb)

Abbreviations

BSW: Brown Swiss Cattle; BTA: Bovine chromosome; BWd: Birth weight direct; BWm: Birth weight maternal; CEd: Calving ease direct; CEm: Calving ease maternal; CNV: Copy number variant; deregBV: Deregressed breeding value; DFS: Days to first service; FPKM: Fragments per kilobase million; GLd: Gestation length direct; GLm: Gestation length maternal; GWAS: Genome-wide association study; HD: High density; IFLC: Interval from first to last insemination in cows; IFLH: Interval from first to last insemination in heifer; MAF: Minor allele frequency; NRC: Non-return rate in cows; NRH: Non-return rate in heifers; PCA: Principal component analysis; QTL: Quantitative trait locus/loci; SBd: Stillbirth direct; SBm: Stillbirth maternal; SNP: Single nucleotide polymorphism; VEP: Variant effect predictor

Acknowledgements

We thank the 1000 Bull Genomes Project for providing the sequencing data, which has been used as a reference for the imputation and for the estimation of minor allele frequencies. We acknowledge Alessandro Bagnato for valuable comments on the manuscript. Furthermore, we thank the Fondation sur la croix and the Swiss Commission for Technology and innovation for financial support of this project.

Funding

This study was partly conducted within the Swiss Low Input Genetics (SLIG) project, which was financially supported by the Swiss Commission for Technology and Innovation, and partly funded by the Fondation sur la Croix.

Authors' contributions

MF, BB, FRS, CF, BG conceived the study, carried out all analyses and wrote the manuscript. HSH, DG, CS, RF, IR, IC, IR, JS, AB and MGS contributed data and helped in the analyses. All authors read and approved the final manuscript.

Ethics approval

No experimental animal studies were conducted for the work in this manuscript. The sequence data for all animals were obtained from the 1000 bulls genome project [14]. The deregressed breeding values were obtained from routine genomic prediction.

Competing interests

The authors declare that they have no competing interests.

Author details

[1]Qualitas AG, Chamerstrasse 56a, 6300 Zug, Switzerland. [2]School of Agricultural, Forest and Food Sciences HAFL, Bern University of Applied Sciences, Länggasse 85, 3052 Zollikofen, Switzerland. [3]Institute of Veterinary, Animal & Biomedical Sciences, Massey University, Hamilton, New Zealand. [4]agn Genetics GmbH, 8b Börtjistrasse, 7260 Davos, Switzerland. [5]Interbull center, SLU - Box 7023, S-75007 Uppsala, Sweden. [6]Technische Universität München, Liesel-Beckmann-Straße 1, 85354 Freising-Weihenstephan, Germany. [7]Tierzuchtforschung e.V, Senator-Gerauer-Str. 23, 85586 Poing, Germany. [8]University of Natural Resources and Life Sciences, Gregor-Mendel-Str 33, 1180 Wien, Austria. [9]Research Institute of Organic Agriculture (FiBL), Ackerstrasse 113, 5070 Frick, Switzerland. [10]Department of Veterinary Medicine, University of Milan, Via Celoria 10, 20133 Milan, Italy.

References

1. Burren A, Alder S. Abgangsursachen und LBE. CHbraunvieh. 2013;3:8–11.
2. Egger-Danner C, Cole JB, Pryce JE, Gengler N, Heringstad B, Bradley A, et al. Invited review: overview of new traits and phenotyping strategies in dairy cattle with a focus on functional traits. Animal. 2015;9:191–207.
3. Berweger M. Nur noch TVD-Daten für die ZWS Geburtsablauf. CHbraunvieh. 2016;4:10–2.
4. Gredler B, Schnyder U. New genetic evaluation of fertility in Swiss Brown Swiss. Interbull Bull. 2013;47:226–9.
5. Pausch H, Flisikowski K, Jung S, Emmerling R, Edel C, Götz KU, et al. Genome-wide association study identifies two major loci affecting calving ease and growth-related traits in cattle. Genetics. 2011;187:289–97.
6. Purfield DC, Bradley DG, Evans RD, Kearney FJ, Berry DP. Genome-wide association study for calving performance using high-density genotypes in dairy and beef cattle. Genet Sel Evol. 2015;47:47.
7. Schulman NF, Sahana G, Iso-Touru T, McKay SD, Schnabel RD, Lund MS, et al. Mapping of fertility traits in Finnish Ayrshire by genome-wide association analysis. Anim Genet. 2011;42:263–9.
8. Höglund JK, Guldbrandtsen B, Lund MS, Sahana G. Identification of genomic regions associated with female fertility in Danish Jersey using whole genome sequence data. BMC Genet. 2015;16:60.
9. Minozzi G, Nicolazzi EL, Stella A, Biffani S, Negrini R, Lazzari B, et al. Genome wide analysis of fertility and production traits in Italian Holstein cattle. PLoS One. 2013;8:e80219.
10. Müller M-P, Rothammer S, Seichter D, Russ I, Hinrichs D, Tetens J, et al. Genome-wide mapping of 10 calving and fertility traits in Holstein dairy cattle with special regard to chromosome 18. J Dairy Sci. 2017;100:1–20.
11. Pausch H, Emmerling R, Schwarzenbacher H, Fries R. A multi-trait meta-analysis with imputed sequence variants reveals twelve QTL for mammary gland morphology in Fleckvieh cattle. Genet Sel Evol. 2016;48:14.
12. Sahana G, Höglund JK, Guldbrandtsen B, Lund MS. Loci associated with adult stature also affect calf birth survival in cattle. BMC Genet. 2015;16:47.
13. Meuwissen THE, Hayes BJ, Goddard ME. Prediction of total genetic value using genome-wide dense marker maps. Genetics. 2001;157:1819–29.
14. Daetwyler HD, Capitan A, Pausch H, Stothard P, van Binsbergen R, Brøndum RF, et al. Whole-genome sequencing of 234 bulls facilitates mapping of monogenic and complex traits in cattle. Nat Genet. 2014;46:858–65.
15. 1000 bull genomes project. Available from: http://www.1000bullgenomes.com/.
16. Pausch H, Macleod IM, Fries R, Emmerling R, Phil J. Evaluation of the accuracy of imputed sequence variants and their utility for causal variant detection in cattle. Genet Sel Evol. 2017;49:24.
17. Jorjani H, Jakobsen J, Nilforooshan MA, Hjerpe E, Zumbach B, Palucci V. Genomic evaluation of BSW populations InterGenomics: results and deliverables. Interbull Bull. 2011;43:5–8.
18. Garrick DJ, Taylor JF, Fernando RL. Deregressing estimated breeding values and weighting information for genomic regression analyses. Genet Sel Evol. 2009;41:55.
19. Khatkar MS, Moser G, Hayes BJ, Raadsma HW. Strategies and utility of imputed SNP genotypes for genomic analysis in dairy cattle. BMC Genomics. 2012;13:538.
20. Sargolzaei M, Chesnais J, Schenkel F. FImpute - an efficient imputation algorithm for dairy cattle populations. J Anim Sci/J Dairy Sci. 2011;89/94:421.
21. Fuchsberger C, Abecasis GR, Hinds DA. Minimac2: faster genotype imputation. Bioinformatics. 2014;31:782–4.
22. Kang HM, Sul JH, Service SK, Zaitlen NA, Kong S-Y, Freimer NB, et al. Variance component model to account for sample structure in genome-wide association studies. Nat Genet. 2010;42:348–54.
23. Flury C, Boschun C, Denzle M, Baps B, Schnyde U, Gredle B, et al. Genome-wide association study for 13 udder traits from linear type classification in cattle. Proc. 10th world Congr. Genet. Appl. To Livest. Prod. 2014;
24. Yang J, Lee SH, Goddard ME, Visscher PMGCTA. A tool for genome-wide complex trait analysis. Am J Hum Genet. 2011;88:76–82.
25. R Core Team. A language and environment for statistical computing http://www.R-project.org/.
26. Reusser D, Francke T. wasim: Visualisation and analysis of output files of the hydrological model WASIM. 2011. Available from: https://cran.r-project.org/package=wasim.
27. McLaren W, Gil L, Hunt SE, Riat HS, Ritchie GRS, Thormann A, et al. The Ensembl variant effect predictor. Genome Biol. 2016;17:122.
28. Chang CC, Chow CC, Tellier LC, Vattikuti S, Purcell SM, Lee JJ. Second-generation PLINK: rising to the challenge of larger and richer datasets. Gigascience. 2015;4
29. Purcell S, Neale B, Todd-Brown K, Thomas L, Ferreira MAR, Bender D, et al. PLINK: a tool set for whole-genome association and population-based linkage analyses. Am J Hum Genet. 2007;81:559–75.
30. Adzhubei IA, Schmidt S, Peshkin L, Ramensky VE, Gerasimova A, Bork P, et al. A method and server for predicting damaging missense mutations. Nat Methods. 2010;7:248–9.
31. Höglund JK, Sahana G, Guldbrandtsen B, Lund MS. Validation of associations for female fertility traits in Nordic Holstein, Nordic red and Jersey dairy cattle. BMC Genet. 2014;15:8.
32. Moore SG, Pryce JE, Hayes BJ, Chamberlain AJ, Kemper KE, Berry DP, et al. Differentially expressed genes in Endometrium and corpus Luteum of Holstein cows selected for high and low fertility are enriched for sequence variants associated with fertility. Biol Reprod. 2016;94:19.
33. Davoodi S, Cooke RF, Fernandes ACC, Cappellozza BI, Vasconcelos JLM, Cerri RLA. Expression of estrus modifies the gene expression profile in reproductive tissues on day 19 of gestation in beef cows. Theriogenology. 2016;85:645–55.
34. Prinsen RTMM, Strillacci MG, Schiavini F, Santus E, Rossoni A, Maurer V, et al. A genome-wide scan of copy number variants using high-density SNPs in Brown Swiss dairy cattle. Livest Sci. 2016;191:153–60.
35. Duran Aguilar M, Roman Ponce SI, Ruiz Lopez FJ, Gonzalez Padilla E, Vasquez Pelaez CG, Bagnato A, et al. Genome-wide association study for milk somatic cell score in holstein cattle using copy number variation as markers. J Anim Breed Genet. 2017;134:49–50.
36. Guo J, Jorjani H, Carlborg Ö. A genome-wide association study using international breeding-evaluation data identifies major loci affecting production traits and stature in the Brown Swiss cattle breed. BMC Genet. 2012;13:82.
37. Olson TA. Reproductive efficiency of cows of different sizes. 1993. Available from: http://animal.ifas.ufl.edu/beef_extension/bcsc/1993/docs/olson.pdf.
38. Martín MA, Lucía A, Arenas J, et al. Glycogen Storage Disease Type V. In: Adam MP, Ardinger HH, Pagon RA, et al, editors. GeneReviews®. Seattle: University of Washington, Seattle; 2006. pp. 1993-2017. Available from: https://www.ncbi.nlm.nih.gov/books/NBK1344/.
39. Quax TEF, Claassens NJ, Söll D, van der Oost J. Codon bias as a means to fine-tune gene expression. Mol Cell. 2015;59:149–61.

40 Eaglen SAE, Coffey MP, Woolliams JA, Wall E. Direct and maternal genetic
    relationships between calving ease, gestation length, milk production,
    fertility, type, and lifespan of Holstein-Friesian primiparous cows. J Dairy Sci.
    2013;96:4015–25.

41 Kalsner L, Chamberlain SJ. Prader-Willi, Angelman, and 15q11-q13
    duplication syndromes. Pediatr Clin N Am. 2015;62:587–606.

42 Driscoll DJ, Miller JL, Schwartz S, et al. Prader-Willi Syndrome. In: Adam MP,
    Ardinger HH, Pagon RA, et al, editors. GeneReviews®. Seattle: University of
    Washington, Seattle; 1998. pp. 1993-2017. Available from: https://www.ncbi.
    nlm.nih.gov/books/NBK1330/.

43 Meguro M, Kashiwagi A, Mitsuya K, Nakao M, Kondo I, Saitoh S, et al. A novel
    maternally expressed gene, ATP10C, encodes a putative aminophospholipid
    translocase associated with Angelman syndrome. Nat Genet. 2001;28:19–20.

44 DuBose AJ, Johnstone KA, Smith EY, Hallett RAE, Resnick JL. Atp10a, a gene
    adjacent to the PWS/AS gene cluster, is not imprinted in mouse and is
    insensitive to the PWS-IC. Neurogenetics. 2010;11:145–51.

45 Schaaf CP, Gonzalez-Garay ML, Xia F, Potocki L, Gripp KW, Zhang B, et al.
    Truncating mutations of MAGEL2 cause Prader-Willi phenotypes and autism.
    Nat Genet. 2013;45:1405–8.

46 Kaneda M, Takahashi M, Yamanaka K, Saito K, Taniguchi M, Akagi S, et al.
    Epigenetic analysis of bovine parthenogenetic embryonic fibroblasts. J
    Repduction Dev. 2017;63:365–75.

47 Schaller F, Watrin F, Sturny R, Massacrier A, Szepetowski P, Muscatelli F. A
    single postnatal injection of oxytocin rescues the lethal feeding behaviour
    in mouse newborns deficient for the imprinted Magel2 gene. Hum Mol
    Genet. 2010;19:4895–905.

48 Schwarzenbacher H, Burgstaller J, Seefried FR, Wurmser C, Hilbe M, Jung S,
    et al. A missense mutation in TUBD1 is associated with high juvenile
    mortality in Braunvieh and Fleckvieh cattle. BMC Genomics. 2016;17:400.

49 Armstrong DL, McGowen MR, Weckle A, Pantham P, Caravas J, Agnew D, et
    al. The core transcriptome of mammalian placentas and the divergence of
    expression with placental shape. bioRxiv. 2017. http://dx.doi.org/10.1101/
    137554.

50 Pasch E, Link J, Beck C, Scheuerle S, Alsheimer M. The LINC complex
    component Sun4 plays a crucial role in sperm head formation and fertility.
    Biol Open. 2015;4:1792–802.

51 Brew O, Sullivan MHF, Woodman A, Dulay A, Nayeri U, Buhimschi C.
    Comparison of normal and pre-Eclamptic placental gene expression: a
    systematic review with meta-analysis. PLoS One. 2016;11:e0161504.

52 Chandra A, van Maldegem F, Andrews S, Neuberger MS, Rada C. Deficiency
    in spliceosome-associated factor CTNNBL1 does not affect ongoing cell
    cycling but delays exit from quiescence and results in embryonic lethality in
    mice. Cell Cycle. 2013;12:732–42.

53 Kappil MA, Green BB, Armstrong DA, Sharp AJ, Lambertini L, Marsit CJ, et al.
    Placental expression profile of imprinted genes impacts birth weight.
    Epigenetics. 2015;10:842–9.

5

Transcriptome analysis of avian reovirus-mediated changes in gene expression of normal chicken fibroblast DF-1 cells

Xiaosai Niu, Yuyang Wang, Min Li, Xiaorong Zhang and Yantao Wu[*]

Abstract

Background: Avian reovirus (ARV) is an important poultry pathogen that can cause immunosuppression. In this study, RNA-Seq technology was applied to investigate the transcriptome-wide changes of DF-1 cells upon ARV infection at the middle stage.

Results: Total RNA of ARV-infected or mock-infected samples at 10 and 18 h post infection (hpi) was extracted to build RNA-Seq datasets. Analysis of the sequencing data revealed that the expressions of numerous genes were altered, and a panel of differentially expressed genes were confirmed with RT-qPCR. At 10 hpi, 104 genes were down-regulated and 64 were up-regulated, while the expressions of 47 genes were increased and only one was down-regulated, which may play a role in retinoic acid biosynthesis, at 18 hpi in the ARV-infected cells. The similar profiles of up-regulated genes between the two groups of infected cells suggest that ARV infection activated a prolonged antiviral response of host cells. Alternative splicing analysis found no significantly changed events altered by ARV infection.

Conclusions: Overall, the differential expression profile presented in this study can be used to expand our understanding of the comprehensive interactions between ARV and the host cells, and may be helpful for us to reveal the pathogenic mechanism on the molecular level.

Keywords: Avian reovirus, RNA-Seq, Transcriptome, Anti-viral responses

Background

Avian reovirus (ARV) is member of the *Orthoreovirus* genus that has recently been classified into the *Spinareovirinae* subfamily, which is one of two subfamilies in the *Reoviridae* family [1]. ARV is an important pathogen of birds and has been impacting poultry for nearly 60 years since it was first detected in 1957 [2, 3], and it is still prevalent in poultry until now, causing considerable economical loss in the global poultry industry [4–6]. Efficient and simple detecting methods may be helpful to control and prevent ARV infection [7]. Horizontal transmission is the main route of infection, with infrequent egg transmission [8]. Though ARV was found to be ubiquitous in poultry flocks, several strains could cause severe diseases [9]. These pathogenic strains can cause tenosynovitis individually [9], and additionally usually cause mixed infections together with other pathogens, such as chicken anemia agent [10, 11]. It has been demonstrated that ARV can replicate in macrophages and cause immunosuppression [8, 12].

The pathogenicity and epizootiology of ARV have been well studied, but the pathogenesis at the molecular level is poorly understood. An excellent review on the structural and biological characteristics of ARV was published 10 years ago [8]. Though many researchers have done brilliant work to reveal the pathogenesis of ARV infection at the molecular level in recent years, several major questions raised in the review remain unresolved. Previous studies showed that σC and P10 can induce apoptosis in different ways [13, 14], and a subsequent study correlated ARV-induced apoptosis with tissue injury [15]. Another study demonstrated that ARV can induce autophagy to promote viral titer [16]. Subsequent studies revealed the connection between ARV-induced

* Correspondence: ytwu@yzu.edu.cn

Jiangsu Co-Innovation Center for Prevention of Animal Infectious Diseases and Zoonoses, College of Veterinary Medicine, Yangzhou University, 48 East Wenhui Road, Yangzhou, Jiangsu 225009, China

autophagy and apoptosis [17, 18]. It was also demonstrated that ARV disrupts many cellular pathways, regulating protein translation, cell proliferation, and cell metabolism [19–22]. However, these results are scattered and hard to reconcile. Some studies applied proteomic analysis and microarray analysis to get a comparatively integrated data set [23, 24]. However, these methods have several disadvantages compared with RNA-Seq. RNA-Seq now provides a way to investigate virus-mediated changes on the transcriptome of host cells, with high accuracy and low background [25, 26]. Additionally, it provides information on alternative splicing events, analyzing single nucleotide polymorphism, and predicting novel transcripts [27–29]. In this study, we tried to build a complete expression profile of ARV-mediated changes at the transcriptional level using RNA-Seq to unveil the complex interactions between ARV and host cells.

Methods
Cell culture and virus inoculation
Chicken embryonic fibroblast cell line DF-1 (CRL-12203, ATCC) cells were cultured with high glucose (4.5 g D-Glucose/L) Dulbecco's Modified Eagle Medium (HG-DMEM) (Basal Media, Shanghai, China) supplemented with 10% (v/v) fetal bovine serum (FBS) (Gibco, Shanghai, China) at 37 °C and 5% CO_2. The ARV strain GX/2010/1, causing severe tenosynovitis and enteritis, was isolated by our lab and propagated in chicken embryo fibroblasts (CEF) cells, and was reported to trigger autophagy in host cells to promote virus production [16]. The sequences of this strain are available in the GenBank database under the accession numbers KJ476699 –KJ476708. The sixth generation of the purified virus was used in this study and the median tissue culture infective dose ($TCID_{50}$) per milliliter (ml) of the virus was determined by the Reed-Muench method in CEF cells [30].

One day before virus inoculation, approximately 2×10^6 DF-1 cells were seeded into 75 cm^2 flasks (Corning, ME, USA). When monolayer was complete (approximately 7×10^6 cells), culture medium was discarded and the cells were rinsed with phosphate buffered saline once. The purified virus was diluted to 10 multiplicity of infection (MOI) per 5 mL with HG-DMEM and applied into each flask of the ARV-infected group and an equal volume of HG-DMEM was added in the mock-infected group. After incubated at 37 °C for 1.5 h, the medium was changed to HG-DMEM supplemented with 2% FBS. Then the cells were continued to be incubated at 37 °C for 2, 10, 18 and 24 h (Fig. 1a).

Total RNA extraction and cDNA library construction
At specified hours post infection (hpi), the medium was discarded and total RNA was extracted from triplicate samples of uninfected or ARV infected groups using the Ultrapure RNA Kit (CWBIO, Beijing, China) according to manufacturer's protocol. RNA degradation and contamination was monitored on 1% agarose gels. RNA purity was checked using the NanoPhotometer® spectrophotometer (IMPLEN, CA, USA). RNA concentration was measured using Qubit® RNA Assay Kit in Qubit® 2.0 Flurometer (Life Technologies, CA, USA). RNA integrity was assessed using the RNA Nano 6000 Assay Kit of the Bioanalyzer 2100 system (Agilent Technologies, CA, USA).

A total amount of 3 µg RNA per sample was used as input material for the RNA sample preparations. Sequencing libraries were generated using NEBNext® Ultra™ RNA Library Prep Kit for Illumina® (NEB, USA) following the manufacturer's recommendations, and index codes were added to attribute sequences to each sample.

Clustering and sequencing
The clustering of the index-coded samples was performed on a cBot Cluster Generation System using TruSeq PE Cluster Kit v3-cBot-HS (Illumia) according to the manufacturer's instructions. After cluster generation, the library preparations were sequenced on an Illumina Hiseq platform and 150 bp paired-end reads were generated. Raw reads of fastq format were first processed through custom written Perl scripts. At the same time, Q20, Q30, and GC content were calculated. All of the downstream analyses were based on clean, high quality data.

Reads mapping and quantification of gene expression level
The index of the chicken reference genome (Ensembl, Galgal4, updated 11-2015) was built using Bowtie v2.2.3 [31] and paired-end clean reads were aligned to the reference genome using TopHat v2.0.12 [32]. And HTSeq v0.6.1 was used to count the number of reads mapped to each gene [33]. Then, the expression level of each gene was calculated by the expected Fragments Per Kilobase of transcript per Million fragments mapped (FPKM) [34].

Differential expression analysis
Differential expression analysis of two groups was performed using the DESeq R package (1.18.0) [35]. DESeq provides statistical routines for determining differential expression in digital gene expression data using a model based on a negative binomial distribution. The resulting P-values were adjusted using the Benjamini and Hochberg's approach for controlling the false discovery rate [36]. Genes with an adjusted P-value < 0.05 found by DESeq were considered to be differentially expressed. Additionally, KOBAS 2.0 software was used to test the statistical enrichment of differentially expressed genes (DEGs) in the Kyoto Encyclopedia of Genes and Genomes (KEGG) pathways [37].

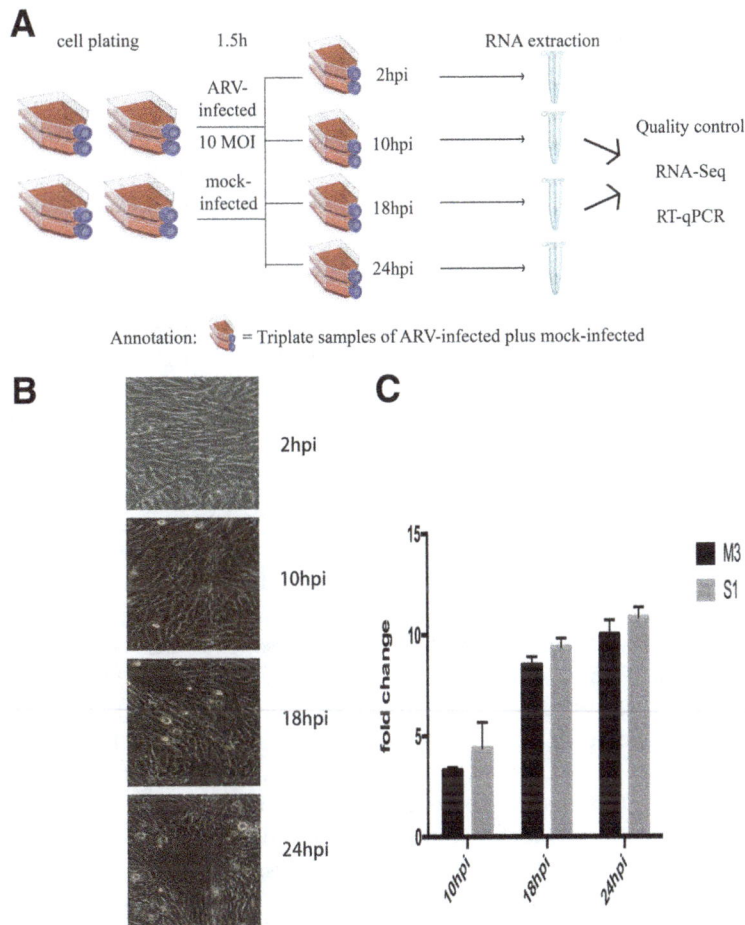

Fig. 1 Overview of RNA-Seq approach. **a** Experimental setup for RNA-Seq datasets. **b** DF-1 cells were infected with ARV and the cytopathic effect was assessed at different time points. **c** The replication of ARV was monitored by RT-qPCR analysis of ARV genes M3 and S1

RT-qPCR

Reverse transcription-quantitative polymerase chain reaction (RT-qPCR) was carried out based on the basic rules of the MIQE guidelines [38]. Briefly, 5 μg of total RNA (described above) was reverse transcribed using M-MLV reverse transcriptase (Transgen, Beijing, China) with a random hexamer primer (Genscript, Nanjing, China). The mixtures were diluted 1:10 with nuclease free water and then used as templates for qPCR. The qPCR analysis was performed using AceQ® qPCR SYBR® Green Master Mix (Vazyme, Nanjing, China) with 250 nM forward and reverse primers (Additional file 1). The reaction was carried out using LightCycler® Nano (Roche) with the following cycling conditions: an initial denaturation at 95 °C for 600 s followed by 45 cycles of 95 °C for 10 s and 60 °C for 30 s. Fold change was determined by the $2^{-\Delta\Delta Ct}$ method [39].

Novel transcripts and alternative splicing prediction

The Cufflinks v2.1.1 Reference Annotation Based Transcript (RABT) assembly method was used to construct and identify both known and novel transcripts from TopHat alignment results [40]. Alternative splicing (AS) events were classified to five major types by the software rMATS (Multivariate Analysis of Transcript Splicing) v3.2.5 [41]. The number of AS events in each sample was estimated separately. Because the chicken genome has been recently updated [42], the differentially expressed novel transcripts were retrieved by the BLAST tool on National Center for Biotechnology Information (NCBI).

Results

ARV infection of DF-1 cells and viral replication dynamics

To further study the molecular mechanism of ARV infection, DF-1 cells were infected with the virus for different time points at 10 MOI. The high dosage of virus was used to overcome the influence of uninfected cells [43, 44]. Infection and mock-infection were performed in biological triplicate for each time point and total RNA was extracted from both groups. The replication of the viral genome was determined by RT-qPCR and the fold change of M3 and S1 showed similar trends (Fig. 1c). Cytopathic effects could be seen at 18 hpi (Fig. 1b). To obtain an obviously changed transcriptome profile and minimize the influence of cell death and lysis, data at 2 hpi and 24 hpi were discarded and the remaining two groups were analyzed by RNA-Seq. One uninfected 10 hpi sample was lost, leaving a final total of 11 samples that were sequenced.

RNA-Seq results

After an overall quality review, mRNA was purified from total RNA using poly-T oligo-attached magnetic beads, and then the cDNA library was constructed, with quality assessment. After cluster generation, the library preparations were sequenced on an Illumina Hiseq platform and 150 bp paired-end reads were generated. The sequence run of each sample yielded at least 42 million clean reads and the lowest value of the reads possessing a Q-score > 20 was 95%, and the bottom line of the reads with a Q-score > 30 was 88% (Table 1). These results meet the requirements that more than 10 million reads are needed to construct a high quality eukaryotic transcriptome for discovering new genes and transcripts [45]. Importantly, all samples had between 71.81% and 77.45% of total reads mapped to the chicken reference genome, and the percentage of uniquely mapped reads was between 70.75% and 76.26% (Table 1).

Then, the mapped data were used to predict the novel transcripts and analyze the five major types of AS events, including SE (skipped exon), A5SS (alternative 5′ splice site), A3SS (alternative 3′ splice site), MXE (mutually exclusive exon), and RI (retained intron). The predicted novel genes were further analyzed together with the known genes. No significantly changed AS events were found between each group. The mapped data was normalized by calculating the FPKM and the distribution of mean FPKM per gene was found to be uniform between the four conditions (Fig. 2a). The correlation of gene expression levels between all of the samples was investigated using the squared Pearson correlation coefficient (R^2), and the minimum value was 0.979 (Fig. 2b). These results indicate that the expression levels of different genes or groups of genes are comparable, suggesting that the treatment is repeatable and has little variation. Therefore, the accuracy of the subsequent analysis of differentially expressed genes is likely to be high. Because the transcripts of the ARV genome do not have a poly-A 3′ tail [8], there are no reads can be used to indicate the replication of ARV.

Differentially expressed genes upon ARV infection

To further investigate the differential expression patterns in DF-1 cells between infected and mock-infected samples, the normalized gene expression level data were analyzed by DESeq. The resulting *P*-values were adjusted after correction for multiple testing and the DEGs were defined by having adjusted P-values (padj) < 0.05. The infected and mock-infected were compared with each other and the outline of the DEGs can be seen in Fig. 3a. Though there were significant alterations in the pairwise comparisons of different time points in the infected samples (Fig. 3b), the corresponding mock-infected samples also had big differences (Fig. 3c). This interference might result from cell culture and so should be discarded in the future studies. All of the DEGs were clustered and the results exhibited a clear time-dependent change in gene expression (Additional file 2).

The distinct effect on gene expression upon ARV infection was carefully examined. Compared with mock infected controls, 168 changes in the transcriptome, with 64 up-regulated and 104 down-regulated DEGs, were observed in response to ARV infection at 10 hpi (Fig. 3d, Additional file 3). Interestingly, only 47 up-regulated DEGs, with 31 genes in accordance with the 10 hpi group, and a single down-regulated novel gene were identified at 18 hpi (Fig. 3e, Additional file 3). These novel genes among the DEGs were retrieved by the BLAST tool. In

Table 1 RNA-Seq overview

Sample name	Raw reads	Clean reads	Q20 (%)	Q30 (%)	Total mapped (%)	Uniquely mapped (%)
V10_1	53,944,714	52,820,194	97.37	93.68	75.82	74.70
V10_2	59,160,706	57,378,292	97.39	93.81	76.44	75.46
V10_3	53,875,186	51,827,434	97.17	93.2	77.45	76.26
NC10_1	57,025,992	54,713,970	96.84	92.51	77.09	75.89
NC10_2	60,174,142	58,164,444	95.01	88.06	71.81	70.75
V18_1	58,591,660	56,688,518	95.1	88.24	72.40	71.34
V18_2	63,030,008	60,813,608	95.05	88.16	71.95	70.89
V18_3	44,002,734	42,770,420	96.75	92.22	73.96	72.89
NC18_1	51,089,720	48,895,588	96.24	91.26	74.02	72.96
NC18_2	61,366,046	59,972,578	97.11	93.2	75.05	73.90
NC18_3	68,023,822	65,597,914	97.36	93.79	76.15	75

Abbreviations: V ARV-infected, NC mock-infected

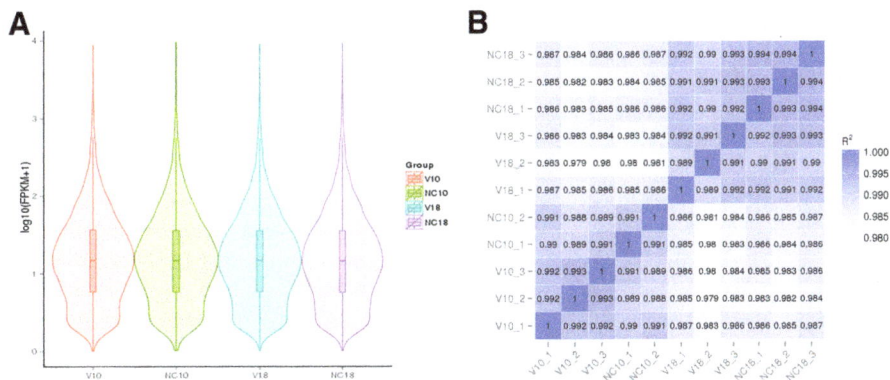

Fig. 2 The quality assessment of RNA-Seq. **a** Violin diagram of the distribution of average FPKM values per sample group. **b** The correlation between all the samples shown as a squared Pearson correlation coefficient (R^2)

addition, the pathway enrichment result can be seen at additional file 4. The DEGs with a fold change larger than 2 are listed in Table 2 with UniProtKB Keywords annotation [46], as a 2-fold threshold is commonly used to indicate biological significance [47].

Four up-regulated DEGs and two down-regulated DEGs were selected to be validated with RT-qPCR. The results show a similar pattern of ARV-mediated changes as was seen in the DEG analysis of RNA-Seq data (Fig. 4).

Discussion

ARV is one of the major pathogens that can cause immunosuppression in poultry [7]. Though the pathology and some molecular characteristics of ARV have been well studied [2, 8, 48], there are only a few reports that

can be used to help us understand the molecular basis of ARV infection. In this study, the DEG listed in Table 2 show that DF-1 cells exerted a prolonged antiviral response upon ARV infection. The up-regulation of RSAD2 (radical S-adenosyl methionine domain containing 2), IFIT5 (interferon induced protein with tetratricopeptide repeats 5), OASL (2′-5′-oligoadenylate synthetase-like), ISG12(2) (interferon-stimulated genes) and Mx (myxovirus resistance) have been reported in the infection of infectious bursal disease virus (IBDV), which is another important pathogen similar to ARV but can cause much higher mortality and much more serious immunosuppression [43]. EPSTI1 (epithelial-stromal interaction 1) does not have a Gene Ontology (GO) annotation. A recent report indicated that EPSTI1 plays a key role in IL-28A

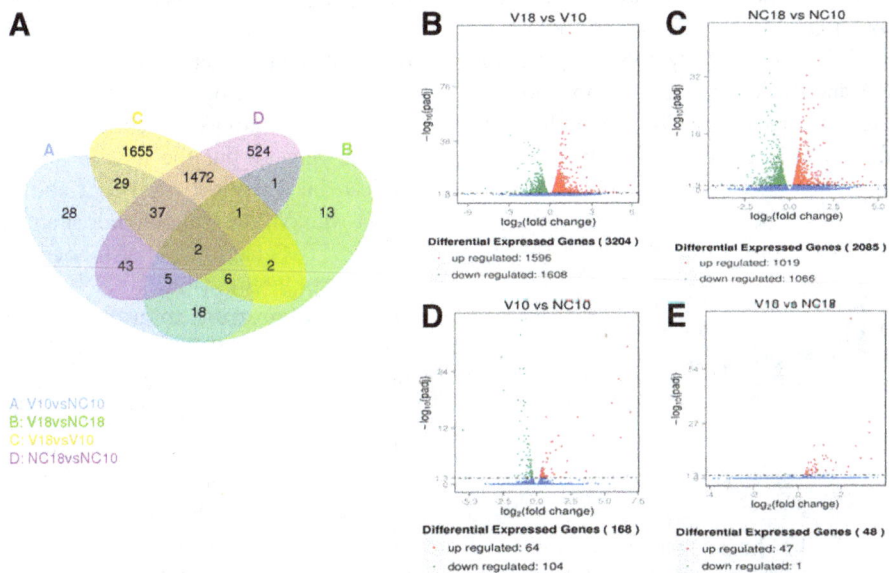

Fig. 3 Overall analysis of ARV-mediated changes in gene expression. **a** Venn diagram of DEGs between all samples. **b**–**e** Volcano plots of DEGs from ARV-infected cells at 10 hpi to 18 hpi (**b**), mock-infected cells at 10 hpi to 18 hpi (**c**), and ARV-infected cells to mock-infected cells at each time point (**d**, **e**)

Table 2 DEGs and the respective biological process in DF-1 cells upon ARV infection. DEGs with fold change > 2 are listed

	Gene-id	Gene-name	Description	Keywords of biological process	Fold change 10 hpi	18 hpi
Up	ENSGALG00000016400	RSAD2	Radical S-adenosyl methionine domain containing 2	antiviral defense, innate immunity, et al.	6.9368	3.2553
	ENSGALG00000006384	IFIT5	Interferon induced protein with tetratricopeptide repeats 5	antiviral defense, innate immunity, et al.	6.7163	3.3529
	ENSGALG00000013723	OASL	2'-5'-oligoadenylate synthetase-like	antiviral defense, innate immunity, et al.	6.2188	3.2579
	ENSGALG00000013575	ISG12(2)	Interferon-stimulated genes 12 (2)	–	6.0596	2.2561
	ENSGALG00000016142	Mx	Myxovirus resistance	antiviral defense, innate immunity, et al.	5.5442	2.653
	ENSGALG00000028982	CMPK2	Cytidine monophosphate (UMP-CMP) kinase 2	Pyrimidine biosynthesis	4.1368	2.3091
	ENSGALG00000009479	SAMD9L	Sterile alpha motif domain containing protein 9-like	endosomal vesicle fusion	3.5798	2.8676
	ENSGALG00000006138	HELZ2	Helicase with zinc finger domain 2	Transcription regulation	2.9797	2.4167
	ENSGALG00000016964	EPSTI1	Epithelial stromal interaction 1	–	2.2115	1.7528
	Novel00328	predicted: TRIM25-like	Tripartite motif containing 25	antiviral defense, innate immunity, et al.	2.1606	–
	Novel00773	predicted: EEA1-like	Early endosome antigen 1-like	endocytosis, vesicle fusion, et al.	2.1431	–
Down	ENSGALG00000010791	LEXM	Lymphocyte expansion molecule	positive regulation of cell proliferation, et al.	−5.4359	–
	ENSGALG00000000607	GPR37L1	G protein-coupled receptor 37 like 1	positive regulation of MAPK cascade, et al.	−5.2343	–
	ENSGALG00000002742	TMEM132B	Transmembrane protein 132B	–	−2.6179	–
	Novel01111	lncRNA	Uncharacterized LOC107053801	–	−2.4302	–

(interferon-λ2) mediated antiviral activity [49]. These interferon- (IFN-) induced genes (ISGs) with high expression levels reflect the stimulation of IFNs. Even though no significant elevation of expression levels of IFN genes were identified in this study, the up-regulation of TLR3 (Toll-like receptor 3), MYD88 (Myeloid differentiation factor 88), IRF1 (IFN regulatory factor 1), and IRF3 (IFN regulatory factor 3) were found. TLR3 plays key roles in detecting virus-derived dsRNA and the TLR3 genes are polymorphic among different chicken breeds [50, 51]. In addition to TLR-induced pathways, members of the RLR family

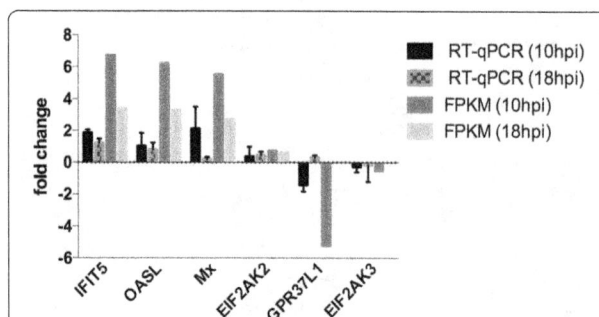

Fig. 4 Confirmation of DEGs by RT-qPCR. Expression of four up-regulated genes and two down-regulated genes was confirmed by RT-qPCR. The results are presented as fold change of ARV-infected cells compared to mock-infected cells at the indicate time point. Fold change using the FPKM was included for comparison of expression pattern

(retinoic acid inducible gene-I like receptor) constitute another TLR-independent anti-virus system. In our results, DHX58 (DEXH-box helicase 58, also known as LGP2, laboratory of genetics and physiology 2, or RLR3, RIG-I like receptor 3), IFIH1 (IFN-induced helicase C domain-containing protein 1, also known as MDA5, melanoma differentiation-associated protein 5), TRIM25 (Tripartite motif-containing protein 25), and a predicted TRIM25-like gene were found to be up-regulated at 10 hpi or at both 10 and 18 hpi time points. Chickens lack RIG-I (retinoic acid-inducible gene I), but the function of sensing viral infections can be performed by LGP2 and MDA5, which can interact with MAVS (mitochondrial antiviral signaling protein) or STING (stimulator of IFN genes) to stimulate the expression of IFNs [52–54].

DF-1 cells construct an antiviral environment through the expression of ISGs, including EIF2AK2 (Eukaryotic translation initiation factor 2-alpha kinase 2, also known as PKR, protein kinase RNA-activated). PKR is IFN-induced dsRNA-dependent enzymes. Active PKR can catalyze Ser-51 phosphorylation of the alpha subunit of EIF2, resulting in inhibition of protein synthesis at the initiation step of translation [55]. However, a previous report demonstrated that σA, an ARV encoded dsRNA binding protein, can block the activation of PKR and restore translation. In that report, the inhibition of vaccinia virus replication might reflect mechanisms other than

OAS and PKR to be responsible for the antiviral effects [56]. Interestingly, another eukaryotic translation initiation factor 2-alpha kinase, EIF2AK3, was found to be down-regulated in our results. EIF2AK3, also known as PERK (PKR-like endoplasmic reticulum-resident kinase), is one of eIF2α kinases regulating gene expression in the unfolded protein response (UPR) and in amino acid starved cells [57]. Protein synthesis can be inhibited during viral infection due to ER stress triggered by UPR, and different viruses may adapt different strategies to interfere with the activity of PERK [58, 59]. The depressed expression of PERK may reflect that ARV can impair the stress response and activate protein translation in DF-1 cells. The regulation of the host cell translation system ensures efficient replication of ARV. There is also a hypothesis that gene expression of ARV is mainly regulated at the translational level, rather than transcriptional level [8]. The replication level of viral genome determined by RT-qPCR in our results is consistent with this hypothesis.

ARV was initially detected from the clinical case of tenosynovitis, and a direct link between the virus and disease had been conclusively demonstrated [2, 8]. Though this virus has been studied for many years, the molecular pathogenesis of the disease remains unclear. In our results, the elevated expression of a gene, WNT9a (also known as Wnt14), was observed, which might play a key role in the development of the disease. A previous report identified that Wnt14 plays a pivotal role in initiating synovial joint formation in the chick limb, but the researchers were unable to determine the specific pathway that is responsible for transducing the Wnt14 signal in joint formation [60]. Later, studies demonstrated that the Wnt/β-catenin signaling pathway is necessary and sufficient to induce early steps of synovial joint formation [61]. Subsequently, a precise expression pattern of various Wnts was analyzed during chick wing development [62]. Continued expression of Wnt14 in the mature joint might be good for the maintenance of joint integrity and was presumed to play a role in the etiology of rheumatoid arthritis in humans [60]. ARV can replicate and perhaps be persistent at hock joint of chicken [48]. The up-regulation of Wnt14, combined with the induction of apoptosis [15], may be responsible for ARV-induced joint damage and more severe tendon rupture.

Conclusions

In conclusion, our results show that ARV infection stimulates a prolonged antiviral response in host cells and interferes with cell growth and cell death pathways. Our results also provide information that may be helpful to further investigate the pathogenesis of ARV infection. Combined with previous studies, we can begin to piece together the interactions between ARV and host cells (Fig. 5). However, the details of these interactions need to be further investigated in future studies.

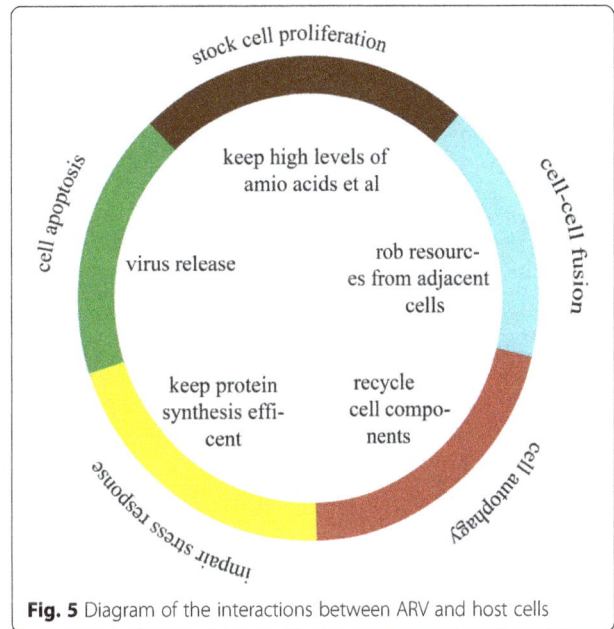

Fig. 5 Diagram of the interactions between ARV and host cells

Additional files

Additional file 1: RT-qPCR Primers. All RT-qPCR primers for detection of the replication of ARV and the DEGs. (DOCX 46 kb)

Additional file 2: Cluster analysis of differentially expressed genes. Heatmap of the DEGs across all datasets based on $\log_{10}$ (FPKM + 1). (PDF 283 kb)

Additional file 3: Differentially expressed genes between sample groups. Sheet 1. The DEGs between mock- or ARV-infected samples at 10 hpi. Sheet 2. The DEGs between mock- or ARV-infected samples at 18 hpi. (XLSX 74 kb)

Additional file 4: KEGG pathway enrichment result. Sheet 1. The KEGG enrichment result of the DEGs between mock- or ARV-infected samples at 10hpi. Sheet 2. The KEGG enrichment result of the DEGs between mock- or ARV-infected samples at 18hpi. (XLSX 60 kb)

Abbreviations
ARV: Avian reovirus; AS: Alternative splicing; Bp: Base pair; CEF: Chicken embryo fibroblasts; DEGs: Differentially expressed genes; FBS: Fetal bovine serum; FPKM: Fragments Per Kilobase of transcript per Million fragments mapped; GO: Gene Ontology; HG-DMEM: High glucose Dulbecco's Modified Eagle Medium; Hpi: Hours post infection; IFN: Interferon; KEGG: Kyoto Encyclopedia of Genes and Genomes; MOI: Multiplicity of infection; Padj: Adjusted P-values; RNA-Seq: RNA sequencing; RT-qPCR: Reverse transcription-quantitative polymerase chain reaction; TCID$_{50}$: Median tissue culture infective dose

Acknowledgements
Thanks to Huaisheng Wu for the assistance in performing RNA-Seq analysis.

Funding
This work was supported by grants from the National Natural Science Foundation of China (31272576), the China Agriculture Research System (CARS-41-K08), and the Priority Academic Program Development of Jiangsu Higher Education Institutions.

Authors' contributions

XN and XZ conceived and designed the experiments. XN performed the experiments. XN and YWa analyzed the data. XZ and ML contributed reagents/materials/analysis tools. XN and YWu wrote the manuscript. All authors read and approved the final manuscript.

Competing interests

The authors declare that they have no competing interests.

References

1. King AM, Adams MJ, Lefkowitz EJ, Carstens EB. Virus taxonomy: ninth report of the international committee on taxonomy of viruses. London: Academic; 2012.
2. Van DHL. The history of avian reovirus. Avian Dis. 2000;44(3):638–41.
3. Van DHL. Viral arthritis/tenosynovitis: a review. Avian Pathol. 1977;6(4):271–84.
4. Woźniakowski G, Samorek-Salamonowicz E, Gaweł A. Occurrence of reovirus infection in Muscovy ducks (Cairina Moschata) in south western Poland. Pol J Vet Sci. 2015;17(2):299–305.
5. Ayalew LE, Gupta A, Fricke J, Ahmed KA, Popowich S, Lockerbie B, Tikoo SK, Ojkic D, Gomis S. Phenotypic, genotypic and antigenic characterization of emerging avian reoviruses isolated from clinical cases of arthritis in broilers in Saskatchewan, Canada. Sci Rep. 2017;7(1):3565.
6. Nham EG, Pearl DL, Slavic D, Ouckama R, Ojkic D, Guerin MT. Flock-level prevalence, geographical distribution, and seasonal variation of avian reovirus among broiler flocks in Ontario. Can Vet J. 2017;58(8):828–34.
7. Woźniakowski G, Niczyporuk JS, Samorek-Salamonowicz E, Gaweł A. The development and evaluation of cross-priming amplification (CPA) for the detection of avian reovirus (ARV). J Appl Microbiol. 2015;118(2):528–36.
8. Schat KA, Skinner MA. Avian immunosuppressive diseases and immunoevasion. In: Davison F, Kaspers B, Schat KA, editors. Avian immunology. 2nd ed. London: Academic; 2014. p. 275–97.
9. Benavente J, Martinez-Costas J. Avian reovirus: structure and biology. Virus Res. 2007;123(2):105–19.
10. Engstrom BE. Blue wing disease of chickens: isolation of avian reovirus and chicken anaemia agent. Avian Pathol. 1988;17(1):23–32.
11. Engstrom BE, Fossum O, Luthman M. Blue wing disease of chickens: experimental infection with a Swedish isolate of chicken anaemia agent and an avian reovirus. Avian Pathol. 1988;17(1):33–50.
12. Mills JN, Wilcox GE. Replication of four antigenic types of avian reovirus in subpopulations of chicken leukocytes. Avian Pathol. 1993;22(2):353–61.
13. Shih WL, Hsu HW, Liao MH, Lee LH, Liu HJ. Avian reovirus σC protein induces apoptosis in cultured cells. Virology. 2004;321(1):65–74.
14. Salsman J, Top D, Boutilier J, Duncan R. Extensive syncytium formation mediated by the reovirus FAST proteins triggers apoptosis-induced membrane instability. J Virol. 2005;79(13):8090–100.
15. Lin HY, Chuang ST, Chen YT, Shih WL, Chang CD, Liu HJ. Avian reovirus-induced apoptosis related to tissue injury. Avian Pathol. 2007;36(2):155–9.
16. Meng S, Jiang K, Zhang X, Zhang M, Zhou Z, Hu M, Yang R, Sun C, Avian WY. Reovirus triggers autophagy in primary chicken fibroblast cells and Vero cells to promote virus production. Arch Virol. 2012;157(4):661–8.
17. Lin PY, Chang CD, Chen YC, Shih WL. RhoA/ROCK1 regulates avian Reovirus S1133-induced switch from autophagy to apoptosis. BMC Vet Res. 2015;11(1):1–12.
18. Duan S, Cheng J, Li C, Yu L, Zhang X. Autophagy inhibitors reduce avian-reovirus-mediated apoptosis in cultured cells and in chicken embryos. Arch Virol. 2015;160(7):1679–85.
19. Ji WT, Wang L, Lin RC, Huang WR, Liu HJ. Avian reovirus influences phosphorylation of several factors involved in host protein translation including eukaryotic translation elongation factor 2 (eEF2) in Vero cells. Biochem Bioph Res Co. 2009;384(3):301–5.
20. Liu HJ, Lin PY, Lee JW, Hsu HY, Shih WL. Retardation of cell growth by avian reovirus p17 through the activation of p53 pathway. Biochem Bioph Res Co. 2005;336(2):709–15.
21. Huang WR, Wang YC, Chi PI, Wang L, Wang CY, Lin CH, Liu HJ. Cell entry of avian reovirus follows a caveolin-1-mediated and dynamin-2-dependent endocytic pathway that requires activation of p38 mitogen-activated protein kinase (MAPK) and Src signaling pathways as well as microtubules and small GTPase Rab5 protein. J Biol Chem. 2011;286(35):30780–94.
22. Chi PI, Huang WR, Lai IH, Cheng CY, Liu HJ. The p17 nonstructural protein of avian reovirus triggers autophagy enhancing virus replication via activation of phosphatase and tensin deleted on chromosome 10 (PTEN) and AMP-activated protein kinase (AMPK), as well as dsRNA-dependent protein kinase (PKR). J Biol Chem. 2013;288(5):3571–84.
23. Lin PY, Liu HJ, Chang CD, Chang CI, Hsu JL, Liao MH, Lee JW, Shih WL. Avian reovirus S1133-induced DNA damage signaling and subsequent apoptosis in cultured cells and in chickens. Arch Virol. 2011;156(11):1917–29.
24. Chen WT, YL W, Chen T, Cheng CS, Chan HL, Chou HC, Chen YW, Yin HS. Proteomics analysis of the DF-1 chicken fibroblasts infected with avian reovirus strain S1133. PLoS One. 2014;9(3):e92154.
25. Kogenaru S, Yan Q, Guo Y, Wang N. RNA-seq and microarray complement each other in transcriptome profiling. BMC Genomics. 2012;13(1):629.
26. Liu JJ, Sturrock RN, Benton R. Transcriptome analysis of Pinus Monticola primary needles by RNA-seq provides novel insight into host resistance to Cronartium Ribicola. BMC Genomics. 2013;14(1):884.
27. Zhao C, Cees W, De WPJGM, Tang D, Theo VDL. RNA-Seq analysis reveals new gene models and alternative splicing in the fungal pathogen Fusarium graminearum. BMC Genomics. 2013;14(1):21.
28. Mcqueen CM, Whitfieldcargile CM, Konganti K, Blodgett GP, Dindot SV, Cohen ND. TRPM2 SNP genotype previously associated with susceptibility to Rhodococcus equi pneumonia in quarter horse foals displays differential gene expression identified using RNA-Seq. BMC Genomics. 2016;17(1):993.
29. Sheynkman GM, Johnson JE, Jagtap PD, Shortreed MR, Onsongo G, Frey BL, Griffin TJ, Smith LM. Using galaxy-P to leverage RNA-Seq for the discovery of novel protein variations. BMC Genomics. 2014;15(1):1–9.
30. Reed LJ, Muench H. A simple method of estimating fifty percent endpoints. Am J Epidemiol. 1938;27(3):493–9.
31. Langmead B, Salzberg SL. Fast gapped-read alignment with bowtie 2. Nat Methods. 2012;9(4):357–9.
32. Kim D, Pertea G, Trapnell C, Pimentel H, Kelley R, Salzberg SL. TopHat2: accurate alignment of transcriptomes in the presence of insertions, deletions and gene fusions. Genome Biol. 2013;14(4):R36.
33. Anders S, Pyl PT, Huber W. HTSeq-a python framework to work with high-throughput sequencing data. Bioinformatics. 2015;31(2):166–9.
34. Trapnell C, Williams BA, Pertea G, Mortazavi A, Kwan G, van Baren MJ, Salzberg SL, Wold BJ, Pachter L. Transcript assembly and abundance estimation from RNA-Seq reveals thousands of new transcripts and switching among isoforms. Nat Biotechnol. 2010;28(5):511–5.
35. Anders S, Huber W. Differential expression analysis for sequence count data. Genome Biol. 2010;11(10):R106.
36. Benjamini Y, Hochberg Y. Controlling the false discovery rate: a practical and powerful approach to multiple testing. J Roy Stat Soc B. 1995;57(1):289–300.
37. Xie C, Mao X, Huang J, Ding Y, Wu J, Dong S, Kong L, Gao G, Li CY, Wei LKOBAS. 2.0: a web server for annotation and identification of enriched pathways and diseases. Nucleic Acids Res. 2011;39:W316–22.
38. Bustin SA, Benes V, Garson JA, Hellemans J, Huggett J, Kubista M, Mueller R, Nolan T, Pfaffl MW, Shipley GL, et al. The MIQE guidelines: minimum information for publication of quantitative real-time PCR experiments. Clin Chem. 2009;55(4):611–22.
39. Livak KJ, Schmittgen TD. Analysis of relative gene expression data using real-time quantitative PCR and the 2(−Delta Delta C(T)) method. Methods. 2001;25(4):402–8.
40. Trapnell C, Roberts A, Goff L, Pertea G, Kim D, Kelley DR, Pimentel H, Salzberg SL, Rinn JL, Pachter L. Differential gene and transcript expression analysis of RNA-seq experiments with TopHat and cufflinks. Nat Protoc. 2012;7(3):562–78.
41. Shen S, Park JW, Lu ZX, Lin L, Henry MD, Wu YN, Zhou Q, Xing Y. rMATS: robust and flexible detection of differential alternative splicing from replicate RNA-Seq data. P Natl Acad Sci USA. 2014;111(51):E5593.
42. Genome database on National Center for Biotechnology Information. https://www.ncbi.nlm.nih.gov/genome/?term=gallus%20gallus. Accessed 30 May 2017.
43. Hui RK, Leung FC. Differential expression profile of chicken embryo fibroblast DF-1 cells infected with cell-adapted infectious Bursal disease virus. PLoS One. 2015;10(6):e0111771.
44. Sessions OM, Tan Y, Goh KC, Liu Y, Tan P, Rozen S, Ooi EE. Host cell transcriptome profile during wild-type and attenuated dengue virus infection. PLoS Negl Trop Dis. 2013;7(3):e2107.

Transcriptome analysis of avian reovirus-mediated changes in gene expression of normal chicken...

69

45. Chen S, Wang A, Sun L, Liu F, Wang M, Jia R, Zhu D, Liu M, Yang Q, Wu Y, et al. Immune-Related Gene Expression Patterns in GPV- or H9N2- Infected Goose Spleens. Int J Mol Sci. 2016;17(12):E1990.

46. Universal Protein Resource databases. The European Bioinformatics Institute (EMBL-EBI), the SIB Swiss Institute of Bioinformatics and the Protein Information Resource (PIR). http://www.uniprot.org/. Accessed 30 May 2017.

47. Lamontagne J, Mell JC, Bouchard MJ. Transcriptome-wide analysis of hepatitis B virus-mediated changes to normal Hepatocyte gene expression. PLoS Pathog. 2016;12(2):e1005438.

48. Jones RC. Avian reovirus infection. Rev Sci Tech OIE. 2000;19(2):614–25.

49. Meng X, Yang D, Yu R, Zhu H. EPSTI1 is involved in IL-28A-mediated inhibition of HCV infection. Mediat Inflamm. 2015;2015:716315.

50. Karpala AJ, Lowenthal JW, Bean AG. Activation of the TLR3 pathway regulates IFNbeta production in chickens. Dev Comp Immunol. 2008;32(4):435–44.

51. Ruan W, An J, Wu Y. Polymorphisms of chicken TLR3 and 7 in different breeds. PLoS One. 2015;10(3):e0119967.

52. Liniger M, Summerfield A, Ruggli N. MDA5 can be exploited as efficacious genetic adjuvant for DNA vaccination against lethal H5N1 influenza virus infection in chickens. PLoS One. 2012;7(12):e49952.

53. Liniger M, Summerfield A, Zimmer G, Mccullough KC, Ruggli N. Chicken cells sense influenza a virus infection through MDA5 and CARDIF signaling involving LGP2. J Virol. 2012;86(2):705–17.

54. Cheng Y, Sun Y, Wang H, Yan Y, Ding C, Sun J, Chicken STING. Mediates activation of the IFN gene independently of the RIG-I gene. J Immunol. 2015;195(8):3922–36.

55. Zykova T, Zhu F, Bode AM, Zhang Y, Dong Z. ERK2 and RSK2 mediate phosphorylation of PKR (Thr451) and PKR directly catalyzes the phosphorylation of eIF2α at Ser51. Cancer Res. 2006;66(Suppl 8):1005.

56. Martínezcostas J, Gonzálezlópez C, Vakharia VN, Benavente J. Possible involvement of the double-stranded RNA-binding Core protein çA in the resistance of avian Reovirus to interferon. J Virol. 2000;74(3):1124–31.

57. Harding HP, Novoa I, Zhang Y, Zeng H, Wek R, Schapira M, Ron D. Regulated translation initiation controls stress-induced gene expression in mammalian cells. Mol Cell. 2000;6(5):1099–108.

58. Pavio N, Romano PR, Graczyk TM, Feinstone SM, Taylor DR. Protein synthesis and endoplasmic reticulum stress can be modulated by the hepatitis C virus envelope protein E2 through the eukaryotic initiation factor 2α Kinase PERK. J Virol. 2003;77(6):3578–85.

59. Minakshi R, Padhan K, Rani M, Khan N, Ahmad F, Jameel S. The SARS Coronavirus 3a protein causes endoplasmic reticulum stress and induces ligand-independent downregulation of the type 1 interferon receptor. PLoS One. 2009;4(12):e8342.

60. Hartmann C, Tabin CJ. Wnt-14 plays a pivotal role in inducing synovial joint formation in the developing appendicular skeleton. Cell. 2001;104(3):341–51.

61. Guo X, Day TF, Jiang X, Garrett-Beal L, Topol L, Yang Y. Wnt/beta-catenin signaling is sufficient and necessary for synovial joint formation. Genes Dev. 2004;18(19):2404–17.

62. Loganathan PG, Nimmagadda S, Huang R, Scaal M, Christ B. Comparative analysis of the expression patterns of Wnts during chick limb development. Histochem Cell Biol. 2005;123(2):195–201.

6

The insect pathogenic bacterium *Xenorhabdus innexi* has attenuated virulence in multiple insect model hosts yet encodes a potent mosquitocidal toxin

Il-Hwan Kim[1,2], Sudarshan K. Aryal[3], Dariush T. Aghai[4], Ángel M. Casanova-Torres[4], Kai Hillman[4], Michael P. Kozuch[4], Erin J. Mans[4,5], Terra J. Mauer[4,5], Jean-Claude Ogier[6], Jerald C. Ensign[4], Sophie Gaudriault[6], Walter G. Goodman[1], Heidi Goodrich-Blair[4,5*][iD] and Adler R. Dillman[3*]

Abstract

Background: *Xenorhabdus innexi* is a bacterial symbiont of *Steinernema scapterisci* nematodes, which is a cricket-specialist parasite and together the nematode and bacteria infect and kill crickets. Curiously, *X. innexi* expresses a potent extracellular mosquitocidal toxin activity in culture supernatants. We sequenced a draft genome of *X. innexi* and compared it to the genomes of related pathogens to elucidate the nature of specialization.

Results: Using green fluorescent protein-expressing *X. innexi* we confirm previous reports using culture-dependent techniques that *X. innexi* colonizes its nematode host at low levels (~3–8 cells per nematode), relative to other *Xenorhabdus-Steinernema* associations. We found that compared to the well-characterized entomopathogenic nematode symbiont *X. nematophila*, *X. innexi* fails to suppress the insect phenoloxidase immune pathway and is attenuated for virulence and reproduction in the Lepidoptera *Galleria mellonella* and *Manduca sexta*, as well as the dipteran *Drosophila melanogaster*. To assess if, compared to other *Xenorhabdus* spp., *X. innexi* has a reduced capacity to synthesize virulence determinants, we obtained and analyzed a draft genome sequence. We found no evidence for several hallmarks of *Xenorhabdus* spp. toxicity, including Tc and Mcf toxins. Similar to other *Xenorhabdus* genomes, we found numerous loci predicted to encode non-ribosomal peptide/polyketide synthetases. Anti-SMASH predictions of these loci revealed one, related to the *fcl* locus that encodes fabclavines and *zmn* locus that encodes zeamines, as a likely candidate to encode the *X. innexi* mosquitocidal toxin biosynthetic machinery, which we designated Xlt. In support of this hypothesis, two mutants each with an insertion in an Xlt biosynthesis gene cluster lacked the mosquitocidal compound based on HPLC/MS analysis and neither produced toxin to the levels of the wild type parent.

Conclusions: The *X. innexi* genome will be a valuable resource in identifying loci encoding new metabolites of interest, but also in future comparative studies of nematode-bacterial symbiosis and niche partitioning among bacterial pathogens.

Keywords: Virulence, Toxin, Symbiosis, Insect, Immunity, NRPS/PKS, Mosquito, Lipopeptide

* Correspondence: hgblair@utk.edu; adlerd@ucr.edu
[4]Department of Bacteriology, University of Wisconsin-Madison, Madison, WI, USA
[3]Department of Nematology, University of California, Riverside, CA, USA
Full list of author information is available at the end of the article

Background

Nematodes in the genus *Steinernema* associate with *Xenorhabdus* bacteria in a mutually beneficial relationship that allows the pair to utilize insect hosts as a reproductive niche. *Steinernema* nematodes have a soil-dwelling stage, known as the infective juvenile (IJ) that carries *Xenorhabdus* bacteria into insect prey that will be killed and used for nutrients that support reproduction. Progeny IJs then emerge from the spent insect cadaver, carrying their *Xenorhabdus* partner, to begin the cycle again. In general, the bacterial symbionts promote nematode fitness by helping kill insect hosts and by contributing to the degradation and protection of the host cadaver from competitors and predators [1]. Because they can be pathogenic to insects when injected without their nematode host, *Xenorhabdus* bacteria and their genes are being exploited for use in agricultural settings to help control important crop pests. For example, certain *X. nematophila* genes can confer resistance to insect pests when expressed transgenically in plants [2, 3]. The potential for insecticidal and natural product discovery has helped spur the sequencing and analysis of multiple *Xenorhabdus* spp. genomes [4–7].

Recently, renewed attention has been placed on the biology of *Steinernema scapterisci*, a nematode first isolated by G.C. Smart and K.B. Nguyen in 1985 from mole crickets found in Uruguay [8–11]. The bacterial symbiont found within these nematodes was later established as a new species, *Xenorhabdus innexi* [12]. The relationship between *S. scapterisci* and *X. innexi* appears to be specific; six species of *Xenorhabdus* have been tested in previous studies and only *X. innexi* colonizes the infective juvenile (IJ) stage of *S. scapterisci* [13].

S. scapterisci is closely related to the well-studied steinernematid nematode *S. carpocapsae* [14–17], but has distinctive characteristics that make it useful for comparative purposes, including its specialization for cricket hosts [9, 11, 18, 19]. While both *S. carpocapsae* and *S. scapterisci* caused death when injected into *A. domesticus* (house cricket), only *S. scapterisci* reproduced to high levels (*S. carpocapsae* produced ~7% the infective juvenile progeny relative to *S. scapterisci*), and fewer (16%) *S. scapterisci* were melanized compared to *S. carpocapsae* (92%), indicating *S. scapterisci* either does not induce an immune response in *A. domesticus* or is resistant to it [19]. A common feature of host-seeking parasitic nematodes is the activation of the IJ stage upon exposure to host tissue [11, 20, 21]. For entomopathogenic nematode (EPN) IJs, this activation process includes morphological changes of the mouth, pharynx, and anterior gut, as well as release of the symbiotic bacteria into the host and secretion of a variety of proteins that are thought to be involved in parasitism [22, 23]. A recent study demonstrated that more than 70% of *S. scapterisci* IJs are activated within 18 h of exposure to cricket tissue while fewer than 30% of the IJs are activated when exposed to *G. mellonella* waxworm tissue for the same period of time [11], supporting the notion that *S. scapterisci* is a cricket specialist.

The specialization of *S. scapterisci* and its symbiont for crickets is in contrast to their attenuated effectiveness against other insects. When injected into *Popillia japonica* (Japanese beetle), *S. carpocapsae* can kill and reproduce, but *S. scapterisci* cannot. Further, although conflicting reports occur in the literature, compared to other *Steinernema* species, *S. scaptersci* appears to have reduced capacity to kill or reproduce in *Galleria mellonella* [13, 18, 24], which is a standard bait host. It has been suggested that the low virulence of *S. scapterisci* in wax worms is due to the relatively low virulence of its associated symbiont *X. innexi* [10, 25], as well as negative impacts from non-*Xenorhabdus* microbes that can be associated with *S. scapterisci* IJs [10].

Generally, *Steinernema* bacterial symbionts are thought to benefit their hosts by contributing to insect death and degradation. Using aposymbiotic *S. scapterisci* nematodes and cultured *X. innexi* symbionts, Bonifassi et al. determined that neither was pathogenic towards *G. mellonella* individually, but were when combined [10]. This indicates that both partners are necessary to kill this insect, in contrast to *S. carpocapsae* and *X. nematophila* each of which can kill insects without the other (see, for example [26, 27]). Later studies demonstrated that *S. scapterisci* could survive, parasitize, and reproduce aposymbiotically in *G. mellonella*, but with reduced overall fitness [24], and that the impact of different *Xenorhabdus* species on *S. scapterisci* fitness is directly correlated with their phylogenetic relatedness to *X. innexi* [13]. Generally, these studies support the idea that *X. innexi* specifically facilitates the establishment of *S. scapterisci* nematode infection and production of progeny IJs in insect hosts. However, Sicard et al. noted that *S. scapterisci* fared better in the absence of its symbiont than did the other nematode species examined, and that it was colonized by fewer bacterial symbionts (~0.07 CFU/IJ average, relative to 43.8 CFU/IJ for *X. nematophila*) as measured using a crushing and plating method [24]. Taken together, these reports suggest that *S. scapterisci* is trending toward decreased dependence of the nematode on its bacterial symbiont.

Although the findings reviewed above hinted that *X. innexi* may be less virulent, at least toward some insect hosts, than other *Xenorhabdus* species, an activity-screening approach revealed that it does secrete a peptide with insecticidal activity effective against the larvae of several mosquito species in the *Aedes*, *Anopheles*, and *Culex* genera [28]. Recent work has indicated the active compound is a lipopeptide, dubbed Xenorhabdus lipoprotein toxin (Xlt) that can create pores in the apical surface of mosquito larval anterior midgut cells [29].

The experiments presented here were geared toward directly testing the nematode colonization and insect virulence properties of *X. innexi*, to provide further insights into the evolution of different symbiotic relationships among *Steinernema-Xenorhabdus* pairings. Further, we sought to identify distinctive virulence determinants that may be encoded by *X. innexi* relative to other *Xenorhabdus* species, predicted based on the specialization of the *X. innexi-S. scapterisci* symbiotic pair for crickets, and the production by *X. innexi* of a mosquitocidal toxin. To pursue these goals we established a laboratory model of *S. scapterisci-X. innexi*-insect symbiosis. We assessed *X. innexi* virulence in several model insects, applied genetic tools to facilitate monitoring its presence and gene function, and used draft genome sequencing and analysis to explore its virulence potential.

Results

S. scapterisci IJ receptacles are colonized by few *X. innexi* cells

To assess the colonization levels of *X. innexi* in *S. scapterisci* nematodes, we added axenic nematodes (see Methods) to lawns of two *X. innexi* strains, one (HGB1681) isolated from *S. scapterisci* nematodes provided by Prof. Grover Smart (FL) and the other isolated from the *S. scapterisci* nematodes being used in this study (provided by Becker Underwood Inc.) (Table 1). IJs emerging from in vitro cultures such as those described above were surface sterilized and subjected to a grinding assay to calculate average colony-forming units (CFU) of bacteria per IJ. Both tested *X. innexi* strains colonized *S. scapterisci* at ~7 CFU/IJ (Table 2) and colonies were confirmed to be *X. innexi* based on distinctive phenotypic traits (catalase negative, characteristic brown color, and distinctive odor). No colonies grew from homogenates of axenic nematodes cultivated on *X. nematophila*, confirming the previous finding that *X. nematophila* does not colonize *S. scapterisci* nematodes (Table 2) [24].

The low colonization level we detected could be due to low frequency of colonization (few nematodes in the population are colonized) or low levels of colonization (the majority of nematodes are colonized by very few bacteria) or a combination of these phenotypes. To address this question, we generated *X. innexi* strains expressing green fluorescent protein (GFP) (Table 1) to facilitate their visualization within IJ receptacles (Fig. 1) [30]. As with non-GFP expressing strains, progeny IJs emerging from lawns of GFP-expressing *X. innexi* were colonized by an average of approximately 7–10 CFU/IJ, as determined by grinding assays (Table 2). Visualization by

fluorescence microscopy revealed GFP-expressing bacterial cell colonization of the *S. scapterisci* IJs (Fig. 1b). *S. carpocapsae* and *S. scapterisci* had visible green-fluorescent bacteria at frequencies of 94.8 ± 0.007 and 92.7 ± 0.016, respectively (mean ± SD of three biological replicates of each nematode species). Like other *Xenorhabdus* spp., including *X. nematophila*, *X. innexi* localized to the receptacle region of the intestine posterior to the basal bulb. However, *X. innexi* appears distinct in that only a few cells (1–5 cells) were visible within the receptacles of individual colonized IJs (Fig. 1b), in contrast to the large number of *X. nematophila* occupying this region in *S. carpocapsae* nematodes (Fig. 1a). These observations support the quantitative data acquired by grinding, and indicate that *X. innexi* colonizes *S. scapterisci* IJs at a high frequency, but at very low levels compared to *X. nematophila* colonization of *S. carpocapsae* (typically ~40 CFU/IJ using this method) [31]. We next examined the growth characteristics of *X. innexi* in laboratory medium, compared to *X. nematophila* and *X. bovienii*, the symbiont of *S. jollieti*, two *Xenorhabdus* bacteria for which complete genomes exist [6]. We found that in LB medium, *X. innexi* displayed a longer lag, a significantly slower growth rate (Additional file 1), and a lower final OD_{600} compared to *X. bovienii* and *X. nematophila* (Fig. 1c).

X. innexi is avirulent at ecologically relevant doses

Given that an individual *S. scapterisci* nematode would inoculate an insect host with few cells of its *X. innexi* symbiont (Fig. 1) [24], we next assessed the contribution of *X. innexi* to the nematode-symbiont complex by injecting quantified doses into several potential insect hosts and the model insect *Drosophila melanogaster*. We compared this to the virulence of *X. nematophila*, the well-characterized bacterial symbiont of *S. carpocapsae*.

Similar to previous studies [32, 33], we found that *X. nematophila* is highly toxic to *G. mellonella* waxworm larvae, rapidly killing these insects even at low doses (Fig. 2a). *X. nematophila* quickly grew in waxworm larvae, reaching over 1 million colony-forming units (CFUs) in less than 24 h, regardless of the inoculating dose (Fig. 2b). We found similar results in adult *D. melanogaster*, where *X. nematophila* rapidly killed the adults and grew to over 1 million CFUs in less than 18 h (Fig. 2c-d; Additional file 2). In contrast to *X. nematophila*, *X. innexi* was nearly avirulent when injected into fruit fly adults (Fig. 2e). We found that all but the highest dose we tried, 100,000 CFUs, proved to have little to no effect on fruit fly survival. We plotted the growth of the bacteria in infected flies over time and found that *D. melanogaster* adults are highly resistant to *X. innexi* (Fig. 2f). The flies reduced bacterial

Table 1 Strains and plasmids used in this study

	Relevant characteristics	Source/Reference
Strain		
HGB800	*Xenorhabdus nematophila* isolated from *Steinernema carpocapsae* All nematodes	ATCC19061
HGB1053	*Xenorhabdus bovienii* SS-2004	[119]
HGB1681	*Xenorhabdus innexi* isolated from *Steinernema scapterisci* nematodes from Grover Smart. Also called *Xenorhabdus* MT, deposited to ATCC in 2005.	G.C. Smart Jr. University of Florida; ATCC PTA-6826
HGB1997	*Xenorhabdus innexi* isolated in 2013 from *Steinernema scapterisci* nematodes obtained from Becker-Underwood	This study
HGB2171	HGB1681 *att*Tn7/Tn7-GFP (from pURR25)	This study
HGB2172	HGB1997 *att*Tn7/Tn7-GFP (from pURR25)	This study
HGB283	*Escherichia coli* S17–1 lambda pir pUX-BF13	[107]
HGB1262	*Escherichia coli* BW29427 pURR25, mini Tn7KS-GFP	B. Lies and D. Newman [108]
TOP10	*E. coli* strain for general cloning	Thermo
Plasmids		
pBluescript II SK (–)	General cloning	Stratagene
pKanWOR	pBluescript KS+ with Km cassette (1 kb) in BamHI site	H. Goodrich-Blair
pCR-Blunt II-TOPO	General cloning vector, Kanr	Thermo
pBlueXIS1_460109Up	XIS1_460109Up inserted in pBluescript SK-	This study
pBlueXIS1_460109UpDn	XIS1_460109Dn inserted in pBlueXIS1_460109Up	This study
pBlueXIS1_460109UpKanDn	Kan cassette from pKanWOR inserted in pBlueXIS1_460109UpDn	This study
pBlueXIS1_460115Up	XIS1_460115Up inserted in pBluescript SK-	This study
pBlueXIS1_460115UpDn	XIS1_460115Dn inserted in pBlueXIS1_460115Up	This study
pBlueXIS1_460115UpKanDn	Kan cassette from pKanWOR inserted in pBlueXIS1_460115UpDn	This study
pKR100	oriR6K suicide vector, Cmr	H. Goodrich-Blair
pKRXIS1_460109	XIS1_460109UpKanDn inserted in pKR100	This study
pKRXIS1_460115	XIS1_460115UpKanDn inserted in pKR100	This study

growth and eventually cleared the bacteria from the system, even when given an initial dose of 10,000 CFUs (Fig. 2f). When we injected 100,000 CFUs, the bacterial cells were able to grow and kill the flies quickly, but this dose would require more than

Table 2 *X. innexi* colonization of *S. scapterisci* nematodes

Strain	Relevant Characteristics	Avg. CFU/IJ ± SE[a]
HGB1681	*X. innexi* (Smart)	6.1 ± 1.1
HGB1997	*X. innexi* (BD)	7.9 ± 1.4
HGB2171	*X. innexi* (Smart) GFP	6.4 ± 1.4
HGB2172	*X. innexi* (BD) GFP	6.7 ± 1.0
HGB800	*X. nematophila*	<0.005

[a]Average colony forming units (CFU) per infective juvenile (IJ) ± standard error (SE) from four independent experiments

14,000 nematode IJs to initiate infection and therefore is not ecologically relevant (Fig. 1) [24]. *X. innexi* was also avirulent in waxworm larvae, except when injected at 100,000 CFUs (Fig. 2g). In *Manduca sexta* larvae, an inoculum of 1000 CFU was sufficient for *X. nematophila* to cause death of 50% of insects by 48 h post-injection, while *X. innexi* only killed 10% of insects toward the end of the experiment (5 d post-injection) (Fig. 2h). The attenuated virulence of *X. innexi* in these various insects supports the idea that the *S. scapterisci*-*X. innexi* complex has a specialized host range. Further, we used the growth data from these experiments to calculate in vivo growth rates in *D. melanogaster*, which, similar to in vitro growth rates, are lower for *X. innexi* than for *X. nematophila* (Additional file 1).

A *S. carpocapsae/X. nematophila*

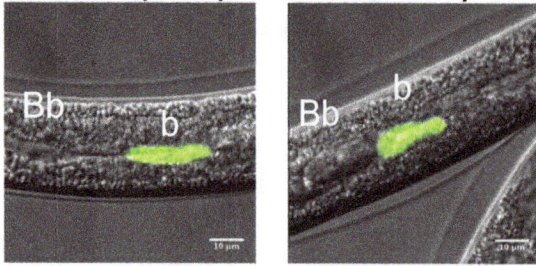

B *S. scapterisci/X. innexi*

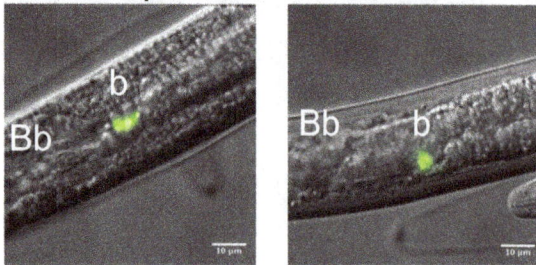

C Growth in LB

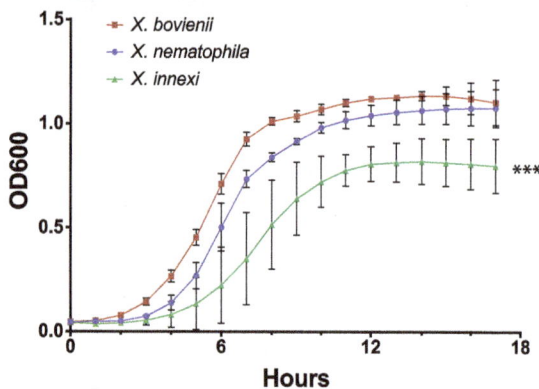

Fig. 1 *X. innexi* nematode colonization levels and in vitro growth rate are lower than other *Xenorhabdus* species. **a** *S. carpocapsae* or (**b**) *S. scapterisci* nematodes were reared on lawns of their respective symbionts, *X. nematophila* and *X. innexi*, engineered to express the green fluorescent protein. Approximately 100 infective juveniles of each nematode species emerging from these lawns were examined by fluorescence microscopy to visualize bacterial colonization of the nematode receptacle and two representative images are shown for each nematode. All colonized *S. scapterisci* nematodes had smaller regions of green fluorescence in the receptacle than did colonized *S. carpocapsae*. When individual bacterial cells could be resolved only 2–3 cells were apparent within *S. scapterisci* nematodes. Both nematode species were colonized at similar frequencies (~92–97%). Bb: basal bulb; b: bacteria. **c** *X. bovienii* (red squares), *X. nematophila* (blue circles), and *X. innexi* (green triangles) bacteria were subcultured into LB medium and monitored for growth based on optical density (OD_{600}).*X. innexi* displayed a longer lag time, slower growth rate, and lower final cell densities than the other two bacterial species, . *X. innexi* density became significantly different from that of *X. nematophila* and *X. bovienii* after 6 h and remained significantly different for the remainder of the experiment (***: $P < 0.002$, 2-way ANOVA at each time point with Tukey's multiple comparisons test), and the overall growth curve was significantly different using Extra sum-of-squares F text ($P = 0.0001$)

X. innexi supernatant does not suppress the *Manduca sexta* phenoloxidase cascade

A common activity associated with *Xenorhabdus* bacteria is the ability to suppress aspects of insect immunity, including the phenoloxidase (PO) system. PO is activated by the cleavage of proPO, which occurs as a result of a serine protease cascade [34]. Several metabolites secreted by *X. nematophila* such as rhabduscin can inhibit the activation of PO [35–37]. To determine if *X. innexi* also secretes immunosuppressive metabolites we isolated cell-free supernatant from it and *X. nematophila* as a control and assessed their abilities to inhibit the activation of PO when incubated with plasma extracted from *M. sexta* insects (Fig. 3). We found that as expected, *X. nematophila* produces heat-tolerant factor(s) that can reduce PO activation to 30% of control reactions. In contrast, *X. innexi* supernatants do not inhibit the activation of PO, indicating that when grown to stationary phase in laboratory culture this bacterium does not secrete immunosuppressive factors at levels sufficient for detection in this assay.

The *X. innexi* genome has a reduced complement of genes predicted to encode virulence determinants, compared to those of other *Xenorhabdus* spp

We have presented data that *X. innexi* is attenuated for virulence in several insect models and for the secretion of immunosuppressive factors. These data and previous publications support a model that *S. scapterisci* is less reliant than other EPNs on its symbiont for fitness. However, *X. innexi* does contribute to *S. scapterisci* success in some insect hosts [10, 13, 24] and also produces several factors of interest, including a mosquitocidal toxin [28]. We predicted that the genome of *X. innexi* might reveal a reduction in the canonical virulence determinants associated with *Xenorhabdus* and related species, while also potentially encoding novel virulence factors that contribute to its specialization for virulence in certain insect hosts.

To further investigate these ideas, we produced a draft genome sequence for *X. innexi* strain HGB1681 (Table 3) (Accession for the whole genome shotgun sequencing project: FTLG00000000.1). The XIS1 draft genome comprises 69 scaffolds (LT699767-LT699835) and 246 contigs (FTLG01000001-FTLG01000246). In total, the genome is similar in size (4,574,778 bp), GC content (43.68%), and coding potential (4418 CDS) and density (83%) to the complete genomes of *X. nematophila* and *X. bovienii* (Table 3) [6]. Due to the draft status of the sequenced genome, only one copy of 16S rRNA, one copy of 23S rRNA and two copies of 5S rRNA were successfully assembled while the completed genomes of both *X. nematophila* and *X. bovienii* have multiple copies of each rRNA gene. Since lower copy numbers of rRNA operons is associated with

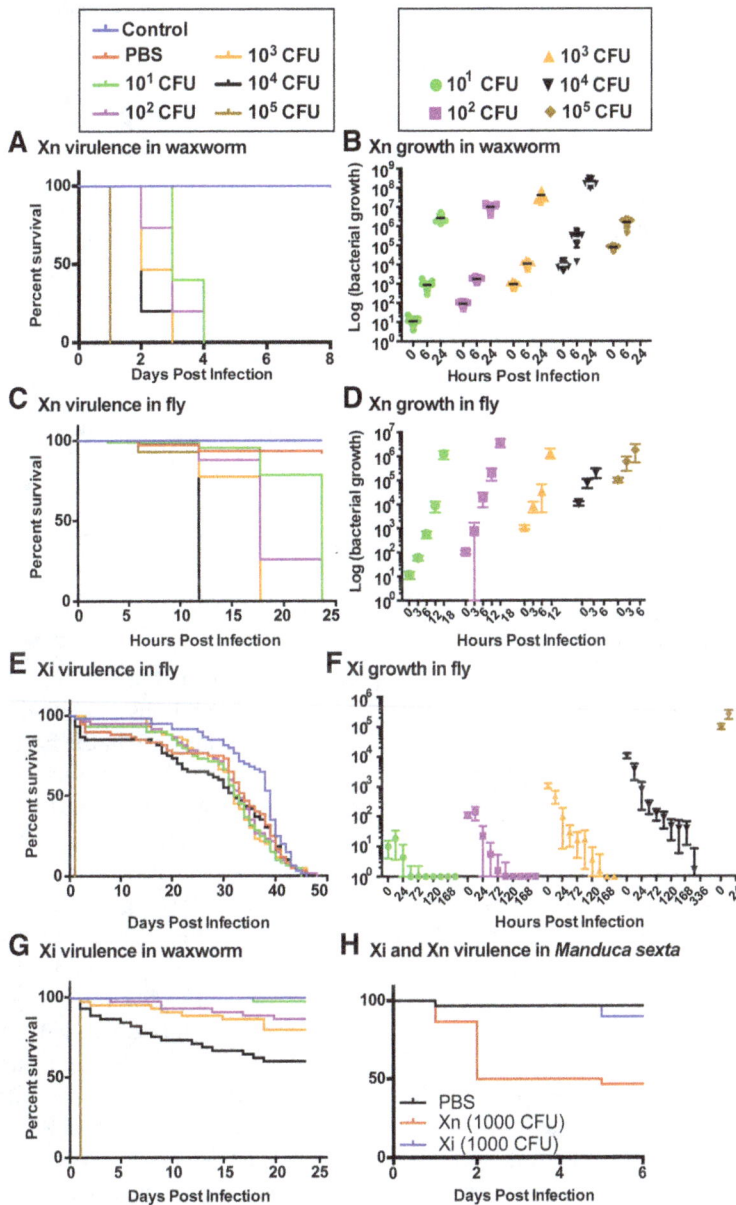

Fig. 2 *X. innexi* is attenuated for virulence in three insect model hosts. *Galleria mellonella* (**a**, **b**, **g**), *Drosophila melanogaster* (**c-f**), or *Manduca sexta* (**h**) insects were injected with *X. nematophila* (**a-d**, **h**) or *X. innexi* (**e-g**, **h**) laboratory-grown bacterial cells at the level indicated in the symbol legend, or with controls as indicated. Over time after injection the insects were monitored for survival (**a**, **c**, **e**, **g**, and **h**) to assess bacterial virulence, or were destructively sampled for bacterial cell number (**b**, **d**, and **f**)

lengthened lag phase and growth rate [38–40], phenotypes we have observed for *X. innexi*, it is possible that *X. innexi* does encode fewer rRNA gene copies, but this conclusion awaits further investigation. The draft genome of *X. innexi* encodes the same number (79) of tRNAs as do the complete genomes of *X. nematophila* and *X. bovienii*.

To investigate the virulence coding potential of the *X. innexi* genome, CDS protein sequences were analyzed using similarity to known virulence factors and conserved protein domains (Table 4, see Methods for details) [41]. In addition to these direct searches, we

used the MicroScope Gene Phyloprofile tool [42] to identify sets of genes specifically absent in *X. innexi* genome (Additional file 3). We used loci present in the completely sequenced genome of the virulent strain *X. nematophila* (ATCC 19061) and identified those with homologs in the genomes of the virulent strains *X. bovienii* SS-2004 and *X. doucetiae* FRM16 [6, 7], but without homologs in the *X. innexi* HGB1681 genome.

Consistent with the reduction of virulence potential and absence of PO inhibition, the draft genome *X. innexi* lacked, or had a reduced complement of virulence

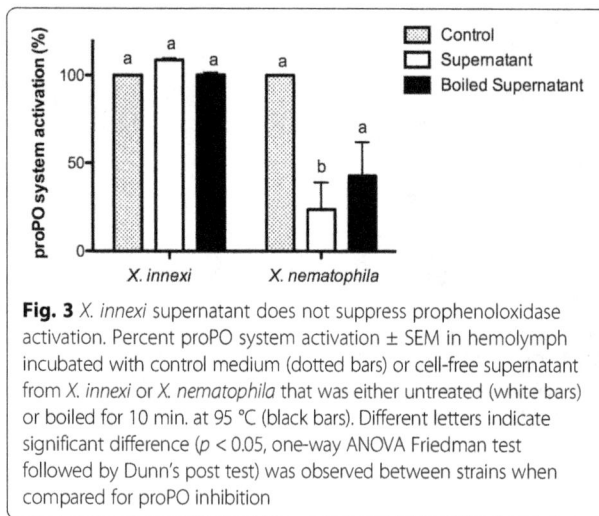

Fig. 3 *X. innexi* supernatant does not suppress prophenoloxidase activation. Percent proPO system activation ± SEM in hemolymph incubated with control medium (dotted bars) or cell-free supernatant from *X. innexi* or *X. nematophila* that was either untreated (white bars) or boiled for 10 min. at 95 °C (black bars). Different letters indicate significant difference ($p < 0.05$, one-way ANOVA Friedman test followed by Dunn's post test) was observed between strains when compared for proPO inhibition

Table 4 Numbers of *X. nematophila* and *X. innexi* genes encoding known virulence factors

Gene family	*X. innexi* HGB1681	*X. nematophila* ATCC 19061
Chitinases	0	2
HIP57 (GroEL)	1	3
MARTX	3[a]	1
Mcf	0	1
Pir toxins	0	2
PrtA	1	1
Rhabduscin	0	3
Tc toxins (A)	0	6[a]
Tc toxins (B)	0	3
Tc toxins (C)	1[a]	3
TPS-Fha	2[a]	0
TPS-Hemolysin	2[a]	1
Xenocin	0	1

[a]indicates at least one fragment

factors typical of other *Xenorhabdus* genomes. For example, *X. innexi* does not encode Tc (or associated chitinases), Mcf, XaxAB, entire Rtx (see below), or Pir toxins [6, 43–46] or rhabduscin-encoding genes [37] (Table 4; Additional file 3).

In silico analysis of select *X. innexi* secretion systems and effectors

Bacteria encode numerous types of secretion systems, many of which allow delivery of virulence factors to the host environment and cells. As with other *Xenorhabdus* bacterial genomes [6], the genome of *X. innexi* lacks a Type III secretion system (T3SS) (determined using *S. enterica* T3S as a model; Additional file 4). Another class of secreted molecules that are often found in pathogens that lack T3SS is the MARTX (Multifunctional Autoprocessing Repeats-in-Toxin Toxins). These polymorphic toxins are very large and comprise an N-terminal region with conserved A and B repeats that appear necessary for delivery of the toxin into host cells, an effector

Table 3 Draft genome statistics

	X. innexi HGB1681	*X. nematophila* ATCC 19061	*X. bovienii* SS-2004
Size of chromosome (bp)	4,575,778	4,432,590	4,225,498
G + C content, %	43.68	44.15	44.97
Coding sequences	4418	4648	4406
Number of scaffolds	69	1	1
Number of contigs	246	1	1
Average CDS length (bp)	885.8	850.81	849.48
Average intergenic length (bp)	179.85	163.62	158.03
Protein coding density %	82.93	82.65	84.07
rRNAs	4	29	29
tRNAs	79	79	83

domain region containing multiple modules with host-modulating functions, a CPD domain that processes the effector domains once in the host cell, and a C-terminal repeat domain necessary for secretion out of the bacterial cell through a Type 1 secretion system encoded by the *rtxEDB* operon and the unlinked *tolC* gene [47, 48] (Fig. 4).

Published literature has established that the *X. nematophila* and *X. bovienii* genomes each contain one complete MARTX-encoding gene, predicted to encode proteins of 4970 aa and 4716 aa respectively, each with canonical A repeats (A1-A10; A11-A14), B repeats (B1–38; B39–41) and C repeats (C1–2; C3–15), and a CPD domain. Both also contain the effector domains DUF1, RID, and MCF, but they are distinct in that *X. nematophila* includes a PMT C1/C2 (now known as RRSP) domain [49, 50]. while that of *X. bovienii* encodes an ABH domain, both immediately following their respective RID domains [51, 52]. Consistent with the genomic context of other organisms, the MARTX-encoding genes of *X. nematophila* and *X. bovienii* are encoded adjacent to those predicted to encode Rtx activating and secretion functions.

The *X. innexi* genome contains 4 contigs with regions that have similarity to MARTX-encoding genes, based on a BLASTp search with XNC1_1381 (Fig. 4). In the assembly, only one gene (XIS1_650005) encodes a full suite of A, B, and C repeats and the effector domains. However, based on alignment with the *X. nematophila* and *X. bovienii* MARTX proteins this protein lacks A repeats 4 through 8 (of 14) (Fig. 4 and Additional file 5). In addition to XIS1_650005, we identified another five regions with one or more MARTX protein-encoding domains. Of these, two (XiS1v1_640001 and XIS1_650001) are predicted to encode an A domain region that, like

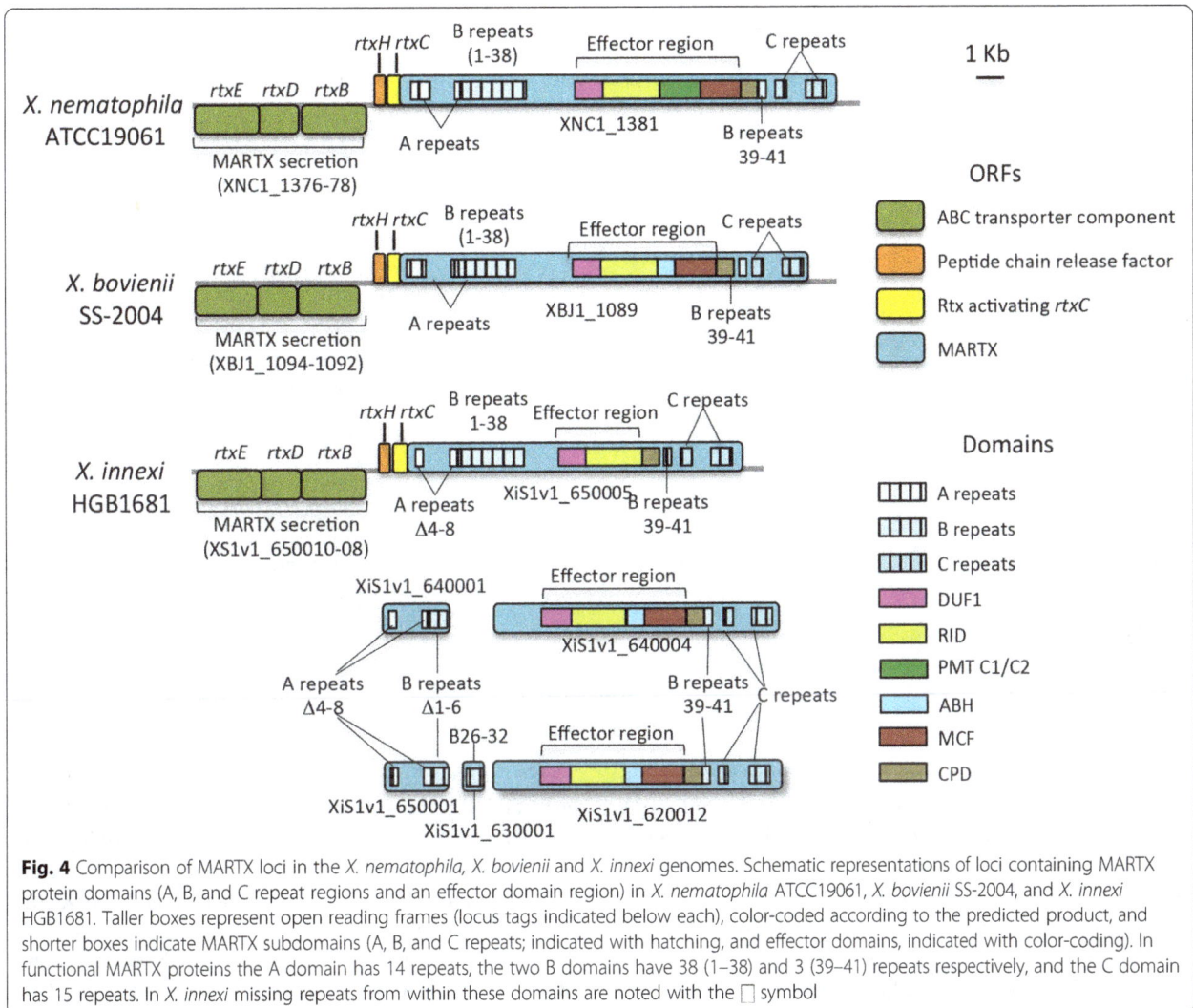

Fig. 4 Comparison of MARTX loci in the *X. nematophila*, *X. bovienii* and *X. innexi* genomes. Schematic representations of loci containing MARTX protein domains (A, B, and C repeat regions and an effector domain region) in *X. nematophila* ATCC19061, *X. bovienii* SS-2004, and *X. innexi* HGB1681. Taller boxes represent open reading frames (locus tags indicated below each), color-coded according to the predicted product, and shorter boxes indicate MARTX subdomains (A, B, and C repeats; indicated with hatching, and effector domains, indicated with color-coding). In functional MARTX proteins the A domain has 14 repeats, the two B domains have 38 (1–38) and 3 (39–41) repeats respectively, and the C domain has 15 repeats. In *X. innexi* missing repeats from within these domains are noted with the ☐ symbol

XIS1_650005, lack repeats 4 through 8, as well as a truncated B domain region (extending up to repeat 16 of 38) that lack repeats B1-B6 (Fig. 4 and Additional file 5). Additional B repeats (corresponding to ~B24 through B33) are found encoded by XIS1_630001. Finally, XIS1_640004 and XIS1_620012 both encode B repeats (38–41) and an effector domain region with the same composition as that of *X. bovienii* (Fig. 4 and Additional file 5). In sum, consistent with the attenuated virulence of *X. innexi*, the genome appears not to encode a complete MARTX protein (since internal repeats appear to be missing from the A domain of XIS1_650005). It will be of interest to determine the functional significance of the absence of these repeats, assuming that the missing repeats are verified and not due to assembly issues.

Another secretion system with implications for virulence are the two-partner secretion (TPS) systems (also known as Type 5 or autotransporters), which encode both the toxin (e.g. hemolysin) and its transport system (Tables 4 and 5) [53]. In these systems one protein forms a beta-barrel pore that facilitates translocation of an exoprotein across the outer membrane. Typically, the beta-barrel protein and the exoprotein are encoded adjacent to each other. A conserved feature of these systems is the TPS domain encoded in the N-terminal 250 aa of the A exoprotein, which is necessary for translocation and contains a conserved NPNL-35aa-NPNGI motif. Generally, this region is conserved across types of secreted exoproteins, while the remaining portions of the protein are distinct. Previous phylogenetic analyses of whole TpsA sequences [6] revealed that Tps proteins are divided into three clusters. The first cluster contains CdiA exoproteins, which are involved in the contact-dependent inhibition systems, playing important roles in inter-strain competition and self/nonself discrimination. CDI systems are mainly distributed among pathogenic Gram-negative bacteria [54–57], and recently described in the entomopathogenic bacterium *Xenorhabdus doucetiae* FRM16 [58].

Table 5 *X. innexi* two partner secretion pathway loci

XIS1_	Gene	Length (aa)	Predicted Function	TPS cluster or relevant features
1110028	$xhlB_{xi}$	558	Beta-barrel	Cluster II (hemolysin)
1110029	$xhlA_{xi}$	1468	Exoprotein	Cluster II (hemolysin), TPS motif
1600025	$xhlB2_{xi}$	558	Beta-barrel	Cluster II (hemolysin)
1600026	$xhlA2_{xi}$ (part 1)	656	Exoprotein fragment	Cluster II (hemolysin), TPS motif
1600027	$xhlA2_{xi}$ (part 2)	801	Exoprotein fragment	Cluster II (hemolysin)
680062	$cdiB_{xi}$	569	Beta-barrel	Cluster I (Cdi)
680061	$cdiA_{xi}$	4029	Exoprotein	Cluster I (Cdi), TPS motif and VENN domain
680060	$cdiI_{xi}$	4029	Immunity protein	Cluster I (Cdi), putative $cdiI_{xi}$
260016	$cdiB\text{-}like_{xi}$	571	Beta-barrel	Cluster I (Cdi)
260017	$cdiA\text{-}like$ (part 1)	1157	Exoprotein fragment	Cluster I (Cdi), TPS motif, lacks VENN motif
270001	$cdiA\text{-}like$ (part 2)	180	Exoprotein fragment	Cluster I (Cdi), lacks TPS motif and VENN domain, contains beta barrel region,
280001	$cdiA\text{-}like$ (part 3)	4062	Exoprotein fragment	Cluster I (Cdi), lacks TPS motif, starts with the beta barrel region, includes VENN domain
1500009	$tpsB_{xi}$	566	Beta-barrel	Cluster III (DUF637 domain)
1500008	$tpsA_{xi}$	1907	Exoprotein	Cluster III (DUF637 domain), TPS motif and DUF637 domain
1500007	$tpsI_{xi}$	110	unknown	potential immunity protein

The second cluster is comprised of active hemolysins, such as PhlA from *Photorhabdus luminescens* and XhlA from *X. nematophila* [59, 60]. A third cluster contains TpsA proteins with unknown functions, which are characterized by the presence of a DUF637 domain.

In the draft *X. innexi* genome we identified a total of five genes predicted to encode proteins with an N-terminal TPS domain including conserved NPNL and NPNGI domains (Table 5). The genomic contexts of these suggest five independent loci encoding Tps systems. XIS1_1110029 and XIS1_1600026 genes encode proteins with TPS domains that belong to the hemolysin phylogenetic cluster (Fig. 5; Additional file 6). The XIS1_1110029-encoding protein displays 67% identity over its entire length (1468 aa) to the *X. nematophila* XhlA hemolysin ($XhlA_{xn}$) [59] and was therefore named $XhlA_{xi}$. $xhlA_{xi}$ is adjacent to a homolog of $xhlB$ predicted to encode the beta-barrel protein component of the TPS system. The genomic locus $xhlBA_{xi}$ is syntenic with that of $xhlBA_{xn}$, and includes genes predicted to encode a Type VI secretion system (T6SS). XIS1_1600026 is contiguous to XIS1_1600027 and they encode putative proteins that have respectively 52% identity with the N-terminal region of $XhlA_{xn}$ and 32% identity with the C-terminal region of $XhlA_{xn}$ These two genes are adjacent to another homolog of $xhlB$ (XIS1_1600025). Overall, it appears that *X. innexi* encodes a second $xhlBA_{xi}$ locus, with $xhlA_{xi}$ truncated in two parts (XIS1_16000276/XIS1_1600027).

XIS1_ 680,061 and XIS1_ 260,017 encode proteins with TPS domains that belong to the Cdi phylogenetic cluster (Fig. 5). The XIS1_680,061-encoding protein displays identity with functionally characterized CdiA proteins (36% identity with the *X. doucetiae* FRM16 CdiA and 34% identity with the *E. coli* EC93 CdiA) and has a VENN domain, which usually separates the conserved N-terminus from the variable C-terminus in many CdiA proteins [61]. Moreover, the adjacent genes XIS1_680062 and XIS1_680060 encode a CdiB ortholog and a potential immunity protein CdiI (based on location of the gene and the small size of the encoded-protein), respectively. We therefore hypothesize that this locus is a *cdiBAI* locus. XIS1_ 260017, XIS1_270001 and XIS1_280001 are on three separate contigs but are contiguous in the assembly of the *X. innexi* genome. They each display partial similarities with sub-regions of CdiA proteins. These three *cdiA*-like genes are adjacent to a CdiB ortholog XIS1_ 260016 (63% identity with CdiB of *X. doucetiae* FRM16), which suggest the presence of a second *cdi* locus, which has been highly shuffled.

The fifth *tps* genomic locus we identified includes XIS1_150008, encoding a 1907 aa protein with a TPS-domain and a DUF637 domain, placing it in the third phylogenetic cluster (Fig. 5), for which no function is described to date. XIS1_150009 encodes a TpsB ortholog. Interestingly, XIS1_150007 displays features of immunity genes due to its location and its small size although the

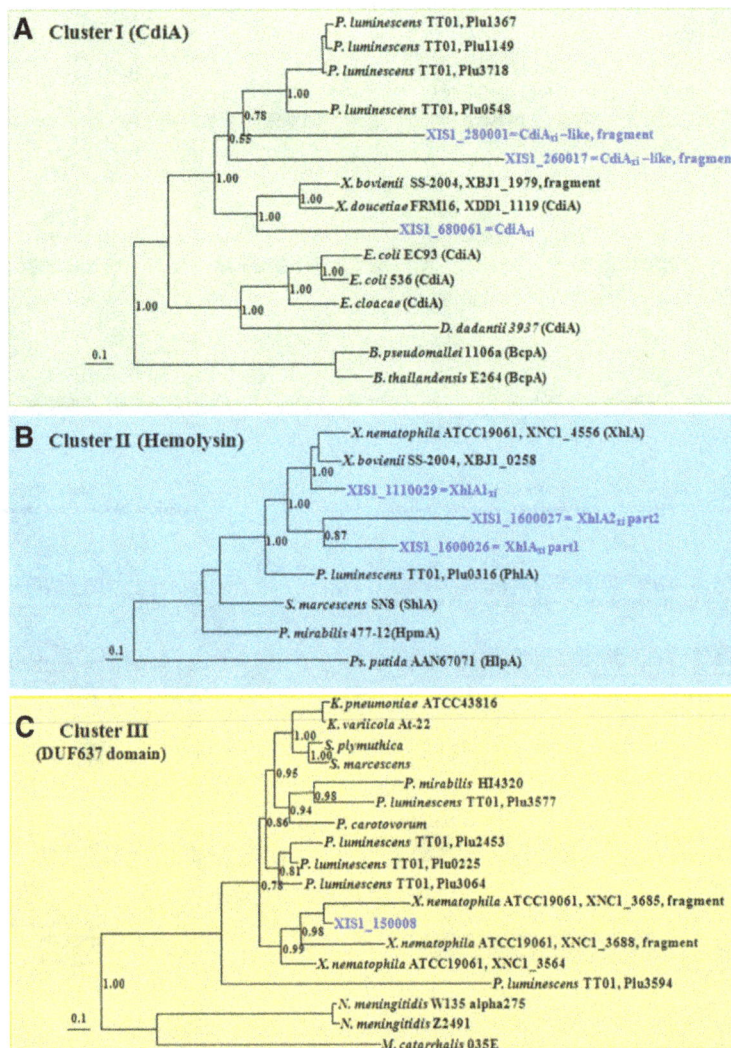

Fig. 5 Phylogenetic analysis of putative TpsA proteins of *X. innexi*. For each family of TpsA proteins, a phylogenetic tree was built by the maximum likelihood (ML) method using the LG substitution model. Branch support values, estimated by the aLRT (SH-like) method, are indicated at the nodes. The branch length scale bar below the phylogenetic tree reflects the number of amino-acid substitutions per site. TpsA proteins fall into three clusters: **a** Cluster I containing CdiA exoproteins, which are involved in contact-dependent inhibition systems, **b** Cluster II containing hemolysins and (**c**) Cluster III containing TpsA proteins with unknown functions, which are all characterized by the presence of DUF637 domain. TpsA are identified by the name of the bacterial strain in each cluster and the label number in the *Photorhabdus* and *Xenorhabdus* genera. The *X. innexi* TpsA proteins are indicated in blue with the label number of their encoding gene. Previous functionally characterized TpsA are named in parentheses. Accession numbers of the sequences are indicated in Additional file 6

tpsA gene does not fall in the Cdi phylogenetic cluster (Table 5).

In summary, the *X. innexii* genome displays *tps* clusters in each of the three phylogenetic clusters, which sets it apart from other *Xenorhabdus* genomes. For instance, in the genome of the highly virulent *Xenorhabdus nematophila* ATCC19061 strain, only hemolysin and DUF637 domain clusters are represented (Fig. 5; Additional file 6).

Another class of secretion systems that can be involved in virulence is the T6SS. These are bacterial nanomachines comprising 13 conserved structural proteins, which deliver toxic effectors into eukaryotic or prokaryotic organisms in a one-step firing mechanism [62]. T6SSs often are associated with roles in virulence and inter-bacterial competition, providing a selective advantage against competitors [63]. To analyze the T6SS content in the draft genome of *X. innexi*, we used a combination of the Magnifying Genomes server (MaGe) and NCBI Conserved Domain Database and identified three T6SS clusters, T6SS-1,2, and 3 (named in order of their appearance in the draft genome) (Fig. 6a, Additional file 7).

T6SS-1 appears to be incomplete, as it lacks the *tssJ, tssG*, and *clpV* components. Although ClpV is dispensable for some T6SSs, TssJ and TssG are required [62] suggesting the

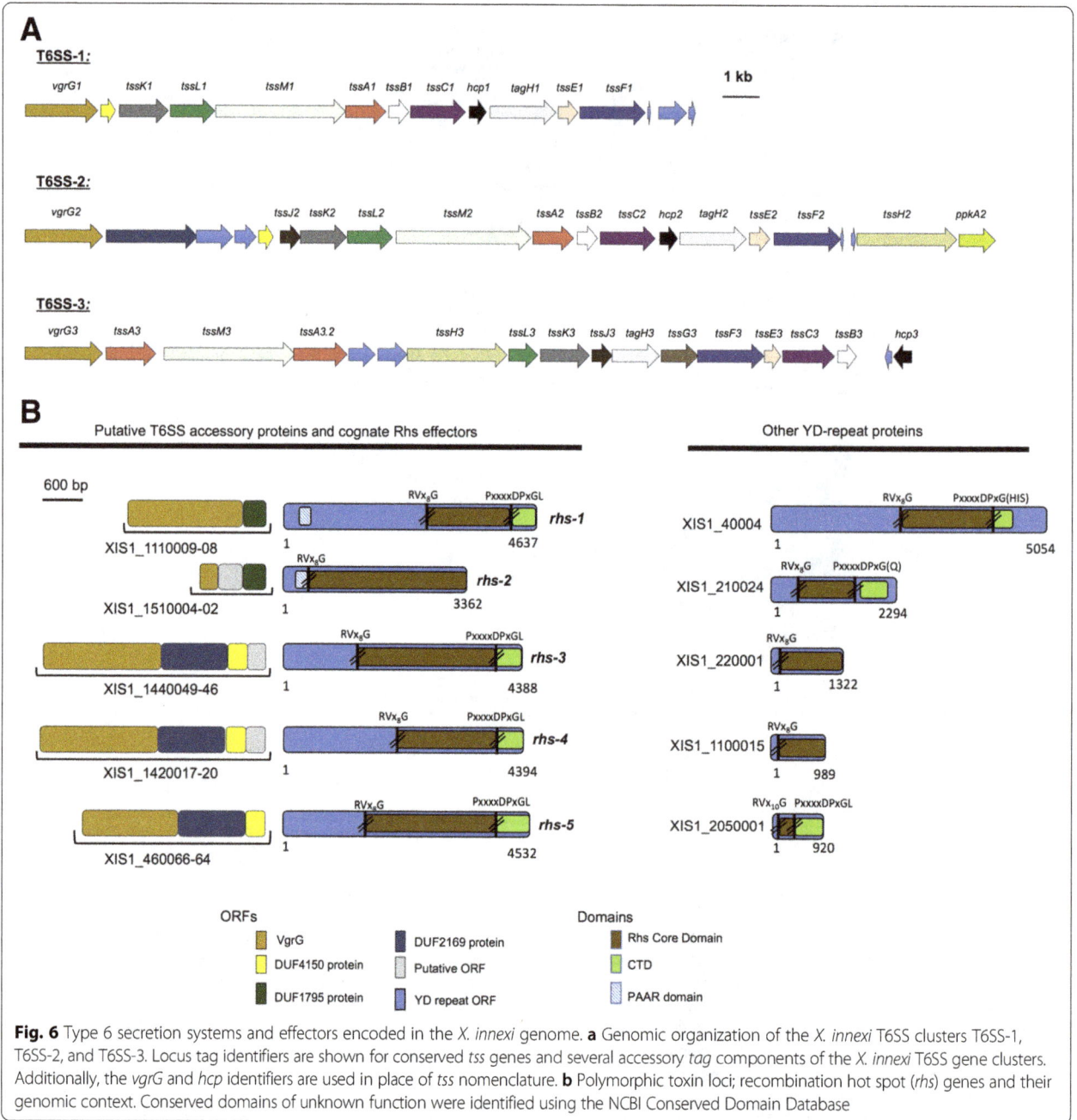

Fig. 6 Type 6 secretion systems and effectors encoded in the *X. innexi* genome. **a** Genomic organization of the *X. innexi* T6SS clusters T6SS-1, T6SS-2, and T6SS-3. Locus tag identifiers are shown for conserved *tss* genes and several accessory *tag* components of the *X. innexi* T6SS gene clusters. Additionally, the *vgrG* and *hcp* identifiers are used in place of *tss* nomenclature. **b** Polymorphic toxin loci; recombination hot spot (*rhs*) genes and their genomic context. Conserved domains of unknown function were identified using the NCBI Conserved Domain Database

X. innexi T6SS-1 system may not be functional. The T6SS-2 cluster lacks *tssG*, but contains all other core components as well as potential effector-immunity (E-I) protein pairs. Additionally, T6SS-2 contains both *tagH*, a forkhead-associated domain-containing protein, and *ppkA*, a transmembrane threonine kinase. TagH and PpkA are components of the threonine phosphorylation pathway (TPP), a post-translational regulatory mechanism for T6SS activity [64]. The T6SS-3 cluster appears complete and includes a duplication of the baseplate protein, *tssA*, and a potential E-I protein pair. Genes encoding putative T6SS E-I pairs can be found clustered with the T6SS

structural genes or scattered about the genome, often linked to a T6SS chaperone. One such group of T6SS effectors is the polymorphic toxins, Rhs proteins.

Rhs proteins containing PAAR domains have been reported as T6SS-dependent antibacterial effectors that mediate both intra- and inter-species competition [65]. Rhs proteins, in human pathogens such as *Pseudomonas aeruginosa* and *Enterobacter cloacae*, mediate bacterial competition under in vitro conditions. For these pathogens, Rhs proteins may play an important role in virulence by establishing a suitable niche for survival during infection of the host [66]. Genes encoding known or putative T6S effectors, including Rhs proteins, are often found near *vgrG* genes and require the cognate Vgr for T6S secretion [67]. *X. innexi* strain HGB1681 encodes ten YD-repeat (PF05593) proteins, five of which are putative Rhs proteins based on the presence of characteristic YD-repeats, a rhs core domain flanked by conserved motifs, and a variable C-terminal 'tip' [68] (Fig. 6b). These five genes can be further categorized into two groups based on their genomic context. Two genes, *Xi_rhs-1* and *Xi_rhs-2* encode an N-terminal PAAR motif (PF05488), though *Xi_rhs-2* is annotated as a truncated ORF, missing its C-terminal-encoding domain. The other three (*Xi_rhs-3* through *rhs-5*) lack the PAAR domain but are encoded next to small open reading frames with a PAAR-like domain, DUF4150/PF13665. Furthermore, the *Xi_rhs-1* and *Xi_rhs-2* are encoded downstream of a DUF1795-containing protein and putative *vgrG* gene, both of which were necessary for Rhs protein translocation in *Serratia marcescens* [69]. The other three *rhs* genes are also encoded downstream of putative *vgrG* genes. The genomic contexts of these three *rhs* genes are distinct from those of *rhs-1* and *rhs-2* in that these gene clusters also encode DUF2169 and DUF4150 domain-containing proteins, which in *Agrobacterium tumefaciens* are demonstrated accessory proteins required for secretion of their cognate T6SS toxin [67].

The highly variable C-terminal domains (CTDs) of Rhs proteins contain the toxic effector activity. An analysis of the CTDs of *X. innexi* Rhs-family proteins (except *Xi_rhs-2* which lacks a CTD) revealed no recognizable CTD function in *rhs-3, −4,* and *−5*. In contrast *rhs-1* contains PF14437, a MafB19 deaminase domain [70]. This domain occurs in the CTDs of several classes of polymorphic toxins, including the recently recognized *Neisseria* MafB toxins, and the Rhs protein putative toxin E1IMF [70]. While our data demonstrate that in several insect hosts, *X. innexi* displays attenuated virulence relative to other *Xenorhabdus* spp. it remains an associate of *S. scapterisci* nematodes (Fig. 1, Table 2), and successfully reproduces within crickets, where it may encounter competing microbes (Additional file 8). Together with the T6SS, the Rhs family proteins encoded by *X. innexi* may play a role in any one of these activities.

The *X. innexi* genome includes numerous loci predicted to encode non-ribosomal peptide and polyketide synthetases

Non-ribosomal peptide synthetase (NRPS) and polyketide synthetase (PKS) clusters encode large molecular weight complexes responsible for the synthesis of small molecules (natural products) with diverse activities, including toxicity against target organisms [71]. To begin to assess the ability of *X. innexi* to produce such compounds we computationally screened for clusters predicted to encode NRPS, PKS, or hybrids. Our initial screening identified one PKS, 12 NRPS and three NRPS-PKS hybrid gene clusters (Table 6). The hybrid genes were

Table 6 NPRS, PKS, and NPRS-PKS hybrid clusters in the *X. innexi* genome

Location	Size (bp)	Type	Number of A or AT domains[a]
XIS1_130014	5139	PKS[b]	1 AT domain
XIS1_250010	13,755	NRPS[c]	6 A domains
XIS1_40005 - XIS1_60002	21,138	NPRS	4 A domains
XIS1_170001 - XIS1_190004	21,111	NPRS	6 A domains
XIS1_370002 - XIS1_370004	15,468	NPRS	3 A domains
XIS1_390007 - XIS1_400001	15,519	NPRS	5 A domains
XIS1_450016	2997	NRPS	1 A domain
XIS1_460014	7323	NPRS	2 A domains
XIS1_460105 - XIS1_460116	44,193	Hybrid[d]	6 A domains and 2 AT domains
XIS1_480023 - XIS1_480027	23,055	Hybrid	3 A domains and 1 AT domain
XIS1_600036	3060	NRPS	1 A domain
XIS1_660020 - XIS1_660029	21,609	Hybrid	3 A domains and 1 AT domain
XIS1_1050018 - XIS1_1050019	3366	NRPS	1 A domain
XIS1_1690009 - XIS1_1690010	23,319	NRPS	7 A domains
XIS1_1700078	11,811	NRPS	3 A domains
XIS1_1750018 - XIS1_1750021	44,826	NRPS	13 A domains

[a]A domain; Adenylation domain, AT domain; Acyltransferase domain predicted by antiSMASH
[b]PKS; Polyketide synthase
[c]NRPS; Non-ribosomal peptide synthetase
[d]Hybrid; NRPS-PKS hybrid gene

further examined by analyzing their amino acid sequences through AntiSMASH or Conserved Domain searches to identify NRPS and PKS domains and to confirm the number of adenylation (A) and acyltransferase (AT) domains, which are responsible for selection and loading of amino acids or carboxylic acids, respectively for incorporation into the product (Table 6).

Identification of the Xlt-encoding NRPS/PKS hybrid gene cluster

X. innexi secretes a small lipopeptide named Xlt with toxicity toward mosquitoes [28, 29]. Previous structural data suggested that Xlt is a cyclic lipopeptide composed of six amino acids and two fatty acids [28] and we hypothesized that it may be synthesized by a hybrid NRPS/PKS cluster [72]. Based on this hypothesis, we predicted that the locus would encode six A- and two AT-domains respectively. Among the three identified hybrid genes in *X. innexi* genome, only one gene cluster, *XIS1_460105* to *_460116* (present in the center of a single contig) has two AT-domains and six A-domains that correspond to the number of fatty acids and amino acids identified in Xlt [28].

The candidate gene cluster encodes 12 ORFs with predicted NRPS or PKS functions based on BLASTp analysis (Table 7). Eight of these had predicted PKS or PKS-related functions: *XIS1_460115* (Type-I PKS), *XIS1_460114* (beta-ketoacyl synthase), *XIS1_460113* (PfaD family protein), *XIS1_460112* (3-oxoacyl-ACP reductase), *XIS1_460111* (thioester reductase), *XIS1_460110* (amidohydrolase), *XIS1_460107* (Type-I PKS) and *XIS1_460105* (acyl-CoA thioesterase) and three had NRPS or NRPS-related

functions: *XIS1_460109* (NRPS), *XIS1_460107* (NRPS) and *XIS1_460106* (condensation protein).

The arrangement of genes from *XIS1_460105* to *_460116* is similar to those of the fabclavine synthesis loci found in *Xenorhabdus budapestensis* DSM 16342 and *Xenorhabdus szentirmaii* DSM 16338 (*fcl*), and pre-zeamine synthesis loci from *Serratia plymuthica* AS 9 (*zmn*) [73, 74]. We compared the *X. innexi* genes predicted to encode Xlt biosynthesis machinery to *fcl* and *zmn* sequences from *X. szentirmaii* and *S. plymuthica* respectively (the sequences of *X. budapestensis fcl* were not available). BLASTp analysis indicated that the predicted function of each gene in Xlt biosynthesis gene cluster is very similar to both Fcl and Zmn coding genes (Table 8). We noted two differences in coding content, both on the flanking edges of the *X. innexi* cluster, relative to *X. szentirmaii*: the first gene in the *X. szentirmaii* locus (*fclA*, predicted to encode a NUDIX hydrolase) is absent in the Xlt-encoding cluster (Tables 7 and 8) [74]. Instead, the flanking genes are predicted to encode a TonB homolog and a cardiolipin synthase. Also, *X. szentirmaii* has cluster genes *fclM* and *fclN*, predicted to encode ABC transporters, immediately downstream of the last condensation domain gene [74]. In contrast, in *X. innexi* the gene following the last condensation domain is predicted to encode an acyl-CoA thioesterase (*XIS1_460105*). This difference may reflect a distinct release mechanism of the final Xlt product relative to fabclavine and zeamine. Acyl-CoA thioesterases are involved in the release of fatty acids [75] and are most active on myristoyl-CoA but also display high activities on palmoityol-CoA, stearoyl-CoA and arachidoyl-CoA [76–78]. Therefore, it is possible that in *X. innexi* the second PKS module produces 3-oxo-saturated fatty acids of the chain length from C14 to C20, consistent with the description of preliminary fatty acid structure data for Xlt [28]. The presence of a distinctive acyl-CoA thioesterase encoding gene within the putative Xlt-biosynthetic cluster provides further support that this cluster is involved in the synthesis of Xlt and that Xlt may have unique characteristics relative zeamine/fabclavine.

Various domain analysis programs were used to verify the predicted biosynthetic activities and specificities of the candidate Xlt synthesis gene cluster (see Methods). As expected based on the similarities noted above, the number of A-domains found from *XIS1_460105* to *_460116* was the same as observed for *fcl* and *zmn* gene clusters [73, 74]. In fact, the predicted Stachelhause codes from Xlt coding genes were nearly identical to that of Fcl and Zmn coding genes (Fig. 7b, Table 2 in [79] and Table 8 in [80]). The peptide moiety incorporated by A-domains in Xlt coding genes closely resembled both Fcl and Zmn synthesis genes, and this further suggested that the candidate Xlt biosynthesis gene

Table 7 Gene location, size and putative function of the candidate Xlt biosynthesis gene cluster from *X. innexi*

Gene location	Size (aa)	Putative function
XIS1_460116	338	Membrane protein of unknown function
XIS1_460115	1974	Type-I PKS
XIS1_460114	1471	Beta-ketoacyl synthase
XIS1_460113	948	PfaD family proteinglutamate-1-semialdehyde 2,1-aminomutase
XIS1_460112	255	3-oxoacyl-(acyl-carrier-protein) reductase
XIS1_460111	412	Thioester reductase/polyketide synthase
XIS1_460110	258	Amidohydrolase/NAD(P)-binding amidase with nitrilase
XIS1_460109	4437	NRPS/glutamate racemase
XIS1_460108	2301	NRPS
XIS1_460107	1644	Type-I PKS/6-deoxyerythronolide-B synthase
XIS1_460106	539	Condensation protein/peptide synthase
XIS1_460105	142	Acyl-CoA thioesterase/acyl-CoA thioester hydrolase

Table 8 Amino acid identities[a] of predicted proteins encoded by *X. innexi* putative Xlt-biosynthetic locus and fabclavine and pre-zeamine biosynthetic clusters encoded by *X. szentirmaii* DSM 16338 and *Serratia plymuthica* AS9 respectively

Locus tag	Identity (%) to *X. szentirmaii fcl* locus		Identity (%) to *S. plymuthica zmn* locus	
XIS1_460116	N/A[b]	–	SerAS9_4283	43.03%
XIS1_460115	FclC	69.11%	SerAS9_4282	59.15%
XIS1_460114	FclD	74.65%	SerAS9_4281	58.37%
XIS1_460113	FclE	79.67%	SerAS9_4280	75.53%
XIS1_460112	FclF	81.57%	SerAS9_4279	70.20%
XIS1_460111	FclG	76.83%	SerAS9_4278	62.72%
XIS1_460110	FclH	79.46%	SerAS9_4277	65.12%
XIS1_460109	FclI	65.72%	SerAS9_4276	50.79%
XIS1_460108	FclJ	70.86%	SerAS9_4275	59.94%
XIS1_460107	FclK	67.36%	SerAS9_4274	55.06%
XIS1_460106	FclL	65.64%	SerAS9_4273	48.05%
XIS1_460105	N/A	–	N/A	–

[a]Based on the protein blast (BLASTp) analysis
[b]N/A- Not available

cluster is homologous to the Fcl and Zmn clusters. In *X. innexi*, NRPSpredictor2 predicted A_1 through A_6 to be A_1: serine, A_2: phenylalanine, A_3: asparagine, A_4: asparagine, A_5: threonine, and A_6: valine (Fig. 7a and b). Also, some programs predicted an epimerization domain, which may indicate that the A_3-domain incorporates a D-asparagine/aspartic acid. Refinement with Stachelhause codes indicated 90% probability that A_3 and A_4 are asparagine and A_5 is threonine (Fig. 7b). However, consistent with the fact that spectral analysis between 260 and 280 nm indicates Xlt lacks phenylalanine (J.C. Ensign, unpublished data), the nearest neighbor scores for this amino acid (as well as serine and valine) were low.

in silico analysis of NRPS and PKS modules in the gene cluster from *XIS1_460105* to *_460116* provided a

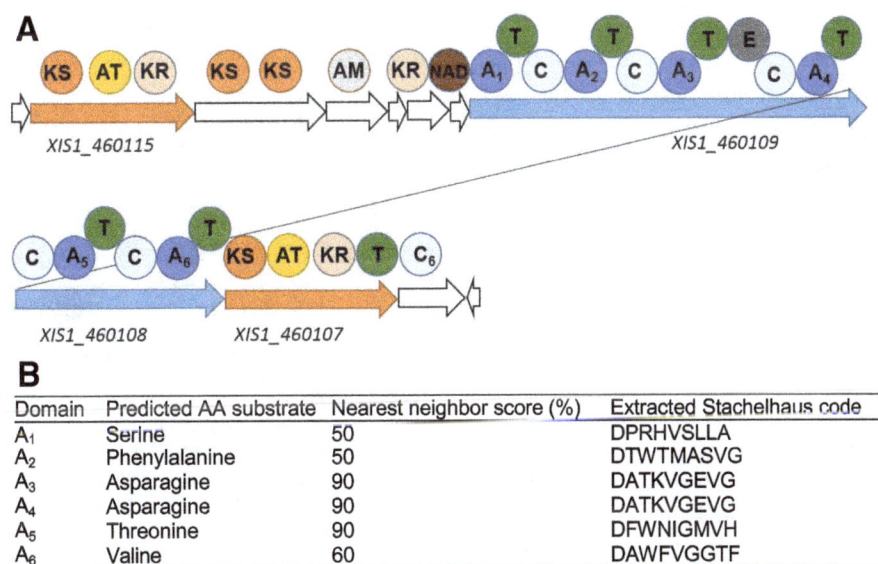

Fig. 7 Predicted NRPS and PKS domains between *XIS1_460105* and *_460116*, and the analysis of adenylation domains. Domains were identified by analyzing translated sequences of each ORF using the Conserved domain search and AntiSMASH. Panel **a** displays the domain annotation based on the AntiSMASH analysis and Conserved Domain search. Aminotransferase (AT) domain containing ORFs are highlighted in orange and adenylation (A) domain containing ORFs are highlighted in blue. Panel **b** represents the predicted amino acid substrate of each adenylation domain from the candidate Xlt synthesis NRPS-PKS hybrid cluster from *X. innexi*. The amino acid substrate prediction was made based on the extracted Stachelhause code by NRPSpredictor2 [120]. KS: ketoacyl-synthase, AT: acyl-transferase, KR: ketoreductase, AM: aminotransferase, NAD: NAD(P)-binding amidase, A: adenylation, T: thiolation/peptide carrier protein, E: epimerization

strong rationale that the selected gene locus is the likely candidate to produce Xlt. This prediction is largely consistent with the preliminary structural analysis of the mosquitocidal toxin, which indicated the presence of serine, asparagine, glycine and at least one oxo-fatty acid of C_8 to C_{20} [28]. The presence of certain amino acid residues of Xlt, including histidine and 2,3-diaminobutyric acid (DAB), could not be explained by the in silico analysis conducted in this study. However, the structural analysis of fabclavine, which is produced by a homologous gene cluster from *X. szentirmaii* and *X. budapestensis*, showed a replacement of phenylalanine by histidine as well as the presence of 2,3-diaminobutyric acid (DAB) in its peptide moiety. The structure of fabclavines, which corresponds to the preliminary structural data of Xlt, provided further support that the selected gene cluster should produce Xlt. Based on our prediction, we next tested if mutation of the gene cluster from *XIS1_460105* to *_460116* would disrupt mosquitocidal toxin activity in *X. innexi*.

Site-directed mutagenesis at *XIS1_460115* or *XIS1_460109* resulted in phenotypic changes

To further explore the possibility that the candidate gene cluster is involved in Xlt mosquitocidal toxin biosynthesis, we used site-directed mutagenesis to mutate two independent genes within the locus: *XIS1_460115* and *XIS1_460109*. Each was individually replaced with a kanamycin cassette (see Methods) and supernatants from the resulting *XIS1_460115::kan* (*ΔXIS1_460115*) and *XIS1_460109::kan* (*ΔXIS1_460109*) mutants were analyzed with MALDI-TOF MS. Consistent with previous preliminary data, which indicated that Xlt has a molecular weight range between 1182 and 1431 Da, with the difference in molecular weights ascribed to varying lengths of fatty acids [28], our MALDI-TOF MS analysis of wild type *X. innexi* (HGB1681) supernatant revealed major peaks between 1348 and 1402 Da (Additional file 9). In contrast the supernatants of *ΔXIS1_460115* and *ΔXIS1_460109* did not have peaks in this region and rather showed either one major peak at 751 Da or three major peaks between 1182 and 1210 Da, respectively (Additional file 9).

Bioassays were conducted to examine if the mutation of *XIS1_460115* or *XIS1_460109* resulted in reduction or loss of the mosquito larvicidal activities, as predicted if the candidate gene cluster locus is necessary for Xlt biosynthesis. Of the mosquito larvae exposed to wild type *X. innexi* supernatant, 100% mortality was observed, up to 25% dilution of the supernatant (Fig. 8). Exposure to dilutions of 12.5% and 6.25% of supernatant resulted over 70% of mortality in 48 h (Fig. 8). However, both *ΔXIS1_460115* and *ΔXIS1_460109* culture supernatants were inactive at dilutions of 25% or lower (Fig. 8), and

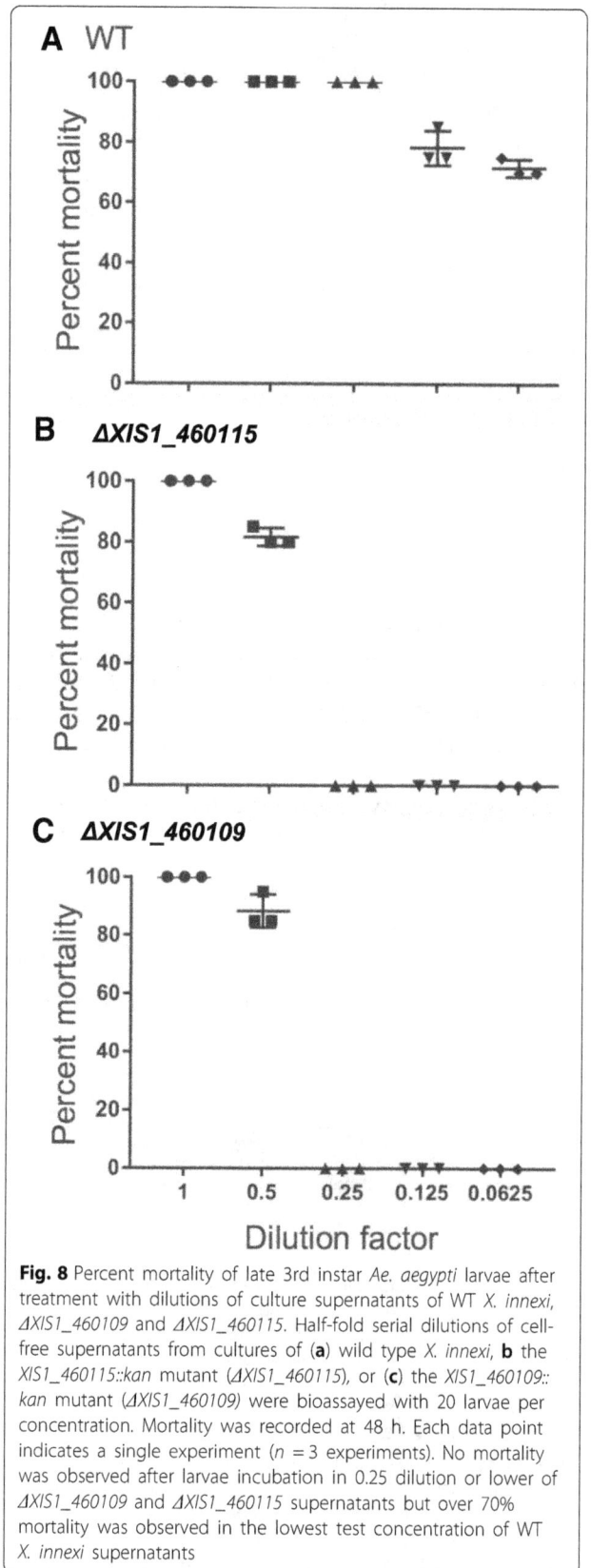

Fig. 8 Percent mortality of late 3rd instar *Ae. aegypti* larvae after treatment with dilutions of culture supernatants of WT *X. innexi*, *ΔXIS1_460109* and *ΔXIS1_460115*. Half-fold serial dilutions of cell-free supernatants from cultures of (**a**) wild type *X. innexi*, **b** the *XIS1_460115::kan* mutant (*ΔXIS1_460115*), or (**c**) the *XIS1_460109:: kan* mutant (*ΔXIS1_460109*) were bioassayed with 20 larvae per concentration. Mortality was recorded at 48 h. Each data point indicates a single experiment (*n* = 3 experiments). No mortality was observed after larvae incubation in 0.25 dilution or lower of *ΔXIS1_460109* and *ΔXIS1_460115* supernatants but over 70% mortality was observed in the lowest test concentration of WT *X. innexi* supernatants

only the undiluted supernatants from these mutants resulted in 100% mortality (Fig. 8).

Discussion

Xenorhabdus bacteria, symbionts of *Steinernema* nematodes, are increasingly exploited for novel products that may be useful in pharmaceutical, agricultural, and industrial settings [81]. Further, exploration of the biology of *Xenorhabdus-Steinernema* associations is yielding new insights into molecular and cellular biology and evolutionary and ecological principles underlying parasitism (e.g. [1, 82, 83]). In this study, we used *X. innexi* and its nematode host *S. scapterisci*, which specializes in parasitism of crickets, to expand our knowledge of potential virulence determinants produced by *Xenorhabdus* bacteria and to discern how *X. innexi* may be impacted by specialization. Our findings that *X. innexi* is an ineffective pathogen of several insects tested, that it does not secrete immunosuppressive factors, and that the *X. innexi* genome lacks many of the canonical virulence determinants encoded by its sister species may indicate that specialization in crickets has led to an erosion of virulence coding potential. However, the specificity of *S. scapterisci* for colonization by *X. innexi*, and our identification of several loci predicted (e.g. T6SS/Rhs) or confirmed (e.g. *xlt*) to be necessary for production of secreted factors indicate that *X. innexi* remains an actively transmitted and biologically active symbiont.

Relative to the well-characterized entomopathogenic nematode symbiont *X. nematophila*, *X. innexi* is attenuated for virulence and reproduction in the lepidopteran hosts *G. mellonella* and *M. sexta*, as well as the dipteran *D. melanogaster*. Unpublished data suggests *X. innexi* is also avirulent towards honeybees (*Apis mellifera)* and Colorado potato beetles (*Leptinotarsa decemlineata*) [29]. This suggests that the toxicity of the *S. scapterisci-X. innexi* pair either relies on the nematode or on an emergent synergism that we did not detect when using the bacteria alone [11].

Common genomic features of Xenorhabdus species with attenuated virulence phenotypes

X. innexi joins a growing list of *Xenorhabdus* species that displays attenuated virulence relative to other members of the genus. Other examples include *X. poinarii* G6, which is attenuated for virulence when injected into *Spodoptera littoralis* and *G. mellonella* insects. Its genome is smaller (3.66 Mbp) than that of either *X. nematophila* (ATCC19061) or *X. bovienii* (SS-2004) and lacks hemolysins, T5SS, Mcf, NRPS, and TA systems [7], suggesting a streamlining of the genome. In contrast, we report here that the genome of *X. innexi* is of similar size (slightly larger) as those of *X. nematophila* and *X. bovienii*, and while it lacks Tc toxins, Mcf, and other canonical *Xenorhabdus* virulence determinants, it does contain genes predicted to encode hemolysins and other T5SS genes and non-ribosomal small molecule biosynthetic machinery, including a locus necessary for production of an extracellular mosquitocidal small molecule. Although caution is necessary when interpreting data based on a draft genome, we propose that in contrast to *X. poinarii*, the *X. innexi* attenuated virulence is due not to genome reduction, but rather to the presence of a distinct repertoire of genes.

In this sense, the *X. innexi* genome may be more similar to *X. bovienii* CS03, the symbiont of *S. weiseri*, another attenuated virulent *Xenorhabdus* bacterium [4]. In this case, rather than genome reduction (as in *X. poinarii*) the attenuated virulence appears to be associated with a genome shift away from virulence determinants and towards inter-bacterial competition. Both *X. bovienii* and *X. innexi* have genomes that are larger than those of *X. nematophila* and *X. bovienii* (SS-2004). Bisch et al. [4] proposed that the *X. bovienii* (CS03) genome had been shaped by the selection for factors mediating inter-microbial competition. A similar phenomenon may be occurring in *X. innexi*, an idea supported by the presence of T6SS and Rhs homologs, which in other systems mediate inter-bacterial competition, concomitant with an absence of canonical insect virulence determinants.

Curiously, *X. poinarii*, *X. bovienii* CS03, and *X. innexi* all lack, or have degraded genes encoding Tc toxins [4, 7]. It should be noted that the production of Tc toxins is not a requirement for virulence, since the Clade C_I bacterium, *X. doucetiae* is virulent in both *S. littoralis* and *G. mellonella*, even though it does not produce Tc toxins [7]. Ogier et al. [7] suggested that the absence of Tc toxins encoded in genomes of members of this clade (C_I) [84] is due to loss of an ancestral component [7]. In the *X. innexi* genome we did not find evidence of fragments or pseudogenized copies of Tc-encoding genes, as are present in the *X. bovienii* CS03 genome [4]. As such, we propose that the apparent lack of these genes in the genome of *X. innexi*, a member of clade C_{IV} indicates a loss event, separate from that proposed to have occurred in clade C_I. Interestingly, in the draft genome of another strain of the C_{IV} clade (*X. cabanillasii*, accession number: GCA_000531755), Tc loci are incomplete, which supports the idea that recent deletions for Tc-encoding genes have occurred in this clade (unpublished data, S. Gaudriault). Regardless, our data combined with those of Ogier et al. [7] and Bisch et al. [4] indicate that the presence of Tc-encoding genes is not a uniformly present trait among *Xenorhabdus* species. It may be that Tc toxins are generalized insecticidal factors that are not of adaptive benefit to *Xenorhabdus* with narrow host ranges. Although not investigated for *S. weiseri- X. bovienii* CS03 pair, both *S. scapterisci-X. innexi* and *S.*

glaseri-*X. poinarii* symbiont pairs appear to have a restricted host range relative to other *Steinernema-Xenorhabdus* pairs [7, 19, 85–87].

A hybrid NRPS/PKS locus is necessary for X. Innexi mosquitocidal toxicity

Despite the ineffectiveness of *X. innexi* as a pathogen when injected into members of various orders of insects, cell-free supernatants from *X. innexi* do exhibit toxicity specifically towards larvae of *Aedes*, *Anopheles* and *Culex* mosquitoes [28]. Our bioinformatic analysis of the *X. innexi* genome revealed a candidate hybrid NRPS/PKS for the biosynthesis of a secreted mosquitocidal toxin. This prediction is supported by our experimental data showing reduction of Xlt lipopeptide synthesis in and mosquitocidal toxicity of the *XIS1_460115* or *XIS1_460109* supernatants. Entomopathogenic bacteria, including *X. innexi*, produce a diversity of secondary metabolites including antibiotics, antifungal and other virulence factors [88] and it is possible that the loss of mosquitocidal toxicity in the *XIS1_460115* and *XIS1_460109* mutants is due to disruption an indirect impact on these other pathways. However, the combined bioinformatic and genetic evidence more strongly support a direct role for the Xlt biosynthetic machinery in the production of the mosquitocidal lipopeptide.

The Xlt biosynthesis gene cluster we have identified is homologous to *fcl* and zmn clusters in the genomes of *X. szentirmaii* and *S. plymuthica* that encode machinery for the synthesis of a class of lipopeptides known as fabclavines and zeamines. Xlt biosynthesis gene cluster also differed from *fcl* and *zmn* cluster by the presence of acyl-CoA thioesterase at the end of the cluster as well as the lack of NUDIX hydrolase gene in the N-terminus and ABC transporter genes in the C-terminus. This genetic similarity and difference, combined with the similar mass to charge ratios of Xlt and fabclavines (~1347) supports the idea that Xlt is a derivative within the fabclavine family. *X. budapestensis* and *X. szentirmaii* produce multiple forms of fabclavine, some of which are distinguished by the presence of either a histidine or phenylalanine moiety at the 2 position. Since Xlt does not absorb at 280 nm (J. Ensign, unpublished data) it is unlikely to contain phenylalanine and thus Xlt is a derivative of a fabclavine Ib [74].

Fabclavines, and the related zeamine have a broad spectrum of bioactivity against bacteria, fungi, nematodes, oomycetes, apicomplexans, and protozoa [79, 80, 89–92]. Similarly, Xlt demonstrated antimicrobial activities towards a broad spectrum of bacteria including *Pseudomonas aeruginosa*, *Salmonella spp.*, *Escherichia coli*, *Listeria monocytogenes*, and *Bacillus cereus* [28].

The results presented here expand the list of fabclavine targets to include mosquito larvae. The literature includes multiple reports of *Steinernema-Xenorhabdus* activities against mosquitoes, which our data suggest could be mediated by bacterially-produced fabclavine and fabclavine derivatives. For instance, *S. carpocapsae* (the nematode host of *X. nematophila*) triggers an immune response in and can kill the larvae of *Aedes aegypti*, a vector of many diseases of humans [73, 93]. Although the mechanism underlying this observation was not investigated, the authors of these studies suggested it could involve a secreted toxin. In support of this concept, recent studies demonstrated toxicity toward *Ae. aegypti* larvae of cell-free supernatants from *X. nematophila*, the symbiont of *S. carpocapsae* [74]. Mosquitoes are unlikely to be natural hosts of *Steinernema-Xenorhabdus* species complexes in nature, raising the question of what the biological function of Xlt may be in the *X. innexi* life history. One possibility may be that it acts in inter-microbial competition, since as a lipopeptide Xlt may be able to disrupt bacterial cell membranes through detergent-like action [94, 95]. Certain bacterial lipopeptides such as surfactins and cyclic lipopeptides (CLPs) from *Bacillus subtilis* have both insecticidal and antimicrobial activity [96–99], although their mode of action against insects is not well understood.

It should be noted that while fabclavines as a class clearly have a broad target spectrum, moiety substitutions within individual fabclavine derivatives could result in varying and specialized activities. In turn, if Xlt and other *Xenorhabdus*-produced fabclavines have non-discriminant broad-spectrum bioactivities, it will be of interest to determine how *Steinernema* nematode hosts associated with the fabclavine-producing *Xenorhabdus* symbionts survive exposure to this generally toxic compound.

Conclusions

As a basis for continued exploration of *X. innexi* in biological studies and biotechnological applications we examined some of its characteristics. We found that unlike other reported EPN/bacterial symbioses, *S. scapterisci* is colonized at very low levels and that *X. innexi* has attenuated virulence compared to other species of *Xenorhabdus*. We have sequenced a draft version of the *X. innexi* genome and reported detailed analyses of several families of known virulence factors. We found no evidence for several key *Xenorhabdus* spp. toxicity genes, including Tc toxins and "makes caterpillars floppy" (Mcf) toxins. However, we also found that the *X. innexi* genome contains two-partner secretion (TPS) system genes from all three TPS clusters, including CdiA exoproteins, active hemolysins, and TpsA proteins. Consistent with other *Xenorhabdus* spp. genomes, we found numerous loci predicted to encode non-ribosomal peptide synthetases, which we explored and identified a locus that putatively encodes a fabclavine derivative with mosquitocidal activity. The *X innexi* genome will be a valuable resource in identifying loci encoding new metabolites of interest, but also in future comparative studies of nematode-bacterial symbiosis and niche partitioning among bacterial pathogens.

Methods
Bacterial strains and growth conditions
Strains and plasmids used in this study are listed in Table 1. Two *X. innexi* strains were tested. One, HGB1681 (a.k.a. PTA-6826), is a lab stock strain acquired by Prof. Jerry Ensign (UW-Madison) from Prof. Grover Smart (University of Florida), the other was isolated from *S. scapterisci* nematodes provided by BD Scientific. In both cases the primary form was isolated as blue colonies on NBTA plates [100]. *Xenorhabdus* strains were incubated at 30 °C in media not exposed to light, or supplemented with 0.1% pyruvate [101]. Permanent stocks of the cultures were stored in broth supplemented with 20% glycerol at −80 °C. Luria Bertani (LB) was used for standard growth, and lipid agar (LA) was used for nematode-bacterium co-culture [102]. When noted, media were supplemented with ampicillin (150 µg/ml), kanamycin (50 µg/ml), streptomycin (150 µg/ml), or diaminopimelic acid (DAP) (80 µg/ml).

To determine the in vitro growth rate of *X. innexi*, we subcultured overnight cultures to an OD600 of 0.1 in LB with limited light exposure and grew them in a 96 well plate (Sarstedt 82.1581.001), 200 µl/well with liquid only (no cultures) in the outermost wells. The plate was incubated in a BioTech plate reader at 30 °C for 17 h constantly shaking in a double orbital pattern, measuring OD_{600} every hour. *X. nematophila* and *X. bovienii* were included for comparison. For each species, three biological replicates were measured, each with three technical replicates within the 96-well plate. The technical replicates were averaged for each biological replicate, and then the biological replicates were plotted with the standard error of the mean. The in vivo growth rates of *X. nematophila* and *X. innexi* in *D. melanogaster* were calculated using the number of CFU (N1) at time 0 (t1) and the number of CFU (N2) recovered at 6 HPI (t2), using the following formula $\ln(N2/N1) = k(t2-t1)$.

Animal sources and husbandry
After purchase from a local vendor (Reptile Rapture, Madison, WI or PetSmart, Knoxville, TN) *A. domesticus* were stored in a large bucket and provided with apple slices and fresh spinach. *S. scapterisci* nematodes were obtained from Becker Underwood Inc. and BD Scientific and established in the laboratory through infection of *Acheta domesticus* house crickets. Typically, 20 crickets were used for infection with *S. scapterisci* nematodes,

while 5 were left uninfected as controls. Crickets were infected within 1–2 days of purchase. Nematodes were propagated every 8 weeks. For infections a 100 mm diameter filter paper was placed in the top of an inverted 100 mm petri dish in which holes had been burned to allow airflow. The filter paper was soaked with 1 ml of *S. scapterisci* IJ stock from the previous infection round, stored in H_2O. In each dish, 3–4 live crickets were placed and provided fresh spinach or apple slices. Infection with ~100 *S. scapterisci* IJs per individual *A. domesticus* cricket yielded $90 \pm 0\%$ mortality (n = 4; 10–20 insects per trial) within 2–3 d of exposure, and some within 1 d. This rapid host killing is a hallmark characteristic of EPNs [103] and reflects efficient release of the bacterial symbiont and/or the release of toxic factors by the nematodes themselves. Once crickets died, the cadavers were placed onto 60 mm filter paper in a 60 mm petri dish, which was then set in a water-filled 100 mm petri dish. After 2–3 days IJs were visible on cadavers and after an additional 4 days IJs emerged from the host and thousands of progeny migrated into the water trap. The nematodes were stored in H_2O for up to 16 weeks. A Stereo Star dissection microscope was used to visually monitor *A. domesticus* infection and collect photos shown in Additional file 8.

Inbred laboratory *Aedes aegypti* (Rockefeller strain) larvae were reared at 26 °C under a 14 L: 10D photoperiod and provided with pellets of fish food [104]. Late 3rd instars were used to bioassay for the presence of mosquito larvicidal lipopeptide, Xlt.

Drosophila melanogaster Oregon-R strain used for infection experiments were kept in standard fly bottles containing dextrose medium (129.4 g dextrose, 7.4 g agar, 61.2 g corn meal, 32.4 g yeast, and 2.7 g tegosept per liter; polypropylene round bottom 8 oz. bottles plugged with bonded dense weave cellulose acetate plugs, Genesee Scientific Cat #49–100) and were housed at 25 °C with 60% relative humidity and a 12 h light and 12 h dark cycle, as previously described [105].

Galleria mellonella waxworms used for infection experiments were purchased from CritterGrub (http://www.critter-grub.com/). Once received, any dead waxworms were discarded and the healthy individuals were kept at 15 °C in the dark until used for experiments. All waxworms were used for experimentation within 14 days.

Tobacco hornworm *Manduca sexta* larvae were raised from eggs (obtained from Carolina Biological Supply Company) on artificial diet (Gypsy moth wheat germ diet, MP Biomedicals, Aurora, OH) with a photoperiod of 16 h.

In vitro colonization assays
After overnight incubation, lawns of *X. innexi* were inoculated with 1 ml of *S. scapterisci* stock and incubated at

room temperature for 72 h or until a large number of adult nematodes were visible. Axenic eggs were isolated from these nematodes as previously described [30] and resuspended in 5 ml LB supplemented with ampicillin. The eggs were used immediately or allowed to hatch into J1 juveniles and stored at room temperature for up to 3 days. The absence of contamination was visually confirmed before use. Approximately 500–1000 axenic eggs and/or J1 nematodes were placed onto lipid agar plates with bacterial lawns and allowed to incubate at room temperature for 3–5 days before placement into White traps to capture emerging IJs [106]. To assess bacterial colonization of IJs, ~1000 IJs were prepared by surface sterilizing in 1.7% sodium hypochlorite solution (5 ml KOH, 32 ml 5.25% sodium hypochlorite [Clorox bleach], and 63 ml ddH20) for 2 min followed by rinsing 6 times in ddH$_2$O. Approximately 200 surface sterilized IJs (in 200 µl) were homogenized for 2 min with a hand-held motor driven grinder and sterile polypropylene pestle (Kontes). The homogenate was dilution plated to observe and quantify CFU.

Construction of X. innexi strains expressing the green fluorescent protein (GFP)

To visualize *X. innexi* within nematodes, we engineered it to express the green-fluorescent protein. pBSL118, a mini Tn5-GFP donor plasmid was used in combination with S17–1 λpir from pUX-BF13, a Tn5 helper strain, to perform GFP conjugations [30, 107, 108]. Briefly, donor, recipient, & helper strain were streaked for single colonies on LB + pyruvate agar plates and grown for 24–48 h at 30 °C without exposure to light. Single colonies were picked grown overnight at 30 °C in liquid LB, with supplementation with 300 µM diaminopimelic acid for the helper and donor strain. Cells were subcultured into fresh medium and grown for an additional 4 h after which 900 µl of *X. innexi* (HGB1681 or HGB1997) and 300 µl each of the helper and donors strains were pelleted separately, washed and re-suspended at their original volumes. The three strains were then mixed together, and plated as a single spot onto a permissive LB pyruvate + DAP plate. After 24 h incubation an inoculation loop was dragged through the spot and the collected cells were re-suspended in LB and plated onto a selective LB pyruvate with ampicillin and kanamycin. After 24–48 h incubation at 30 °C the resulting colonies were analyzed for the expression of GFP with a Nikon Eclipse TE300 inverted fluorescent microscope.

Bacterial infection of insects

Injections into *D. melanogaster* adults were performed as previously described [105]. Briefly, different colony forming unit (CFU) doses were injected into CO$_2$ anesthetized adult male flies aged 5–7 days old with control flies being injected with PBS. Each fly received a total volume of 50 nl injections in the anterior abdomen. Injections were performed using a MINJ-FLY high-speed pneumatic injector (Tritech Research, CA) and a pulled glass needle. After each injection all flies were maintained at 25 °C and 60% humidity. The bacteria were grown to log phase and then diluted to obtain the desired CFU count in a 50 nl volume. To determine CFUs in infected flies, individual flies were homogenized in 200 µl of PBS, diluted serially, and spotted 50 µl onto LB plates supplemented with 0.1% sodium pyruvate. Plates were kept overnight at 28 °C and total CFUs were then determined. For each virulence experiment we injected ≥60 flies, per dose of bacteria. Each experiment was repeated three times. For each in vivo growth assay, we injected and homogenized ≥10 flies, per dose at each time point. These experiments were repeated in triplicate.

Injections into *G. mellonella* larvae were performed as previously described [33]. Briefly, different colony forming unit (CFU) doses were injected into CO$_2$ anesthetized 6th instar larvae. The larvae weighed between 0.19 and 0.30 g. We injected 10 µl in to the hindmost left proleg using a 27-gauge needle. After injections, all insects were kept in 60 mm petri dishes in the dark at 25 °C. Mortality was checked every 12 h. To determine CFUs in infected waxworms, we extracted approximately 10 µl of hemolymph from individual larvae and diluted this with 190 µl of PBS. The diluted hemolymph was then diluted serially, and 50 µl was spotted onto LB plates supplemented with 0.1% sodium pyruvate. Plates were kept overnight at 28 °C and total CFUs were then determined. For each virulence experiment we injected ≥10 larvae, per dose of bacteria. These experiments were repeated experiments in triplicate. For each growth assay, we injected and bled ≥10 larvae, per dose. These experiments were repeated in triplicate.

For injections into *M. sexta*, fifth-instar insect larvae were incubated on ice for approximately 10 min prior to injection. Ten microliters of the diluted culture were injected behind the first set of prolegs of each of 10 insect larvae per treatment using a 30-gauge syringe (Hamilton, Reno, NV). Dilution plating of the inoculum confirmed that for each treatment, an individual insect received 10^4 CFU.

Activation of the proPO system in insect plasma

Supernatants of *X. innexi* and *X. nematophila* strains were used to test their proPO inhibitory activity. Bacterial cultures were grown in LB broth for ~18 h at 30 °C and bacterial supernatant was isolated by spinning cells for 5 min at 8000 x g and filtering through a 0.2 µm syringe filter. Filtered supernatants were heat-treated for 10 min at 95 °C to inactivate heat-labile factors in the supernatant.

Hemolymph (plasma) from wounded fifth instar *M. sexta* larvae was harvested as described previously [109]. In vitro activation of the proPO system was assessed by combining the following in wells of a 96-well plate: 150 µl PBS (phosphate-buffered saline; 137 mM NaCl, 2.7 mM KCL, 10 mM Na_2HPO_4, 1.8 mM KH_2PO_4, pH 7.4), 10 µl plasma, and 20 µl of bacterial supernatant (filtered through a 0.20 µm syringe filter). Fresh LB was used as a negative control. This reaction was incubated at room temperature with constant shaking for 30 min to allow time for inhibition of proPO activation. Immediately following incubation, 20 µl of L-dihydroxyphenylalanine (L-DOPA) (4 mg/ml PBS) were added to the reaction. A microplate reader was used to monitor absorbance at 490 nm every min for 1 h. proPO activation was measured by calculating the rate of synthesis of dopachrome (a melanin intermediate) from L-DOPA. Data are presented as the percentage of each treatment against a negative control for proPO inhibition.

DNA extraction, genome sequencing and annotation

The *X. innexi* genomic DNA was isolated using a standard protocol [110] and submitted for Roche (454) pyrosequencing and assembly at the University of Wisconsin Biotechnology Center. The assembled genome sequence was annotated using the Magnifying Genomes server (MaGe) from MicroScope Microbial Genome Annotation and Analysis Platform. Sequences are available through accession numbers: FTLG01000001-FTLG01000246.

Identification of putative toxin genes in *X. nematophila* and *X. innexi*

The *X. nematophila* ATCC19061 genome was used as a reference to identify the various toxin gene families that we evaluated [104]. We determined the presence or absence of genes encoding putative toxins in *X. innexi* in three ways: using *X. nematophila* sequences as BLAST queries (E ≤ 0.00005) [111], performing Pfam analyses to identify the presence of Pfam domains associated with the various toxin proteins, and using the MicroScope Gene Phyloprofile tool [42] to identify sets of genes specifically absent in *X. innexi* genome. For BLAST analyses, we used the following *X. nematophila* genes as queries: MARTX (XNC1_1376, 1377, 1378, 1380, 1381); Mcf (XNC1_2265); Pir toxins (XNC1_1142, and XNC1_1143); PrtA (XNC1_4025); Tc toxins A (XNC1_2333 + 2334, XNC1_2560 + 2561, XNC1_2566, XNC1_2569, XNC1_3020 + 3021 + 3022 + 3023 + 3024, and XNC1_2187); B (XNC1_2186, XNC1_2335, XNC1_2568); and C (XNC1_2188, XNC1_2336, XNC1_2567); chitinases (XNC1_2562 and XNC1_2569); Txp40 (XNC1_1129); XaxAB (XNC1_3766 and XNC1_3767); Xenocin (XNC1_1221–1223). For Pfam searches we used hmmscan from the latest version of HMMER (3.0) software package, which implements probabilistic profile hidden Markov models. We set our threshold E-value criterion at 10^{-6}, to reduce the probability of false-positive matches. For MaGe analyses we used loci present in the completely sequenced genome of the virulent strain *X. nematophila* (ATCC 19061) and identified those with homologs in the genomes of the virulent strains *X. bovienii* SS-2004 and *X. doucetiae* FRM16 [6, 7], but without homologs in the *X. innexi* HGB1681 genome. The following homology constraints were used: bidirectional best hit, minimal alignment coverage of 0.8, and amino acid sequence identity of 30%.

Identification and analysis of Tps genes in *X. innexi*

TpsA proteins sequences were aligned using the CLUSTAL W program implemented in SEAVIEW [112], and alignments were cleaned using Gblocks [113]. The phylogenetic trees were built by the maximum likelihood (ML) method using the LG substitution model, and branch support values, estimated by the aLRT (SH-like) method, are indicated at the nodes.

Search for type III secretion system homologs in *X. innexi*

The Type III Secretion (T3S) genes (Additional file 4) of *Salmonella enterica* (NCBI Reference Sequence NC_003197.2) were used to search for homologs in *X. innexi*. The nucleotide sequence of the genes in Additional file 4 were used as query sequences in a nucleotide BLAST performed with the Magnifying Genomes server (MaGe) from MicroScope Microbial Genome Annotation and Analysis Platform. Consistent with other examined species of *Xenorhabdus*, *X. innexi* did not contain homologs for any T3S genes.

NRPS-PKS hybrid cluster domain analysis and identification of a candidate Xlt biosynthetic gene cluster in *X. innexi*

X. innexi genome was screened to locate NRPS, PKS and NRPS-PKS hybrid gene clusters. The initial screening was conducted by analyzing protein sequences of each coding DNA sequence (CDS) through a conserved domains search in National Center for Biotechnology Information (NCBI). If a conserved domain search recognized the candidate gene sequence as NRPS, PKS or NRPS-PKS hybrid, the number of A- or AT- domains were examined. NRPS, PKS and NRPS-PKS hybrid genes identified were further analyzed by submitting the corresponding protein sequences into the antibiotic and secondary metabolite analysis shell (AntiSMASH) to identify NRPS and PKS domains [114]. The data file generated by AntiSMASH analyses of the candidate gene cluster is available in Additional file 10.

One candidate gene cluster predicted to encode the Xlt biosynthetic machinery was identified based on

Table 9 Primers used in this study

Primers	5' to 3' sequence[a]	Use
XIS1_460109ApaUpF	NNNNNNGGGCCCCAGGATATGCCATTCAGC	Mutant construction
XIS1_460109BamUpR	NNNNNNGGATCCCAATGACATCAGGCACAC	Mutant construction
XIS1_460109BamDnF	NNNNNNGGATCCGAACCATCGCAGATTGAG	Mutant construction
XIS1_460109XbaDnR	NNNNNNTCTAGAGCCCAATCGCTTCATATC	Mutant construction
XIS1_460115ApaUpF	NNNNNNGGGCCCGAATCGCCCTGGATTATG	Mutant construction
XIS1_460115BamUpR	NNNNNNGGATCCCCCTCTGGCTGATAATAG	Mutant construction
XIS1_460115BamDnF	NNNNNNGGATCCCTCAGGCTCGATTATTGG	Mutant construction
XIS1_460115XbaDnR	NNNNNNTCTAGACTGAATGTACTCCTGCTG	Mutant construction
NilBF	NNNCATATGAGGAAAACGCCACATTCCGG	Confirmation PCR
NilBR	NNNGGGCCCTTGCATGGTTTGGTTG	Confirmation PCR
M13F (−20)	GTAAAACGACGGCCAG	Sequencing PCR
M13R	CAGGAAACAGCTATGAC	Sequencing PCR

[a]N represents A, T, G or C. Engineered restriction enzyme sites are underlined

preliminary chemical data on Xlt structure and composition. Additional in silico analyses were conducted to further test this prediction. Protein sequences of each ORF in the cluster were examined through protein BLAST to predict the putative function, and then analyzed through the conserved domain search to identify PKS, NRPS and non-PKS/NRPS domains. Protein sequences of A-domains in NRPS modules were analyzed through NRPSpredictor2 [115] and AT- domains identified in PKS modules were analyzed using I-TASSER server [116].

Construction of *XIS1_460115* and *XIS1_460109* mutants

To provide a functional test of the role of the candidate *xlt* gene cluster in Xlt biosynthesis, we used allelic exchange site-directed mutagenesis to replace the PKS (*XIS1_460115*) or NRPS (*XIS1_460109*) genes with a kanamycin cassette [117] and tested relevant phenotypes of the resulting mutants. Briefly, upstream and downstream regions of *XIS1_460115* or *XIS1_460109* were amplified using restriction-site-containing primers (Table 9). Amplified fragments were cloned individually into pBluescript SK (–) plasmids; the kanamycin resistant cassette from pKanWor plasmid was cloned into the BamHI site of pBlueXIS1_460109UpDn or pBlueXIS1_460115UpDn (Table 1). The pBlueXIS1_460109UpDn or pBlueXIS1_460115UpDn construct was cloned into a pKR100 suicide vector; the resulting pKRXIS1_460115 and pKRXIS1_460109 constructs (Table 1) were separately conjugated into the WT *X. innexi* using *E. coli* S-17 λpir donor strain. The resulting mutants were first verified by PCR amplification of *nilB*, which is a *Xenorhabdus*-specific gene [118]. The position of mutation was also confirmed by PCR amplification of the flanking regions of the inserted kanamycin cassette.

Mosquito larvicidal bioassays

Mosquito larval bioassays were conducted to determine if mutation at *XIS1_460115* or *XIS1_460109* resulted in the loss of mosquito larvicidal activity. WT *X. innexi*, *ΔXIS1_460115* and *ΔXIS1_460109* were grown in liquid LB media overnight at 30 °C. Samples of overnight cultures were transferred to fresh liquid LB media and were grown at 30 °C until they reached an optical density of 1.0 at 600 nm. Bacterial cultures were centrifuged at 6000 rpm for 10 min and only supernatants were used for bioassays. Various dilutions of the supernatants were made in water and then 2 ml of each dilutions were pipetted into 24- well plastic plates (Becton Dickinson Labware, Franklin Lakes, NJ). Five *Ae. aegypti* larvae were transferred into each well with four replications in each treatment. The experiment was repeated five times and the percent mortality in each concentration of the bacterial supernatant was calculated.

MALDI-TOF MS analysis of WT *X. innexi*, *ΔXIS1_460115* and *ΔXIS1_460109* culture supernatants: *ΔXIS1_460115*, *ΔXIS1_460109* and WT *X. innexi* were cultured in liquid LB media for 24 h at 30 °C, and then centrifuged at 6000 rpm to collect supernatants. Supernatants were submitted for Matrix-assisted laser desorption/ionization time-of-flight mass spectrometry (MALDI-TOF MS) analysis to examine the potential mass profile differences between WT *X. innexi* and mutants (Biotechnology Center, University of Wisconsin-Madison).

Additional Files

Additional file 1: Growth rates of *X. innexi*, *X. nematophila* and *X. bovienii* in vitro and in vivo (PDF 77 kb)

Additional file 2: Percent survival over 50 days of *D. melanogaster* flies injected with controls or *X. nematophila.* (PDF 500 kb)

Additional file 3: List of *X. nematophila* ATCC19061 genes present in *X. bovienii* SS-2004, *X. doucetiae* FRM16, but absent in *X. innexi* HGB1681. (XLSX 42 kb)

Additional file 4: ORFs used for T3SS BLASTp analysis of *X. innexi* draft genome. (PDF 104 kb)

Additional file 5: Repeat domains in MARTX-like genes of *X. innexi* (PDF 141 kb)

Additional file 6: Accession numbers of the sequences used in the phylogenetic analyses of TpsA proteins (PDF 85 kb)

Additional file 7: *X. innexi* loci with genes predicted to encode T6SS components. (PDF 72 kb)

Additional file 8: *A. domesticus* infected with *S. scapterisci.* (PDF 5505 kb)

Additional file 9: MALDI-TOF MS of WT *X. innexi,* ΔXIS1_460109 and ΔXIS1_460115. (PDF 330 kb)

Additional file 10: AntiSMASH analysis of NRPS and PKS genes from *XIS1_460105* to *XIS1_460116.* (ZIP 2696 kb)

Abbreviations

A-: Adenylation-; AA: Amino acid; AntiSMASH: Antibiotics and secondary metabolite analysis shell; AT-: Acyltransferase-; BLAST: Basic local alignment search tool; BLASTp: Protein basic local alignment search tool; CDS: Coding DNA sequence; Cfu: Colony forming unit; CTDs: C-terminal domains; Da: Dalton; DAB: Diaminobutyric acid; DAP: Diaminopimelic acid; DNA: Deoxyribonucleic acid; EPN: Entomopathogenic nematode; GC: Guanine-cytosine; GFP: Green fluorescent protein; HPI: Hours post infection; HPLC/MS: High performance liquid chromatograph/mass spectrometry; IJ: Infective juvenile; LA: Lipid agar; LB: Luria bertani; L-DOPA: L-dihydroxyphenylalanine; MaGe: Magnifying genomes server; MALDI-TOF MS: Matrix assisted laser desorption ionization-time of flight mass spectrometry; MARTX: Multifunctional autoprocessing repeats-in-toxin toxins; Mbp: Megabase pair; Mcf: Makes caterpillars floppy; NBTA: Nutrient bromothymol blue agar; NCBI: National center for biotechnology information; NRPS: Nonribosomal peptide synthetase; OD_{600}: Optical density at a wavelength of 600 nm; ORF: Open reading frame; PBS: Phosphate buffered saline; PCR: Polymerase chain reaction; PKS: Polyketide synthase; PO: Phenoloxidase; Rpm: Revolutions per minute; rRNA: Ribosomal ribonucleic acid; SD: Standard deviation; T3S: Type III secretion; T5SS: Type V secretion system; T6SS: Type VI secretion system; TA: Toxin-antitoxin; TPP: Threonine phosphorylation pathway; TPS: Two-partner secretion; tRNA: Transfer ribonucleic acid; Xlt: Xenorhabdus lipoprotein toxin

Acknowledgements
The authors wish to thank Melinda Hauser and Xiaofei Bai for their assistance with microscopy.

Funding
Work in the Goodrich-Blair lab was supported by a grant (IOS-1353674) from the National Science Foundation, the UW-Madison United States Department of Agriculture (USDA) Hatch Multi-state research formula fund (WIS01582), and the University of Tennessee-Knoxville. TJM and ÁMCT were supported by a grant from the National Institutes of Health Research Service Award T32-GM07215 and ÁMCT was also supported by an Advance Opportunity Fellowship through the Science and Medicine Graduate Research Scholars Program at UW-Madison. SKA and ARD were supported by a National Institutes of Health K22 award from the National Institute of Allergy and Infectious Diseases (AI119155) awarded to ARD. None of these funding agencies had any role in the design of the study or the collection, analysis, and interpretation of data or in writing the manuscript.

Authors' contributions
I-HK carried out mosquitocidal toxin experiments and wrote portions of the manuscript. SKA conducted virulence assays in *Drosophila* and *Galleria,* while AMC-T and MPK conducted virulence assays in *Manduca.* DTA established *X. innexi/S. scapterisci* rearing mechanisms in crickets and calculated *Xenorhabdus* colonization levels. EJM and TJM evaluated colonization phenotypes of *X. innexi* in *S. scapterisci* and TJM also conducted bacterial growth analyses and T3SS comparisons and helped write and edit the manuscript. KH conducted T6SS and Rhs comparisons and phenoloxidase assays. J-CO, and SG conducted comparisons of TPS and helped edit the manuscript. WGG helped supervise I-HK at UW-Madison and edited the manuscript. JCE identified and completed preliminary characterization of the mosquitocidal toxin and acquired funding to obtain the genome sequence. HGB was responsible for supervision of the research at UW-Madison and UTK, acquisition of funding, analysis of TPS, MARTX, and other genomic elements, and writing of the manuscript. ARD was responsible for supervising research at UC Riverside, acquisition of relevant funding, writing the manuscript, and analyzing genomic content. All authors read and approved the final manuscript.

Competing interests
JCE declares pursuit of a patent pertaining to the activity of the mosquitocidal toxin (see bibliography for reference). All other authors declare that they have no competing interests.

Author details
[1]Department of Entomology, University of Wisconsin-Madison, Madison, WI, USA. [2]Present address: Laboratory of Malaria and Vector Research, National Institute of Allergy and Infectious Diseases, Rockville, MD, USA. [3]Department of Nematology, University of California, Riverside, CA, USA. [4]Department of Bacteriology, University of Wisconsin-Madison, Madison, WI, USA. [5]Department of Microbiology, University of Tennessee-Knoxville, Knoxville, TN, USA. [6]DGIMI, INRA, Université de Montpellier, 34095 Montpellier, France.

References
1. Herbert EE, Goodrich-Blair H. Friend and foe: the two faces of *Xenorhabdus nematophila.* Nat Rev Microbiol. 2007;5(8):634–46. doi: 10.1038/nrmicro1706.
2. Kumari P, Mahapatro GK, Banerjee N, Sarin NB. Ectopic expression of GroEL from *Xenorhabdus nematophila* in tomato enhances resistance against *Helicoverpa armigera* and salt and thermal stress. Transgenic Res. 2015;24(5): 859–73. doi: 10.1007/s11248-015-9881-9.
3. Zhang H, Mao J, Liu F, Zeng F. Expression of a nematode symbiotic bacterium-derived protease inhibitor protein in tobacco enhanced tolerance against *Myzus persicae.* Plant Cell Rep. 2012;31(11):1981–9. doi: 10.1007/s00299-012-1310-4.
4. Bisch G, Ogier JC, Medigue C, Rouy Z, Vincent S, Tailliez P, et al. Comparative genomics between two *Xenorhabdus bovienii* strains highlights differential evolutionary scenarios within an entomopathogenic bacterial species. Genome Biol Evol. 2016;8(1):148–60. doi: 10.1093/gbe/evv248.
5. Challinor VL, Bode HB. Bioactive natural products from novel microbial sources. Ann N Y Acad Sci. 2015;1354:82–97. doi: 10.1111/nyas.12954.
6. Chaston JM, Suen G, Tucker SL, Andersen AW, Bhasin A, Bode E, et al. The entomopathogenic bacterial endosymbionts *Xenorhabdus* and *Photorhabdus:* convergent lifestyles from divergent genomes. PLoS One. 2011;6(11):e27909. doi: 10.1371/journal.pone.0027909.
7. Ogier JC, Pages S, Bisch G, Chiapello H, Medigue C, Rouy Z, et al. Attenuated virulence and genomic reductive evolution in the entomopathogenic bacterial symbiont species, *Xenorhabdus poinarii.* Genome Biol Evol. 2014;6(6):1495–513. doi: 10.1093/gbe/evu119.
8. Nguyen KB, Smart GCJ. *Steinernema scapterisci,* new species (Rhabditida: Steinernematidae). J Nematol. 1990;22(2):187–99.
9. Nguyen KB. A new nematode parasite of mole crickets: its taxonomy, biology and potential for biological control. [Ph.D.]. Gainesville: University of Florida; 1988.

10. Bonifassi E, Fischer-Le Saux M, Boemare N, Lanois A, Laumond C, Smart G. Gnotobiological study of infective juveniles and symbionts of *Steinernema scapterisci*: a model to clarify the concept of the natural occurrence of monoxenic associations in entomopathogenic nematodes. J Invertebr Pathol. 1999;74:164–72.

11. Lu D, Sepulveda C, Dillman AR. Infective juveniles of the entomopathogenic nematode *Steinernema scapterisci* are preferentially activated by cricket tissue. PLoS One. 2017;12(1):e0169410. doi: 10.1371/journal.pone.0169410.

12. Lengyel K, Lang E, Fodor A, Szallas E, Schumann P, Stackebrandt E. Description of four novel species of *Xenorhabdus*, family Enterobacteriaceae: *Xenorhabdus budapestensis* sp. nov., *Xenorhabdus ehlersii* sp. nov., *Xenorhabdus innexi* sp. nov., and *Xenorhabdus szentirmaii* sp. nov. Syst Appl Microbiol. 2005;28:115–22.

13. Sicard M, Ramone H, Le Brun N, Pages S, Moulia C. Specialization of the entomopathogenic nematode *Steinernema scaptersci* with its mutualistic *Xenorhabdus* symbiont. Naturwissenschaften. 2005;92:472–6.

14. Spiridonov SE, Reid AP, Podrucka K, Subbotin SA, Moens M. Phylogenetic relationships within the genus *Steinernema* (Nematoda: Rhabditida) as inferred from analyses of sequences of the ITS1-5.8S-ITS2 region of rDNA and morphological features. Nematology. 2004;6:547–66.

15. Nadler SA, Bolotin E, Stock SP. Phylogenetic relationships of *Steinernema* Travassos, 1927 (Nematoda: Cephalobina: Steinernematidae) based on nuclear, mitochondrial and morphological data. Syst Parasitol. 2006;63(3): 161–81. doi: 10.1007/s11230-005-9009-3.

16. Lee MM, Stock SP. A multilocus approach to assessing co-evolutionary relationships between *Steinernema* spp. (Nematoda: Steinernematidae) and their bacterial symbionts *Xenorhabdus* spp. (gamma-Proteobacteria: Enterobacteriaceae). Syst Parasitol. 2010;77(1):1–12. doi: 10.1007/s11230-010-9256-9.

17. Dillman AR, Macchietto M, Porter CF, Rogers A, Williams B, Antoshechkin I, et al. Comparative genomics of *Steinernema* reveals deeply conserved gene regulatory networks. Genome Biol. 2015;16(1):200. doi: 10.1186/s13059-015-0746-6.

18. Nguyen KB, Smart GCJ. Pathogenicity of *Steinernema scapterisci* to selected invertebrates. J Nematol. 1991;23(1):7–11.

19. Wang Y, Gaugler R, Cui L. Variations in immune response of *Popillia japonica* and *Acheta domesticus* to *Heterorhabditis bacteriophora* and *Steinernema* species. J Nematol. 1994;26(1):11–8.

20. Bonner TP. Changes in the structure of *Nippostrongylus brasiliensis* intestinal cells during development from the free-living to the parasitic stages. J Parasitol. 1979;65(5):745–50.

21. Hawdon JM, Schad GA. Serum-stimulated feeding in vitro by third-stage infective larvae of the canine hookworm *Ancylostoma caninum*. J Parasitol. 1990;76(3):394–8.

22. Balasubramanian N, Hao YJ, Toubarro D, Nascimento G, Simoes N. Purification, biochemical and molecular analysis of a chymotrypsin protease with prophenoloxidase suppression activity from the entomopathogenic nematode *Steinernema carpocapsae*. Int J Parasitol. 2009;39(9):975–84. doi: 10.1016/j.ijpara.2009.01.012.

23. Toubarro D, Lucena-Robles M, Nascimento G, Costa G, Montiel R, Coelho AV, et al. An apoptosis-inducing serine protease secreted by the entomopathogenic nematode *Steinernema carpocapsae*. Int J Parasitol. 2009; 39(12):1319–30. doi: 10.1016/j.ijpara.2009.04.013.

24. Sicard M, Le Brun N, Pages S, Godelle B, Boemare N, Moulia C. Effect of native *Xenorhabdus* on the fitness of their *Steinernema* hosts: contrasting types of interactions. Parasitol Res. 2003;91:520–4.

25. Grewal PS, Matsuura M, Converse V. Mechanisms of specificity of association between the nematode *Steinernema scapterisci* and its symbiotic bacterium. Parasitology. 1997;114(5):483–8.

26. Mitani DK, Kaya HK, Goodrich-Blair H. Comparative study of the entomopathogenic nematode, *Steinernema carpocapsae*, reared on mutant and wild-type *Xenorhabdus nematophila*. Biol Control. 2004;29:382–91.

27. Hussa E, Goodrich-Blair H. Rearing and injection of *Manduca sexta* larvae to assess bacterial virulence. J Vis Exp. 2012;70:e4295. doi: 10.3791/4295.

28. Ensign JC, Lan Q, Dyer DH, inventors; Mosquitocidal *Xenorhabdus*, lipopeptide and methods. 2014 US Patent US20140274880 A1.

29. Kim IH, Ensign J, Kim DY, Jung HY, Kim NR, Choi BH, et al. Specificity and putative mode of action of a mosquito larvicidal toxin from the bacterium *Xenorhabdus innexi*. J Invertebr Pathol. 2017;149:21–8. doi: 10.1016/j.jip.2017.07.002.

30. Murfin KE, Chaston J, Goodrich-Blair H. Visualizing bacteria in nematodes using fluorescence microscopy. J Vis Exp. 2012;68:e4298. doi: 10.3791/4298.

31. Veesenmeyer JL, Andersen AW, Lu X, Hussa EA, Murfin KE, Chaston JM, et al. NilD CRISPR RNA contributes to *Xenorhabdus nematophila* colonization of symbiotic host nematodes. Mol Microbiol. 2014;93(5):1026–42. doi: 10.1111/mmi.12715.

32. Dunphy GB. Interaction of mutants of *Xenorhabdus nematophilus* (Enterobacteriaceae) with antibacterial systems of *Galleria mellonella* larvae (Insecta: Pyralidae). Can J Microbiol. 1994;40(3):161–8.

33. Blackburn D, Wood PL Jr, Burk TJ, Crawford B, Wright SM, Adams BJ. Evolution of virulence in *Photorhabdus* spp., entomopathogenic nematode symbionts. Syst Appl Microbiol. 2016;39(3):173–9. doi: 10.1016/j.syapm.2016.02.003.

34. Cerenius L, Söderhäll K. The prophenoloxidase-activating system in invertebrates. Immunol Rev. 2004;198:116–26.

35. Seo S, Lee S, Hong Y, Kim Y. Phospholipase A2 inhibitors synthesized by two entomopathogenic bacteria, *Xenorhabdus nematophila* and *Photorhabdus temperata* subsp. temperata. Appl Environ Microbiol. 2012;78(11):3816–23. doi: 10.1128/AEM.00301-12.

36. Song CJ, Seo S, Shrestha S, Kim Y. Bacterial metabolites of an entomopathogenic bacterium, *Xenorhabdus nematophila*, inhibit a catalytic activity of phenoloxidase of the diamondback moth, *Plutella xylostella*. J Microbiol Biotechnol. 2011;21(3):317–22.

37. Crawford JM, Portmann C, Zhang X, Roeffaers MB, Clardy J. Small molecule perimeter defense in entomopathogenic bacteria. Proc Natl Acad Sci U S A. 2012;109(27):10821–6. doi: 10.1073/pnas.1201160109.

38. Condon C, Liveris D, Squires C, Schwartz I, Squires CL. rRNA operon multiplicity in *Escherichia coli* and the physiological implications of *rrn* inactivation. J Bacteriol. 1995;177(14):4152–6.

39. Asai T, Condon C, Voulgaris J, Zaporojets D, Shen B, Al-Omar M, et al. Construction and initial characterization of *Escherichia coli* strains with few or no intact chromosomal rRNA operons. J Bacteriol. 1999;181(12):3803–9.

40. Gyorfy Z, Draskovits G, Vernyik V, Blattner FF, Gaal T, Posfai G. Engineered ribosomal RNA operon copy-number variants of *E. coli* reveal the evolutionary trade-offs shaping rRNA operon number. Nucleic Acids Res. 2015;43(3):1783–94. doi: 10.1093/nar/gkv040.

41. Castagnola A, Stock SP. Common virulence factors and tissue targets of entomopathogenic bacteria for biological control of lepidopteran pests. Insects. 2014;5(1):139–66. doi: 10.3390/insects5010139.

42. Vallenet D, Labarre L, Rouy Z, Barbe V, Bocs S, Cruveiller S, et al. MaGe: a microbial genome annotation system supported by synteny results. Nucleic Acids Res. 2006;34(1):53–65. doi: 10.1093/nar/gkj406.

43. Waterfield N, Bowen DJ, Fetherston JD, Perry RD, ffrench-Constant RH. The toxin complex genes of Photorhabdus: a growing gene family. Trends Microbiol. 2001;9:185–91.

44. Waterfield N, Dabord PJ, Dowling AJ, Yang G, Hares M. ffrench-Constant RH. The insecticidal toxin makes caterpillars floppy 2 (Mcf2) shows similarity to HrmA, an avirulence protein from a plant pathogen. FEMS Microbiol Lett. 2003;229:265–70.

45. Waterfield N, Kamita SG, Hammock BD, ffrench-Constant R. The Photorhabdus Pir toxins are similar to a developmentally regulated insect protein but show no juvenile hormone esterase activity. FEMS Microbiol Lett. 2005;245:47–52.

46. Vigneux F, Zumbihl R, Jubelin G, Ribeiro C, Poncet J, Baghdiguian S, et al. The *xaxAB* genes encoding a new apoptotic toxin from the insect pathogen *Xenorhabdus nematophila* are present in plant and human pathogens. J Biol Chem. 2007;282:9571–80.

47. Gavin HE, Satchell KJ. MARTX toxins as effector delivery platforms. Pathog Dis. 2015;73(9):ftv092. doi: 10.1093/femspd/ftv092.

48. Kim BS, Gavin HE, Satchell KJ. Distinct roles of the repeat-containing regions and effector domains of the Vibrio vulnificus multifunctional-autoprocessing repeats-in-toxin (MARTX) toxin. MBio. 2015;6(2) doi: 10.1128/mBio.00324-15.

49. Antic I, Biancucci M, Zhu Y, Gius DR, Satchell KJ. Site-specific processing of Ras and Rap1 switch I by a MARTX toxin effector domain. Nat Commun. 2015;6:7396. doi: 10.1038/ncomms8396.

50. Biancucci M, Rabideau AE, Lu Z, Loftis AR, Pentelute BL, Satchell KJF. Substrate recognition of MARTX Ras/Rap1-specific Endopeptidase. Biochemist. 2017;56(21):2747–57. doi: 10.1021/acs.biochem.7b00246.

51. Satchell KJ. Structure and function of MARTX toxins and other large repetitive RTX proteins. Annu Rev Microbiol. 2011;65:71–90. doi: 10.1146/annurev-micro-090110-102943.

52. Satchell KJ. Multifunctional-autoprocessing repeats-in-toxin (MARTX) toxins of Vibrios. Microbiol Spectr. 2015;3(3) doi: 10.1128/microbiolspec.VE-0002-2014.

53. Jacob-Dubuisson F, Locht C, Antoine R. Two-partner secretion in gram-negative bacteria: a thrifty, specific pathway for large virulence proteins. Mol Microbiol. 2001;40(2):306–13.

54. Nikolakakis K, Amber S, Wilbur JS, Diner EJ, Aoki SK, Poole SJ, et al. The toxin/immunity network of Burkholderia pseudomallei contact-dependent growth inhibition (CDI) systems. Mol Microbiol. 2012;84(3):516–29. doi: 10.1111/j.1365-2958.2012.08039.x.

55. Aoki SK, Poole SJ, Hayes CS, Low DA. Toxin on a stick: modular CDI toxin delivery systems play roles in bacterial competition. Virulence. 2011;2(4):356–9.

56. Aoki SK, Pamma R, Hernday AD, Bickham JE, Braaten BA, Low DA. Contact-dependent inhibition of growth in Escherichia coli. Science. 2005;309(5738):1245–8. doi: 10.1126/science.1115109.

57. Aoki SK, Diner EJ, de Roodenbeke CT, Burgess BR, Poole SJ, Braaten BA, et al. A widespread family of polymorphic contact-dependent toxin delivery systems in bacteria. Nature. 2010;468(7322):439–42. doi: 10.1038/nature09490.

58. Ogier JC, Duvic B, Lanois A, Givaudan A, Gaudriault S. A new member of the growing family of contact-dependent growth inhibition systems in Xenorhabdus doucetiae. PLoS One. 2016;11(12):e0167443. doi: 10.1371/journal.pone.0167443.

59. Cowles KN, Goodrich-Blair H. Expression and activity of a Xenorhabdus nematophila haemolysin required for full virulence towards Manduca sexta insects. Cell Microbiol. 2005;2:209–19.

60. Brillard J, Ribeiro C, Boemare N, Brehélin M, Givaudan A. Two distinct hemolytic activities in Xenorhabdus nematophila are active against immunocompetent insect cells. Appl Environ Microbiol. 2001;67:2515–25.

61. Ruhe ZC, Low DA, Hayes CS. Bacterial contact-dependent growth inhibition. Trends Microbiol. 2013;21(5):230–7. doi: 10.1016/j.tim.2013.02.003.

62. Ho BT, Dong TG, Mekalanos JJ. A view to a kill: the bacterial type VI secretion system. Cell Host Microbe. 2014;15(1):9–21. doi: 10.1016/j.chom.2013.11.008.

63. Hachani A, Allsopp LP, Oduko Y, Filloux A. The VgrG proteins are "a la carte" delivery systems for bacterial type VI effectors. J Biol Chem. 2014;289(25):17872–84. doi: 10.1074/jbc.M114.563429.

64. Mougous JD, Gifford CA, Ramsdell TL, Mekalanos JJ. Threonine phosphorylation post-translationally regulates protein secretion in Pseudomonas aeruginosa. Nat Cell Biol. 2007;9(7):797–803. doi: 10.1038/ncb1605.

65. Cianfanelli FR, Monlezun L, Coulthurst SJ. Aim, load, fire: the type VI secretion system, a bacterial nanoweapon. Trends Microbiol. 2016;24(1):51–62. doi: 10.1016/j.tim.2015.10.005.

66. Whitney JC, Beck CM, Goo YA, Russell AB, Harding BN, De Leon JA, et al. Genetically distinct pathways guide effector export through the type VI secretion system. Mol Microbiol. 2014;92(3):529–42. doi: 10.1111/mmi.12571.

67. Bondage DD, Lin JS, Ma LS, Kuo CH, Lai EM. VgrG C terminus confers the type VI effector transport specificity and is required for binding with PAAR and adaptor-effector complex. Proc Natl Acad Sci U S A. 2016;113(27):E3931–40. doi: 10.1073/pnas.1600428113.

68. Jackson AP, Thomas GH, Parkhill J, Thomson NR. Evolutionary diversification of an ancient gene family (rhs) through C-terminal displacement. BMC Genomics. 2009;10:584. doi: 10.1186/1471-2164-10-584.

69. Alcoforado Diniz J, Liu YC, Coulthurst SJ. Molecular weaponry: diverse effectors delivered by the type VI secretion system. Cell Microbiol. 2015;17(12):1742–51. doi: 10.1111/cmi.12532.

70. Jamet A, Nassif X. New players in the toxin field: polymorphic toxin systems in bacteria. MBio. 2015;6(3):e00285-15. doi: 10.1128/mBio.00285-15.

71. Nikolouli K, Mossialos D. Bioactive compounds synthesized by non-ribosomal peptide synthetases and type-I polyketide synthases discovered through genome-mining and metagenomics. Biotechnol Lett. 2012;34(8):1393–403. doi: 10.1007/s10529-012-0919-2.

72. Du L, Sánchez C, Shen B. Hybrid peptide–polyketide natural products: biosynthesis and prospects toward engineering novel molecules. Metab Eng. 2001;3(1):78–95.

73. Beresky MA, Hall DW. The influence of phenylthiourea on encapsulation, melanization, and survival in larvae of the mosquito Aedes aegypti parasitized by the nematode Neoaplectana carpocapsae. J Invertebr Pathol. 1977;29(1):74–80.

74. da Silva OS, Prado GR, da Silva JL, Silva CE, da Costa M, Heermann R. Oral toxicity of Photorhabdus luminescens and Xenorhabdus nematophila (Enterobacteriaceae) against Aedes aegypti (Diptera: Culicidae). Parasitol Res. 2013;112(8):2891–6. doi: 10.1007/s00436-013-3460-x.

75. Benning MM, Wesenberg G, Liu R, Taylor KL, Dunaway-Mariano D, Holden HM. The three-dimensional structure of 4-hydroxybenzoyl-CoA thioesterase from Pseudomonas sp. strain CBS-3. J Biol Chem. 1998;273(50):33572–9.

76. Wilcke M, Alexson SE. Characterization of acyl-CoA thioesterase activity in isolated rat liver peroxisomes. FEBS J. 1994;222(3):803–11.

77. Svensson LT, Alexson SE, Hiltunen JK. Very long chain and long chain acyl-coA thioesterases in rat liver mitochondria. Identification, purification, characterization and induction by peroxisome proliferators. J Biol Chem. 1995;270(20):12177–83.

78. Hunt MC, Solaas K, Kase BF, Alexson SE. Characterization of an acyl-coA thioesterase that functions as a major regulator of peroxisomal lipid metabolism. J Biol Chem. 2002;277(2):1128–38. doi: 10.1074/jbc.M106458200.

79. Masschelein J, Mattheus W, Gao LJ, Moons P, Van Houdt R, Uytterhoeven B, et al. A PKS/NRPS/FAS hybrid gene cluster from Serratia plymuthica RVH1 encoding the biosynthesis of three broad spectrum, zeamine-related antibiotics. PLoS One. 2013;8(1):e54143. doi: 10.1371/journal.pone.0054143.

80. Fuchs SW, Grundmann F, Kurz M, Kaiser M, Bode HB. Fabclavines: bioactive peptide-polyketide-polyamino hybrids from Xenorhabdus. Chembiochem. 2014;15(4):512–6. doi: 10.1002/cbic.201300802.

81. Pidot SJ, Coyne S, Kloss F, Hertweck C. Antibiotics from neglected bacterial sources. Int J Med Microbiol. 2014;304(1):14–22. doi: 10.1016/j.ijmm.2013.08.011.

82. Bashey F, Hawlena H, Lively CM. Alternative paths to success in a parasite community: within-host competition can favor higher virulence or direct interference. Evolution. 2013;67(3):900–7. doi: 10.1111/j.1558-5646.2012.01825.x.

83. Murfin KE, Lee MM, Klassen JL, McDonald BR, Larget B, Forst S, et al. Xenorhabdus bovienii strain diversity impacts coevolution and symbiotic maintenance with Steinernema spp. nematode hosts. MBio. 2015;6(3):e00076. doi: 10.1128/mBio.00076-15.

84. Tailliez P, Laroui C, Ginibre N, Paule A, Pages S, Boemare N. Phylogeny of Photorhabdus and Xenorhabdus based on universally conserved protein-coding sequences and implications for the taxonomy of these two genera. Proposal of new taxa: X. vietnamensis sp. nov., P. luminescens subsp. caribbeanensis subsp. nov., P. luminescens subsp. hainanensis subsp. nov., P. temperata subsp. khanii subsp. nov., P. temperata subsp. tasmaniensis subsp. nov., and the reclassification of P. luminescens subsp. thracensis as P. temperata subsp. thracensis comb. nov. Int J Syst Evol Microbiol. 2010;60(Pt 8):1921–37. doi: 10.1099/ijs.0.014308-0.

85. Converse V, Grewal PS. Virulence of entomopathogenic nematodes to the western masked chafer Cyclocephala hirta (Coleoptera: Scarabaeidae). J Econ Entomol. 1998;91(2):428–32.

86. Rosa JS, Cabral C, Simoes N. Differences between the pathogenic processes induced by Steinernema and Heterorhabditis (Nemata: Rhabditida) in Psudaletia unipuncta (Insecta: Lepidoptera). J Invertebr Pathol. 2002;80:46–54.

87. Fallon DJ, Solter LF, Bauer LS, Miller DL, Cate JR, McManus ML. Effect of entomopathogenic nematodes on Plectrodera scalator (Fabricius) (Coleoptera: Cerambycidae). J Invertebr Pathol. 2006;92(1):55–7. doi: 10.1016/j.jip.2006.01.006.

88. Bode HB. Entomopathogenic bacteria as a source of secondary metabolites. Curr Opin Chem Biol. 2009;13(2):224–30. doi: 10.1016/j.cbpa.2009.02.037.

89. Boszormenyi E, Ersek T, Fodor A, Fodor AM, Foldes LS, Hevesi M, et al. Isolation and activity of Xenorhabdus antimicrobial compounds against the plant pathogens Erwinia amylovora and Phytophthora nicotianae. J Appl Microbiol. 2009;107(3):746–59. doi: 10.1111/j.1365-2672.2009.04249.x.

90. Fuchs SW, Sachs CC, Kegler C, Nollmann FI, Karas M, Bode HB. Neutral loss fragmentation pattern based screening for arginine-rich natural products in Xenorhabdus and Photorhabdus. Anal Chem. 2012;84(16):6948–55. doi: 10.1021/ac300372p.

91. Hellberg JE, Matilla MA, Salmond GP. The broad-spectrum antibiotic, zeamine, kills the nematode worm Caenorhabditis elegans. Front Microbiol. 2015;6:137. doi: 10.3389/fmicb.2015.00137.

92. Masschelein J, Clauwers C, Stalmans K, Nuyts K, De Borggraeve W, Briers Y, et al. The zeamine antibiotics affect the integrity of bacterial membranes. Appl Environ Microbiol. 2015;81(3):1139–46. doi: 10.1128/AEM.03146-14.

93. Andreadis TG, Hall DW. Neoaplectana carpocapsae:encapsulation in Aedes aegypti and changes in host hemocytes and hemolymph proteins. Exp Parasitol. 1976;39(2):252–61.

94. Carrillo C, Teruel JA, Aranda FJ, Ortiz A. Molecular mechanism of membrane permeabilization by the peptide antibiotic surfactin. Biochim Biophys Acta. 2003;1611(1–2):91–7. doi: 10.1016/S0005-2736(03)00029-4.

95. Straus SK, Hancock REW. Mode of action of the new antibiotic for gram-positive pathogens daptomycin: comparison with cationic antimicrobial peptides and lipopeptides. Biochim Biophys Acta. 2006;1758(9):1215–23. doi: 10.1016/j.bbamem.2006.02.009.

96. Assie LK, Deleu M, Arnaud L, Paquot M, Thonart P, Gaspar C, et al. Insecticide activity of surfactins and iturins from a biopesticide *Bacillus subtilis* Cohn (S499 strain). Meded Rijksuniv Gent Fak Landbouwkd Toegep Biol Wet. 2002;67(3):647–55.

97. Das K, Mukherjee AK. Assessment of mosquito larvicidal potency of cyclic lipopeptides produced by *Bacillus subtilis* strains. Acta Trop. 2006;97(2):168–73. doi: 10.1016/j.actatropica.2005.10.002.

98. Ongena M, Henry G, Thonart P. The Roles of Cyclic Lipopeptides in the Biocontrol Activity of *Bacillus subtilis*. In: Gisi U, Chet I, Gullino M. (eds) Recent Developments in Management of Plant Diseases. Plant Pathology in the 21st Century (Contributions to the 9th International Congress), Springer, Dordrecht. 2010;1.

99. Singh P, Cameotra SS. Potential applications of microbial surfactants in biomedical sciences. Trends Biotechnol. 2004;22(3):142–6. doi: 10.1016/j.tibtech.2004.01.010.

100. Boemare NE, Akhurst RJ. Biochemical and physiological characterization of colony form variants in *Xenorhabdus* spp. (Enterobacteriaceae). J Gen Microbiol. 1988;134:751–61.

101. Xu J, Hurlbert RE. Toxicity of irradiated media for *Xenorhabdus* spp. Appl Environ Microbiol. 1990;56:815–8.

102. Vivas EI, Goodrich-Blair H. *Xenorhabdus nematophilus* as a model for host-bacterium interactions: *rpoS* is necessary for mutualism with nematodes. J Bacteriol. 2001;183(16):4687–93.

103. Dillman AR, Chaston JM, Adams BJ, Ciche TA, Goodrich-Blair H, Stock SP, et al. An entomopathogenic nematode by any other name. PLoS Path. 2012;8(3):e1002527. doi: 10.1371/journal.ppat.1002527.

104. Krebs K, Lan Q. Isolation and expression of a sterol carrier protein-2 gene from the yellow fever mosquito, *Aedes aegypti*. Insect Mol Biol. 2003;12(1):51–60.

105. Pham LN, Dionne MS, Shirasu-Hiza M, Schneider DS. A specific primed immune response in *Drosophila* is dependent on phagocytes. PLoS Path. 2007;3(3):e26. doi: 10.1371/journal.ppat.0030026.

106. White GFA. Method for obtaining infective nematode larvae from cultures. Science. 1927;66:302–3.

107. Bao Y, Lies DP, Fu H, Roberts GP. An improved Tn7-based system for the single-copy insertion of cloned genes into chromosomes of gram-negative bacteria. Gene. 1991;109:167–8.

108. Teal TK, Lies DP, Wold BJ, Newman DK. Spatiometabolic stratification of *Shewanella oneidensis* biofilms. Appl Environ Microbiol. 2006;72(11):7324–30. doi: 10.1128/AEM.01163-06.

109. Orchard SS, Goodrich-Blair H. Identification and functional characterization of a *Xenorhabdus nematophila* oligopeptide permease. Appl Environ Microbiol. 2004;70(9):5621–7.

110. Sambrook J, Fritsch EF, Maniatis T. Molecular cloning: a laboratory manual. 2nd ed. Cold Spring Harbor: Cold Spring Harbor Laboratory Press; 1989.

111. Altschul SF, Madden TL, Schaffer AA, Zhang J, Zhang Z, Miller W, et al. Gapped BLAST and PSI-BLAST: a new generation of protein database search programs. Nucleic Acids Res. 1997;25(17):3389–402.

112. Galtier N, Gouy M, Gautier C. SEAVIEW and PHYLO_WIN: two graphic tools for sequence alignment and molecular phylogeny. Comput Appl Biosci. 1996;12(6):543–8.

113. Castresana J. Selection of conserved blocks from multiple alignments for their use in phylogenetic analysis. Mol Biol Evol. 2000;17(4):540–52.

114. Blin K, Medema MH, Kazempour D, Fischbach MA, Breitling R, Takano E, et al. antiSMASH 2.0–a versatile platform for genome mining of secondary metabolite producers. Nucleic Acids Res. 2013;41(Web Server issue):W204-12. doi: 10.1093/nar/gkt449.

115. Röttig M, Medema MH, Blin K, Weber T, Rausch C, Kohlbacher O. NRPSpredictor2—a web server for predicting NRPS adenylation domain specificity. Nucl Ac Res. 2011;39:W362-7.

116. Zhang Y. I-TASSER server for protein 3D structure prediction. BMC Bioinformat. 2008;9(1):40.

117. Richards GR, Goodrich-Blair H. Examination of *Xenorhabdus nematophila* lipases in pathogenic and mutualistic host interactions reveals a role for *xlpA* in nematode progeny production. Appl Environ Microbiol. 2010;76(1):221–9. doi: 10.1128/AEM.01715-09.

118. Bhasin A, Chaston JM, Goodrich-Blair H. Mutational analyses reveal overall topology and functional regions of NilB, a bacterial outer membrane protein required for host association in a model of animal-microbe mutualism. J Bacteriol. 2012;194(7):1763–76. doi: 10.1128/JB.06711-11.

119. Sugar DR, Murfin KE, Chaston JM, Andersen AW, Richards GR, Deleon L, et al. Phenotypic variation and host interactions of *Xenorhabdus bovienii* SS-2004, the entomopathogenic symbiont of *Steinernema jollieti* nematodes. Env Microbiol. 2012;14(4):924–39. doi: 10.1111/j.1462-2920.2011.02663.x.

7

Lactation-related metabolic mechanism investigated based on mammary gland metabolomics and 4 biofluids' metabolomics relationships in dairy cows

Hui-Zeng Sun, Kai Shi, Xue-Hui Wu, Ming-Yuan Xue, Zi-Hai Wei, Jian-Xin Liu and Hong-Yun Liu[*] ⓘ

Abstract

Background: Lactation is extremely important for dairy cows; however, the understanding of the underlying metabolic mechanisms is very limited. This study was conducted to investigate the inherent metabolic patterns during lactation using the overall biofluid metabolomics and the metabolic differences from non-lactation periods, as determined using partial tissue-metabolomics. We analyzed the metabolomic profiles of four biofluids (rumen fluid, serum, milk and urine) and their relationships in six mid-lactation Holstein cows and compared their mammary gland (MG) metabolomic profiles with those of six non-lactating cows by using gas chromatography-time of flight/mass spectrometry.

Results: In total, 33 metabolites were shared among the four biofluids, and 274 metabolites were identified in the MG tissues. The sub-clusters of the hierarchical clustering analysis revealed that the rumen fluid and serum metabolomics profiles were grouped together and highly correlated but were separate from those for milk. Urine had the most different profile compared to the other three biofluids. Creatine was identified as the most different metabolite among the four biofluids (VIP = 1.537). Five metabolic pathways, including gluconeogenesis, pyruvate metabolism, the tricarboxylic acid cycle (TCA cycle), glycerolipid metabolism, and aspartate metabolism, showed the most functional enrichment among the four biofluids (false discovery rate < 0.05, fold enrichment >2). Clear discriminations were observed in the MG metabolomics profiles between the lactating and non-lactating cows, with 54 metabolites having a significantly higher abundance ($P < 0.05$, VIP > 1) in the lactation group. Lactobionic acid, citric acid, orotic acid and oxamide were extracted by the S-plot as potential biomarkers of the metabolic difference between lactation and non-lactation. The TCA cycle, glyoxylate and dicarboxylate metabolism, glutamate metabolism and glycine metabolism were determined to be pathways that were significantly impacted ($P < 0.01$, impact value >0.1) in the lactation group. Among them, the TCA cycle was the most up-regulated pathway ($P < 0.0001$), with 7 of the 10 related metabolites increased in the MG tissues of the lactating cows.

Conclusions: The overall biofluid and MG tissue metabolic mechanisms in the lactating cows were interpreted in this study. Our findings are the first to provide an integrated insight and a better understanding of the metabolic mechanism of lactation, which is beneficial for developing regulated strategies to improve the metabolic status of lactating dairy cows.

Keywords: Biofluids relationship, Dairy cow, Lactation, Mammary gland, Metabolic function, Metabolomics

* Correspondence: hyliu@zju.edu.cn
Institute of Dairy Science, MoE Key Laboratory of Molecular Animal Nutrition, College of Animal Sciences, Zhejiang University, Hangzhou 310058, People's Republic of China

Background

Dairy cows contribute the most important source of dairy foods for humans, and this relies largely on the cows' lactation performance [1]. The inherent metabolic status during lactation and the difference from the non-lactation period are the most important metabolic biological processes for determining milk production and sustainability[2], and they depend on the development of the mammary gland (MG) and the overall coordinated metabolism and physiology.

Metabolomics, a vital part of systemic biology, can be utilized to explore the disease, diet or environment-related comprehensive metabolism by analyzing endogenous small molecules in the biofluids or tissues [3]. To date, biofluid metabolomics combined with tissue metabolomics has provided valuable information on the overall or partial metabolic mechanisms [4]. Biofluid metabolomics has been extensively applied in disease diagnosis [5], biomarker discovery [6] and novel pathway identification [7] in human clinical studies. Recently, biofluid metabolomics has been widely used to investigate the relationship between the rumen fluid metabolites and cow health [8], the effect of heat stress on milk metabolites [9], and the urinary biomarkers under different quality forage diets [6]. However, most biofluid metabolomics studies in dairy cows have been limited to a single biofluid and/or focused on the changes in the physiological conditions under different treatments. Multiple biofluid metabolomics and their relationships under the same treatment allow new insights into the inherent metabolic pattern and global metabolism, which could contribute to improving the metabolic status for better productivity and sustainability. Tissue metabolomics can capture the subtle metabolic variation of the specific organs or parts of the body and reflect the actual biological processes that affect gene expression, transcription and translation [10]. It has been revealed that MG metabolomics is an effective way to explore the initiation, maintenance and regulation of lactation [11, 12] and can provide direct connections with the milking phenotypes. Therefore, identifying the metabolites in the MG during lactation and comparing their key pathways with non-lactating cows could enhance our understanding of the lactation mechanism.

In this study, we hypothesize about the inherent metabolic characteristics during lactation and their differences from the non-lactation period, which can be addressed by biofluid metabolomics combined with tissue metabolomics. The relationships among four biofluids' (rumen fluid, serum, milk and urine) metabolomics profiles from the lactating cows were analyzed to investigate the inherent metabolic patterns and possible effects on lactation sustainability. The MG metabolomics profiles were compared between lactating and non-lactating cows to identify the key metabolites and pathways and their potential regulatory roles in lactation.

Methods

The experimental procedures were approved by the Animal Care Committee at Zhejiang University (Hangzhou, China) and were in accordance with the university's guidelines for animal research.

Study design and sample collection

In total, 12 Holstein dairy cows, 6 lactating and 6 non-lactating, were used in this study. Among them, 6 multiparous dairy cows with similar milk yields (30.4 ± 2.29 kg/d, mean $\pm$ SD) and at similar lactation stages (days in milk = 164 ± 19.6 d, mean $\pm$ SD) were fed a diet with a forage-to-concentrate ratio of 45:55, and they were provided 16.7% (DM basis) crude protein and 1.57 Mcal/kg net energy for lactation, following the protocol described in our previous study [13]. The diet ingredients and nutrient composition are commonly applied in state-of-the-art farms throughout China (Additional file 1: Table S1). Approximately 1000 mg of MG tissues was collected immediately after slaughter according to previously described methods [14] from each cow in both the lactation and non-lactation groups. The tissues were then stored at −80 °C until metabolite extraction. Biofluid samples, including 50 mL of rumen fluid, 50 mL of milk, 10 mL of blood and 10 mL of urine, were collected from the lactating cows before the morning feeding at 6:00 AM. Rumen fluid was collected using an oral stomach tube following the standard procedures [15], milk samples were collected using a milk sampling device (Waikato Milking Systems NZ Ltd., Waikato, Hamilton, New Zealand), blood samples were collected from the jugular vein using pro-coagulation 10-mL tubes, and urine samples were collected using the vulval stimulation method. To minimize any possible degradation of the metabolites, rumen fluid, milk and urine samples were infused into a 15-mL sterilized centrifuge tube and immediately placed in liquid nitrogen. Before storage in a − 80 °C freezer, the rumen fluid, milk and urine samples were centrifuged at 6000×g, 3000×g, and 3000×g, respectively, and incubated at 4 °C for 15 min. Blood samples were centrifuged at 4 °C, 3000×g for 15 min within 20 min after sample collection. More details can be obtained from our previous study [16].

Extraction of compounds

The biofluid metabolomics was examined using an Agilent 7890 gas chromatography system equipped with a Pegasus 4D time of flight mass spectrometer (LECO, St. Joseph, MI, USA) [6]. The tissue metabolomics procedures were performed as follows. First, 100 mg of MG tissue from each sample was added to a 2-mL Eppendorf tube with 0.4 mL of methanol-chloroform ($V_{methanol}$: $V_{chloroform} = 3:1$) and 30 μL of L-2-chlorophenylalanine

(1 mg/mL, stored in dH$_2$O) and was mixed by vortexing for 10 s. Second, steel balls were placed in the tube and milled for 5 min at 55 Hz. The sample was then centrifuged at 4 °C at 12,000 rpm for 15 min. Third, approximately 0.4 mL of supernatant was transferred into a 2 mL silylated vial. An equal volume (10 μL) of each sample was placed in a new 2 mL silylated vial as a mixed sample for the quality control of the stability of the equipment system, the standard deviation of the beginning, middle and ending retention time of the mixed samples was less than 0.2, which indicates good stability.

Metabolite derivatization

The extracts were dried in a vacuum concentrator at 30 °C for 1.5 h. Methoxymethyl amine salt (80 μL; dissolved in pyridine to a final concentration of 20 mg/mL) was added to the dried metabolites, mixed and gently sealed. The solution was then incubated at 37 °C for 2 h in an oven. Then, 100 μL of bis trifluoroacetamide (containing 1% trimethylchlorosilane, v/v) was added to each sample, which was sealed again and incubated at 70 °C for 1 h. In addition, 10 μL of fatty acid methyl esters (a standard mixture of fatty acid methyl esters, C8-C16: 1 mg/mL; C18-C30: 0.5 mg/mL in chloroform) was added to the mixed sample and cooled to room temperature. Finally, the samples were mixed well and subjected to gas chromatography-time of flight/mass spectrometry (GC-TOF/MS) testing.

Metabolite identification by GC-TOF/MS

GC-TOF/MS analysis of the tissue samples was performed on an Agilent 7890 gas chromatograph system in cooperation with a Pegasus HT time-of-flight mass spectrometer (LECO, St. Joseph, MI, USA). The system used a DB-5 MS capillary column coated with 5% cross-linked diphenyl and 95% dimethyl polysiloxane (30 m × 250 μm inner diameter, 0.25 μm film thickness; J&W Scientific, Folsom, CA, USA). One μL aliquot of the analyte was added in splitless mode. Helium was used as the carrier gas. The flow of the front inlet purge was 3 mL min^{-1}, and the gas flow rate through the column was 20 mL min^{-1}. The original temperature was set at 50 °C and was maintained for 1 min. The temperature was increased to 330 °C at a speed of 10 °C min^{-1} and was maintained for 5 min at 330 °C. Temperatures of 280 °C, 280 °C, and 220 °C were used for the injection, transfer line, and ion source, respectively. The energy was −70 eV in electron influence mode. The full-scan mode of the mass spectrometry data was 85 m/z - 600 m/z at a rate of 20 spectra per second after a solvent delay of 366 s.

Identification of differentially expressed metabolites

Using the interquartile range denoising method, missing values of the raw data were filled by half of the minimum value and valid peaks were detected and only the

metabolites remained. Additionally, an internal standard (L-2-chlorobenzene alanine) log normalization method was applied. The resulting three-dimensional data, including the peak number, sample name and normalized peak area, were uploaded to the online analysis platform - Metaboanalyst 3.0 (http://www.metaboanalyst.ca/) for univariate and multivariate analyses. Both one-way ANOVA and post hoc analysis were used to identify the important metabolites of the four biofluids. When the P value was less than 0.05, the metabolite was characterized as differentially expressed. Univariate analyses provided a preliminary overview of the features that were potentially significant. The false discovery rate (FDR) was used to conduct multiple comparisons testing, with an FDR value less than 0.05 indicating significance. Multivariate analyses included hierarchical cluster analysis (HCA), principal component analysis (PCA), partial least squares discriminant analysis (PLS-DA), and orthogonal projections to latent structures (OPLS). PCA was used to visualize the dataset of multitudinous variables in a 2 or 3-dimensional figure and display the general similarity and difference. The first principal component of the PLS (variable importance projection, VIP value) was obtained to identify the differentially expressed metabolites. Variables with VIP values exceeding 1.0 and P values less than 0.05 were selected as differentially expressed metabolites. An S-plot was generated to further identify the statistically significant and potentially biochemically significant metabolites in both their contributions to the module variables and the reliability of the module.

Pathway characterization

The compound names and relative concentrations of the shared metabolites were imported into Metaboanalyst 3.0 (http://www.metaboanalyst.ca) to perform functional enrichment and impact pathway analyses (integrates pathway enrichment analysis and pathway topology analysis). The functional enrichment analysis was based on several libraries containing approximately 6300 groups of biologically meaningful metabolite sets. Functional enrichment analysis was performed using the metabolite set enrichment analysis (MSEA) to investigate the enrichment of the predefined groups of functionally related metabolites, which are usually associated with biological pathways [17]. Significant functional pathways were selected using a fold enrichment (FE) threshold greater than 2.0 and an FDR value less than 0.05 [18]. Pathway topological analysis was used to calculate the importance of the metabolites based on the out-of-degree centrality and the relative betweenness measures of a metabolite in a given metabolic network [19]. The impact pathway value was calculated as the sum of the matched important metabolites of all the metabolites in each pathway [20]. Impact values greater than 0.10 and *P* values less than 0.05 were defined as significant impact pathways. Extremely different pathways were defined with *P* values less than 0.01. Additionally, commercial databases, including the Kyoto Encyclopedia of Genes and Genomes (KEGG, http://www.genome.jp/kegg/) and the Small Molecular Pathway Database (SMPDB, http://smpdb.ca), were used to search for metabolites and for the integrated pathway analysis.

Results
Lactation performance
The 4% fat-corrected milk yield of the six lactating cows was 26.29 ± 2.66 kg/d (mean ± SD), and the milk protein, milk fat and lactose contents were $3.29 \pm 0.16\%$, $4.17 \pm 0.37\%$ and $4.91 \pm 0.14\%$ (mean ± SD), respectively. Detailed results can be found in our previous study [21]. No adverse events occurred in any animal.

Metabolites identified in biofluids using GC-TOF/MS
From the metabolomics datasets of the four biofluids in six lactating cows, 33 shared metabolites were identified in the rumen fluid, serum, milk and urine, including noradrenaline, methylmalonic acid, glycine, lyxose, L-malic acid, thymol, creatine, 5-methoxytryptamine, oxoproline, glycerol, L-threose, m-cresol, aminomalonic acid, 2-hydroxybutanoic acid, alanine, hydroxylamine, phosphate, fumaric acid, glucose, prostaglandin E2, isoleucine, lactose, malonic acid, N-methyl-L-glutamic acid, phenylethylamine, 2,4-diaminobutyric acid, oxalic acid, lactic acid, 4-androsten-1-beta-ol-3,17-dione, norleucine, asparagine, conduritol-b-epoxide, and 5-aminovaleric acid. The relative abundance of each metabolite is presented in Additional file 1: Table S2.

When we compared the MG metabolomics and milk metabolomics of the 6 same lactating dairy cows in the lactation group, a total of 118 shared metabolites, 156 MG-specific metabolites and 67 milk-specific metabolites were identified (Fig. 1, Additional file 1: Table S3).

Metabolites identified in the MG tissue using GC-TOF/MS
In total, 394 valid peaks were detected from the analytes of the MG tissues. Based on the LECO/Fiehn Metabolomics Library, 274 metabolites were characterized and quantified (Additional file 1: Table S4). Among them, stigmasterol, sarcosine, palatinitol, oxamide, lyxose, L-threose, D-erythronolactone, cholesterol, carnitine and carbamoyl-aspartic acid were identified in the lactation group only and mainly belong to lipid or lipid-like molecules, carbohydrates and amino acids. Conversely, terephthalic acid, phenyl beta-D-glucopyranoside, octadecanol and alpha-ketoisocaproic acid were found only in the non-lactation dairy cow group.

Multivariate statistical analysis
The 3D–PCA score plot of the metabolic profiles showed significantly separated clusters among the four biofluids (Fig. 2a). All score plots for the rumen fluid, serum, milk, and urine sample were in the Hotelling T^2 ellipse with 95% confidence. As shown in the PLS-DA score map, the samples were clearly separated into four parts, indicating the differential metabolic profiling of the four biofluids samples (Fig. 2b). HCA revealed sub-clusters containing varying numbers of metabolites, with different metabolomics profiles from each biofluid being clustered together (Fig. 2c). The metabolites from the rumen and serum samples were clustered together, which then clustered with the milk samples, but the cluster derived from the urine samples was separated from the other three biofluids. Several metabolites with higher relative concentrations were identified in each biofluid: glycerol, noradrenaline, lyxose, 5-methoxytryptamine and glucose in rumen fluid; lactic acid, glycine, N-methyl-L-glutamic acid, phenylethylamine, aminomalonic acid and glucose in the serum; glycerol, oxoproline, oxalic acid, L-malic acid, lyxose and L-threose in the milk; and isoleucine, m-cresol, 2-hydroxybutanoic acid, phosphate, creatine and lactose in the urine. HCA also revealed the presence of one cluster that included an amino acid and its derivatives (e.g., glycine, aminomalonic acid, N-methyl-L-glutamic acid, 5-methoxytryptamine, noradrenaline and phenylethylamine) and an organic acid and its derivatives (e.g., glucose, lactic acid and malonic acid). Another cluster consisted of aliphatic compounds, such as oxoproline, L-threose, oxalic acid, L-malic acid, glycerol and lyxose.

For the MG metabolomics, the heat map revealed the 6 lactating and 6 non-lactating dairy cows were assigned to 2 significantly separated clusters (Fig. 3a). The overview of the heatmap showed that the concentrations of most MG metabolites in the lactation group were higher than those in MG of non-lactating dairy cows. The 3-D PCA score map of GC-TOF/MS metabolic profiles of MG tissue showed significant discrimination between the two animal groups (Fig. 3b). All score plots for the MG samples in the 2 groups were in the 95% Hotelling T^2 ellipse in the PLS-DA score map (Fig. 3c). In summary, clear separation in metabolomic profiles was found between the lactating and non-lactating dairy cow groups.

Identification of differentially expressed metabolites
The differentially expressed metabolites identified from the four biofluids are listed in Additional file 1: Table S5. Twenty-six metabolites showed obvious differences among the four biofluids, with four shared metabolites being differentially expressed in each comparison (FDR value <0.01): oxoproline, L-threose, creatine and lactose. As illustrated in Fig. 4, the VIP

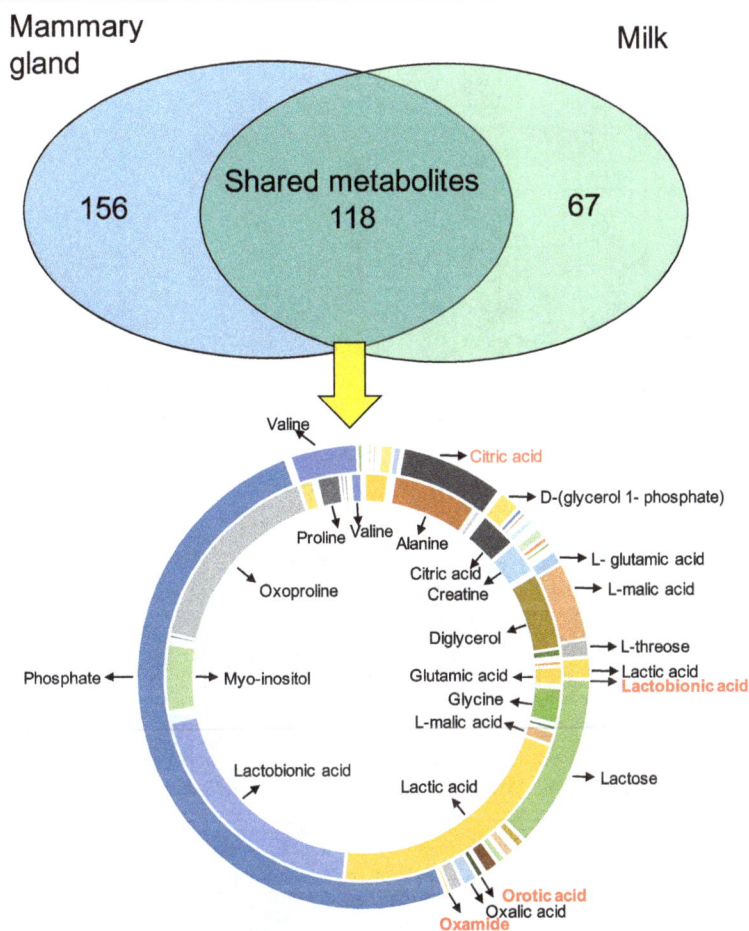

Fig. 1 Metabolite comparison profiling of mammary gland tissue and milk from the same 6 lactating dairy cows. In total, 118 shared metabolites, 156 mammary gland-specific metabolites and 67 milk-specific metabolites were identified. The 118 shared metabolites from the mammary gland tissue (inside) and milk (outside) are displayed in the ring chart. The potential biomarkers in the mammary gland metabolomics study were also identified in milk and are marked in red

values of the PLS-DA model >1 included creatine, L-malic acid, oxalic acid, lyxose, lactose, phenylethylamine, glycerol, L-threose, m-cresol and oxoproline, with creatine exhibiting maximum integrated classification performance (VIP = 1.537).

In total, 73 differentially expressed metabolites were identified to in MG tissues between the lactating and non-lactating cows (VIP > 1, $P < 0.05$ & $Q < 0.10$), with 54 significantly higher and 19 significantly lower relative concentration metabolites in the lactating cows (Additional file 1: Table S4). The S-plot showed the key metabolites (Fig. 5). On the left-hand side of the S-plot, 4 variables with strong model contribution and high statistical reliability were explored as potential biomarkers to characterize the metabolic discriminations between lactation and non-lactation: lactobionic acid ($P_{cov} = -4.669$, $P_{corr} = -0.806$), citric acid ($P_{cov} = -2.983$, $P_{corr} = -0.793$), orotic acid ($P_{cov} = -2.786$, $P_{corr} = -0.887$) and oxamide ($P_{cov} = -2.491$, $P_{corr} = -0.975$).

KEGG pathway analysis

Overall, 20 pathways were obtained when the 33 shared metabolites from the four biofluids were imported into the KEGG analysis. Figure 6 shows the functional enrichment of the different pathways. The most enriched functional pathways included gluconeogenesis (FDR < 0.001, FE = 3.79), pyruvate metabolism (FDR = 0.003, FE = 2.68), the citric acid cycle (FDR = 0.007, FE = 2.63), glycerolipid metabolism (FDR = 0.017, FE = 2.01), and aspartate metabolism (FDR = 0.021, FE = 3.24). For the MG metabolomics, significantly higher abundance metabolites were involved in the 22 pathways in the lactation group (Additional file 1: Table S6). The key functional impact of the pathways is illustrated in Fig. 7. The results of the enrichment and impact pathways demonstrated that there were four significantly up-regulated pathways in the lactating cows: the tricarboxylic acid cycle (TCA), glyoxylate and dicarboxylate metabolism, glutamate metabolism, and glycine biosynthesis and degradation

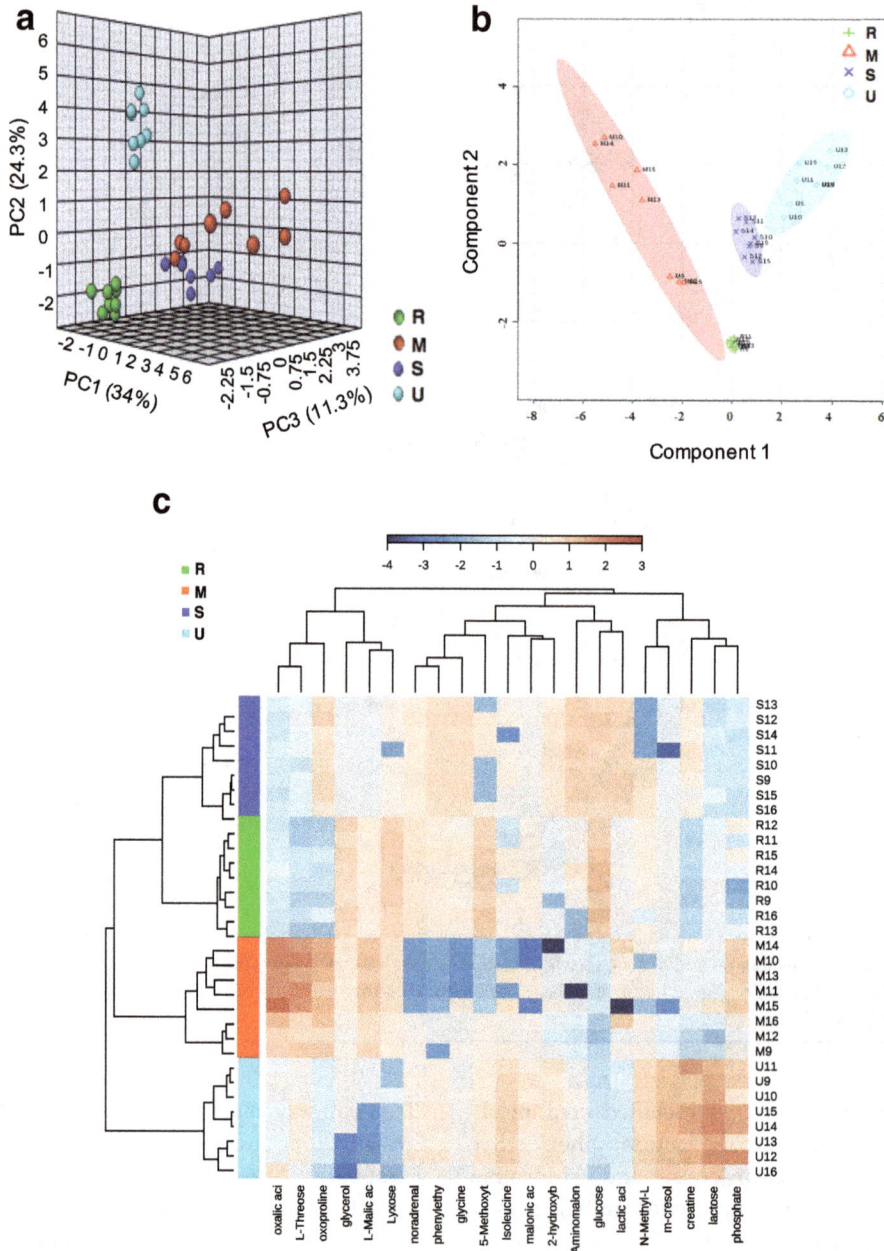

Fig. 2 The 3-D principal component analysis (PCA) score map, 2-D partial least squares discriminant analysis (PLS-DA) score map, and hierarchical clustering analysis (HCA) for the shared metabolites in the rumen fluid, serum, milk and urine derived from gas chromatography-time of flight/mass spectrometry. Subgraph **a**, **b** and **c** represent the PCA, PLS-DA and HCA, respectively. The green, dark blue, red and light blue represent the samples from the rumen fluid, serum, milk and urine, respectively. The patterns shown in each row were categorized using an average linkage hierarchical clustering program. The light blue boxes indicate an expression ratio that is less than the mean, and dark red boxes denote an expression ratio that is greater than the mean. The tree clusters and their shorter Euclidean distances indicate higher similarities. The similarity between the two metabolites is represented by the branch height; the lower a node is vertically, the more similar its subtree is

($P < 0.05$, impact value >0.10). The integrated overview pathway map combined the above four key cellular pathways with the corresponding most significant and relevant metabolites (Fig. 8). Among these were seven metabolites that are involved in the TCA cycle, glyoxylate and dicarboxylate pathway, three metabolites that are involved in the glutamate

metabolism pathway and three metabolites that are involved in the glycine biosynthesis and degradation pathway.

Discussion

Remarkable progress has been made in our understanding of dairy cow metabolism, especially of the nutrient

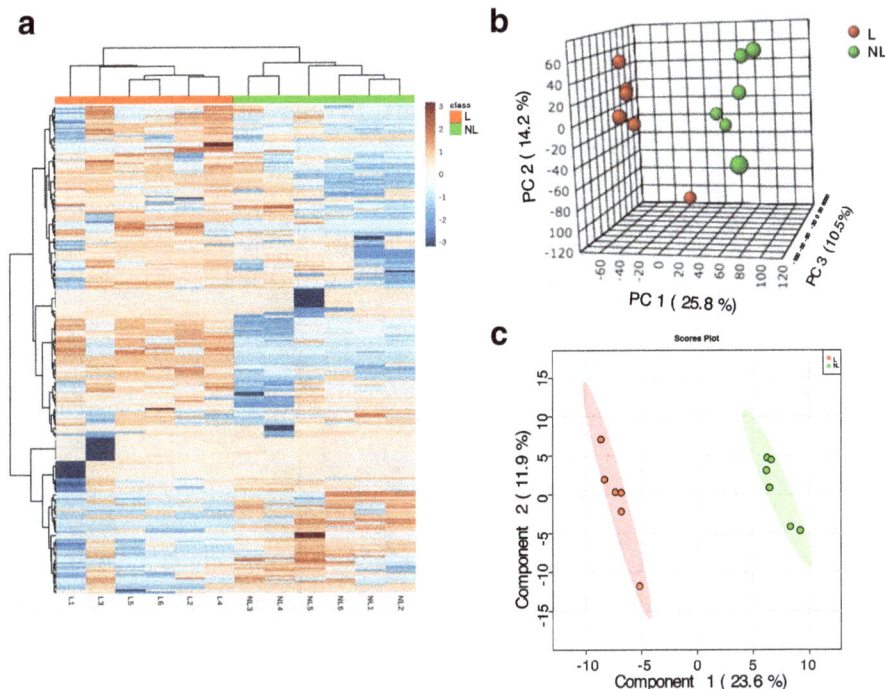

Fig. 3 Heatmap, 3-D principal component analysis (PCA) score map and 2-D partial least squares discriminant analysis (PLS-DA) score map of the mammary gland tissue from 6 lactating cows and 6 non-lactating cows. Subgraph **a**, **b** and **c** represent the Heatmap, PCA, and PLS-DA, respectively. Red and green color bars represent lactating and non-lactating cows, respectively. The blue boxes indicate an expression ratio that is less than the mean, and the red boxes denote an expression ratio that is greater than the mean. The darker the color is, the larger the difference there is from the mean value

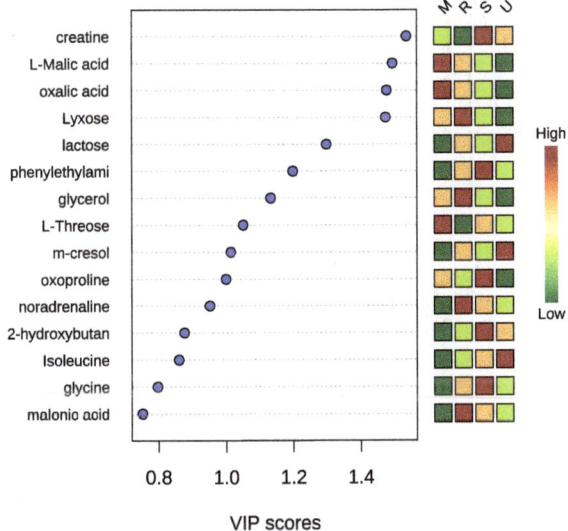

Fig. 4 Differentially expressed metabolites in the rumen fluid, serum, milk and urine of the lactating dairy cows identified by partial least squares discriminant analysis (PLS). The colored boxes on the right indicate the relative concentrations of the corresponding metabolite in each biofluid. R: rumen fluid; M: milk; S: serum; U: urine. Darker green indicates a lower relative concentration and darker red represents a higher relative concentration. The x-axis represents the first principal component of the PLS (variable importance projection, VIP value), and the blue dots represent the different metabolites

requirements of the biochemical reactions and functions [22]. However, the complex metabolic mechanisms that occur during lactation are still considered a "black box" due to our lack of in-depth understanding [23]. The increasing application of the -omics technologies (i.e., metagenomics, transcriptomics, proteomics and metabolomics) have become a powerful tool leading to the understanding of this "black box" [24]. Indeed, these -omics technologies are being used to evaluate the physiological and molecular changes of dairy cows under different environments, such as varied diets and management. The inherent metabolic status during lactation and the difference in the non-lactation period play a vital role in the lactation physiology, which determines the milk production, udder health and dairy sustainability [2]. Very little information is available on the metabolomics profiles of these two stages in the dairy cows. To our knowledge, this paper is the first to address the lactation-related metabolic profiles by using tissue and biofluid metabolomics. Both the overall lactation-related metabolism (variation and coordinated metabolic mechanism based on four biofluids' metabolomics) and partial metabolism (comparison metabolomics in MGs between lactating and non-lactating cows) provide novel insights and a greater understanding of the metabolic mechanism of lactation.

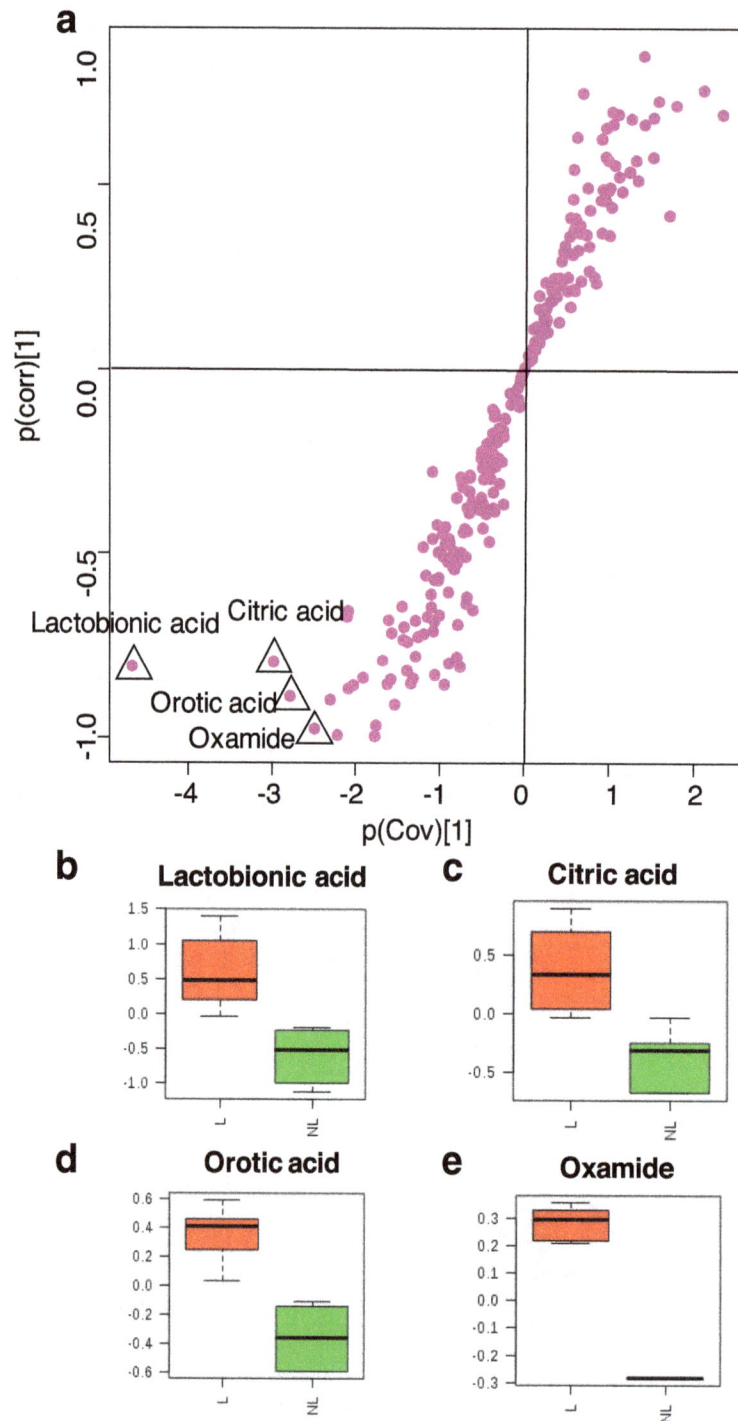

Fig. 5 Potentially differentially expressed metabolites extracted by the S-plot between the lactation and non-lactation groups. **a** S-Plot, 4 metabolites are highlighted. These are shown in **b-e**. **b** Log normalized relative concentration of the 2 groups for lactobionic acid; **c** Log normalized relative concentration of the 2 groups for citric acid; **d** Log normalized relative concentration of the 2 groups for orotic acid; **e** Log normalized relative concentration of the 2 groups for oxamide; L: lactation group, NL: non-lactation group. The x-axis, p(cov), in figure a is a visualization of the contribution (covariance) to the module variables, and the y-axis, p(corr), in figure a is a visualization of the reliability (correlation) of the module. The preferred selection of the potential biomarkers is high covariance combined with high correlation. The red bar represents the lactation group, and the green bar represents the non-lactation group

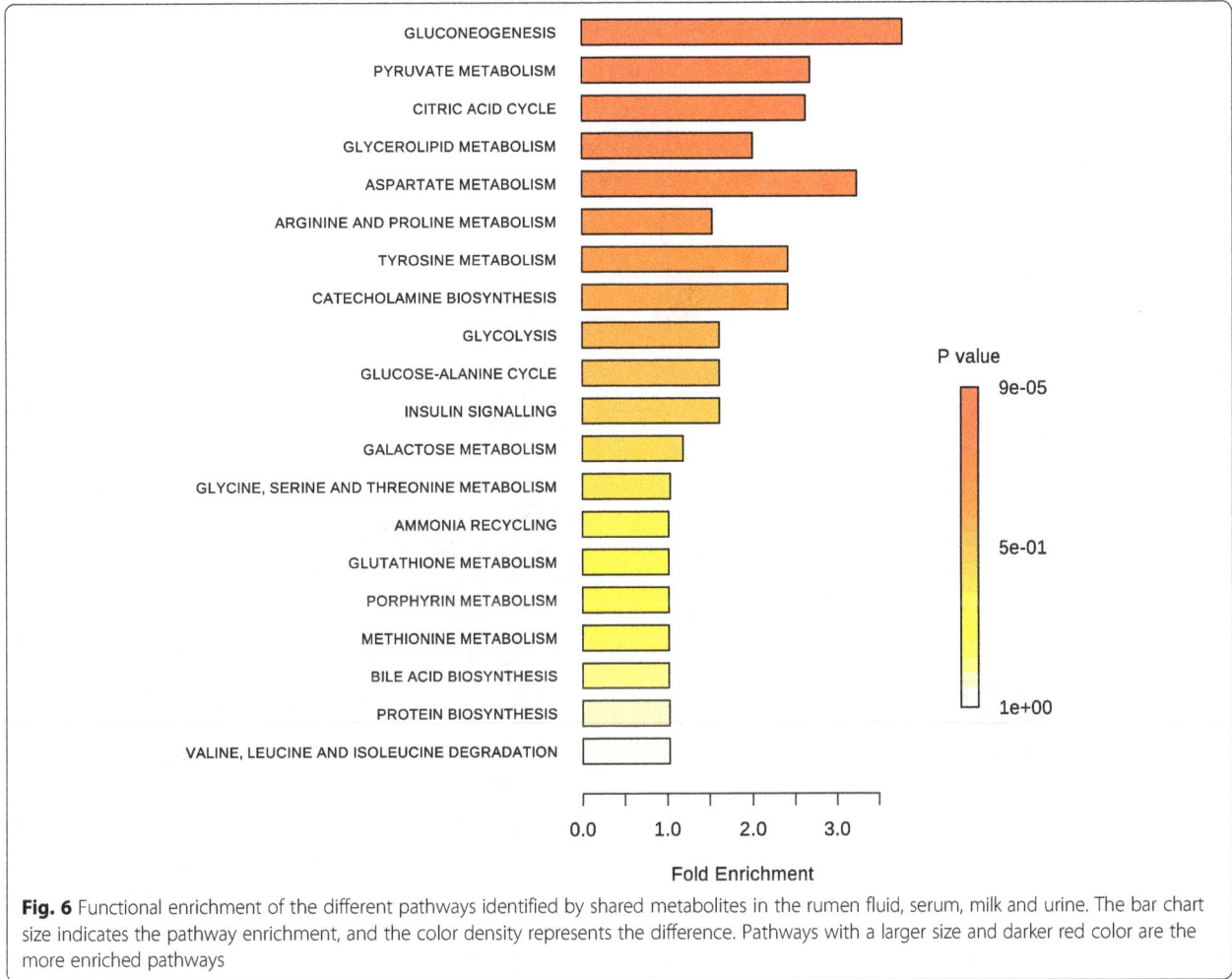

Fig. 6 Functional enrichment of the different pathways identified by shared metabolites in the rumen fluid, serum, milk and urine. The bar chart size indicates the pathway enrichment, and the color density represents the difference. Pathways with a larger size and darker red color are the more enriched pathways

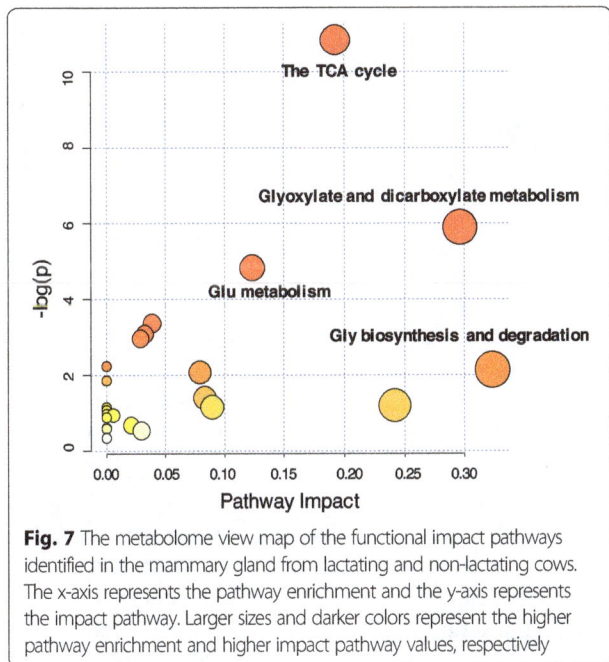

Fig. 7 The metabolome view map of the functional impact pathways identified in the mammary gland from lactating and non-lactating cows. The x-axis represents the pathway enrichment and the y-axis represents the impact pathway. Larger sizes and darker colors represent the higher pathway enrichment and higher impact pathway values, respectively

For dairy cows, common biofluids include the rumen fluid, blood, milk and urine. The rumen fluid metabolites are critical for microbial metabolism and growth, host nutrient utilization and health, all of which can influence milk production [25]. Blood is an intermediary biofluid that consists of metabolites that are excreted from the body and links different organs and tissues [26]. The milk metabolomics reflect the product quality, metabolic processes and status of MG [27]. Urine is rich in metabolites and contains many biochemical pathway signals [28]. It is easy to distinguish the rumen fluid, serum, milk and urine based on their physical characteristics, such as the color and odor. However, the metabolic relationships, which may contain subtle metabolic profiling for the whole body, are not well understood. According to the metabolic flow in the dairy cow, the nutrients in the diet are first ingested and digested in the rumen, absorbed or passed through the rumen, transferred into blood in the small intestines, and then allocated to and utilized in the different organs, such as the MG. Waste materials are then extracted from the

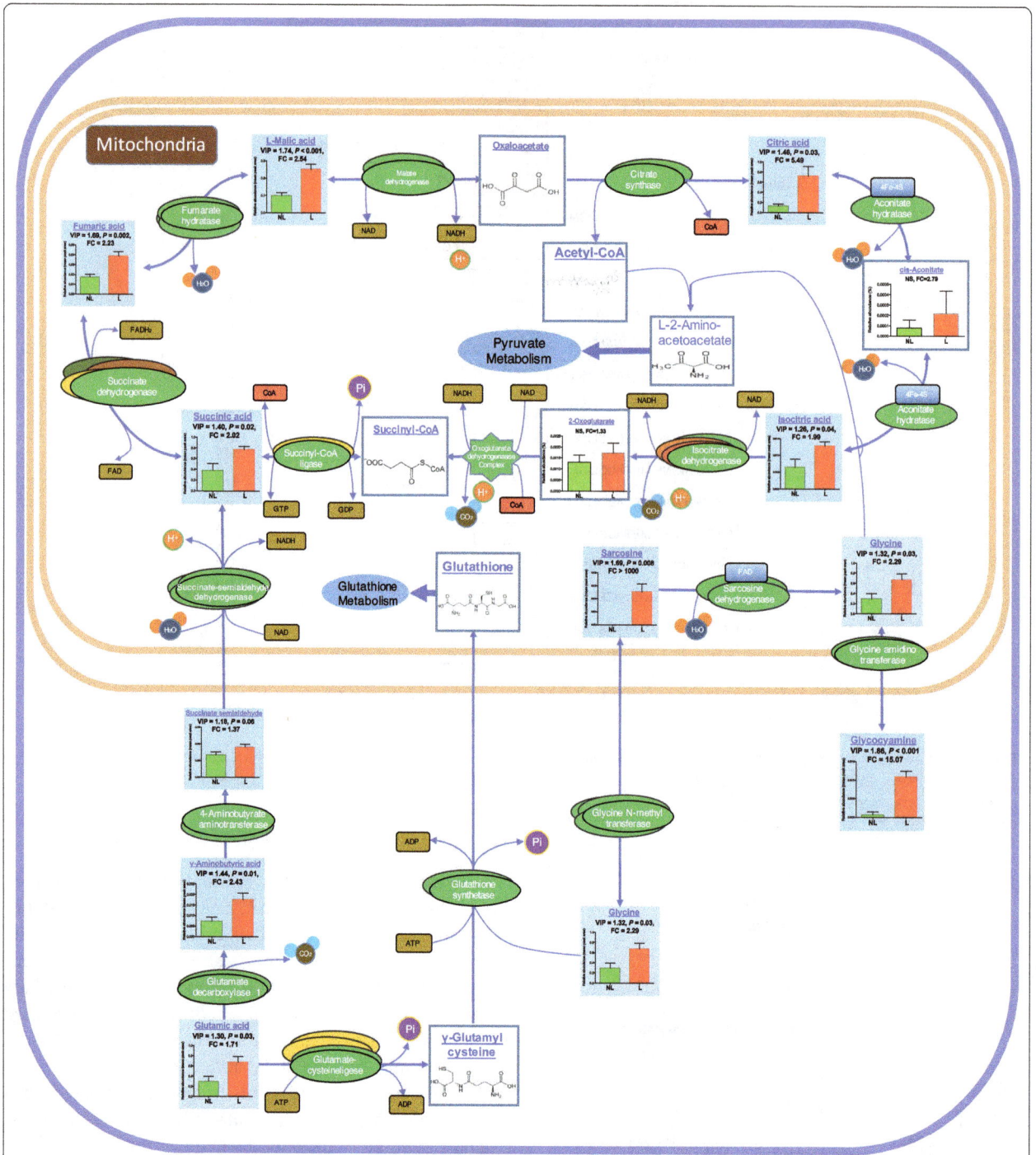

Fig. 8 Integrated metabolomics of the lactating vs non-lactating mammary gland tissue. The figure combines the TCA cycle, glutamate metabolism, glycine biosynthesis and degradation pathways together with the detailed reactions, compounds, enzymes and energy carriers. The differentially expressed metabolites are illustrated by a column chart of 6 lactating cows and 6 non-lactating cows with the VIP (variable importance projection) and P values and fold changes between the groups

body through the urine [24]. It is reasonable to consider that the metabolites in the rumen fluid and serum are similar, whereas those in the urine are the most different from the other biofluids. In this study, creatine was characterized as an important metabolite that crosses four

biofluids that have potential associations with lactation performance and health in dairy cows. Creatine is an important intermediate metabolite in the energy reactions, and its phosphorylated form is an important metabolite that is part of the energy shuttle [29]. In the reverse reaction, a high-energy phosphate group, such as ATP, is transferred to creatine to form phosphocreatine and ADP. There are two ways in which creatine may accumulate in animals due to the continual replacement of lost creatinine [30]: diet and de novo synthesis. Creatine is mainly synthesized in the liver from three amino acids: arginine, glycine and methionine. It is generally used by tissues, absorbed into blood and excreted in urine [31]. The relative concentration across the four biofluids from high to low were serum > urine > milk > rumen fluid, which represented the basic metabolic distribution of this metabolite. Milk creatine can help infants who suffer from defects in the arginine:glycine amidinotransferase and glycine amidinotransferase [32]. It is reported that more creatine is needed to ease the serious negative energy balance in dairy cows [33]. The creatine content is decreased in cows with ovarian inactivity, which may be related to the increased energy consumption in inactive cows [34]. Additionally, creatine was identified as a potential biomarker in diagnosing the heat stress status in dairy cows, which may be attributed to the phosphocreatine in the muscle tissue that have been mobilized for energy supply [35].

MSEA is a group-based approach that does not require the pre-selection of metabolites by using an arbitrary threshold. The key idea behind MSEA is to directly investigate the enrichment of pre-defined groups of functionally related metabolites (or metabolites sets) instead of individual metabolites, which have proven to be successful in deriving new information from the untargeted metabolomics studies [18]. In this study, all of the shared metabolites in the four biofluids were evaluated together, and their related biological information was incorporated into the results. For the mid-lactation dairy cows, the most active and important pathways, including gluconeogenesis, pyruvate metabolism, the TCA cycle, glycerolipid metabolism and aspartate metabolism, which were identified, reflected and confirmed by the overall metabolomics profiles of the four biofluids. Among them, gluconeogenesis, pyruvate metabolism and the TCA cycle are the key pathways of energy metabolism. Gluconeogenesis generates glucose from lactate, glycerol and glucogenic amino acids [36]. Glucose is extremely important for milk synthesis. In dairy cows, gluconeogenesis mostly occurs in the liver and provides up to 90% of the glucose required for host maintenance and production [37]. Pyruvate is the starting point of gluconeogenesis and the end product for glycolysis, and it can also be generated by the transamination of alanine [38]. Almost all of the dietary carbohydrates are fermented to volatile fatty acids (acetate, propionate and butyrate) in the rumen of the dairy cows, with propionate as the predominant substrate for gluconeogenesis [39]. Pyruvate is an important intermediate metabolite for the generation of propionate from the succinic pathway or the lactate pathway [40]. It can be converted by the pyruvate dehydrogenase complex into acetyl-CoA, which can then enter the TCA cycle. The TCA cycle plays a central role in cellular respiration and the supply of energy to all living cells [41], which is of paramount significance to the cell's metabolic efficiency and, therefore, to the cow's metabolism and production [42]. Insufficient energy is highly associated with ketosis (a severe metabolic disease) and can lead to decreased milk production [43]. In the glycerolipid metabolism pathway, glycerol can be converted into glucose by the liver and provide energy for cellular metabolism and lactation maintenance [44]. Aspartate is a precursor of many compounds that are involved in cellular signaling, such as N-acetyl-aspartate. It is also a metabolite in the urea cycle and participates in gluconeogenesis in dairy cows [45]. Based on the overview map of the KEGG pathway, these 5 key metabolic pathways were closely integrated together, suggesting that the carbohydrate pathways and energy pathways play vital roles in regulating lactation maintenance for the dairy cow.

Compared to the increase in the human breast metabolomics studies, to our knowledge, only 1 study has reported the goat MG secretory tissue metabolomics using NMR and identified 46 metabolites, with lactose, glutamate, glycine and lactate being the most abundant [46]. In our study, the most abundant metabolites in the lactating dairy cow MGs were lactic acid, lactobionic acid, oxoproline, alanine, diglycerol, glycine, citric acid, creatine and glutamic acid, of which lactobionic acid and oxoproline were first identified following other studies [27, 47]. The milk metabolomics of the dairy cows mainly focused on the composition and phenotype-related analyses, including distinguishing the formula milk from breast milk [48], comparing the colostrum and milk [49], discerning Holstein cows and other minor dairy animals [27], and analyzing the relationship with heat stress [9] or methane emission [42]. In this study, we also compared the similarities and differences between MG and milk metabolomics from the same lactating dairy cows. However, the four potential biomarkers (citric acid, lactobionic acid, oxamide, and orotic acid), including those metabolites that were found specifically in the MG from lactating cows, were also identified in the milk, which indicates that milk better reflects the physiological status of the MG compared to the non-lactating cows.

In this study, the S-plot combined with the impact pathway analysis was proven to be an effective and easy way to screen most statistically significant metabolites and functional impact pathways. The S-plot indicated that lactobionic acid, citric acid, orotic acid and oxamide (identified only in the lactation group) play vital roles in discriminating between lactating and non-lactating cows.

The TCA cycle, glutamate metabolism and glycine metabolism pathways are the most important and work together in the MG for lactation initiation. Milk component synthesis and secretion is the main transition in lactation initiation. Lactobionic acid is the intermediate metabolite in the lactose biosynthesis pathway, is produced by lactose oxidation and has strong mineral complex properties that serve as promising bioactive ingredients in human nutrition [50]. Citric acid is formed in the TCA cycle or from the diet and participates in the intermediate metabolism of carbohydrate oxidation in animal tissues [51]. The level of citric acid is significantly higher in the lactating dairy cows and may enhance energy by participating in the TCA cycle. Orotic acid is found at a high concentration in bovine milk and almost exclusively originates in the MG cells [52]. Orotic acid has been applied to improve athletic performance and body composition, and it has proven to enhance ATP [53]. Oxamide is a diamide that is derived from oxalic acid and participates in the TCA cycle. In this study, oxamide and other secondary metabolites were significantly higher in the lactating dairy cows, indicating that the TCA cycle is more active during lactation initiation. Oxamide was also identified in the milk in the lactation group, suggesting that the TCA cycle is up-regulated in the lactation group.

Moreover, the functional impact pathway analysis showed that the TCA cycle pathway had the highest P value between the groups, which was mainly attributed to the 7 up-regulated metabolites of the 10 main substrates in the TCA cycle. It has been reported that a critical energy output is required during the initiation of lactation to support milk synthesis and secretion of mammary epithelial cells [54]. Figure 8 depicts the integration of the most significant pathways and the most significant and relevant metabolites, which suggests the different metabolic mechanisms in the MG of the lactating cows compared to the non-lactating cows. Therefore, the downstream metabolites in the glutamic acid pathway entered the TCA cycle to generate succinate in the mitochondria through succinate semialdehyde, thereby releasing NADH. It is well known that NADH is used by the oxidative phosphorylation pathway to generate ATP [55], which can transport chemical energy within the cells for metabolism and provide a larger amount of energy for lactation maintenance and related biological processes. In addition to supplying energy sources, the TCA cycle integrates many other pathways to unify carbohydrate, protein and fat metabolism [56]. For example, citric acid can be transported out of the mitochondrion to produce cytosolic acetyl-CoA for fatty acid and cholesterol (which further synthesizes steroid hormones and vitamin D) synthesis, which is also very important for lactation

maintenance [57]. Another intermediate of glutamate metabolism, γ-glutamyl-cysteine, is used to synthesize glutathione, which is vital for preventing damage to important cellular components of the mammary epithelial cells [58]. As a prime metabolic source of glutathione, creatine, purines and serine and a protein constituent in the lactating mammary epithelial cells, glycine plays a significant role in various biological processes [59]. With respect to the Gly pathway, two metabolites were highlighted: sarcosine and glycocyamine. Sarcosine is an intermediate and byproduct in glycine synthesis and degradation [60]. It can be rapidly degraded into glycine in the mitochondria, which serves as a substrate along with acetyl-CoA to generate L-2-aminoacetoacetate and interacts with the pyruvate metabolism pathway. Glycocyamine is a precursor of creatine, which serves as an essential substrate for muscle energy metabolism and has been used in the feed additive industry to improve production quality [61]. The above results suggest that major energy-related metabolic changes occur in the MG of lactating cows at the initiation of lactation to accommodate the demand for increased milk synthesis nutrients.

Conclusions

Using MG tissue and multi-biofluid metabolomics, the metabolic mechanisms of inherent lactation metabolic patterns and differences with non-lactation periods were addressed in this study. Based on the HCA of the 33 shared metabolites, the rumen fluid and serum were highly correlated and grouped together with milk. Creatine was characterized as the key metabolite to explain the biological variation among the four biofluids. For the lactating cows, lactobionic acid, citric acid, orotic acid and oxamide were identified as potential biomarkers among 54 differentially expressed metabolites. Gluconeogenesis, pyruvate metabolism, the TCA cycle, glycerolipid metabolism and the aspartate metabolism pathways were the most functionally enriched pathways. Integrated analysis of the differentially expressed metabolites involved in the TCA cycle, glutamate metabolism and glycine biosynthesis and degradation pathways revealed the probable key metabolic mechanism in the MG during lactation. Overall, our results provide a better physiological understanding of the lactation metabolism of mid-lactation dairy cows, which can help elucidate the regulated metabolic strategies for the lactating dairy cows in the future. More importantly, the combined application of the multi-biofluids and the tissue metabolomics in this study provide new insights into addressing complex biological questions. Further studies are required to validate the potential biomarkers and pathways identified in this study.

Additional file

Additional file 1: Table S1. Ingredients and nutrient composition of the experimental diet. **Table S2.** The relative abundance of mutual metabolites in rumen fluid, serum, milk and urine. **Table S3.** Mutual metabolites identified in the mammary gland and milk in lactation group. **Table S4** The metabolites identified in the lactation (N) and non-lactation (NL) groups. **Table S5.** Differentially expressed metabolites among the 4 biofluids identified by one-way ANOVA and post-hoc analysis. **Table S6.** Significantly up-regulated pathways in lactation group. (DOCX 70 kb)

Abbreviations

FDR: False discovery rate; FE: Fold enrichment; GC-TOF/MS: Gas chromatography-time of flight/mass spectrometry; HCA: Hierarchical cluster analysis; MG: Mammary gland; MSEA: metabolite set enrichment analysis; OPLS: Orthogonal projections to latent structures; PCA: Principal component analysis; PLS-DA: Partial least squares discriminant analysis; TCA: Tricarboxylic acid cycle; VIP: Variable importance projection

Acknowledgements

The authors gratefully thank the personnel of Hangjiang Dairy Farm (Hangzhou, China) for their assistance in the feeding and care of the animals. We would like to thank B. Wang from the Institute of Dairy Science of Zhejiang University (Hangzhou, China) for assistance in animal trial and sampling.

Funding

This study was supported by the National Natural Science Foundation of China (No. 31472121) and the China Agriculture Research System (No. CARS-37). The funding body has not participated in or interfered with the research.

Authors' contributions

HS, JL and HL contributed to the design of the study and discussed the results. SK helped analysis mammary gland metabolomics data. XW and M X investigated the relationship analysis of four biofluid metabolomics. HS and ZW performed rumen fluid, serum, urine and milk samples collection and collected mammary gland tissue samples. All authors read and approved the final manuscript.

Ethics approval

The procedures of this study were approved by the Animal Care and Use Committee of Zhejiang University, Hangzhou, P. R. China and were in accordance with the university's guidelines for animal research. All the animals used in this study were selected from Hangjiang Dairy Farm (Hangzhou, China).

Competing interests

The authors declare that they have no competing interests.

References

1. Ferreira AM, Bislev SL, Bendixen E, Almeida AM. The mammary gland in domestic ruminants: a systems biology perspective. J Proteome. 2013;94: 110–23.
2. Sejrsen K, Hvelplund T, Nielsen MO. Ruminant physiology: digestion, metabolism and impact of nutrition on gene expression, immunology and stress. The Netherlands: Wageningen Academic Pub; 2006.
3. Johnson CH, Ivanisevic J, Siuzdak G. Metabolomics: beyond biomarkers and towards mechanisms. Nat Rev Mol Cell Biol. 2016;17(7):451–9.
4. Mörén L, Bergenheim AT, Ghasimi S, Brännström T, Johansson M, Antti H. Metabolomic screening of tumor tissue and serum in glioma patients reveals diagnostic and prognostic information. Meta. 2015;5(3):502–20.
5. Ryan D, Newnham ED, Prenzler PD, Gibson PR. Metabolomics as a tool for diagnosis and monitoring in coeliac disease. Metabolomics. 2015; 11(4):980–90.
6. Sun H, Wang B, Wang J, Liu H, Liu J. Biomarker and pathway analyses of urine metabolomics in dairy cows when corn stover replaces alfalfa hay. J Anim Sci Biotechnol. 2016;7(1):49.
7. Menni C, Graham D, Kastenmüller G, Alharbi NH, Alsanosi SM, McBride M, Mangino M, Titcombe P, Shin S-Y, Psatha M. Metabolomic identification of a novel pathway of blood pressure regulation involving hexadecanedioate. Hypertension. 2015;66(2):422–9.
8. Saleem F, Ametaj B, Bouatra S, Mandal R, Zebeli Q, Dunn S, Wishart D. A metabolomics approach to uncover the effects of grain diets on rumen health in dairy cows. J Dairy Sci. 2012;95(11):6606–23.
9. Tian H, Zheng N, Wang W, Cheng J, Li S, Zhang Y, Wang J. Integrated Metabolomics study of the milk of heat-stressed lactating dairy cows. Sci Rep. 2016;6:24208.
10. Huang Q, Tan Y, Yin P, Ye G, Gao P, Lu X, Wang H, Xu G. Metabolic characterization of hepatocellular carcinoma using nontargeted tissue metabolomics. Cancer Res. 2013;73(16):4992–5002.
11. Rawson P, Stockum C, Peng L, Manivannan B, Lehnert K, Ward HE, Berry SD, Davis SR, Snell RG, McLauchlan D. Metabolic proteomics of the liver and mammary gland during lactation. J Proteome. 2012;75(14):4429–35.
12. Li Z, Liu H, Jin X, Lo L, Liu J. Expression profiles of microRNAs from lactating and non-lactating bovine mammary glands and identification of miRNA related to lactation. BMC Genomics. 2012;13(1):1.
13. Wang B, Mao S, Yang H, Wu Y, Wang J, Li S, Shen Z, Liu J. Effects of alfalfa and cereal straw as a forage source on nutrient digestibility and lactation performance in lactating dairy cows. J Dairy Sci. 2014;97(12):7706–15.
14. Apelo SA, Singer L, Lin X, McGilliard M, St-Pierre N, Hanigan M. Isoleucine, leucine, methionine, and threonine effects on mammalian target of rapamycin signaling in mammary tissue. J Dairy Sci. 2014;97(2):1047–56.
15. Shen J, Song L, Sun H, Wang B, Chai Z, Chacher B, Liu J. Effects of corn and soybean meal types on rumen fermentation, nitrogen metabolism and productivity in dairy cows. Asian austral. J Anim Sci. 2015;28(3):351.
16. Sun HZ, Wang DM, Wang B, Wang JK, Liu HY, Guan le L, Liu JX. Metabolomics of four biofluids from dairy cows: potential biomarkers for milk production and quality. J Proteome Res. 2015;14(2):1287–98.
17. Xia J, Sinelnikov IV, Han B, Wishart DS. MetaboAnalyst 3.0—making metabolomics more meaningful. Nucleic Acid Res. 2015;43(W1):W251–W57.
18. Xia J, Wishart DS. MSEA: a web-based tool to identify biologically meaningful patterns in quantitative metabolomic data. Nucleic Acid Res. 2010;38(suppl 2):W71–7.
19. Abbasi A, Hossain L, Leydesdorff L. Betweenness centrality as a driver of preferential attachment in the evolution of research collaboration networks. J Inf Secur. 2012;6(3):403–12.
20. Xia J, Wishart DS. MetPA: a web-based metabolomics tool for pathway analysis and visualization. Bioinformatics. 2010;26(18):2342–4.
21. Wang D, Liang G, Wang B, Sun H, Liu J, Guan LL. Systematic microRNAome profiling reveals the roles of microRNAs in milk protein metabolism and quality: insights on low-quality forage utilization. Sci Rep. 2016;6:21194.
22. Drackley J, Donkin S, Reynolds C. Major advances in fundamental dairy cattle nutrition. J Dairy Sci. 2006;89(4):1324–36.
23. Garnsworthy PC: Nutrition and lactation in the dairy cow: Elsevier; 2013.
24. Loor JJ, Bionaz M, Invernizzi G. Systems biology and animal nutrition: insights from the dairy cow during growth and the lactation cycle. Syst biol. Livest Sci. 2011:215–46.
25. Mao SY, Huo WJ, Zhu WY. Microbiome–metabolome analysis reveals unhealthy alterations in the composition and metabolism of ruminal microbiota with increasing dietary grain in a goat model. Environ Microbiol. 2016;18(2):525–41.
26. Kenéz Á, Dänicke S, Rolle-Kampczyk U, von Bergen M, Huber K. A metabolomics approach to characterize phenotypes of metabolic transition from late pregnancy to early lactation in dairy cows. Metabolomics. 2016;12(11):165.

27. Yang Y, Zheng N, Zhao X, Zhang Y, Han R, Yang J, Zhao S, Li S, Guo T, Zang C. Metabolomic biomarkers identify differences in milk produced by Holstein cows and other minor dairy animals. J Proteome. 2016;136:174–82.

28. Wang X, Zhang A, Han Y, Wang P, Sun H, Song G, Dong T, Yuan Y, Yuan X, Zhang M. Urine metabolomics analysis for biomarker discovery and detection of jaundice syndrome in patients with liver disease. Mol Cell Proteomics. 2012;11(8):370–80.

29. Mercimek-Mahmutoglu S, Stoeckler-Ipsiroglu S, Adami A, Appleton R, Araújo HC, Duran M, Ensenauer R, Fernandez-Alvarez E, Garcia P, Grolik C. GAMT deficiency features, treatment, and outcome in an inborn error of creatine synthesis. Neurol. 2006;67(3):480–4.

30. Wyss M, Kaddurah-Daouk R. Creatine and creatinine metabolism. Physiol Rev. 2000;80(3):1107–213.

31. Poortmans JR, Kumps A, Duez P, Fofonka A, Carpentier A, Francaux M. Effect of oral creatine supplementation on urinary methylamine, formaldehyde, and formate. Med Sci Sport Exer. 2005;37(10):1717.

32. Schulze A. Creatine deficiency syndromes. Mol Cell Biochem. 2003;244(1-2): 143–50.

33. Wang Y, Gao Y, Xia C, Zhang H, Qian W, Cao Y. Pathway analysis of plasma different metabolites for dairy cow ketosis. Ital J Anim Sci. 2016; 15(3):545–51.

34. Xu C, Xia C, Sun Y, Xiao X, Wang G, Fan Z, Shu S, Zhang H, Xu C, Yang W. Metabolic profiles using 1H-NMR spectroscopy in postpartum dairy cows with ovarian inactivity. Theriogenology. 2016;86(6):1475–81.

35. Tian H, Wang W, Zheng N, Cheng J, Li S, Zhang Y, Wang J. Identification of diagnostic biomarkers and metabolic pathway shifts of heat-stressed lactating dairy cows. J Proteome. 2015;125:17–28.

36. Nelson DL, Lehninger AL, Cox MM: Lehninger principles of biochemistry: Macmillan; 2008.

37. Aschenbach JR, Kristensen NB, Donkin SS, Hammon HM, Penner GB. Gluconeogenesis in dairy cows: the secret of making sweet milk from sour dough. IUBMB Life. 2010;62(12):869–77.

38. Denton R, Halestrap A. Regulation of pyruvate metabolism in mammalian tissues. Essays Biochem. 1978;15:37–77.

39. Zhang Q, Koser SL, Bequette BJ, Donkin SS. Effect of propionate on mRNA expression of key genes for gluconeogenesis in liver of dairy cattle. J Dairy Sci. 2015;98(12):8698–709.

40. Jeyanathan J, Martin C, Morgavi D. The use of direct-fed microbials for mitigation of ruminant methane emissions: a review. Animal. 2014; 8(02):250–61.

41. Grassian AR, Parker SJ, Davidson SM, Divakaruni AS, Green CR, Zhang X, Slocum KL, Pu M, Lin F, Vickers C. IDH1 mutations alter citric acid cycle metabolism and increase dependence on oxidative mitochondrial metabolism. Cancer Res. 2014;74(12):3317–31.

42. Antunes-Fernandes E, van Gastelen S, Dijkstra J, Hettinga K, Vervoort J. Milk metabolome relates enteric methane emission to milk synthesis and energy metabolism pathways. J Dairy Sci. 2016;99(8):6251–62.

43. Bravo D, Wall E. The rumen and beyond: nutritional physiology of the modern dairy cow1. J Dairy Sci. 2016;99(6):4939–40.

44. Doran AG, Berry DP, Creevey CJ. Whole genome association study identifies regions of the bovine genome and biological pathways involved in carcass trait performance in Holstein-Friesian cattle. BMC Genomics. 2014;15(1):1.

45. Piccioli-Cappelli F, Loor J, Seal C, Minuti A, Trevisi E. Effect of dietary starch level and high rumen-undegradable protein on endocrine-metabolic status, milk yield, and milk composition in dairy cows during early and late lactation. J Dairy Sci. 2014;97(12):7788–803.

46. Palma M, Hernández-Castellano LE, Castro N, Arguëllo A, Capote J, Matzapetakis M, de Almeida AM. NMR-metabolomics profiling of mammary gland secretory tissue and milk serum in two goat breeds with different levels of tolerance to seasonal weight loss. Mol BioSyst. 2016;12:2094–107.

47. Pisano MB, Scano P, Murgia A, Cosentino S, Caboni P. Metabolomics and microbiological profile of Italian mozzarella cheese produced with buffalo and cow milk. Food Chem. 2016;192:618–24.

48. Scano P, Murgia A, Demuru M, Consonni R, Caboni P. Metabolite profiles of formula milk compared to breast milk. Food Res Int. 2016;87:76–82.

49. Curtasu M, Theil P, Hedemann M. Metabolomic profiles of colostrum and milk from lactating sows. J Anim Sci. 2016;94(7supplement3):272–5.

50. Schaafsma G. Lactose and lactose derivatives as bioactive ingredients in human nutrition. Int Dairy J. 2008;18(5):458–65.

51. Krebs HA, Johnson W. The role of citric acid in intermediate metabolism in animal tissues. FEBS Lett. 1980;117(S1):2–10.

52. Saidi B, Warthesen J. Analysis and stability of orotic acid in milk. J Dairy Sci. 1989;72(11):2900–5.

53. Loeffler M, Carrey EA, Zameitat E. Orotic acid, more than just an intermediate of Pyrimidine de novo synthesis. J Genet Genomics. 2015;42(5):207–19.

54. Chen C-C, Stairs DB, Boxer RB, Belka GK, Horseman ND, Alvarez JV, Chodosh LA. Autocrine prolactin induced by the Pten–Akt pathway is required for lactation initiation and provides a direct link between the Akt and Stat5 pathways. Genes Dev. 2012;26(19):2154–68.

55. Malloy CR, Sherry AD, Jeffrey F. Evaluation of carbon flux and substrate selection through alternate pathways involving the citric acid cycle of the heart by 13C NMR spectroscopy. J Biol Chem. 1988;263(15):6964–71.

56. He W. Miao FJ-P, Lin DC-H, Schwandner RT, Wang Z, Gao J, Chen J-L, Tian H, Ling L: citric acid cycle intermediates as ligands for orphan G-protein-coupled receptors. Nature. 2004;429(6988):188–93.

57. Akers RM. Lactation and the mammary gland. Ames: Wiley; 2016.

58. Kannan N, Nguyen LV, Makarem M, Dong Y, Shih K, Eirew P, Raouf A, Emerman JT, Eaves CJ. Glutathione-dependent and-independent oxidative stress-control mechanisms distinguish normal human mammary epithelial cell subsets. P Natl Acad Scd USA. 2014;111(21):7789–94.

59. Shennan D, McNeillie S, Curran D. The effect of a hyposmotic shock on amino acid efflux from lactating rat mammary tissue: stimulation of taurine and glycine efflux via a pathway distinct from anion exchange and volume-activated anion channels. Exp Physiol. 1994;79(5):797–808.

60. Khan AP, Rajendiran TM, Bushra A, Asangani IA, Athanikar JN, Yocum AK, Mehra R, Siddiqui J, Palapattu G, Wei JT. The role of sarcosine metabolism in prostate cancer progression. Neoplasia. 2013;15(5):491–IN13.

61. Ostojic SM. Advanced physiological roles of guanidinoacetic acid. Eur J Nutr. 2015;54(8):1211–5.

8

Differences in global gene expression in muscle tissue of Nellore cattle with divergent meat tenderness

Larissa Fernanda Simielli Fonseca[1*], Daniele Fernanda Jovino Gimenez[1], Danielly Beraldo dos Santos Silva[1], Roger Barthelson[2], Fernando Baldi[1], Jesus Aparecido Ferro[1] and Lucia Galvão Albuquerque[1]

Abstract

Background: Meat tenderness is the consumer's most preferred sensory attribute. This trait is affected by a number of factors, including genotype, age, animal sex, and pre- and post-slaughter management. In view of the high percentage of Zebu genes in the Brazilian cattle population, mainly Nellore cattle, the improvement of meat tenderness is important since the increasing proportion of Zebu genes in the population reduces meat tenderness. However, the measurement of this trait is difficult once it can only be made after animal slaughtering. New technologies such as RNA-Seq have been used to increase our understanding of the genetic processes regulating quantitative traits phenotypes. The objective of this study was to identify differentially expressed genes related to meat tenderness, in Nellore cattle in order to elucidate the genetic factors associated with meat quality. Samples were collected 24 h postmortem and the meat was not aged.

Results: We found 40 differentially expressed genes related to meat tenderness, 17 with known functions. Fourteen genes were up-regulated and 3 were down-regulated in the tender meat group. Genes related to ubiquitin metabolism, transport of molecules such as calcium and oxygen, acid-base balance, collagen production, actin, myosin, and fat were identified. The PCP4L1 (Purkinje cell protein 4 like 1) and BoLA-DQB (major histocompatibility complex, class II, DQ beta) genes were validated by qRT-PCR. The results showed relative expression values similar to those obtained by RNA-Seq, with the same direction of expression (i.e., the two techniques revealed higher expression of PCP4L1 in tender meat samples and of BoLA-DQB in tough meat samples).

Conclusions: This study revealed the differential expression of genes and functions in Nellore cattle muscle tissue, which may contain potential biomarkers involved in meat tenderness.

Keywords: RNA-Seq, Transcriptome, Meat quality

Background

Meat quality traits in Brazilian animal breeding programs have not been fully explored because of the late expression of these attributes and the complex evaluation that can only be made after slaughter. Furthermore, on the domestic market, producers are generally not paid for meat quality, a fact that diminishes interest in improving meat quality traits and has hindered their inclusion in traditional selection objectives. In contrast, on international markets, meat tenderness is one of the most valued traits [39], a fact that highlights the importance towards improving this trait since Brazil is one of the world's largest beef exporters.

Meat tenderness is the preferred sensory attribute of consumers [7]. According to Scollan et al. [46], the European food industry has sought to improve this trait to gain market share over other types of food. In Brazil, about 80% of the cattle herd consists of Zebu animals or their crossbreeds [1]. In this respect, the improvement of meat tenderness becomes important since Ferguson et al. [17] have shown that the higher the proportion of Zebu genes in a population, mainly Nellore cattle, the less tender the meat. Meat tenderness can only be measured after slaughter making this trait more complex to select animals.

* Correspondence: la_simielli@yahoo.com.br
[1]Faculty of Agricultural and Veterinary Sciences, São Paulo State University, FCAV/UNESP, Jaboticabal, São Paulo, Brazil
Full list of author information is available at the end of the article

Thus, alternative tools are useful to include meat tenderness in animal breeding programs [9].

Modern recently developed large-scale RNA sequencing technologies (RNA-Seq) have been useful in understanding the genetic and physiological processes that regulate the phenotype of quantitative traits in a certain situation [34]. RNA-Seq permits analysis of the transcriptional profiles of cells, tissues or organs in a certain situation and the discovery of known and unknown genes involved in a given cellular process [57]. This new technique can be used to identify novel potential molecular markers that permit more accurate and early genetic predictions [51], with a consequent reduction in the generation interval that would contribute to the improvement of difficult-to-measure traits such as meat tenderness.

RNA-Seq has been widely used in recent studies to investigate differentially expressed genes related to meat tenderness in different species. For example, genes related to the degradation of filamins, lipogenesis and collagen synthesis have been identified in a study on meat tenderness in broiler chickens [40]. Gonçalves [20] found genes related to metabolic pathways involved in apoptosis, calcium transport, proteolysis and ribosome synthesis in castrated Nellore cattle, classified as extreme for tenderness based on estimated breeding values for shear force measured after 14 days of aging. Bongiorni et al. [8], who studied gene expression in longissimus dorsi muscle of animals of two Italian beef breeds (Maremmana and Chianina) representing the extremes for meat tenderness, detected differentially expressed genes related to growth and sodium-potassium pumps, among others.

Despite the above-mentioned publications, studies investigating differentially expressed genes related to meat tenderness in cattle are rare. In this respect, the better understanding and identification of the transcripts and biological processes, associated with this complex and economically important trait, will permit to highlight genes that could contain potential biomarkers involved in meat tenderness.

The objective of this study was to identify genes differentially expressed in muscle tissue (longissimus dorsi) of Nellore cattle with divergent meat tenderness using RNA-Seq in order to obtain data that increase our understanding of the genetic and metabolic mechanisms underlying this trait.

Results
RNA sequencing, alignment, and assembly of the transcripts
The TopHat2 program identified a total of 942 million reads (2×100 bp) and the sequencing coverage was 63X (coverage for all transcripts of all samples). An average of almost 24 million reads were obtained per sample and 88.3% of the reads were mapped. For tender meat group, an average of 24,928,506 (89%) million

reads were mapped, while for tough meat group, an average of 22,170,021 million reads were mapped (89%) (Additional file 1: Table S1).

We found transcript for 28,059 genes and 103,309 potential new isoforms.

To evaluate the quality of sequencing, the expression profiles of the Glucuronidase Beta (GUSB), erythrocyte hydroxymethylbilane synthase (HMBS), Hypoxanthine Phosphoribosyltransferase 1 (HPRT1), phosphoglycerate kinase 1 (PGK1) and TATA-Box Binding Protein (TBP) genes were analyzed, which exhibited a similar expression profile in the tender and tough meat groups (Additional file 2: Figure S1).

A box plot (Additional file 3: Figure S2) containing the transformed FPKM values ($\log_{10}$) for each group and the plot of principal component analysis (PCA) (Additional file 4: Figure S3) were constructed using the cummerRbund package. As can be seen in the box plot, the distribution of quartiles was consistent between groups, indicating high quality of the data. In addition, the medians were similar in the two groups and close to -1, indicating that the level of sequencing coverage permitted the identification of low-expressed genes [11, 51]. PCA showed the formation of different groups (tender and tough meat), indicating differences in the expression of genes between the tender and tought meat groups'.

Analysis of differentially expressed genes
Analysis of differential expression in the tender and tough meat groups revealed 40 differentially expressed genes (q-value <0.05) (Table 1). Seventeen of these genes have a known function. The $\log_2$ signal (fold change) was used was used to partition the DE genes into up- and down-regulated groups. In this analysis, 35 genes were found to be up-regulated and 5 were down-regulated in relation to the tough meat group. Among the genes with known function, 14 were up-regulated and 3 were down-regulated.

Combined functional annotation using all differentially expressed (up- and down-regulated) genes for meat tenderness was performed with the DAVID v6.7 database using *Bos taurus* as a reference. This analysis permitted the identification of seven functional groups (annotation clusters; Additional file 5: Table S2). These genes were classified according to their function: cell fraction (GO: 000267), cell junction (GO: 0030054), intrinsic component of membrane (GO: 0031224), regulation of cell communication (GO: 0010646), catalytic activity (GO: 0003824), organelles (GO: 0043226), and binding (GO: 0005488), among others.

Using the ClueGO plug-in, the differentially expressed transcripts HMOX1, AT2, CLDN19, CLEC4G, CLEC12A, PNP and SYP were found to be inter-related through biological processes (cell communication, regulation of response to stimuli), molecular function (binding proteins), or cell components (integral membrane component)

Table 1 Differentially expressed genes detected in the samples divergent for meat tenderness

Gene	Locus	Tender	Tough	log2 (fold change)[a]	p-value	q-value
XLOC_002455	10:2,270,299–2,271,226	213.386	0.108797	429.375	0.0001	0.0078
XLOC_014455	21:61,587,088–61,588,096	194.026	0.110438	413.493	0.0001	0.0078
XLOC_002491	10:20,795,303–20,795,839	49.187	0.291183	407.828	0.0001	0.0078
IQCG	1:70,792,451–70,838,105	0.975199	0.0683124	383.548	0.0001	0.0078
CLEC4G	7:17,785,971–17,789,026	213.916	0.152397	381.114	0.0002	0.0251
EXOSC2	11:101,010,350–101,010,501	146.306	109.334	374.217	0.0001	0.0078
DMGDH	10:9,994,009–10,066,954	0.700546	0.0710559	330.145	0.0001	0.0078
XLOC_029184	GJ060137.1:2576–3597	194.018	0.224785	310.957	0.0001	0.0078
CLDN19	3:104,091,609–104,096,449	112.997	0.133369	308.279	0.0001	0.0078
XLOC_006476	14:27,940,730–27,941,920	173.549	0.233603	289.322	0.0001	0.0078
USP32	19:13,087,794–13,088,565	293.099	0.401415	286.822	0.0001	0.0078
CTNNB1	22:13,897,328–13,898,816	129.849	0.179967	285.103	0.0001	0.0078
AT2	X:1,024,840–1,027,901	112.057	0.163197	277.955	0.0001	0.0078
XLOC_022848	4:115,985,386–115,986,391	206.906	0.302608	277.346	0.0001	0.0078
XLOC_028948	9:32,650,968–32,653,108	0.903283	0.135688	273.489	0.0001	0.0078
CLEC12A	5:100,388,588–100,401,353	205.909	0.31943	268.844	0.0001	0.0078
XLOC_024373	5:47,857,813–47,858,488	368.063	0.591044	263.861	0.0001	0.0078
ZKSCAN2	25:23,091,496–23,104,117	0.689172	0.130967	239.566	0.0001	0.0078
XLOC_022865	4:119,488,532–119,490,812	0.857893	0.171544	232.222	0.0001	0.0078
XLOC_022836	4:114,608,102–114,609,417	166.248	0.334759	231.214	0.0001	0.0078
SYP	X:92,308,615–92,320,415	0.814972	0.169198	226.804	0.0001	0.0078
XLOC_029021	9:76,153,645–76,154,023	105.826	23.362	217.945	0.0001	0.0078
XLOC_002519	10:28,447,880–28,449,134	173.421	0.402111	210.862	0.0001	0.0136
XLOC_021066	3:20,773,659–20,775,848	100.205	0.252969	198.593	0.0001	0.0078
XLOC_001046	1:150,301,045–150,303,606	0.781722	0.211096	188.876	0.0001	0.0136
ASAH1	27:18,312,322–18,314,792	0.924241	0.265766	179.811	0.0001	0.0136
XLOC_018808	28:17,838,746–17,847,124	120.255	0.353753	176.528	0.0001	0.0136
XLOC_001185	1:91,928,092–91,930,927	0.775044	0.232207	173.887	0.0001	0.0078
TCF7L1	11:49,714,024–49,715,693	931.924	299.392	163.818	0.0001	0.0078
XLOC_022736	4:59,964,007–59,966,831	0.820647	0.270615	160.052	0.0002	0.0198
XLOC_001280	1:151,442,513–151,443,449	221.613	0.163023	37.649	0.0001	0.0078
ENSBTAG00000022947	21:21,965,065–21,966,315	156.027	0.257492	25.992	0.0001	0.0078
LOC511981	14:58,850,047–58,851,086	300.014	108.315	14.698	0.0001	0.0136
PCP4L1	3:8,249,980–8,276,724	179.755	0.413335	2.120	0.0001	0.0078
BoLA-DQB	23:25,855,145 25,863,052	122.213	682.692	−0.840093	0.0001	0.0078
HMOX1	5:73,980,699–73,987,841	115.512	197.874	−0.776539	0.0002	0.0251
PNP	6:87,555,287–87,556,157	163.788	543.206	−159.226	0.0001	0.0078
XLOC_004829	13:42,396,473–42,396,753	241.093	676.471	−183.349	0.0001	0.0078
XLOC_029070	9:102,893,791–102,894,054	21.503	101.088	−441.086	0.0001	0.0078

Symbol of the differentially expressed gene (Gene), location of the gene in the *Bos taurus* genome, FPKM values obtained for tender and tough meat, relative expression (log$_2$ fold change), p-value, and q-value
[a] The fold-change estimates (relative expression) refer to the tender meat group

(Fig. 1). The gene HMOX1 was expressed more in tough meat group, while the other five genes were expressed more in tender meat group. The proteins encoded by these transcripts are involved in the transport of molecules such as sodium, potassium, calcium, and oxygen [15, 29, 36, 43].

Using the same programs, the DMGDH gene (dimethylglycine dehydrogenase) was identified as a member of the "glycine, serine and threonine metabolism" pathway (Fig. 2). Glycine makes up about one-third of the helical polypeptide chains of collagen [30]. On the other hand, according to Bailey [2], collagen is degraded by serine proteases, with serine also being part of the glycine metabolic pathway, and by cysteine proteases whose metabolic pathway ("cysteine and methionine metabolism") is associated with the DMGDH pathway. In the present study, the transcript of this gene was expressed more in tender meat.

Figure 3 illustrates the interrelationships between the TCF7L1, EXOSC2, DMGDH and ASAH1 transcripts obtained by enrichment analysis. This analysis shows that the main link between these genes is the cell component called "intracellular membrane-bound organelle". This category refers to structures found inside the cell such as the nucleus and mitochondria [10]. Gene expression analysis in Angus cattle also showed a relationship between meat tenderness and this cell component category [59]. The genes identified in this study are related to actin-myosin assembly, collagen synthesis, lipid accumulation, and serine and glycine metabolic pathways [2, 22, 30, 38].

Validation of differentially expressed genes

The relative expression values ($\log_2$) of the transcripts were similar for the two techniques used, RNA-Seq and qRT-PCR, with values of 2.12 and 2.03 (standard deviation = 0.89) for PCP4L1 and of −0.84 and −0.644 (standard deviation = 0.44) for BoLA-BQD, respectively (Fig. 4). Similar to the RNA-Seq analysis, higher expression of the PCP4L1 and BoLA-BQD genes was observed in the tender and tough meat groups, respectively. Thus, these transcripts showed similar patterns of mRNA abundance in the RNA-Seq and qRT-PCR analyses, with the same direction of expression (i.e., up-regulated and down-regulated, respectively, in relation to the tender meat group).

Discussion

A higher proportion of Zebu genes in cattle herds considerably reduces meat tenderness when compared to taurine breeds. In Brazil, the herd mostly consists in Zebu cattle, mainly Nellore, then improve meat tenderness is very important, because for the beef export market, in which Brazil plays an important role, tenderness is paramount in determining the value of the product.

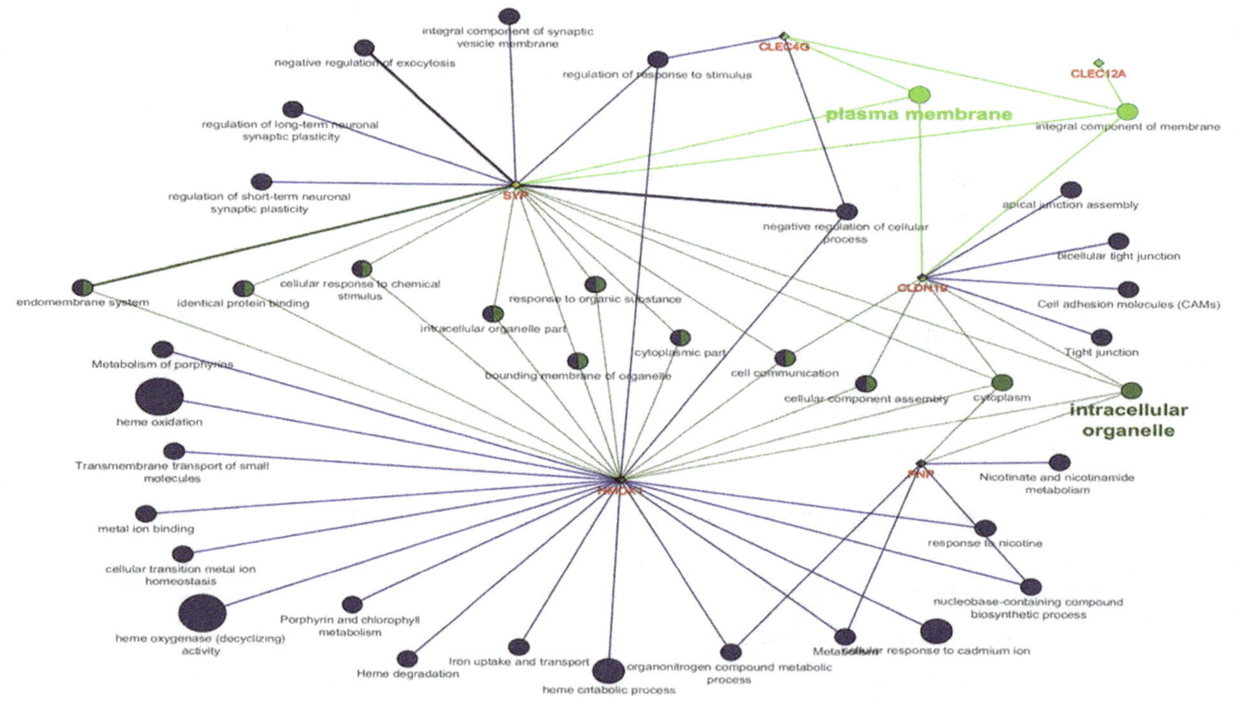

Fig. 1 Enrichment analysis of the HMOX1, CLDN19, CLEC4G, CLEC12A, PNP and SYP genes using the ClueGO plug-in of the Cytoscape program. Note the interrelationships between these genes, which are related to the transport of molecules

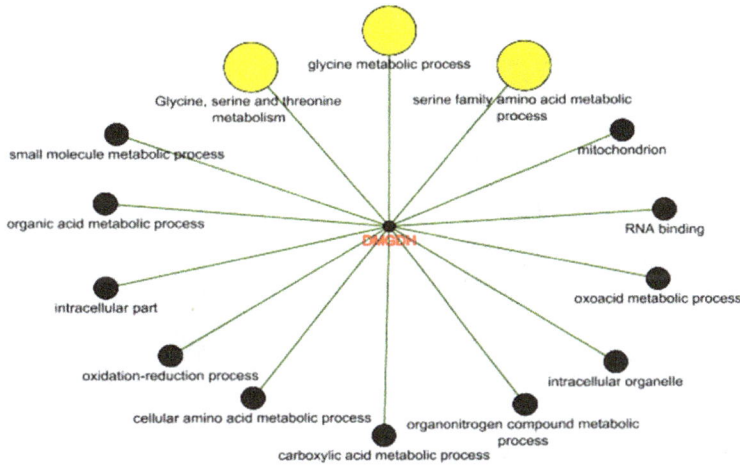

Fig. 2 Enrichment analysis of the DMGDH gene with the ClueGO plug-in. The yellow circlels highlight the biological processes and serine and glycine metabolic pathway in which this gene is involved

Gene expression studies have been used as a tool to identify gene candidates and metabolic pathways related to traits of economic interest. In the present study, the USP32 (ubiquitin specific peptidase 32) transcript was expressed more in tender meat. Members of the ubiquitin-proteasome system are important during the transformation of muscle into meat. These proteins are involved in proteolysis, causing the degradation of myofibrillar proteins in muscle cells [47].

In a genome-wide association study (GWAS) of Nellore cattle using different meat tenderness measures, Tizioto et al. [52] identified genes of the USP family, including USP32. Another study on cattle also associated genes of the USP family with meat tenderness. In Wagiu cattle, the USP2 gene was strongly associated with meat tenderness [12] and gene expression analysis in Nellore cattle showed that the USP2 gene was expressed more in tender meat samples [20].

The functional categories cell junction, regulation of cell communication and intrinsic component of membrane are related to the binding, communication and transport of molecules between cells [10]. Among the transcripts related to these categories, CTNNB1 (catenin - cadherin-associated protein beta 1), which was expressed more in tender meat, is involved in the same metabolic pathway as actin and myosin. Actin and myosin are the proteins found in thin and thick myofilaments, respectively, which form the myofibril that is responsible for muscle contraction. These proteins are the most abundant in the mechanism of muscle contraction, accounting for 52 to 56% of all muscle proteins [48].

Each actin filament binds to the plasma membrane of the cell through a structure, called focal contact. This structure consists of binding proteins and of a transmembrane protein that are products of the "focal adhesion" pathway to which the CTNNB1 and TCF7L1 (transcription factor 7

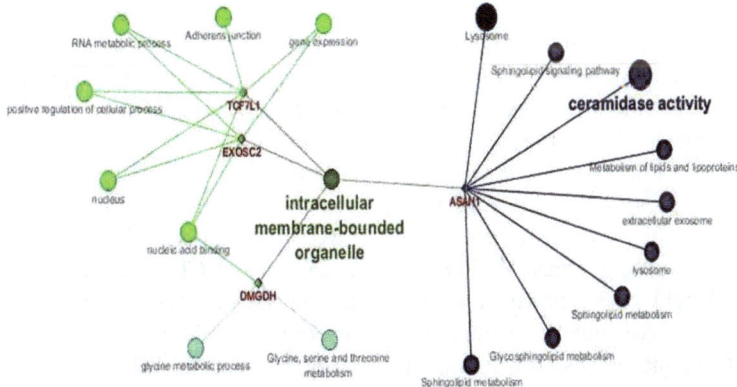

Fig. 3 Enrichment analysis of the TCF71, EXOSC2, DMGDH and ASAH1 genes with the ClueGO plug-in

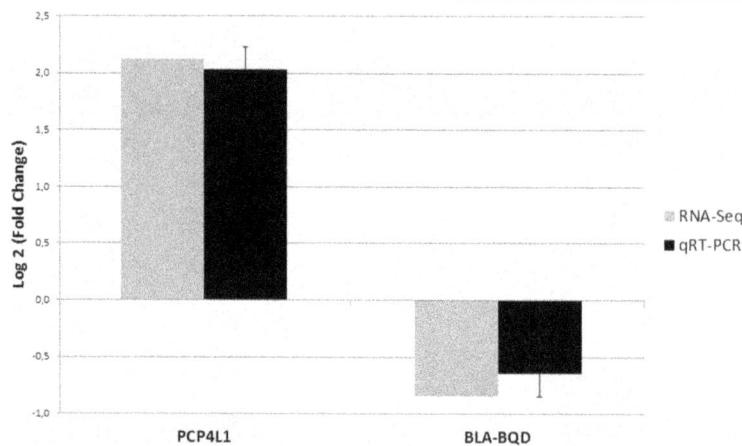

Fig. 4 Comparison of the relative expression values of two differentially expressed transcripts obtained by RNA-Seq and qRT-PCR

like 1) genes belong. On the outer side of the cell, in the extracellular matrix, the transmembrane protein binds to a collagen fiber [14, 23]. According to Bailey [2], a direct association exists between collagen content and the toughening of meat. However, in the present study, the CTNNB1 and TCF7L1 transcripts were expressed more in tender meat.

The SYP (synaptophysin) transcript, which was expressed more in tender meat, encodes an integral membrane protein found in small synaptic vesicles. In a study on rats, [44] showed that the phosphorylation of synaptophysin is calcium dependent. The authors observed a four-fold increase in serine phosphorylation of synaptophysin in the presence of the calmodulin-calcium complex. According to Bailey et al. [2], serine proteases are responsible for the degradation of collagen, which, in turn, directly influences meat tenderness. In addition, calcium is essential for muscle contraction by acting as a catalyst of enzymatic proteolytic activity, which is directly related to the process of meat tenderization [37].

The AT2 transcript, which encodes angiotensin II, was expressed more in tender meat. This protein is involved in vasoconstriction and regulates the secretion of aldosterone, which, in turn, stimulates the reabsorption of sodium by the kidneys. In this respect, after slaughter and during bleeding, angiotensin is activated to restore blood pressure. The result of these stimuli is the depolarization of the cell membrane, altering the distribution of sodium and potassium, in addition to permitting the flow of calcium ions [43]. In a study on crossbred cattle (Luxi-Simmental), Zhong-Liang et al. [60] observed a decline in shear force after the injection of angiotensin II into the carcass for 7 days after slaughter. Bongiorni et al. [8], studying gene expression in longissimus dorsi muscle of Italian Maremmana and Chianina breeds, also found the differential expression of genes to be related to sodium and potassium flow.

The functional category "catalytic activity" is related to increases in the velocity of a biochemical reaction at physiological temperatures [10]. Some reactions that occur during the postmortem period depend on calcium and cellular pH, which decrease in the first 24 h after slaughter [25]. A member of this functional category is ASAH1 (N-acylsphingosine amidohydrolase (acid ceramidase) 1), which belongs to a family of hydrolases that catalyze the synthesis and degradation of ceramide into sphingolipid and free fatty acid and are acid pH dependent [32]. A genetic deficiency in ASAH1 that reduces its catalytic activity causes a lysosomal sphingolipid storage disorder characterized by the accumulation of lipids in cells and tissues throughout the organism [38]. ASAH1 also belongs to the "sphingolipid signaling pathway" and "sphingolipid metabolism" categories in which serine is also involved, with serine protease degrading collagen [2]. Thus, ASAH1, which was expressed more in tender meat, may be related to the process of meat tenderization.

Another member of the "catalytic activity" category is HMOX1 (heme oxygenase 1), which was expressed more in tough meat. This gene encodes a protein involved in the metabolism of porphyrins, molecules whose catalytic activity is activated by iron [35]. Porphyrins are precursors of hemes, the main components of hemoglobin, myoglobin and cytochromes which are responsible for the transport of oxygen and electrons in tissues [36].

The C-type lectin (CLEC) family comprises calcium-dependent carbohydrate-binding protein domains that are involved in cell-cell adhesion [15]. In the present study, the CLEC4G and CLEC12A transcripts were expressed more in tender meat. GWAS in Nellore cattle demonstrated an association of the CLEC12A gene with different meat tenderness measures [52].

The IQCG transcript (IQ motif containing G), which was expressed more in tender meat, encodes a protein that functions as a binding site for different proteins, including myosin light chains and calmodulins. Calmodulin phosphorylates myosin, a process that permits the sliding of fibers and muscle contraction. In this case, calcium present in the reaction, binds to calmodulin, attached to IQ motif, and stimulates the ATPase activity of myosin [42]. According to Duston [16], in addition to factors such as collagen content, the structure and state of contraction of myofibrils (which mainly consists of myosin and actin) directly affect meat tenderness.

The protein encoded by the PNP transcript (purine nucleoside phosphorylase), which was expressed more in tough meat, plays a role in nicotinate and nicotinamide metabolism. Nicotinate (niacin or vitamin B3) is a precursor of NAD^+ and $NADP^+$ coenzymes, which are essential for the production of ATP in the cell [28].

Numerous structural changes and biochemical events occur in the first 24 h after slaughter of the animal, which are responsible for the conversion of muscle into meat [25]. In the early postmortem stages, ATP levels are maintained constant by the conversion of ADP plus phosphocreatine into ATP and oxygen supply ceases because of the cessation of blood circulation. At this stage, slow production of lactate is observed and the onset of rigor mortis occurs (slow phase). The decrease in phosphocreatine levels characterizes the rapid phase, which consists of a rapid decline in available ATP that is used as an energy reserve after the consumption of glycogen and other carbohydrates and is therefore hydrolyzed again to ADP. The scarcity of ATP during this phase is accompanied by the release of calcium ions into the myofibrillar space, which causes muscle shortening with a direct influence on meat tenderness [5].

Another event that occurs during this phase is the anaerobic conversion of glycogen into glucose, producing lactate and reducing the pH of the medium. In addition, the transport of sodium and potassium across the cell membrane, which uses the energy released by the hydrolysis of ATP into ADP, is impaired because it occurs against the concentration gradient. The protons generated during the hydrolysis of ATP into ADP cause a significant decline in intracellular pH [3]. According to Darrel et al. [13], this drop in pH directly influences the final tenderness of meat, especially during the process of aging.

According to Koohmaraie et al. [26], calcium is responsible for the activation of calpains and calpastatins (calcium-dependent cysteine proteases) and calpain I has been shown to be the main enzyme responsible for postmortem tenderization of meat by degrading cytoskeletal proteins that confer the structural integrity of the myofibrillar matrix. Nevertheless, in the present study, the calpain and calpastatin genes were not differentially expressed in the tender and tough meat groups. This finding might be explained by the fact that the amount of calpastatin in cells is higher 24 h after slaughter [43]

and in this study the samples were collected immediately after cleaning the carcasses. Other GWAS and gene expression studies of muscle tissue in Nellore cattle also found no relationship between meat tenderness and calpain or calpastatin [20, 52].

The EXOSC2 transcript, which encodes exosome component 2, was expressed more in tender meat. According to Jong et al. [22], this gene is related with collagen activity in humans. This found could indicated a relationship between this genes and collagen activity in bovines, because there is a direct association exists between collagen content and the toughening of meat [2].

The ZKSCAN2 transcript (zinc finger with KRAB and SCAN domains 2), which was expressed more in tender meat, is vertebrate specific and synthesizes zinc finger proteins that bind through an N-terminus to the SCAN domain (dimerization motif). The function of this gene is not well known, but zinc finger proteins have been associated with the regulation of growth factor transcription and lipid metabolism [45].

In cattle, the main histocompatibility complex class II is called BoLA-DQB (bovine leukocyte antigen) [24]. In the present study, the BoLA-DQB transcript was expressed more in tough meat. We found no studies investigating the association of this gene with meat tenderness. However, this gene has been associated with growth traits in Holstein and beef cattle (Angus, Charolais, Hereford, Limousin, Simmental); [4, 49] and, according to Koohmaraie et al. [27], animals with higher growth rates have more palatable and more tender meat.

When we compared this study with a GWAS study for meat tenderness using the same Nellore population, we do not found common genes between them, but there were some shared functions related to phosphorylation and catalytic activity [33]. These functions are related with oxygen and calcium transport, and collagen degradation, important processes for the the toughening of meat, especially after slaughter. In a GWAS study using another Nellore cattle population, Tizioto et al. [52] identified regions that influence tenderness at three different time points (24 h and 7 and 14 days after slaughter). Some of the genes reported by these authors were also identified in the present study, such as CLDN19, CLEC12A and USP32. In addition to these genes, the authors reported an association of genes belonging to the family of BoLA-BQD, CTNNB1, EXOSC2 and IQCG transcripts and meat tenderness.

Conclusions

Global gene expression analysis in animals phenotypically divergent for meat tenderness identified genes related to ubiquitin metabolism, transport of molecules such as calcium and oxygen, acid-base balance, collagen synthesis, actin and myosin, and fat accumulation. These results contribute to the understanding of the molecular mechanisms involved in the meat tenderization process, at the time of slaughter, and to the development of strategies to select animals with more tender meat.

Methods
Animals and sample collection
Meat samples were collected from 132 intact male (non castrated) Nellore animals belonging to the same contemporary group (i.e., animals that remained together from birth to slaughter). The animals were from the Capivara Farm that participates in the Qualitas Nelore Breeding Program. All animals were finished in feedlots for, approximately, 90 days and slaughtered at an average age of 731 ± 81 days on the same day and under the same conditions.

The slaughter occurred in a commercial plant, under usual process in Brazilian beef industry: the animals are slaughtered and the half-carcasses are refrigerated by 24 h. After that, the carcass is deboned, frozen and commercialized. All samples were frozen and none of them was aged.

For RNA, muscle tissue (*longissimus dorsi*) samples were collected immediately after slaughtering and stored in 15-mL Falcon tubes containing 5 mL RNA holder (BioAgency, São Paulo, SP, Brazil) at –80 °C until the time for total RNA extraction. Additionally, for shear force measurements, a *longissimus* muscle sample was removed during deboning, after 24 h in a cold chamber, between the 12th and 13th rib of each left half-carcass.

Transcriptome studies show the genes expressed in a specific time for a specific cell, i.e. it shows which genes are been expressed at the moment of the sample collection. So, we have chosen to study the gene expression related with tenderness using the phenotype measured closest to the sample collection for RNA extraction, that is, after 24 h postmortem.

Analysis of shear force
Longissimus dorsi samples measuring 2.54 cm in thickness were obtained for analysis of tenderness. The standardized procedure proposed by Wheeler et al. [58] was used for shear force determination in a mechanical Salter Warner-Bratzler Shear Force device. The samples analyzed were not submitted to any type of aging process. From this analysis ($n = 132$), 40 samples derived from animals extreme for meat tenderness (20 with tender meat and 20 with tough meat) were selected. The Student t-test implemented in the R environment [41] was applied to verify differences between the tender and tough meat groups (Table 2).

RNA sequencing
Total RNA was extracted from the samples obtained from the extreme animals selected ($n = 40$). Muscle tissue (longissimus dorsi) samples that were collected immediately after slaughter and stored in 15-mL Falcon tubes containing 5 mL RNA holder (BioAgency, São Paulo, SP, Brazil) at –80 °C were used to extract total RNA. An average of 50 mg of the muscle tissue previously stored in RNA holder (BioAgency, São Paulo, SP, Brazil) was used for extraction with the RNeasy Lipid Tissue Mini Kit (Qiagen, Valencia, CA, USA) according to manufacturer recommendations. The purity of the extracted RNA was determined by reading absorbance in a NanoDrop 1000 spectrophotometer (Thermo Fisher Scientific, Santa Clara, CA, USA, 2007). The quality of the total RNA extracted was evaluated in an Agilent 2100 Bioanalyzer (Agilent, Santa Clara, CA, USA, 2009) and its concentration and contamination with genomic DNA were measured in a Qubit® 2.0 Fluorometer (Invitrogen, Carlsbad, CA, USA, 2010).

Sequencing (RNA-Seq) was performed on an Illumina HiSeq 2500 System. Messenger RNA was obtained from the total RNA extracted and libraries containing 200 bp fragments were constructed and pooled to perform multiplexing sequencing. The reads obtained were paired-end of 2×100 bp.

Sequence processing and alignment
The sequence data generated with the Illumina HiSeq 2500 System were converted into FastQ format, using the Casava software (https://support.illumina.com/downloads/casava_18_changes.html). The computational analyses were performed on CyVerse platform [19].

First, sequenced fragments (reads) of low quality were trimmed using the Sickle program (github.com/najoshi/sickle). The TopHat2 v2.0.9 program [54] was then used to map the fragments and to align them with the bovine reference genome (UMD3.1) available in the NCBI database (http://www.ncbi.nlm.nih.gov/genome/?term=bos+taurus). For each library, a .bam file containing the aligned reads in relation to the reference genome was generated.

Assembly and quantification of the transcripts
The Cufflinks2 v2.1.1 program [55] was used to assemble the aligned reads of each sample and to estimate the number of transcripts, expressed as fragments per kilobase of transcript per million reads mapped (FPKM). The Cufflinks2 result per sample was concatenated in a single file with the Cuffmerge2 v2.1.1 program, which was used as a reference in the differential gene expression analysis.

Differential gene expression analysis
Using the Cuffdiff2 v 2.1.1 program [53, 55], the sequence alignment files generated (.bam) were divided into two contrasting groups according to meat tenderness. The FPKM values of each transcript were calculated for each

Table 2 Number of animals (N), mean, standard error, minimum and maximum of meat tenderness measured by shear force (kgf/cm^2)

	N	Mean	Standard error	Minimum	Maximum	P-value
Tender meat	20	4.36	0.34	3.51	4.80	8.40e-17
Tough meat	20	8.22	1.13	7.33	11.15	

sample. The Cuffdiff2 program uses a *t*-test for the calculation of *p*-values. False discovery rates (FDR) were controlled by the Benjamini-Hochberg procedure considering a q-value of less than 5%.

The CummeRbund package [55], implemented in the R environment [41], was used for exploration and visualization of the data obtained and generate PCA and boxplot graphics.

Annotation of differentially expressed genes
The Database for Annotation, Visualization, and Integrated Discovery (DAVID) v6.7, which consists of an integrated system of biological databases and analytical tools designed to systematically extract the biological meaning from a large list of genes and/or proteins [21], was used to annotate and interpret the lists of differentially expressed genes. The Functional Annotation Tool was used for this purpose, which determines the most relevant Gene Ontology (GO) terms for each list of genes. The Functional Annotation Clustering algorithm was applied to generate annotations of functional groups. DAVID pathway mapping was used to identify metabolic pathways in which the differentially expressed genes are involved.

The ClueGo plug-in of the Cytoscape program was used to visualize non-redundant biological terms for genes in functionally grouped networks [6].

Validation of differentially expressed genes
Real-time quantitative PCR (qRT-PCR) was used to validate the differential expression of the genes identified by RNA-Seq analysis. All the 40 RNA samples used in the RNA-Seq analyses was used to validate the data by qRT-PCR. Two

differentially expressed genes were chosen randomly for this purpose: bovine leukocyte antigen (BoLA-DQB) and Purkinje cell protein 4-like 1 (PCP4L1). In addition to these genes, three reference genes were chosen and quantified by qRT-PCR, as proposed by Vandesompele et al. [56], to normalize the data. The RNA-Seq technique detected no differences in the expression of the beta-glucuronidase (GUSB), hypoxanthine phosphoribosyltransferase 1 (HPRT1) and TATA box binding protein (TBP) genes between the groups studied and these genes were therefore chosen as housekeeping genes and were tested by qRT-PCR.

The method (conditions and equipment) described by Fonseca et al. [18] was used for validation of the differentially expressed genes by qRT-PCR: One μg total RNA was used to synthetize the first complementary DNA (cDNA) strand using SuperScript III First-Strand Synthesis SuperMix for qRT-PCR (Invitrogen). To design the primers (Table 3), the Primer Express 3.0 software (Applied Biosystems, 2004) was used and the GenBank database (http://www.ncbi.nlm.nih.gov) was accessed to obtain the mRNA nucleotide sequences. The primers specificity was tested by NCBI BLAST algorithm (https://blast.ncbi.nlm.nih.gov/Blast.cgi). Genorm (https://genorm.cmgg.be/) and Expression Suite softwares v1.0 (Applied Biosystems, Foster, CA, USA, 2012) were used to test the expression stability of the housekeeping genes.

All qRT-PCR reactions were done with 7500 Real-Time PCR (Applied Biosystems, 2009). For these reactions we used: 0.1 μg cDNA; 1X SYBR Green Master Mix and forward and reverse primers. Primers concentrations were determined by titration: 600 nanoMolar (nM) forward and

Table 3 Sequence of the forward (F) and reverse (R) primers used in the qRT-PCR assays

Gene	Locus	Sequence (5' – 3')	Tm (° C)[a]	Amplicon size (bp)
BoLA-DQB F	23:25,855,145–25,863,052	CATCACAGGAGCCAGAAG	80	64
BoLA-DQB R		GCAAACACCAATCCCAAAAT		
PCP4L1 F	3:8,249,980–8,276,724	ATCTCCAGCAACCAACCAGG	82	119
PCP4L1 R		TCTCTGTTTCTGGTGCCGTC		
GUSB F	25:28,101,456–28,233,126	CAGGGCGGGATGCTCTA	84	93
GUSB R		GTTGTCGGAGAAGTCGGC		
HPRT1 F	3:103,318,856–103,327,860	TGATGAAGGAGATGGGTGGC	83	81
HPRT1 R		CCAACAGGTCGGCAAAGAAC		
TBP F	9:105,686,118–105,690,814	ATAGTGTGCTGGGGATGCTC	80	114
TBP R		GTGGGAGGCTGTTGTTCTGA		

[a]Denaturation temperature of the amplicon

reverse primers (600/600) for BoLA-BQD and GUSB; 300 nanoMolar (nM) forward and reverse primers (300/300) for HPRT1 and 100 nanoMolar (nM) forward and reverse primers (100/100) for PCP4L1 and TBP. The analyses were performed in triplicate. For each gene (target and housekeeping), we included a negative and a positive control in every reaction.

Serial dilutions of cDNA (1:5) were used to build a standard curve and to calculate the qRT-PCR efficiency for each gene. Only PCR primers showing an efficiency of 90–110% were used [31].

The amplification conditions were: 40 cycles at 50 °C for 2 min, 95 °C for 10 min, and 60 °C for 1 min. Dissociation analyzes were performed to monitor the reactions specificity.

For the housekeeping genes, the geometric means of Ct values were calculated [56]. For the analysis of relative expression, a mixed linear model was fitted [50]:

$$Y_{gikr} = T_{ig} + D_{ik} + e_{gikr}$$

where: Y_{gikr} is the Ct obtained from the thermocycler software for gene g, in the rth well of the plate (referring to the technical replicate) in a sample obtained from animal k of treatment i (low or high meat tenderness group). T_{ig} is the group of animals effect i (low or high meat tenderness group) on the expression of gene g; D_{ik} is a random sampling specific effect which captures differences between samples shared by genes, particularly those affecting RNA concentration such as different extraction and amplification efficiency, and e_{gikr} is a residual effect.

Additional files

Additional file 1: Table S1. Samples number (N), classification of the sample, shear force (kgf/cm²), number of transcripts aligned in pairs (N reads), and percentage of transcripts aligned in pairs (% reads). (DOCX 16 kb)

Additional file 2: Figure S1. Expression profile of reference genes in the experimental groups (tender and tough meat). (TIFF 176 kb)

Additional file 3: Figure S2. Box plot of expression values (log10 FPKM) obtained for the groups studied (tender and tough meat). (TIFF 162 kb)

Additional file 4: Figure S3. Principal component analysis (PCA) of the transcripts found in the tender (red) and tough (blue) meat groups. (TIFF 215 kb)

Additional file 5: Table S2. Enriched GO terms obtained with the DAVID software for differentially expressed genes. (XLSX 14 kb)

Abbreviations
ADP: Adenosine Diphosphate; ASAH1: N-Acylsphingosine Amidohydrolase (Acid Ceramidase) 1; AT2: Angiotensin II Receptor Type 2; ATP: Adenosine Triphosphate; BoLA-DQB: Major Histocompatibility Complex, Class II, DQ Beta; cDNA: Complementary DNA; CLDN19: Claudin 19; CLEC: C-Type Lectin Family; CLEC12A: C-Type Lectin Domain Family 12 Member A; CLEC4G: C-

Type Lectin Domain Family 4 Member G; Ct: Threshold Cycle; CTNNB1: Catenin - Cadherin-Associated Protein Beta 1; DMGDH: Dimethylglycine Dehydrogenase; EXOSC2: Exosome Component 2; FDR: False Discovery Rates; FPKM: Fragments Per Kilobase Of Transcript Per Million Reads Mapped; GO: Gene Ontology; GUSB: Glucuronidase Beta; GWAS: Genome-Wide Association Study; HMBS: Erythrocyte Hydroxymethylbilane Synthase; HMOX1: Heme Oxygenase 1; HPRT1: Hypoxanthine Phosphoribosyltransferase 1; IQCG: IQ Motif Containing G; NAD⁺: Nicotinamide Adenine Dinucleotide; NADP⁺: Nicotinamide Adenine Dinucleotide Phosphate; nM: nanoMolar; PCA: Principal Component Analysis; PCP4L1: Purkinje Cell Protein 4 Like 1; PGK1: Phosphoglycerate Kinase 1; PNP: Purine Nucleoside Phosphorylase; qRT-PCR: Quantitative Real Time Polymerase Chain Reaction; RNA-Seq: RNA Sequencing; SYP: Synaptophysin; TBP: TATA-Box Binding Protein; TCF7L1: Transcription Factor 7 Like 1; USP: Ubiquitin Specific Peptidase Family; USP2: Ubiquitin Specific Peptidase 2; USP32: Ubiquitin Specific Peptidase 32; ZKSCAN2: Zinc Finger With KRAB And SCAN Domains 2

Acknowledgments
We thank the Qualitas Nelore breeding program company for providing the tissue samples and database used in this study.

Funding
The RNA sequencing was fund by the project "Genomic tools for the genetic improvement of traits of direct economic importance in Nelore cattle". This was finance also by the São Paulo Research Foundation – FAPESP. (FAPESP grant # 2009/16118–5). The LFSF scholarship was fund by the São Paulo Research Foundation – FAPESP (FAPESP grant # 2013/09190–7).

Authors' contributions
LFSF, DFJG, LGA, JAF and FB conceived and designed the experiment; LFSF and DFJG performed the experiments; LFSF, LGA, DBSS and RB analyzed and interpreted the results; LFSF, LGA, JAP and RB drafted and revised the manuscript. All authors read and approved the final version of the manuscript.

Ethics approval
All experimental procedures were approved by Ethics Committee of the Faculty of Agrarian and Veterinary Sciences of Sao Paulo State, Jaboticabal, São Paulo (protocol number 18,340/16). The animals were provideded by Qualitas Nelore breeding program company and they were slaughtered in commercial slaughterhouses. These slaughterhouses have animal welfare departments staffed by professionals trained by WAG (World Animal Protection) to ensure that the animals are killed humanely using a captive bolt pistol for stunning.

Competing interests
The authors declare that they have no competing interests.

Author details
[1]Faculty of Agricultural and Veterinary Sciences, São Paulo State University, FCAV/UNESP, Jaboticabal, São Paulo, Brazil. [2]CyVerse, University of Arizona, Tucson, USA.

References
1. ABIEC, Associação Brasileira das Indústrias Exportadoras de Carne, 2016. http://www.abiec.com.br/. Accessed 8 Feb 2016.
2. Bailey AJ, Paul RG, Knott L. Knott. Mechanisms of maturation and aging of collagen. Mech Ageing Dev.1998, 106:1–56.
3. Bate-Smith EC, Bendall JR. Factors determining the time course of rigor mortis. J Physiol 1949, 110:47–65.
4. Batra TR, Lee AJ, Gavora JS, Stear MJ. CLASS I alleles of the bovine major histocompatibility system and their association with economic traits. J Dairy Sci. 1989;72:2115–2124.

5. Bendall JR. Posmortem changes in muscle. In: GH BOURNE, editor. The structure and function of muscle, vol. 2. New York: Academic Press; 1973. p. 244–309.
6. Bindea G, Mlecnik B, Hackl H, Charoentong P, Tosolini M. ClueGO: a Cytoscape plug-in to decipher functionally grouped gene ontology and pathway annotation networks. J Bioinformatics. 2009;25(8):1091–3.
7. Boleman SJ, Boleman SL, Miller RK, Taylor JF, Cross HR, Wheeler TL, Koohmaraie M, Shackelford SD, Miller MF, West RL, Johnson DD, Savell JW. Consumer evaluation of beef of known categories of tenderness. J Anim Sci. 1997;75:1521–4.
8. Bongiorni S, Gruber CEM, Bueno S, Chillemi G, Ferr F, Failla S, Moioli B, Valentini A. Transcriptomic investigation of meat tenderness in two Italian cattle breeds. Anim Genet. 2016. doi:10.1111/age.12418.
9. Campo MM, Sañudo C, Panea B, Alberti P, Santolaria P. Breed type and aging time effects on sensory characteristics of beef strip loin steaks. Meat Sci. 1999;51:383–91.
10. Carbon S, Ireland A, Mungall CJ, Shu S, Marshall B, Lewis S. AmiGO: online access to ontology and annotation data. Bioinformatics. 2009;2009(25):288–9.
11. Chapple RH, Tizioto PC, Wells KD. Characterization of the rat developmental liver transcriptome. Physiol Genomics. 2013;45:301–11.
12. CRC: cooperative Resersh Centre for Beef Genetic Technologies. Annual Report of CRC for Beef Genetics Tecnologies. High Quality beef for Global Consumers. Armidale, Australia, 2008.
13. Darrel E, Goll DE, Valery F, Thompson VF, Li H. The calpain system. Physiol Rev. 2003;83:731–801.
14. De Robertis E. Bases da Biologia Celular e Molecular. Rio de Janeiro: Editora Guanabara Koogan; 2010.
15. Drickamer K. C-Type lectin-like domains. Curr Opin Struct Biol. 1999;5:585–90.
16. Duston TR, Hostetler RL, Carpenter ZL. Effect of collagen levels and sarcomere shortening on muscle tenderness. J Food Sci. 1976;41:863–6.
17. Ferguson DM, et al. Effect of electrical stimulation on protease activity and tenderness of M. Longissimus from cattle with different proportions of Bos Indicus content. Meat Sci. 2000;55:265–72.
18. Fonseca LFS, Gimenez DF, Mercadante ME, Bonilha SF, Ferro JA, Baldi F, Souza FRP, Albuquerque LG. Expression of genes related to mitochondrial function in Nellore cattle divergently ranked on residual feed intake. Mol Biol Rep. 2015;42:559–65.
19. Goff SA, Vaughn M, Mckay S. The iPlant collaborative: cyberinfrastructure for plant biology. Front Plant Sci. 2011;2:34. doi:10.3389/fpls.2011.00034.
20. Gonçalves TM. Dissertation. In: differential expression of genes related with meat tenderness in Nellore cattle. ESALQ, USP; 2015. p. 97. http://www.teses.usp.br/teses/disponiveis/11/11139/tde-12052015-165345/pt-br.php. Accessed 16 Mar 2016.
21. Huang W, Sherman BT, Lempick RA. Systematic and integrative analysis of large gene lists using DAVID bioinformatics resources. Nat Protoc. 2009;4:44–57.
22. Jong OG, Balkom BWM, Gremmels H, Verhaar MC. Exosomes from hypoxic endothelial cells have increased collagen crosslinking activity through up-regulation of lysyl oxidase-like 2. J Cell Mol Med. 2015;XX:1–9.
23. Junqueira LCU, Carneiro J. Biologia Celular e Molecular. 8th ed. Guanabara; 2005.
24. Klein J, Bontrop RE, Dawkins RL, Erlich HA, Gyllensten UB, Heise ER, Jones PP, Parham P, Wakeland EK, Watkins DI. Nomenclature for the major histocompatibility complexes of different species: a proposal. Immunogenetics. 1990;4:217–9.
25. Koohmaraie M. The biological basis of meat tenderness and potential genetic approaches for its control and prediction. Proc Recip Meat Conf. 1995;48:69–75.
26. Koohmaraie M. Biochemical factors regulating the toughening and tenderization process of meat. Meat Sci. 1996;43:S193–201.
27. Koohmaraie M, Kent MP, Shackelford SD, Veiseth E, Wheeler TL. Meat tenderness and muscle growth: is there any relationship? Meat Sci. 2002;62:345–52.
28. LAMP: Library of Apicomplexan Metabolic Pathways. Nicotinate and nicotinamide metabolism. http://www.llamp.net/?q=Nicotinate%20metabolism. Accessed 29 Dec 2015.
29. Lee NP, Tong MK, Leung PP. Kidney claudin-19: localization in distal tubules and collecting ducts and dysregulation in polycystic renal disease. FEBS Lett. 2006;580:923–31.
30. Lehninger AL, Nelson LD, Cox MM. Princípios de bioquímica. 3rd ed. São Paulo: SARVIER; 2002. p. 1009.
31. Livak KJ, Schmittgen TD. Analysis of relative gene expression data using real-time quantitative PCR and the 2(−Delta Delta C(T)) method. Methods. 2001;25:402–8.
32. Lucki NC, Sewer MB. Genistein stimulates MCF-7 breast cancer cell growth by inducing acid ceramidase (ASAH1) gene expression. JBC. 2011;286: 19399–409.
33. Magalhães AFB, de Camargo GMF, Junior FGA, Gordo DGM, Tonussi RL, et al. Genome-wide association study of meat quality traits in Nellore cattle. PLoS One. 2016;11(6):e0157845. doi:10.1371/journal.pone.0157845
34. Malone JH, Oliver B. Microarrays, deep sequencing and the true measure of the transcriptome. BMC Biol. 2011;9:34.
35. Manso CMCP, Neri CR, Vidoto EA, Sacco HC, Ciuffi KJ, Iwamoto LS, Iamamoto Y, Nascimento OR, Serra OA. Characterization of iron(III) porphyrin-hidroxo complexes in organic media through Uv-Vis and EPR spectroscopies. J. Inorg. Biochem. 1999;73:85–93.
36. Otterbein LE, Soares MP, Yamashita K, Bach FH. Heme oxygenase-1: unleashing the protective properties of heme. Trends Immunol. 2003;24:449–55.
37. Ouali A, Gagaoua M, Boudida Y, Becila S, Boudjellal A, Herrera-Mendez CH, Sentandreu MA. Biomarkers of meat tenderness: presente knowledge and perspectives in regards to our current understanding of the mechanisms involved. Meat Sci. 2013;95:854–70.
38. Park JH, Schuchman EH. Acid ceramidase and human disease. Biochim Biophys Acta. 2006;1758:2133–8.
39. Paz CCP, Luchiari Filho A. Melhoramento genético e diferenças de raças com relação à qualidade da carne bovina. Pecuária de corte. 2000;101:58–63.
40. Piorkowska K, Żukowski K, Nowak J, Połtowicz K, Ropka-Molik K, Gurgul A. Genome-wide RNA-Seq analysis of breast muscles of two broiler chicken groups differing in shear force. Anim Genet. 2015;47(1):68–80.
41. R Core Team. R: a language and environment for statistical computing. Vienna: R Foundation for Statistical Computing; 2015. https://www.R-project.org/.
42. Rhoads AR, Friedberg F. Sequence motifs for calmodulin recognition. FASEB J Official Publ Federation of American Societies for Exp Biol. 1997;11:331–40.
43. Rubensam JM, Transformações Post Mortem E Qualidade Da Carne Suína. 1ª Conferência Internacional Virtual sobre Qualidade de Carne Suína. 2000. http://www.cnpsa.embrapa.br/sgc/sgc_publicacoes/anais00cv_jane_pt.pdf. Accessed 30 Dec 2015.
44. Rubenstein JL, Greengard P, Czernik AJ. Calcium-dependent serine phosphorylation of synaptophysin. Synapse. 1993;13:161–72.
45. Sander TL, Stringer KF, Maki JL, Szauter P, Stone JR, Collins T. The SCAN domain defines a large family of zinc finger transcription factors. Gene. 2003;310:29–38.
46. Scollan N, Hocquette J, Nuernberg K, Dannenberger D, Ian R, Moloney A. Innovations in beef production systems that enhance the nutritional and health value of beef lipids and their relationship with meat quality. Meat Sci. 2006;74:17–33.
47. Sekikawa M, Seno K, Mikami M. Degradation of ubiquitin in beef during storage. Meat Sci. 1998;48:201–4.
48. Sgarbieri VC. Proteínas em alimentos protéicos. São Paulo: Varela; 1996. p. 517.
49. Stear MJ, Pokorny TS, Muggli NE, Stone RT. The relationships of birth weight, preweaning gain and postweaning gain with the bovine major histocompatibility system. J Anim Sci. 1989;67:641–9.
50. Steibel JP, Poletto R, Coussens PM, Rosa JMG. A powerful and flexible linear mixed model framework for the analysis of relative quantification RT-PCR data. Genomics. 2009;94:146–52.
51. Tizioto PC, Coutinho LL, Decker JE, Schnabel RD, Rosa KO, Oliveira PSN, Souza MM, Mourão GB, Tullio RR, Chaves AS, Lannad PD, Zerlotini-Neto A, Mudadu MA, Taylor JF, Regitano LCA. Global liver gene expression differences in Nelore steers with divergent residual feed intake phenotypes. BMC Genomics. 2015;16:216.
52. Tizioto PC, Decker JE, Taylor JF, Schnabel RD, Mudadu MA. Genome scan for meat quality traits in Nelore beef cattle. Physiol Genomics. 2013;45:1012–20.
53. Trapnell C, Hendrickson DG, Sauvageau M, Goff L, Rinn JL, Pachter L. Differential analysis of gene regulation at transcript resolution with RNA-seq. Nat Biotechnol. 2013;31:46–53.
54. Trapnell C, Pachter L, Salzberg SL. TopHat: discovering splice junctions with RNA-Seq. Bioinformatics. 2009;25:1105–11.
55. Trapnell C, Roberts A, Goff L, Pertea G, Kim D, Kelley DR. Differential gene and transcript expression analysis of RNA-seq experiments with TopHat and cufflinks. Nat Protoc. 2012;7(3):562–78.
56. Vandesompele J, De Preter K, Pattyn F, Poppe B, Van Roy N, De Paepe A, Speleman F. Accurate normalization of real-time quantitative RT-PCR data by geometric averaging of multiple internal control genes. Genome Biol. 2002;3(7):RESEARCH0034.

57.   Wang Z, Gerstein M, Snyder M. RNA-Seq: a revolutionary tool for transcriptomics. Advance Online Plublication: Nat. Rev. Genet; 2008.

58.   Wheeler TL, Koohmaraie M, Shackelford SD. Standardized Warner-Bratzler shear force procedures for meat tenderness measurement. Clay Center: Roman L. Hruska U. S. MARC. USDA, 1995.

59.   Zhao C, Tian F, Yu Y, Liu G, Zan L, Scott M, Song J. miRNA-dysregulation associated with tenderness variation induced by acute stress in Angus cattle. J Anim Sci Biotechnol. 2012;3:12.

60.   Zhong-liang HU, Xing JL, Xiao-feng YE. Effects of Angiotensin II on Beef Quality. Acta Agric Jiangxi. 2009:11.

9

Comparative analysis of avian poxvirus genomes, including a novel poxvirus from lesser flamingos (*Phoenicopterus minor*), highlights the lack of conservation of the central region

Olivia Carulei[1], Nicola Douglass[1] and Anna-Lise Williamson[1,2,3*] (iD)

Abstract

Background: Avian poxviruses are important pathogens of both wild and domestic birds. To date, seven isolates from subclades A and B and one from proposed subclade E, have had their genomes completely sequenced. The genomes of these isolates have been shown to exhibit typical poxvirus genome characteristics with conserved central regions and more variable terminal regions. Infection with avian poxviruses (APVs) has been reported in three species of captive flamingo, as well as a free-living, lesser flamingo at Kamfers dam, near Kimberley, South Africa. This study was undertaken to further characterise this virus which may have long term effects on this important and vulnerable, breeding population.

Results: Gene content and synteny as well as percentage identities between conserved orthologues was compared between Flamingopox virus (FGPV) and the other sequenced APV genomes. Dotplot comparisons revealed major differences in central regions that have been thought to be conserved. Further analysis revealed five regions of difference, of differing lengths, spread across the central, conserved regions of the various genomes. Although individual gene identities at the nucleotide level did not vary greatly, gene content and synteny between isolates/species at these identified regions were more divergent than expected.

Conclusion: Basic comparative genomics revealed the expected similarities in genome architecture but an in depth, comparative, analysis showed all avian poxvirus genomes to differ from other poxvirus genomes in fundamental and unexpected ways. The reasons for these large genomic rearrangements in regions of the genome that were thought to be relatively conserved are yet to be elucidated. Sequencing and analysis of further avian poxvirus genomes will help characterise this complex genus of poxviruses.

Keywords: Poxvirus, Avipoxvirus, Flamingopox, Genome sequence

* Correspondence: Anna-Lise.Williamson@uct.ac.za
[1]Division of Medical Virology, Department of Pathology, Faculty of Health Sciences, University of Cape Town, Cape Town, South Africa
[2]Institute of Infectious Disease and Molecular Medicine, University of Cape Town, Cape Town, South Africa
Full list of author information is available at the end of the article

Background

Avian poxviruses are important pathogens of both wild and domestic birds. In domestic poultry, avian poxvirus infection can cause significant economic losses due to transient decrease in egg production, impaired fertility, reduced growth in young birds and increased mortality [1]. In wild bird populations, including endangered and endemic species, poxvirus infection may lead to secondary bacterial or fungal infections, decrease ability to care for young, and affect vision and/or the ability to feed making them prone to predation [2–4].

As of 2007, poxvirus infections had been reported in 278 species of wild and domestic birds from 70 families and 20 orders [5, 6]. Since then, the total number of avian species reported to be infected with a poxvirus has risen to at least 329 across 76 families and it is likely that many more species are susceptible (Unpublished data).

Genetic, phylogenetic and genomic analyses are increasingly being used to classify and characterise members of this large genus of viruses. To date, isolates from a chicken (*Gallus gallus*) (FWPV and Fp9) [7, 8], canary (*Serinus canaria*) (CNPV) [9], African penguin (*Spheniscus demersus*) (PEPV) [10], feral pigeon (*Columba livia*) (FeP2) [10], turkey (*Melleagris gallopavo*) (TKPV) [11] and two Pacific shearwaters (*Ardenna carneipes*) (SWPV-1); (*Ardenna pacificus*) (SWPV-2) [12], have had their genomes fully sequenced. Because of the relative lack of complete genome sequences of avian poxviruses, construction of phylogenies has to date relied on single gene analyses, with the P4b gene (fwpv167; cnpv240; vacv A3L) [13], one of the 49 genes conserved in all poxviruses, being the most commonly used [14–20]. These analyses have shown that the *Avipoxvirus* genus is divided into 3 clades, A (Fowlpox like viruses), B (Canarypox like viruses) and C (Parrotpox (PRPV) like viruses) as well as two proposed clades, D (Quailpox virus (QPV)) and E (Turkeypox virus(TKPV)). Clades A and B are further divided into several subclades which differ slightly in their composition depending on the genetic locus used for analysis and as such, are still being resolved.

Gene content and synteny have also been used to help elucidate evolutionary relationships between poxviruses. Alignment and comparison of the central genomic regions of viruses from eight *Chordopoxvirinae* (ChPV) genera (using the Vaccinia Virus (VACV) genome as the reference) showed that Molluscum contagiosum virus and FWPV were most divergent, encoding 40 and 33 unique genes in their central genomic regions respectively. In contrast, myxoma virus (MYXV), Yaba-like disease virus (YLDV), lumpy skin disease virus (LSDV) and swinepox virus (SWPV) contained three or less unique genes in this region. In terms of gene order, FWPV also showed major differences with large blocks of genes being translocated and/or inverted compared to

VACV and the other ChPVs [21]. These findings are in accordance with what has been shown through phylogenetic analysis.

Poxvirus infections have been documented in four species of flamingo to date. The first documented case occurred in Chilean flamingos (*Phoenicopterus chilensis*) that were housed at a zoo in Hino City, Tokyo, Japan [22]. Two separate cases of infections in American flamingos (*Phoenicopterus ruber*) have been reported with the first occurring in a bird housed at the National Zoological Park in Washington DC [23]. In this case, a 4.5 kb *Hind*III fragment ranging from the equivalent of fwpv193–203, was reported to show 99.7% nucleotide identity to an isolate from an Andean condor (*Vultur gryphus*) which groups in clade B phylogenetically. A second case was reported in a young American flamingo housed at the Lisbon zoo. Phylogenetic analysis based on P4b and the CNPV 186–187 fragment showed this isolate to group in clade B2 with the highest identity to isolates from various species of bustard [24]. Another case occurred at a zoo in Japan but the infection was noted in two, young Greater flamingos (*Phoenicopterus roseus*). Based on analysis of the P4b gene this isolate was shown to group with two isolates from pigeons (PPV-B7 and CVL950), also in clade B2. All of the above cases occurred in captive flamingos [25].

Documented cases of poxvirus infections in birds in South Africa date back to the early 1960's when infections were noted in Cape turtle doves (*Streptopelia capicola*) and a Cape thrush (*Turdus olivaceus*) [26]. Infections were later noted in ostriches (*Struthio camelus australis*) [27], an African penguin (penguinpox virus (PEPV) - *Spheniscus demersus*) [10, 17, 20, 28], a feral pigeon (pigeonpox virus (FeP2) - *Columba livia*) [10, 20], and a speckled (rock) pigeon (*Columba guinea*) [29].

The flamingopox virus (FGPVKD09) isolate further characterised in this study, was obtained during a poxvirus outbreak in 2008, that occurred in a permanent, breeding population of lesser flamingos (*Phoenicopterus minor*), living at Kamfers Dam, a perennial wetland near Kimberley, South Africa. All lesions seen in this population were of the cutaneous form (present on the legs and faces) and regressed over time showing little effect to their overall health [30]. Approximately 30% of the juvenile population was estimated to have developed lesions over the observation period from January to June 2008. The sample analysed was taken from a lesion near the tibiotarsal joint of a juvenile flamingo which was euthanised and examined further. Phylogenetic analysis based on the alignment of partial P4b nucleotide sequences showed that this isolate grouped in subclade A3 with 99–100% nt identity to the other A3 isolates from various species [18, 20, 30]. Due to this similarity, it was suggested that the flamingos were infected with a virus that was already in circulation in wild birds. This was the first reported case of poxvirus infection in free-living flamingos as well as the first reported case of a flamingo infected by a clade A virus.

Flamingos are gregarious in nature and therefore vulnerable to infectious disease. Due to the declining population and human induced threats to the already small number of breeding sites, the lesser flamingo is listed as near-threatened in both South Africa and internationally. This study was undertaken to further characterise this virus which may have long term effects on this important, breeding population.

Methods

Virus isolation

The Flamingopox virus (FGPV) sample was collected on post mortem from flamingo at Kamfers Dam by Dr. David Zimmermann and Dr. Mark Anderson as part of an investigation into the cause of the dermal lesions. The sample was donated by Dr. Emily Lane of National Zoological Gardens as scab tissue and stored at −20 °C until further processing. A portion of the scab was diced with a scalpel and added to a Dounce homogeniser in 1 ml of McIlvains buffer (pH 7.4) containing penicillin (500 U/ml), streptomycin (100 μg/ml) and fungin (1 μg/ml). The homogenate was centrifuged at 800 rpm for 5 mins and the supernatant used for inoculation onto the chorioallantoic membranes (CAMs) of embryonated hens' eggs as described previously [31].

DNA sequencing

DNA was extracted as described previously [10], and sent to the Central Analytical Facility (CAF) at the University of Stellenbosch, in Stellenbosch, South Africa for full genome sequencing. DNA was sheared ultrasonically using the Covaris S2 sample preparation system (Covaris Inc., USA). One 316 chip was used followed by use of one half of a 318 chip on the Ion Torrent Personal Genome Machine (PGM) (Life Technologies), according to the manufacturer's instructions.

Basic quality control was performed by Anelda van der Walt (CAF, University of Stellenbosch, Stellenbosch, South Africa) using Torrent Suite software (version 3.2.1). Reads were trimmed of adaptor sequences and further trimmed if average base quality values (Q value) were <25 with window size = 11. Reads were discarded if read length was <50 nt and filtered to remove polyclonal reads. All reads that passed the above quality control filters were

mapped to the chicken genome (*Gallus gallus* WASHUC2) using Newbler 2.6 and BLASTed to the chicken genome using CLC Genomics Workbench 4.7.1 (Qiagen) to filter out reads of host origin. Unmapped reads were used as input data for de novo assembly using the CLC Genomics Workbench 4.7.1.

DNA analysis

CLC Genomics Workbench 4.7.1 was used for all analysis unless otherwise stated. Open reading frames (ORFs) longer than 90 nt with a methionine start codon (ATG) were identified. These ORFs were annotated as potential genes and numbered from left to right if alignment to the NCBI nr database using BLASTn and/or BLASTp and/or BLASTx gave BLAST expect values of E ≤ 1e-5. ORFs were annotated as described by Hendrickson et al., [32]. ORFs were annotated as intact (I) if the 5′ end is intact and the ORF is ≥80% the length of the closest homologue. If the 5′ end is intact but the ORF is <80% the length of the closest homologue, it was annotated as truncated (T). If the 5′ end is not intact the ORF was annotated as a fragment (F). If the ORF is ≥20% the length of the closest homologue it was annotated as extended (E) if the 5′ and 3′ ends were intact or as extended at the 5′ or 3′ end. Expression studies and functional analysis would be needed to determine whether fragmented and truncated ORFs are expressed and/or functional. The left most nucleotide was nominated as base 1, as the ITRs and terminal hairpin loops were not resolved. The FGPV sequence can be accessed from Genbank with accession number MF678796.

Dotplots were created using Gepard software, with word length = 10 [33]. Pairwise and multiple sequence alignments were created using MAFFT version 7, with default settings. Genome sequences of each of the fully sequenced avian poxvirus genomes (FWPV - AF198100; Fp9 - AJ581527; CNPV - AY318871; FeP2 - KJ801920; PEPV - KJ859677; TKPV - KP728110; SWPV1 - KX857216; SWPV2 - KX857215) were analysed to determine the number of conserved ORFs and copy numbers of gene family proteins. VACV (strain Copenhagen) - M35027 was also used in dotplot analysis.

The PEPV and FeP2 gene annotations in Genbank have been changed relative to the annotations found in the journal article describing the sequences [10]. When referring to PEPV and FeP2 ORFs in this study, the

Table 1 Genome statistics of FGPV compared with each of the fully sequenced avian poxvirus genomes

Statistic	FWPV	FP9	FGPV	PEPV	FeP2	TKPV	SWPV1	SWPV2	CNPV
Length (kbp)	289	266	293	307	282	189	327	351	360
A + T (%)	69.1	69.2	70.5	70.5	70.5	70.2	72.4	69.8	69.6
# of ORFs	260	244	285	285	271	171	310	312	328

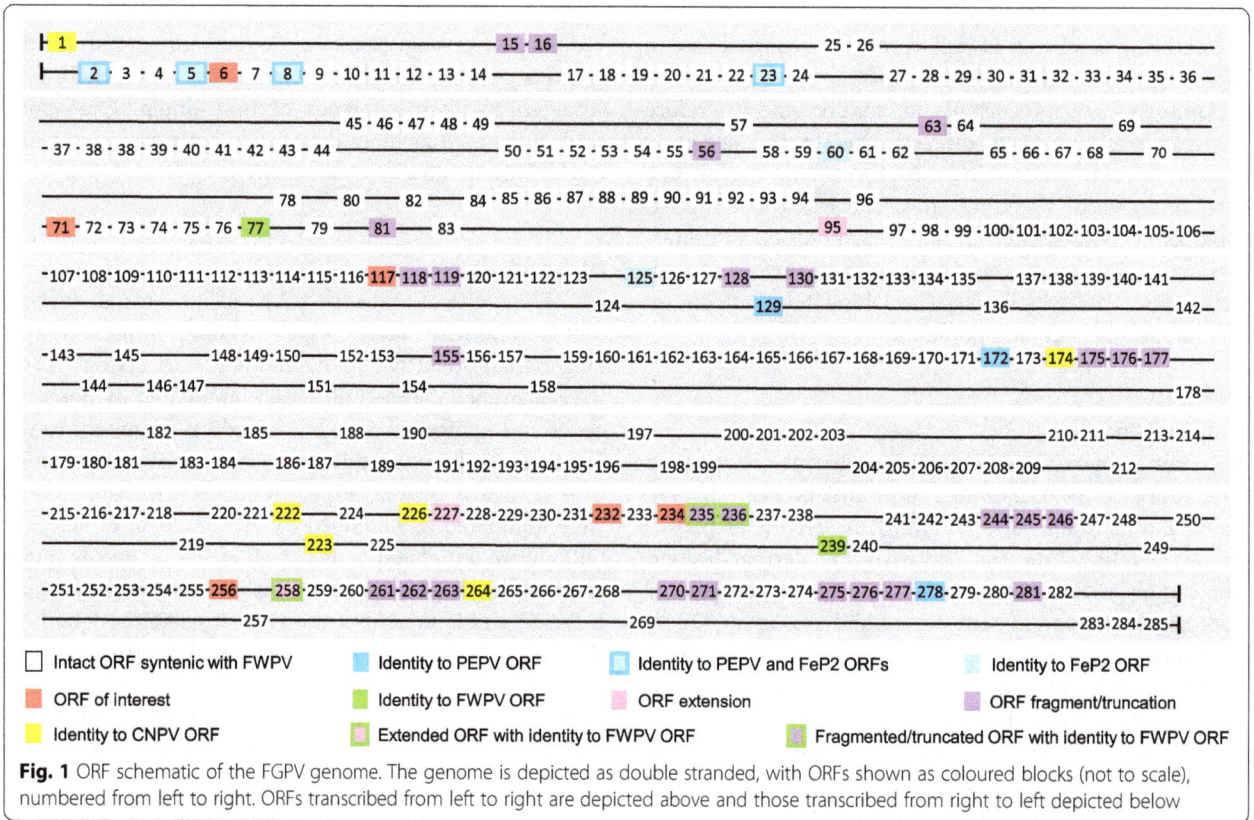

Fig. 1 ORF schematic of the FGPV genome. The genome is depicted as double stranded, with ORFs shown as coloured blocks (not to scale), numbered from left to right. ORFs transcribed from left to right are depicted above and those transcribed from right to left depicted below

original annotations from the publication are used which are referred to in Genbank as "old locus tag".

Results

Ion torrent sequencing on the 316 chip resulted in 3,745,381 reads with a mean read length of 180 bp. After performing quality control (QC) and filtering reads of chicken origin, 1,309,385 (35%) poxvirus specific reads remained. On the 318 chip, 2,148,517 reads were generated with a mean read length of 251 bp. Only 570,143 (27%) remained after QC and filtering of reads of host origin. Read assembly in CLC Genomics workbench

resulted in one contiguous sequence of 293,130 bp with an average of 1090× coverage. Basic genome statistics are shown in Table 1 below. The A + T content across the whole sequenced region was found to be 70.5% and to contain 285 potential ORFs encoding proteins ranging from 37 to 1984 amino acids in length, representing a coding density of 91% (Table 1). Relative to their closest orthologues, 259 ORFs have been annotated as intact, 21 as truncated and/or fragmented, and 4 as extended. One ORF, fgpv256 showed no similarity to any ORFs in Genbank but was identified to have a P-type ATPase motif and was therefore annotated as a

Table 2 Table of ORFs of interest identified in the FGPV genome as well as their location in eight other avian poxvirus genomes if present

ORF	FWPV	FGPV	PEPV	FeP2	TKPV	SWPV1	SWPV2	CNPV	Notes
IL-10	–	006	014	014	–	–	014	018	Found in similar genomic locations
UBQ	Frag.	071	074	Frag.	–	086	091	096	Found in the same genomic location
ANK	–	117	–	–	–	–	–	–	Ankyrin repeat family with limited (30%) identity to avipoxvirus ANKs
C-type lectin	–	232	–	–	–	–	–	–	C-type lectin / Brevican core protein/ NK receptor like
C7L	–	234	231	223	159	–	–	–	Host range gene thought to have been restricted to orthopoxviruses until 2014 [10]
P-type ATPase	–	256	–	–	–	–	–	–	No significant identity to known ORFs in Genbank

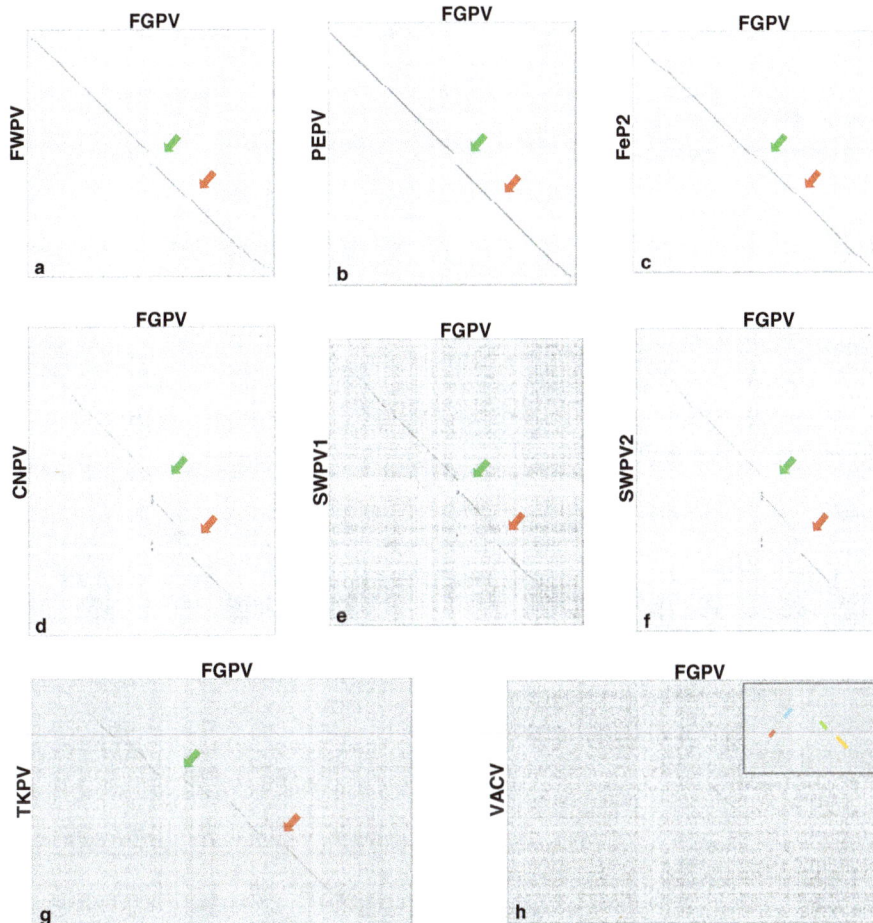

Fig. 2 Dotplots of the FGPV genome (x axis) vs other sequenced avian poxvirus genomes (y axis). **a**) FGPV vs FWPV **b**) FGPV vs PEPV **c**) FGPV vs FeP2 **d**) FGPV vs CNPV **e**) FGPV vs SWPV1 **f**) FGPV vs SWPV2 **g**) FGPV vs TKPV **h**) FGPV vs VACV (Copenhagen) H inset) Conserved areas of the 2H) dotplot highlighted in colours corresponding to Fig. 5. Green arrows indicate the first region of difference (fwpv114–126) and red arrows indicate the second region of difference (fwpv146–165). Plots are not to scale. Window size = 10

hypothetical ORF. All FGPV ORFs have been listed and compared to their closest orthologues in an additional table (Additional file 1).

Comparison of the FGPV genome to other avian poxvirus, clade A genomes shows the central region to be relatively highly conserved in gene content and synteny while the terminal regions are more variable. Several ORFs in the terminal regions of the FGPV genome show greater nucleotide identity to CNPV ORFs than FWPV or other clade A ORFs and many are also truncated or fragmented compared to their closest orthologues (Fig. 1). Of the 25 FGPV ORFs noted to be truncated, fragmented or extended, eight (33%) encode hypothetical proteins, and 12 (46%) encode proteins belonging to gene families (ankyrin repeat, CC-chemokine and V-type Ig domain). ORFs of interest have been highlighted (Fig. 1, red blocks) that have either been identified in poxviruses for the first time

(fgpv 232 and 256), appear to be novel members of gene families (fgpv117) or were previously identified and noted in other avian poxvirus genomes (fgpv006, fgpv071, fgpv234) (Table 2).

Brevican core protein

FGPV encodes an orthologue fragment (fgpv232) of a C-type lectin with similarity to brevican core proteins encoded by eukaryotes. Amino acid identity is relatively low at 30–40%. In humans this protein is involved in nervous system development. To our knowledge, C-type lectins of this variety have not been identified in any viral species to date.

Ankyrin repeat family

The FGPV genome was noted to have significantly more ankyrin repeat family proteins than what has been

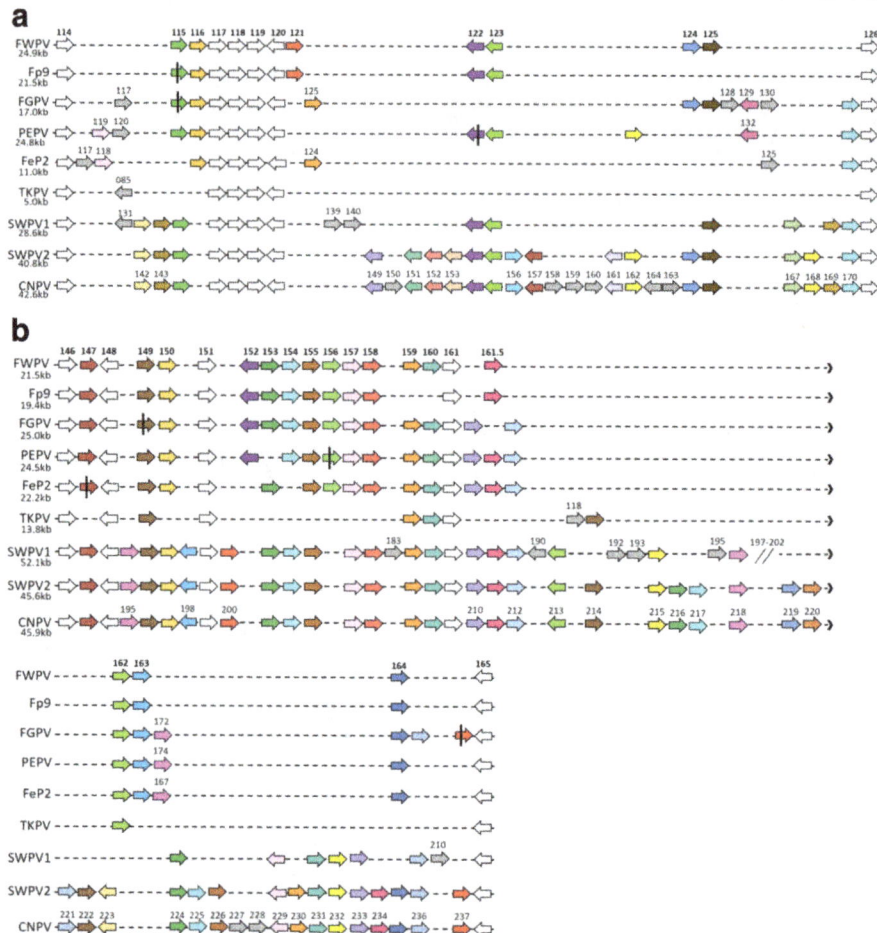

Fig. 3 Schematic representing the ORFs present in the rearranged regions **a**) fwpv114–126 and **b**) fwpv146–165 in nine avian poxvirus genomes. ORFs are represented as arrows pointing in the direction of transcription. Numbers below the virus labels show the length of each region in kilobase pairs. (White = present in all genomes; Grey = unique to one genome; Coloured = present in 2–6 genomes or present in all genomes but with one or more orthologues not intact. Homologous, syntenic ORFs are shaded in the same colour across genomes (some colours have been repeated across the length of the genomes but do not indicate synteny or homology – only ORFs of the same colour (excluding grey) and directly above or below each other are syntenic homologues); Black vertical bar = fragmented and/or truncated ORF). Alignment is not to scale and ORF colours do not correspond between figures

reported in other clade A avian poxviruses but if truncated and fragmented ORFs are assumed to be non-functional and therefore excluded, the number of ankyrin repeat family proteins decreases to levels previously noted in other, clade A, avian poxvirus genomes (FWPV and PEPV). The large number of disrupted ankyrin repeat proteins found in the FGPV genome could be due to gradual loss of these ORFs that were once the result of genomic accordion gene expansions. In the left hand, terminal region are four ORFs containing ankyrin repeats (fgpv002, fgpv005, fgpv008 and fgpv0023) that are only found in the genomes of South African isolates and five of the six FGPV ORFs that are homologues of ORFs only found in CNPV, are found in the terminal regions of the FGPV genome. FGPV also contains an ORF encoding an ankyrin repeat protein (fgpv275) with

similarity to a serine/threonine protein phosphatase from various species including trichomonas vaginalis, and various insects and birds but not to any poxvirus ankyrin repeat proteins in Genbank. It is possible that this ORF was horizontally transferred from a host at some point in the evolutionary history of this virus.

IL-10

FGPV encodes an IL-10 like protein (fgpv006) with identity to homologues found in the CNPV, PEPV and Fep2 genomes. This ORF is found in the same location as the copies found in PEPV and Fep2, between ORF equivalents pepv13-pepv15 and fep14-fep16. This region is highly conserved between the three African isolates.

Ubiquitin

FGPV encodes an ubiquitin homologue (fgpv071) at the same genomic location as the CNPV and PEPV homologues, that shows 100% amino acid identity to homologues found in eukaryotes.

C7L

Like PEPV, FeP2 and TKPV, FGPV contains an orthologue of orthopoxvirus C7L (fgpv234) which is found in an equivalent genome position between orthologues of

fwpv216 and fwpv217. These orthologues are highly conserved with 96–98% aa identity. Although there is no ORF present, the equivalent region in the FWPV genome shows 67–68% nt identity to the above ORFs suggesting that FWPV may have once contained a C7L orthologue.

Dotplots were created comparing FGPV with other sequenced avian poxvirus genomes and VACV (Copenhagen) to compare overall genomic synteny. This analysis showed the FGPV genome to be highly syntenic with FWPV, PEPV and FeP2 genomes overall

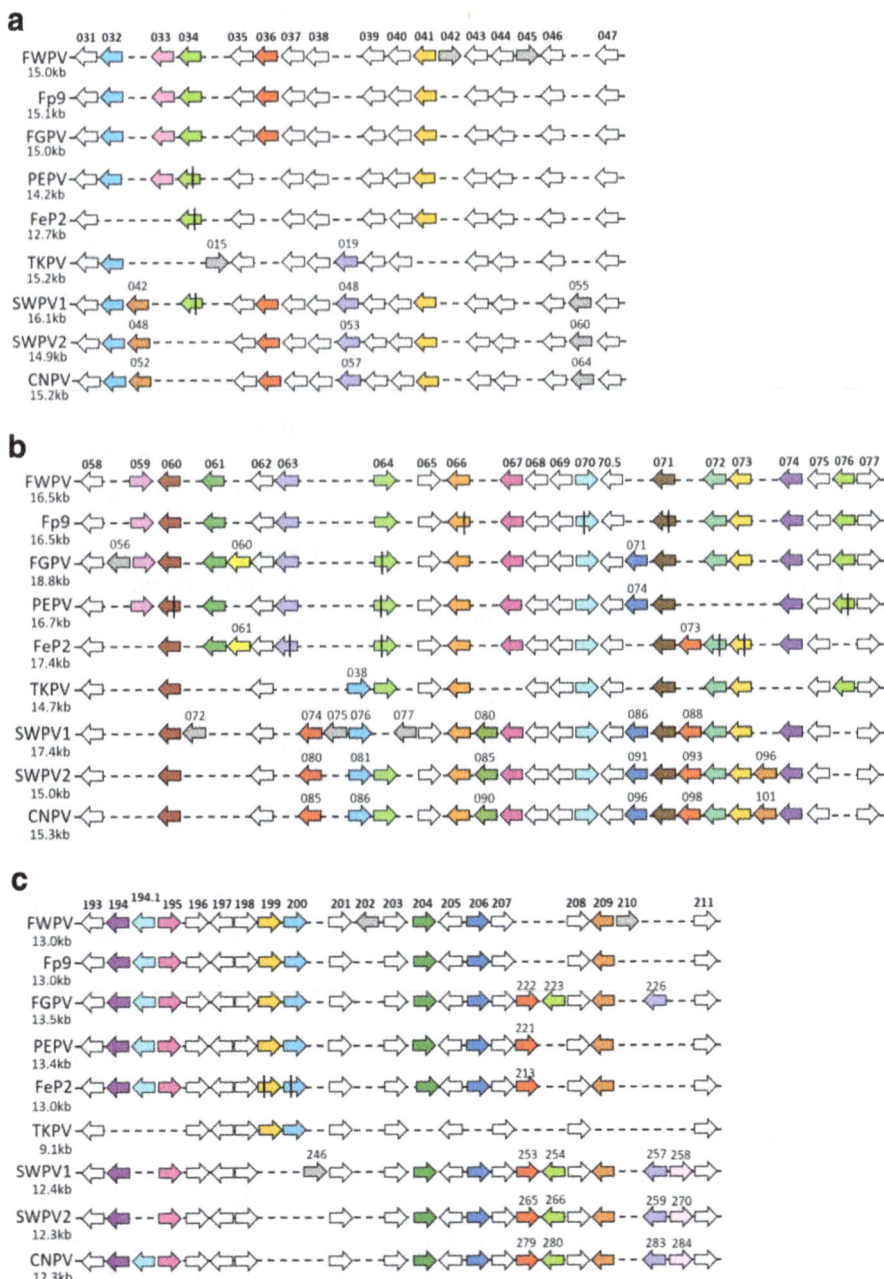

Fig. 4 Schematic representing the ORFs present in the rearranged regions a) fwpv031–047, b) fwpv058–077 and c) fwpv193–211 in nine avian poxvirus genomes. Annotations are depicted as in Fig. 3

and to show major differences compared to CNPV, TKPV, and the two, SWPV genomes, as expected, due to the large differences in genome size. Also notable are the two large breaks in synteny located in the central regions of the dotplots, indicated by arrows (Fig. 2). As was seen in the FeP2 and TKPV genomes [10, 11], a large, rearranged region is present in the FGPV genome between fgpv116–132 (fwpv114–126; cnpv141–171 – green arrows), shown as an alignment schematic in fig. 3a. A second area of rearrangement is also noted between fgpv152–178 (fwpv146–165; cnpv192–238 – red arrows), shown as an alignment schematic in fig. 3b. Regions of rearrangement are referred to from here onward by the FWPV gene annotations as it is the prototype of the genus.

Analysis of full genome alignments revealed three more regions of difference closer to the boundaries of the cores and the termini. These regions fwpv031–047 (fgpv030–044; cnpv050–065), fwpv058–077 (fgpv055–078; cnpv082–104) and fwpv193–211 (fgpv207–227; cnpv267–285) (Fig. 4a–c) were more similar in size across genomes and therefore not as easily visible in the dotplots. Figure 2H) shows a dotplot comparison of FGPV and VACV (strain Copenhagen). Regions of identity have been highlighted as coloured lines in the inset image which correspond to the conserved regions depicted in Fig. 5.

Figure 5 shows a genome schematic of VACV compared to FWPV with each of the four conserved segments previously identified [21] in different colours as in Fig. 2h). The regions of difference identified in this study are shown below these segments, with ORF numbers and size in kilobases for FWPV. The regions of difference largely correspond with the definitions of the conserved regions identified previously although some overlap is present.

Core/conserved ORFs
Ninety ORFs have been noted to be conserved in all ChPV genomes [21, 34, 35]. VACV F16 L was erroneously added to the above list, as this ORF is not present in avian poxvirus genomes; and two ORFs (fwpv194 and fwpv194.1) and their equivalents in the other genomes, previously considered conserved among ChPV, have been excluded from the list as they are not present in the TKPV genome. A fourth ORF, fwpv103 was also removed as it is truncated/fragmented in the TKPV genome (tkpv074). ORF fwpv095 and the relative equivalents were also removed as no orthologue is present in SWPV2. Lastly, orthologues fwpv168 (288aa) and cnpv241 (215aa) differ in length by 25% and were excluded. Therefore, in this study, 83 ORFs are considered to be conserved amongst ChPV. It was also previously, noted that a further 89 ORFs were conserved between FWPV, CNPV, FeP2 and PEPV [10]. This list has been updated with the addition of the four, more recent genomes (TKPV, SWPV-1, SWPV-2 and FGPV) and exclusion of ORFs that differ in length by more than 20% bringing the total to 47 (Table 3) and bringing the total number of conserved ORFs in the sequenced avian poxvirus genomes to 130. The TKPV genome is considerably smaller than the other avian poxvirus genomes and as such, is the only genome missing an ORF that would otherwise be conserved in 32 cases (data not shown). Of these 32 ORFs, 28 are either hypothetical proteins or members of gene families suggesting that they are unlikely to be essential to the viral life cycle. Several ORFs were identified that were unreported in the study describing the TKPV genome [11]. Eight of these (tkpv63.1, tkpv86.1, tkpv121.1, tkpv127.1 and tkpv130.1–130.4) are of the 83 ORFs considered to be conserved in all ChPV genomes. A further two ORFs (tkpv1.1 and tkpv151.1) in common amongst avian poxviruses and ORF tkpv60.1, which is not conserved, were also identified on further inspection.

A concatenated nucleotide alignment of the 130 conserved ORFs from each of the sequenced virus genomes shows FGPV to have the greatest degree of identity to the other South African isolates, PEPV and FeP2 (~96%) followed by FWPV (~90%), the clade B isolates (68–70%) and lastly TKPV (~65%) (Table 4). An amino acid alignment showed percentage identities to be very

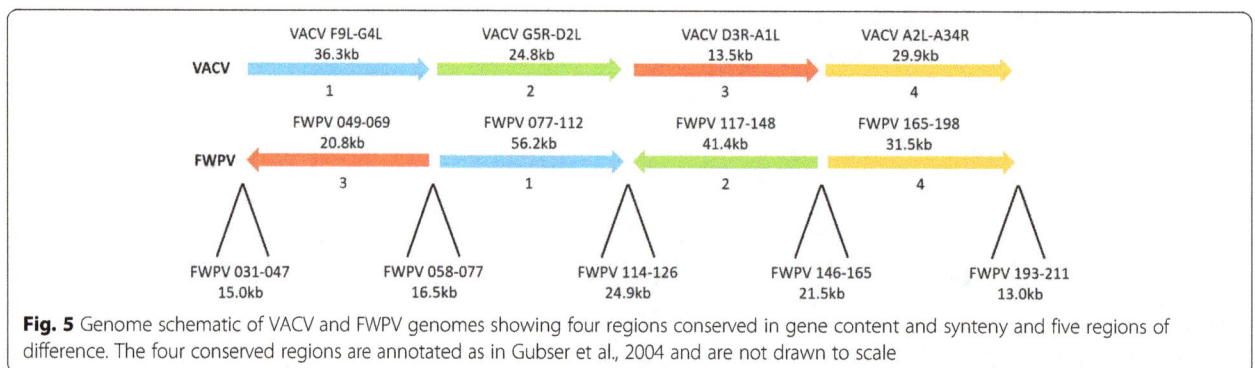

Fig. 5 Genome schematic of VACV and FWPV genomes showing four regions conserved in gene content and synteny and five regions of difference. The four conserved regions are annotated as in Gubser et al., 2004 and are not drawn to scale

Table 3 47 ORFs found to be uniquely conserved in each of the fully sequenced avian poxvirus genomes

FWPV	CNPV	PEPV	FeP2	FGPV	TKPV	SWPV1	SWPV2	Function
016	032	019	019	011	001.1a	024	028	Ig-like domain
017	033	020	020	012	002	025	029	V-type Ig domain
020	038	024	024	017	005	028	034	C4L/C10L protein
021	039	025	025	018	006	029	035	GPCR
022	040	026	026	019	007	030	036	Ankyrin repeat
023	041	027	027	020	008	031	037	Ankyrin repeat
024	042	028	028	021	009	032	038	Ankyrin repeat
030	048	035	035	029	012	038	044	Alkaline phosphodiesterase
031	050	036	036	030	013	040	046	Ankyrin repeat
035	053	040	038	034	016	044	049	Hypothetical protein
037	055	041	039	036	017	046	051	Hypothetical protein
039	058	043	041	038	020	049	054	B-cell lymphoma 2 (Bcl-2)
040	059	044	042	039	021	050	055	Serpin
043	061	046	044	041	022	052	057	DNA ligase
044	062	047	045	042	023	053	058	Serpin family
046	063	048	046	043	024	054	059	Hydroxysteroid dehydrogenase
047	065	049	047	044	025	056	061	Semaphorin
048	068	050	048	045	026	059	064	GNS1/SUR4
054	076	056	054	051	032	066	072	mutT motif
065	088	067	065	064	040	078	083	Hypothetical protein
068	092	070	068	067	042	082	087	Hypothetical protein
070	094	072	070	069	044	084	089	T10-like protein
071	097	075	072	072	046	087	092	Hypothetical protein
075	103	078	077	076	050	092	098	N1R/p28
086	113	089	087	087	060	102	108	Thymidine kinase
091	118	095	093	093	065	107	113	Hypothetical protein
092	119	096	094	094	066	108	114	Hypothetical virion core protein
104	131	108	106	106	075	120	126	Hypothetical protein
105	132	109	107	107	076	121	127	Hypothetical protein
110	137	114	112	112	080	126	132	Hypothetical protein
113	140	117	115	115	083	129	135	Hypothetical protein
145	191	153	146	151	109	167	179	Hypothetical protein
151	199	159	153	157	113	175	187	Deoxycytidine kinase
190	264	203	195	204	140	237	250	A-type inclusion protein
191	265	204	196	205	141	238	251	A-type inclusion protein
196	270	210	202	211	144	243	256	Hypothetical protein
201	273	215	207	216	149	247	259	Hypothetical protein
203	274	216	208	217	150	248	260	Tyrosine kinase
205	276	218	210	219	151	250	262	Hypothetical protein
207	278	220	212	221	151.1a	252	264	Hypothetical protein
208	281	222	214	224	152	255	267	Hypothetical protein
211	285	225	216	227	153	259	271	Epidermal Growth Factor
212	286	226	217	228	154	260	272	Serine/threonine protein kinase
213	287	227	218	229	155	261	273	Hypothetical protein

Table 3 47 ORFs found to be uniquely conserved in each of the fully sequenced avian poxvirus genomes *(Continued)*

FWPV	CNPV	PEPV	FeP2	FGPV	TKPV	SWPV1	SWPV2	Function
214	289	228	219	230	156	263	275	Putative 13.7 kDa protein
219	296	234	226	238	161	272	282	Ankyrin repeat
232	304	248	238	251	164	283	290	Ankyrin repeat

[a]tkpv 001.1 and 151.1 as well as [a]swpv1 241.1 and [a]swpv2 254.1 were not reported in the literature [11, 12], but were identified on inspection of the sequences deposited in Genbank

similar with a maximum difference of 3% compared to the equivalent nucleotide identity (data not shown).

Multigene families

Avian poxviruses contain several, large, multigene families with immune related functions that can make up close to 50% of the genome. Table 5 below outlines the copy numbers of each of the 14 multigene families identified in the FGPV genome compared to that of the other sequenced avian poxvirus genomes. Overall FGPV has a similar complement of multi-gene families but has significantly more ankyrin repeat family genes than are found in other clade A viruses.

Reticuloendotheliosis virus (REV)

REV insertions are typically found between fwpv201–203. The entire FGPV genome was searched for the presence of REV LTRs as well as the *gag, pol* and *env* genes. None of the REV elements were found anywhere in the FGPV genome.

Discussion

Overall, the FGPV genome was found to be similar to other avian poxvirus genomes in terms of genome size, AT content and number of ORFs but was also found to be distinct from these genomes in several ways. FGPV was found to have a unique complement of gene family proteins as well as several ORFs that were truncated, fragmented or extended relative to their closest orthologues. The majority of these altered ORFs were found in

the terminal regions and encode hypothetical or gene family proteins which is expected as they are largely involved in virus-host interactions which are host specific. Also present in the terminal regions are the majority of the genes of interest that are only present in a subset of avian poxvirus genomes, if at all. Three FGPV ORFs were discovered (fgpv 117, 232 and 256) that have yet to be identified in other avian poxvirus genomes and the FGPV genome differed from all other avian poxvirus genomes at the regions of difference identified in this analysis. A nucleotide alignment of 130 conserved ORFs showed FGPV to be most closely related to South African isolates PEPV and FeP2, followed by FWPV, the clade B isolates and lastly TKPV.

This paper confirms the differences between avipoxviruses and orthopoxviruses in gene synteny. The FWPV genome was first shown to exhibit major organisational differences compared to the genome of Vaccinia virus (VACV) using restriction enzyme mapping. It was shown that large segments of the FWPV genome had been reversed and/or translocated relative to VACV although gene content appeared to be largely maintained [36]. Sequencing of the FWPV genome and other ChPV genomes allowed for more detailed comparisons which showed that the core region forms a continuous block in all ChPVs except parapoxviruses and avian poxviruses due to various genome rearrangements. It was specifically noted that the core region of avian poxviruses has broken into four segments two of which have been reversed and one of which has been translocated [7, 21]. Sequencing of the CNPV genome allowed comparison of the regions

Table 4 Pairwise comparison of the % identity and number of differences between nucleotide alignments of 130 conserved genes in eight avian poxvirus genomes

	FWPV	FeP2	PEPV	FGPV	CNPV	SWPV1	SWPV2	TKPV
FWPV		12,486	12,550	11,763	40,283	39,648	40,454	44,591
FeP2	90.2		4967	4636	39,532	39,000	39,705	44,412
PEPV	90.2	96.1		4548	39,558	39,041	39,733	44,468
FGPV	90.8	96.4	96.5		39,719	39,138	39,893	44,602
CNPV	68.8	69.4	69.4	69.3		27,136	1228	45,260
SWPV1	69.3	69.8	69.8	69.7	78.9		27,405	44,385
SWPV2	68.7	69.3	69.3	69.2	99.0	78.7		45,466
TKPV	65.6	65.8	65.8	65.8	65.2	65.8	65.0	

% identities are shown in the lower left and the number of nucleotide differences shown in the upper right

Table 5 Copy number of ORFs in each of the 14 multi-gene families identified in each of the fully sequenced avian poxvirus genomes

Gene family	FWPV	FP9	CNPV	PEPV	FeP2	TKPV	FGPV	SWPV1	SWPV2
Ankyrin Repeat	31	22	51	33	26	16	45	50	46
B22R	6	5	6	5	4	1	4	6	7
N1R/p28	10	8	26	11	11	3	13	20	20
C4L/C10L	3	3	3	2	2	2	2	2	3
CC chemokine	4	4	5	1	4	2	6	6	5
C-type lectin	9	6	11	7	4	2	4	13	11
GPCR	3	2	4	3	2	2	3	4	4
HT motif	6	6	5	5	4	1	7	4	4
Ig-like domain	5	4	9	6	4	3	9	9	8
Serpin	5	5	5	4	4	3	5	5	5
EFc	3	2	2	1	1	1	1	2	2
TGF-β	1	1	5	1	1	1	1	3	4
B-NGF	2	2	2	0	0	2	3	2	2
IL-18 BP	1	1	3	1	0	2	0	3	3
TOTAL	89	71	137	80	67	41	103	129	124
% of TOTAL ORFs	34	29	42	28	25	24	36	42	40

found between these four segments, to the equivalent FWPV regions, and showed major differences in gene content. At the time, it was unclear if these differences were due to subclade specificities or were a feature of all avian poxviruses.

Although the overall genome architecture of avian poxviruses is largely conserved, with the expected variability in the termini, a pattern is emerging with all sequenced isolates exhibiting major differences in multiple, defined, central regions. FP9 was the first clade A isolate noted to be somewhat different to FWPV in the fwpv114–126 region with the truncation of fp9 115 and deletion of fp9 125 and fp9 126 [8]. FeP2 was then noted to have a large deletion of over 10 kb and although this region in the PEPV genome was of similar length to FWPV, several inserted and deleted ORFs were noted [10]. In the TKPV genome, ORF tkpv085 (fwpv114) was identified as being affected by genomic rearrangement [11]. In this study, alignment of this region in all clade A viruses shows a large variation in length from ~11 kb in Fep2, to over 24 kb in FWPV and PEPV. In FGPV this region spans 16.5 kb and encodes 16 ORFs (Fig. 3a). If this comparison is expanded to avian poxviruses in other clades, the difference is much larger with a variation in length from ~5 kb (encoding seven ORFs) in TKPV to over 42.6 kb (encoding 31 ORFs) in CNPV. Four ORFs in this region (fwpv117 - fwpv120 and the relative equivalents) are conserved among all viruses and syntenic as would be expected, as they are of the 83 genes conserved and considered essential among all ChPVs. What is unexpected is the placement and retention of this pocket of four essential genes in a region of highly divergent gene content and synteny. ORF fwpv117 encodes a putative nuclease involved in viral DNA replication [37], fwpv118 encodes RNA polymerase subunit RPO7 fwpv119 is of unknown function and fwpv120 encodes a virion core protein involved in several stages of virion morphogenesis [38]. This region also contains several ORFs unique to avian poxviruses.

The second region of difference found in the genome core shows less difference in length among clade A viruses from ~19 kb in Fp9 to ~25 kb in FGPV but as above, when including viruses in other clades this difference in length increases considerably from ~14 kb in TKPV (encoding 17 ORFs) to ~52 kb in SWPV1 (encoding 42 ORFs) (Fig. 3b). Only one ORF (fwpv148) conserved in all ChPV is present in this region and encodes a virion protein involved in immature virion formation [39].

The three regions of difference found closer to the termini (Fig. 4a), b) and c)) are more similar in size across genomes and contain several more conserved ORFs. Again, this is unusual for poxviruses as we would expect these regions to be less conserved compared to the two, central regions of difference. The three clade B viruses (CNPV, SWPV1 and SWPV2) are more similar to each other in terms of gene content at these locations than the clade A viruses and CNPV and SWPV2 are nearly identical as expected based on the conserved ORF identities and phylogenetic analysis.

Poxviruses have been known to use gene duplication and subsequent, mutationally driven, diversification of paralogues to their advantage to combat host immune responses. Elde et al., 2012, specifically looked at the ability of VACV to adapt to growth in human cells where the host range factor K3 L is non-functional. E3L functions similarly to K3 L and is functional in human cells. It was found that when E3L was deleted, leaving the virus susceptible to host antiviral responses, the K3 L gene was recurrently amplified, with each of the paralogues able to explore mutational space until an adaptive substitution was found. Effective copies of the K3 L gene were retained and the others lost over generations. It was also noted that duplications other than K3 L all occurred in the terminal regions of the VACV genome [40].

In the case of avian poxviruses, it is interesting to note that several ORFs in the fwpv114–126; cnpv141–171 region of difference of CNPV, SWPV1 and SWPV2 are present as repeats/paralogues of gene family proteins: cnpv143–144 = ANK repeat; cnpv150–151 = ANK repeat; cnpv154–155 = B22R; cnpv157–158 = TGF-β; cnpv159–160 = N1R/p28; cnpv161–162 = TGF-β; cnpv166–167 = Ig-like domain; cnpv168–169 = N1R/p28.In the second region of difference (fwpv146–165; cnpv192–238)) found in the central region, the CNPV genome has 18 copies of N1R/p28 like proteins in this region while SWPV1 and SWPV2 contain differing subsets of these, which may be the result of genomic accordions at work earlier in their evolutionary histories.

Conclusions

Genome sequencing and comparative genomics are the gold standards in terms of determining phylogenetic and evolutionary relationships among viral species and explaining differences in host range and pathogenicity. Several important bird species and commercial flocks have been shown to be severely affected by avian poxvirus infection. This study provides the genome sequence of a novel, South African isolate from lesser flamingos and provides insight into overall genome architecture that appears to be unique to avian poxviruses. Given the relative conservation of the central region of other poxvirus genomes, the regions of difference identified here are particular areas of interest in avian poxvirus genomics, but it is currently unclear why these regions would be so susceptible to rearrangement. The mechanisms responsible for such large-scale rearrangements are also yet to be elucidated. As more avian poxvirus genomes are sequenced, exploration and confirmation of these intriguing differences in these important pathogens can be conducted.

Acknowledgements

Thank you to Anelda Philine van der Walt for assistance with QC of the sequence data and to Dr. David Zimmerman of South African National Parks: Veterinary Wildlife Services and Dr. Mark Anderson of Northern Cape Nature Conservation for post mortal collection of the samples and Dr. Emily Lane of National Zoological Gardens for donation of the sample.

Funding

This work is based on research supported by the South African Research Chairs Initiative of the Department of Science and Technology and National Research Foundation (NRF), South Africa. Any opinion, finding and conclusion or recommendation expressed in this material is that of the authors and the NRF does not accept any liability in this regard. Financial support was also provided by the Clinical Infectious Diseases Research Initiative (CIDRI).

Authors' contributions

OC, ND and ALW designed the study. OC performed the experiments and bioinformatics analysis and drafted the manuscript. All authors read, edited and approved the final manuscript.

Author's information

Not applicable.

Competing interests

The authors declare that they have no competing interests

Author details

[1]Division of Medical Virology, Department of Pathology, Faculty of Health Sciences, University of Cape Town, Cape Town, South Africa. [2]Institute of Infectious Disease and Molecular Medicine, University of Cape Town, Cape Town, South Africa. [3]National Health Laboratory Service, Cape Town, South Africa.

References

1. Tripathy DN, Reed WM. Pox. In: Saif YM, Fadly AM, Glisson JR, McDougald LR, Nolan LK, Swayne DE, editors. . Dis. Poult. 12th ed. Ames, Iowa: Blackwell pub. Professional; 2008. p. 291–308.
2. Tripathy DN, Schnitzlein WM, Morris PJ, Janssen DL, Zuba JK, Massey G, et al. Characterization of poxviruses from forest birds in Hawaii. J Wildl Dis. 2000; 36:225–30.
3. Lachish S, Bonsall MB, Lawson B, Cunningham AA, Sheldon BC. Individual and population-level impacts of an emerging poxvirus disease in a wild population of great tits. PLoS One. 2012;7:e48545.
4. Niemeyer C, Favero CM, Kolesnikovas CKM, Bhering RCC, Brandão P, Catão-Dias JL. Two different avipoxviruses associated with pox disease in Magellanic penguins (Spheniscus magellanicus) along the Brazilian coast. Avian Pathol. 2013;42:546–51.
5. Bolte AL, Meurer J, Kaleta EF. Avian host spectrum of avipoxviruses. Avian Pathol. 1999;28:415–32.
6. van Riper C, Forrester DJ. Avian pox. In: Thomas NJ, Hunter DB, Atkinson CT, editors. Infect. Dis. Wild birds. Blackwell publishing professional; 2007. p. 131–76.
7. Afonso CL, Tulman ER, Lu Z, Zsak L, Kutish GF, Rock DL. The genome of fowlpox virus. J Virol. 2000;74:3815–31.
8. Laidlaw SM, Skinner MA. Comparison of the genome sequence of FP9, an attenuated, tissue culture-adapted European strain of Fowlpox virus, with those of virulent American and European viruses. J Gen Virol. 2004; 85:305–22.
9. Tulman ER, Afonso CL, Lu Z, Zsak L, Kutish GF, Rock DL. The genome of canarypox virus. J Virol. 2004;78:353–66.
10. Offerman K, Carulei O, van der Walt AP, Douglass N, Williamson A-L. The complete genome sequences of poxviruses isolated from a penguin and a pigeon in South Africa and comparison to other sequenced avipoxviruses. BMC Genomics. 2014;15:463.
11. Bányai K, Palya V, Dénes B, Glávits R, Ivanics É, Horváth B, et al. Unique genomic organization of a novel Avipoxvirus detected in turkey (Meleagris gallopavo). Infect Genet Evol. 2015;35:221–9.
12. Sarker S, Das S, Lavers JL, Hutton I, Helbig K, Imbery J, et al. Genomic characterization of two novel pathogenic avipoxviruses isolated from pacific shearwaters (Ardenna spp.). BMC Genomics. 2017;18:298.

13. Binns MM, Boursnell ME, Tomley FM, Campbell J. Analysis of the fowlpoxvirus gene encoding the 4b core polypeptide and demonstration that it possesses efficient promoter sequences. Virology. 1989;170:288–91.

14. Lüschow D, Hoffmann T, Hafez HM. Differentiation of avian poxvirus strains on the basis of nucleotide sequences of 4b gene fragment. Avian Dis. 2004; 48:453–62.

15. Weli SC, Traavik T, Tryland M, Coucheron DH, Nilssen Ø. Analysis and comparison of the 4b core protein gene of avipoxviruses from wild birds: evidence for interspecies spatial phylogenetic variation. Arch Virol. 2004;149:2035–46.

16. Jarmin S. Avipoxvirus phylogenetics: identification of a PCR length polymorphism that discriminates between the two major clades. J Gen Virol. 2006;87:2191–201.

17. Carulei O, Douglass N, Williamson A-L. Phylogenetic analysis of three genes of Penguinpox virus corresponding to vaccinia virus G8R (VLTF-1), A3L (P4b) and H3L reveals that it is most closely related to Turkeypox virus, Ostrichpox virus and Pigeonpox virus. Virol J. 2009;6:52.

18. Manarolla G, Pisoni G, Sironi G, Rampin T. Molecular biological characterization of avian poxvirus strains isolated from different avian species. Vet Microbiol. 2010;140:1–8.

19. Gyuranecz M, Foster JT, Dan A, Ip HS, Egstad KF, Parker PG, et al. Worldwide phylogenetic relationship of avian poxviruses. J Virol. 2013;87:4938–51.

20. Offerman K, Carulei O, Gous TA, Douglass N, Williamson A-L. Phylogenetic and histological variation in avipoxviruses isolated in South Africa. J Gen Virol. 2013;94:2338–51.

21. Gubser C, Hué S, Kellam P, Smith GL. Poxvirus genomes: a phylogenetic analysis. J Gen Virol. 2004;85:105–17.

22. Arai S, Arai C, Fujimaki M, Iwamoto Y, Kawarada M, Saito Y, et al. Cutaneous tumour-like lesions due to poxvirus infection in Chilean flamingos. J Comp Pathol. 1991;104:439–41.

23. Mondal SP, Lucio-Martínez B, Buckles EL. Molecular characterization of a poxvirus isolated from an American flamingo (*Phoeniconais ruber rubber*). Avian Dis. 2008;52:520–5.

24. Henriques AM, Fagulha T, Duarte M, Ramos F, Barros SC, Luís T, et al. Avian poxvirus infection in a flamingo (*Phoenicopterus ruber*) of the lisbon zoo. J Zoo Wildl Med. 2016;47:161–74.

25. Terasaki T, Kaneko M, Mase M. Avian poxvirus infection in flamingos (*Phoenicopterus roseus*) in a zoo in Japan. Avian Dis. 2010;54:955–7.

26. Middlemiss E. Avian pox in South Africa. Ostrich. 1961;32:20–2.

27. Allwright DM, Burger WP, Geyer A, Wessles J. Avian pox in ostriches. J S Afr Vet Assoc. 1994;65:23–5.

28. Stannard LM, Marais D, Kow D, Dumbell KR. Evidence for incomplete replication of a penguin poxvirus in cells of mammalian origin. J Gen Virol. 1998;79:1637–46.

29. Bwala DG, Fasina FO, Duncan NM. Avian poxvirus in a free-range juvenile speckled (rock) pigeon (*Columba guinea*). J S Afr Vet Assoc. 2015;86:1259.

30. Zimmermann D, Anderson MD, Lane E, Van Wilpe E, Carulei O, Douglass N, et al. Avian poxvirus epizootic in a breeding population of lesser flamingos (*Phoenicopterus minor*) at Kamfers dam, Kimberley, South Africa. J Wildl Dis. 2011;47:989–93.

31. Kotwal G, Abrahams M-R. Growing poxviruses and determining virus titer. In: Isaacs S, editor. Vaccinia virus Poxvirology. Humana press; 2004. p. 101–12.

32. Hendrickson RC, Wang C, Hatcher EL, Lefkowitz EJ. Orthopoxvirus genome evolution: the role of gene loss. Viruses. 2010;2:1933–67.

33. Krumsiek J, Arnold R, Rattei T. Gepard: a rapid and sensitive tool for creating dotplots on genome scale. Bioinforma Oxf Engl. 2007;23:1026 8.

34. Upton C, Slack S, Hunter AL, Ehlers A, Roper RL. Poxvirus orthologous clusters: toward defining the minimum essential poxvirus genome. J Virol. 2003;77:7590–600.

35. Lefkowitz EJ, Wang C, Upton C. Poxviruses: past, present and future. Virus Res. 2006;117:105–18.

36. Mockett B, Binns MM, Boursnell ME, Skinner MA. Comparison of the locations of homologous fowlpox and vaccinia virus genes reveals major genome reorganization. J Gen Virol. 1992;73:2661–8.

37. Senkevich TG, Koonin EV, Moss B. Predicted poxvirus FEN1-like nuclease required for homologous recombination, double-strand break repair and full-size genome formation. Proc Natl Acad Sci U S A. 2009;106:17921–6.

38. Mercer J, Traktman P. Genetic and cell biological characterization of the vaccinia virus A30 and G7 phosphoproteins. J Virol. 2005;79:7146–61.

39. Szajner P, Jaffe H, Weisberg AS, Moss BA. Complex of seven vaccinia virus proteins conserved in all chordopoxviruses is required for the association of membranes and viroplasm to form immature virions. Virology. 2004;330:447–59.

40. Elde NC, Child SJ, Eickbush MT, Kitzman JO, Rogers KS, Shendure J, et al. Poxviruses deploy genomic accordions to adapt rapidly against host antiviral defenses. Cell. 2012;150:831–41.

10

Single-cell RNA-Seq analysis reveals dynamic trajectories during mouse liver development

Xianbin Su[1†], Yi Shi[1†], Xin Zou[1†], Zhao-Ning Lu[1†], Gangcai Xie[2], Jean Y. H. Yang[3], Chong-Chao Wu[1], Xiao-Fang Cui[1], Kun-Yan He[1], Qing Luo[1], Yu-Lan Qu[1], Na Wang[1], Lan Wang[1] and Ze-Guang Han[1,4*] (ID)

Abstract

Background: The differentiation and maturation trajectories of fetal liver stem/progenitor cells (LSPCs) are not fully understood at single-cell resolution, and a priori knowledge of limited biomarkers could restrict trajectory tracking.

Results: We employed marker-free single-cell RNA-Seq to characterize comprehensive transcriptional profiles of 507 cells randomly selected from seven stages between embryonic day 11.5 and postnatal day 2.5 during mouse liver development, and also 52 Epcam-positive cholangiocytes from postnatal day 3.25 mouse livers. LSPCs in developing mouse livers were identified via marker-free transcriptomic profiling. Single-cell resolution dynamic developmental trajectories of LSPCs exhibited contiguous but discrete genetic control through transcription factors and signaling pathways. The gene expression profiles of cholangiocytes were more close to that of embryonic day 11.5 rather than other later staged LSPCs, cuing the fate decision stage of LSPCs. Our marker-free approach also allows systematic assessment and prediction of isolation biomarkers for LSPCs.

Conclusions: Our data provide not only a valuable resource but also novel insights into the fate decision and transcriptional control of self-renewal, differentiation and maturation of LSPCs.

Keywords: Liver stem/progenitor cells, Single-cell RNA-Seq, Developmental trajectory, Cholangiocyte, Fate decision

Background

The two major epithelial cell types of liver, hepatocytes and cholangiocytes, are differentiated from hepatoblasts during embryonic liver development [1–3]. Liver stem/progenitor cells (LSPCs) have been suggested to exist in fetal and adult liver and are generally defined as cells with the potential to differentiate into both hepatocytes and cholangiocytes [4]. Hepatoblasts are considered a type of LSPC during liver development; however, liver development involves many different cell types derived from endoderm and mesoderm and their reciprocal interactions. Thus, it is possible that fetal LSPCs could

exhibit more complex genetic programs and developmental controls [5].

Previous studies involving the cell fate determination of LSPCs relied on isolation with surface membrane proteins such as EpCAM [6, 7], DLK1 [8], E-cadherin [9], CD13 [10, 11], and CD133 [11]. However, the cell fate determination of LSPCs isolated from embryonic livers, by employing different surface markers is controversial, reflecting the complexity of liver organogenesis and the heterogeneity of LSPCs [4]. Liver development is a dynamic process with the possibility of changing cell markers within LSPC populations. Thus, isolation of LSPCs based on the a priori knowledge about the limited surface markers may restrict the recognition of LSPCs and their functional features to some extent. The best way to perform the clonogenicity and repopulation assays is to assess the differentiation potential of single cells from embryonic livers; however, the practical difficulty of in vivo or ex vivo experiments hinders

* Correspondence: hanzg@sjtu.edu.cn
†Equal contributors
[1]Key Laboratory of Systems Biomedicine (Ministry of Education) and Collaborative Innovation Center of Systems Biomedicine, Shanghai Center for Systems Biomedicine, Shanghai Jiao Tong University, 800 Dongchuan Road, Shanghai 200240, China
[4]Shanghai-MOST Key Laboratory for Disease and Health Genomics, Chinese National Human Genome Center at Shanghai, Shanghai, China
Full list of author information is available at the end of the article

observation of the trajectory of cell lineage differentiation and maturation of these naïve LSPCs at single-cell resolution.

Recently, the development of single-cell sequencing-based technology has provided a unique chance to address many longstanding questions, such as cell lineage relationships and heterogeneity in a given cell population [12–14]. Single-cell transcriptomic analysis, such as RNA-Seq, would supplant the coarse notions of the marker-based cell types and uncover new cell types by the unbiased sampling of single cells [15]. For liver research, the gene expression profiles of zonation and spatial division of hepatic lobule in adult mouse liver were revealed at single-cell resolution [16], and multi-lineage communication is also shown to be important for human liver bud development [17]. Currently, the definition and molecular state of LSPCs during liver development are still obscure, and the developmental trajectories of LSPCs, including self-renewal, differentiation and maturation, are not fully understood at single-cell resolution.

To address the above questions, in this study, we applied single-cell RNA-Seq and quantitative RT-PCR (qPCR) to analyze ~ 800 single cells from eight different stages during mouse liver development. The transcriptomic analysis of LSPCs reconstructed their stepwise differentiation and maturation process at single-cell resolution. In addition to cell surface molecules, we also uncovered the contiguous but discrete genetic control by transcription factors and signaling pathways. To further understand the fate decision and differentiation process of LSPCs, we also analyzed the single-cell transcriptomic profiles of Epcam-positive cholangiocytes. Our single-cell transcriptomic analysis of LSPCs during mouse liver development provides insights into the transcriptional control of their self-renewal, differentiation and maturation and is a useful resource for future research, including research on isolation methods designed for LSPCs.

Results
Overview of single-cell qPCR and RNA-Seq of developing mouse livers

To comprehensively understand the transcriptional program during liver development, we carried out single-cell transcriptomic analysis, including qPCR, on 722 cells and RNA-Seq on 559 cells derived from mouse fetal livers at eight developmental stages, including embryonic day (E) 11.5, 12.5, 13.5, 14.5, 16.5, 18.5 and postnatal day (P) 2.5 and P3.25 (Fig. 1a-b). We first randomly selected 467 single cells and then assessed them via single-cell qPCR with genes related to cell types and liver development (Fig. 1c and Additional file 1: Figure S1). We observed that hepatic marker genes such as *Afp*, *Alb*, *Ttr* and *Serpina1a* were highly expressed in some cells from E11.5 to E16.5 livers, which were later identified as hepatoblasts. However, a similar gene expression pattern was rarely observed in single cells from E18.5 and P2.5 livers (Additional file 1: Figure S1). After removing low quality libraries, we performed RNA-Seq on 415 single cells using the same cDNA libraries as qPCR. We proposed the molecular patterns for putative LSPCs after analysis of these cells and then collected 255 single cells from another batch of fetal livers as biological replicates, and 92 single cells were chosen for RNA-Seq (Fig. 1b). We also used flow cytometry to isolate Epcam$^+$ cells from P3.25 livers, which were likely to be cholangiocytes [7, 18], and then sequenced 52 these Epcam$^+$ single cells (Fig. 1b).

In this study, the median mapping rates of sequencing reads within each developmental stage ranged from 57% to 78%. The median numbers of unique mapped reads ranged from 1.1 to 3.8 million per cell. The median numbers of genes detected with confidence of fragments per kilobase of exon model per million (FPKM) > 1 ranged from approximately 3000 to 6000 for all stages except Epcam$^+$ cells from P3.25 livers, which only showed a median number of around 2000 genes despite similar sequencing depth and mapping rate (Additional file 1: Figure S2a and Additional file 2: Table S1). The decreased number of genes expressed in Epcam$^+$ cells from P3.25 livers could be due to their more differentiated status. We introduced ERCC RNA Spike-ins as technical controls, and high correlation coefficients among single cells at each stage based on the 92 Spike-ins were observed (Additional file 1: Figure S2b), indicating low technical noise in our data. We further quantitatively evaluated the correlation between RNA-Seq and qPCR data from the same single cells, and they were positively correlated with each other (Additional file 1: Figure S2c). Here, the median correlation coefficients between single-cell RNA-Seq and qPCR were approximately 0.9 for all stages (Additional file 1: Figure S2c).

Identification of LSPCs in developing mouse livers via marker-free transcriptomic profiling

Limited markers may lead to the incorrect identification of cell populations, and single-cell transcriptomic profiling facilitates ab initio cell-type characterization. Because a very large portion of E18.5 and P2.5 cells were mature hepatocytes (Additional file 1: Figs. S1, S5), we focused on single cells from E11.5 to E16.5 for cell type identification. To reduce the disturbance in the gene expression-based cell clustering analysis, the transcripts severely corrupted by technical noise were filtered via a statistical model (Additional file 1: Figure S2d). Subsequently, hierarchical clustering (HC) enabled decomposition of these cells into six groups which were later confirmed as endothelial cells, erythrocyte, hepatoblast,

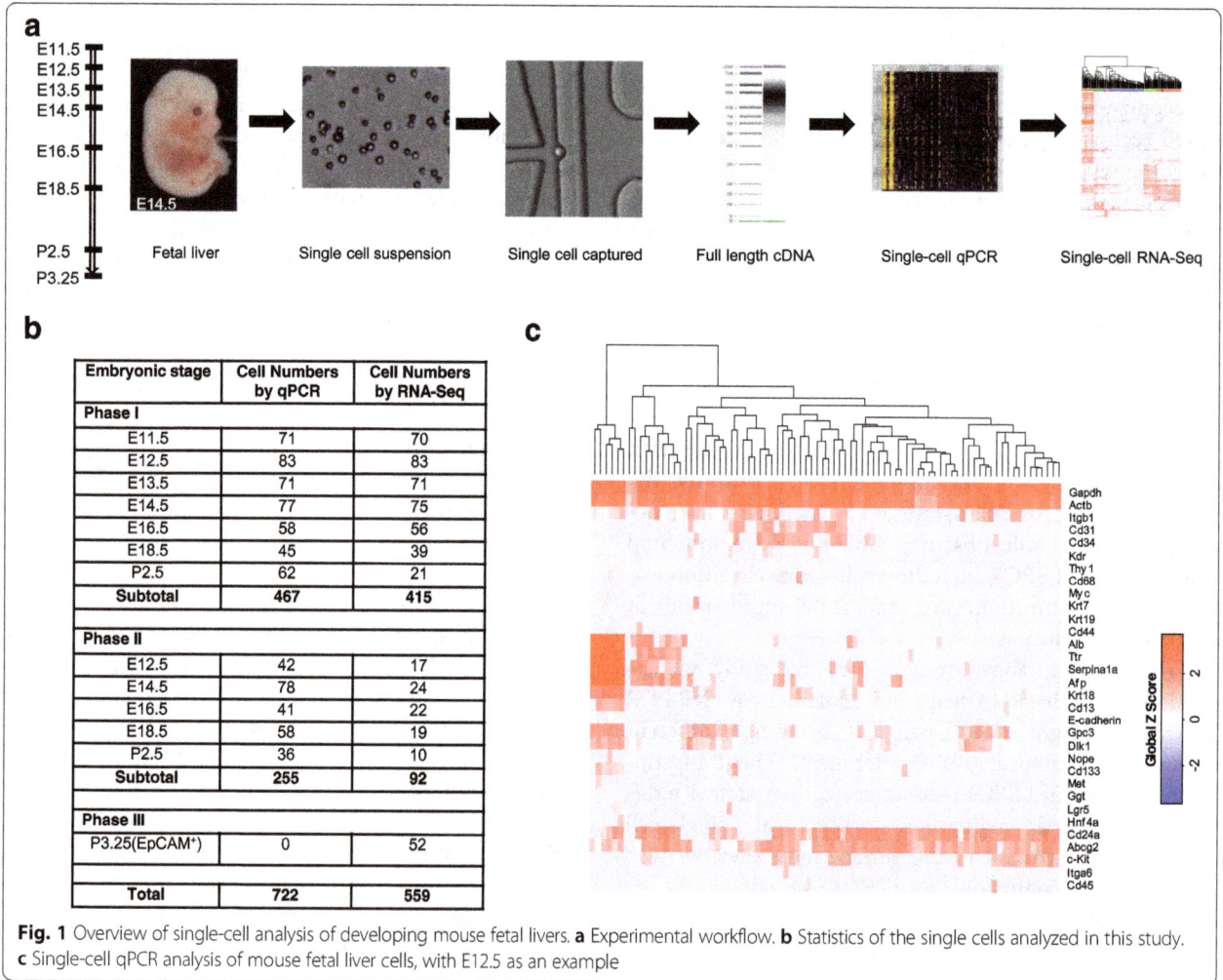

Fig. 1 Overview of single-cell analysis of developing mouse fetal livers. **a** Experimental workflow. **b** Statistics of the single cells analyzed in this study. **c** Single-cell qPCR analysis of mouse fetal liver cells, with E12.5 as an example

macrophage, megakaryocyte and mesenchymal cells (Fig. 2a and Additional file 1: Figure S3a-b), where the classification also could be clearly visualized with t-distributed stochastic neighbor embedding (t-SNE) plot (Fig. 2b).

Among the six groups, one group was classified as the endothelial lineage, while three were assigned to myeloid lineages of the hematopoietic system, including erythrocytic, megakaryocytic, monocytic or macrophagic (likely Kupffer cells) lineages based on specific markers (Fig. 2c and Additional file 1: Figure S3b) and the Gene Ontology (GO) enrichment analysis (Additional file 1: Figure S3c), consistent with the fact that fetal liver is the main hematopoietic organ during this developmental period. These single cells expressed known lineage-specific marker genes; for example, endothelial cells express *Lyve1* and *Kdr*; erythrocytes highly express *Hba-a1* and *Hbb-bt*; macrophages express *Ptprc* (*Cd45*), *Cd68* and *Cd52*; and megakaryocytes express *Itga2b* and *Itgb3*.

One of the remaining two groups belonged to hepatoblasts that express hepatic markers such as *Afp* and *Alb*,

stem/progenitor-related genes such as *Gpc3* and *Dlk1* (Fig. 2c and Additional file 1: Figure S3a), and genes related to liver functions such as lipid metabolism and blood coagulation (Additional file 1: Figure S3c). Interestingly, the other group expressed some stem/progenitor-related genes, such as *Gpc3*, *Dlk1*, *Sox9* and *Sox11*, but had no or low expression of genes related to hepatic lineages (Additional file 1: Figure S3b).

As known, the liver bud is expanding in septum transversum mesenchyme at E11.5 stage, so we speculated that this un-identified group may be related to the mesenchymal phenotype. By checking the expression profiles of certain marker genes closely related to LSPCs and mesenchymal cells (Fig. 2d), we figured out the distinct signatures that can distinguish hepatoblasts from mesenchymal cells. Significantly, hepatoblasts highly expressed many hepatic lineage-specific markers, such as *Afp*, *Alb*, *Hnf4a*, *Krt18*, *Krt8*, *Hnf1b*, *Hhex* and *Met* (*c-Met*), as well as *Anpep* (*Cd13*) and *Cdh1* (E-cadherin), while the un-identified group expressed significantly higher levels of mesenchymal-related marker genes such as *Vim*,

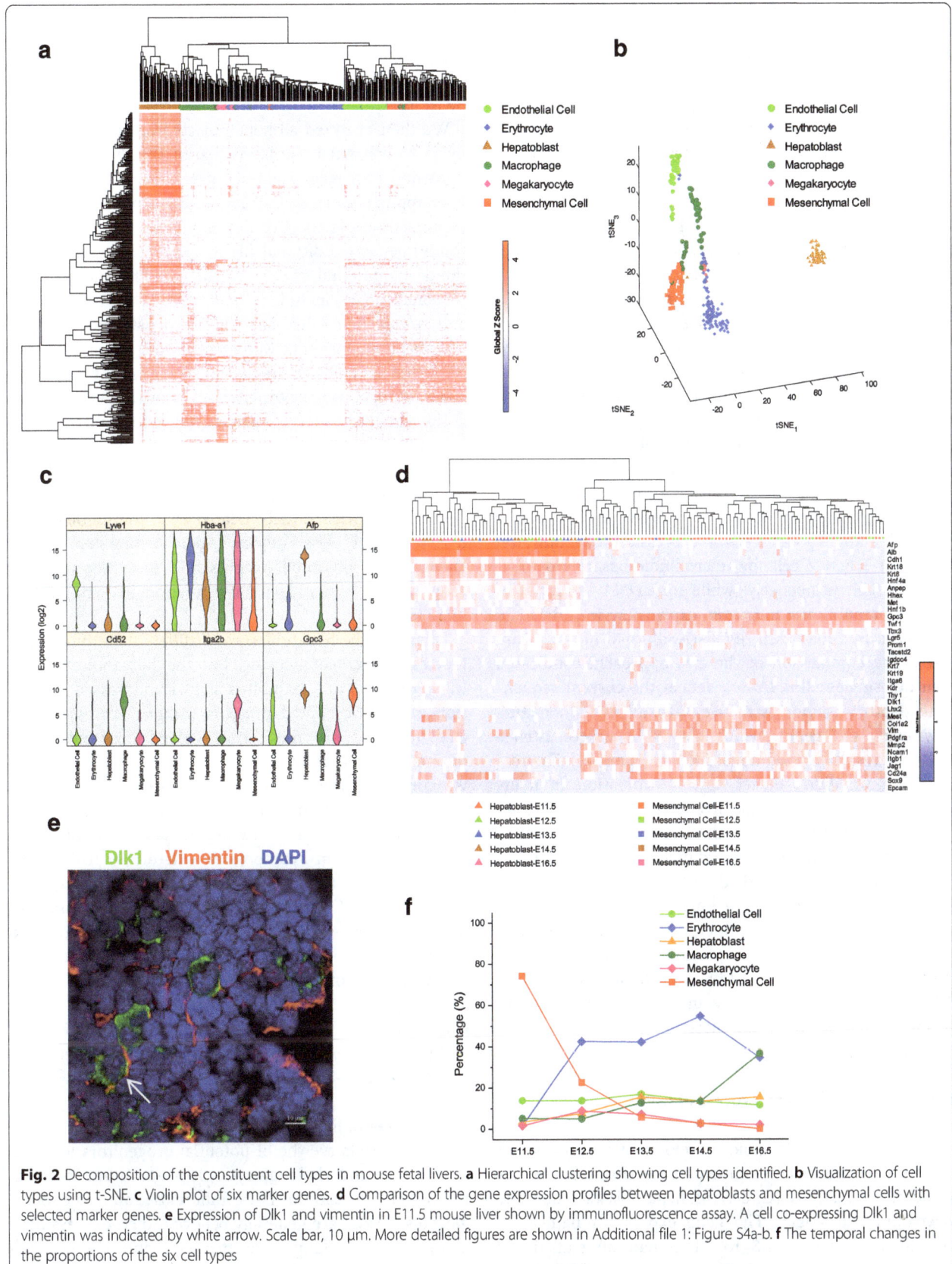

Fig. 2 Decomposition of the constituent cell types in mouse fetal livers. **a** Hierarchical clustering showing cell types identified. **b** Visualization of cell types using t-SNE. **c** Violin plot of six marker genes. **d** Comparison of the gene expression profiles between hepatoblasts and mesenchymal cells with selected marker genes. **e** Expression of Dlk1 and vimentin in E11.5 mouse liver shown by immunofluorescence assay. A cell co-expressing Dlk1 and vimentin was indicated by white arrow. Scale bar, 10 μm. More detailed figures are shown in Additional file 1: Figure S4a-b. **f** The temporal changes in the proportions of the six cell types

Col1a2, *Mest*, *Mmp2*, *Pdgfra*, *Ncam1* and *Lhx2*. We thereby named the group as mesenchymal cells. The interesting thing is that *Dlk1*, a marker of hepatic stem cells, was expressed in the group classified as mesenchymal cells. Here we thus employed immunofluorescence assay and confirmed the existence of such cells co-expressing both Dlk1 and vimentin, a mesenchymal marker (Fig. 2e and Additional file 1: Figure S4a-b).

Epcam is a marker closely related to the differentiation of LSPC into cholangiocytes [4, 18]. Our single-cell RNA-Seq data showed that *Epcam* expression was detected in 1 of 2 cells of E11.5 hepatoblasts and 5 of 43 cells of E11.5 mesenchymal cells, not detected in E12.5 ~ E14.5 cells, but was detected again in 2 of 12 cells of E16.5 hepatoblasts (Fig. 2d). This temporal change of *Epcam* expression indicated our single-cell RNA-seq data was consistent with the previous observation via immunofluorescence assay or flow cytometric analysis [7]. As single-cell RNA-Seq may suffer drop-out issue, here we selected the 8 cells with *Epcam* transcript and 8 cells without *Epcam* transcript form E11.5 and E16.5 for further qPCR validation. For cDNA library before Nextera amplification, *Epcam* expression was only detected in 1 and 2 cells by microfluidic-based and tube-based qPCR, respectively; while for cDNA library after Nextera amplification, *Epcam* expression was detected in 4 cells by tube-based qPCR (Additional file 1: Figure S4c). The comparison of the two approaches indicated that our single-cell RNA-Seq data at the current sequencing depth was more sensitive than qPCR in detecting low expressed genes, which also showed consistence with the results in Additional file 1: Figure S2c. The reason that we didn't detect *Epcam* expression in many cells in these developing livers is probably because *Epcam* is expressed in a small fraction of LSPCs, where the marker-free approach was difficult to detect them.

We then characterized the distribution of the six groups at each stage. Interestingly, the fetal livers from E11.5 to E16.5 all had six cell types despite different proportions (Fig. 2f). The proportion of erythrocytes increased from E11.5 to E14.5 and then decreased at E16.5, and a stepwise decrease in mesenchymal cells and a slight increase in hepatoblasts from E11.5 to E16.5 were observed.

Dynamic developmental process of LSPCs at single-cell resolution

To decipher the dynamic developmental process of LSPCs during liver development, we performed gene set enrichment analysis (GSEA) of hepatoblasts along the developmental stages. The gene expression pattern of hepatoblasts from E12.5 to E14.5 had no statistically meaningful differences, but the comparison of E14.5 and E16.5 hepatoblasts revealed that the cell cycle and mitosis-related genes were significantly enriched in the E14.5 stage (Additional file 1: Figure S5a). This finding suggested that the transition from E14.5 to E16.5 may be the critical differentiation switch for hepatoblasts via cell division.

We then analyzed all hepatoblasts from E11.5 to E16.5 livers to construct the landscape of the dynamic developmental processes. Analysis of variance (ANOVA) was first applied to rank the genes that were differentially expressed across the five developmental stages, and the top 30 genes (Additional file 3: Table S2) were used for the developmental track construction. The gradually upregulated genes included *Apoh*, *Ahsg*, *Alb*, *Kng2*, *Adh1* and *Aldob*, which are specific for hepatocyte function; whereas the genes that were decreased stepwise included *Mdk* that is highly expressed in mid-gestation [19], and *Hhat* (hedgehog acyltransferase) that is required for SHH signaling which is closely related to embryonic development, liver regeneration and liver cancer stem cells [20–23] (Fig. 3a-b and Additional file 1: Figure S5b). In the developmental track, it is clear that E12.5 ~ E14.5 hepatoblasts were closely related in a continuous manner (Fig. 3c). The stepwise gene expression profile changes also provided insights into the transcriptional programs of hepatoblast differentiation and maturation (Fig. 3d).

Theoretically, hepatoblasts will differentiate into hepatocytes and cholangiocytes, but the hepatoblasts from E11.5 to E16.5 only exhibited stepwise increased expression of hepatic-related proteins without expressing biliary markers. Are these cells indeed hepatoblasts or differentiated immature hepatocytes? To figure this out, we compared these embryonic hepatoblasts with P2.5 hepatic cells. Only one hepatic cell from the P2.5 stage was closely related to E16.5 hepatoblasts, and all others were distinct from hepatoblasts, where they had low or no expression of stem/progenitor-related markers, such as *Gpc3*, *Dlk1* and *Afp*, but higher expression of hepatocyte-related genes, such as *Fbp1*, *Mat1a* and *Sult1a1* (Additional file 1: Figure S5c-d). The data indicated that the majority of hepatic cells from P2.5 livers were hepatocytes, although there were a few hepatoblasts, possibly representing precursors for oval cells within adult livers. Here, our data indicated that hepatoblasts in E11.5 ~ E16.5 are authentic hepatoblasts.

LSPC differentiation into cholangiocytes

As hepatoblasts are the bi-potential progenitors for both hepatocytes and cholangiocytes, it is important to reveal the cell fate decision of LSPC differentiation into cholangiocytes. LSPCs are generally believed to co-express hepatocyte and cholangiocyte markers, and we checked such a possibility in these single liver cells from E11.5 to E16.5. Only 4 cells from E11.5, E12.5 and E14.5 livers

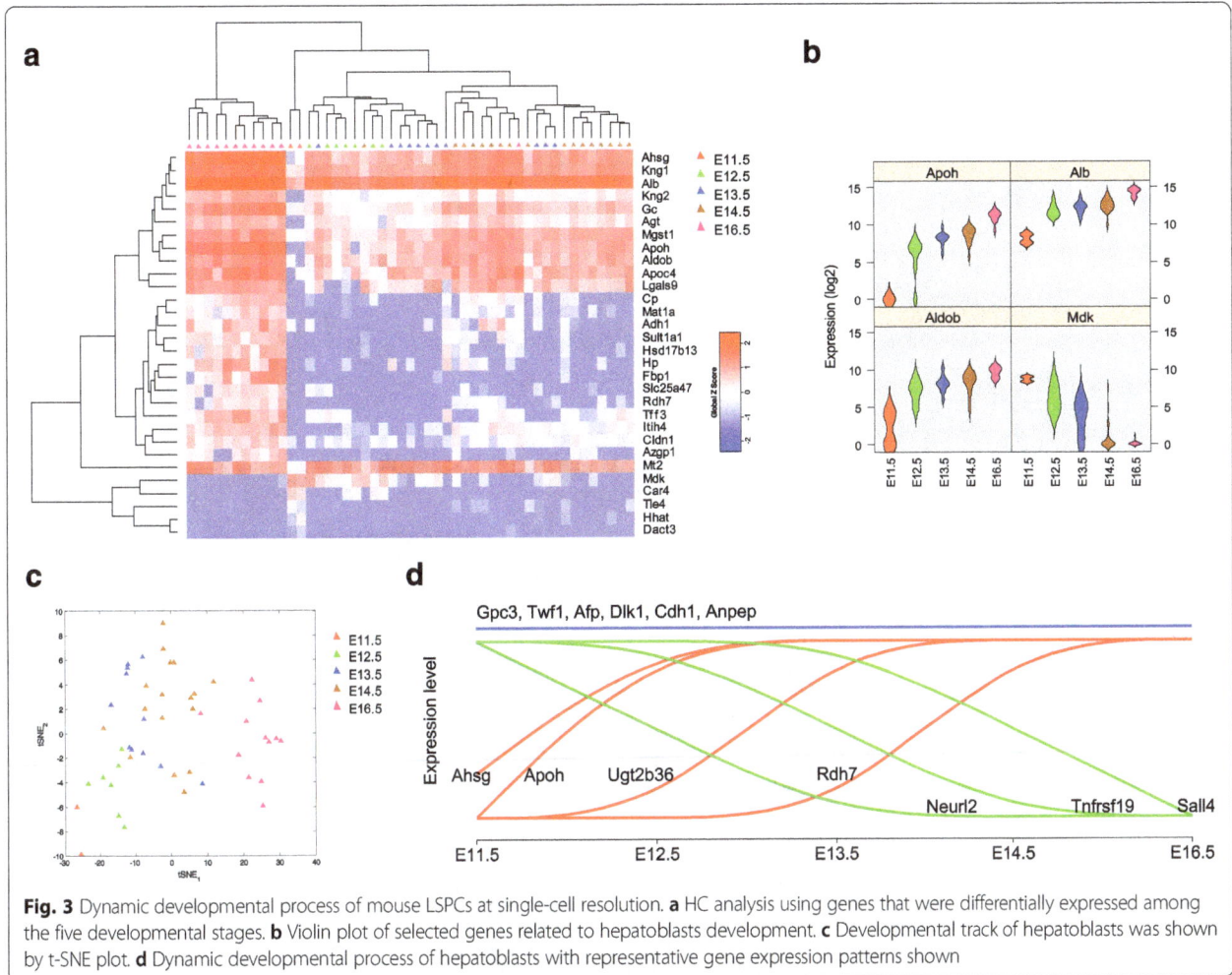

Fig. 3 Dynamic developmental process of mouse LSPCs at single-cell resolution. **a** HC analysis using genes that were differentially expressed among the five developmental stages. **b** Violin plot of selected genes related to hepatoblasts development. **c** Developmental track of hepatoblasts was shown by t-SNE plot. **d** Dynamic developmental process of hepatoblasts with representative gene expression patterns shown

co-expressed hepatocyte-specific markers, such as *Krt8* and *Krt18*, and cholangiocyte-specific markers, such as *Krt7* and *Krt19* (Fig. 2d). The results indicated that only a minority of hepatoblasts from mouse fetal livers co-express hepatocyte and cholangiocyte markers at this developmental period. Our randomly selected single cells contain few cells showing markers of cholangiocytes, probably due to the asymmetric lineage fate of hepatoblast differentiation into hepatocyte and cholangiocyte. We thus employed flow cytometry to enrich the relative rare mouse cholangiocytes using the well-known marker Epcam [6, 7, 18], which will enable single-cell transcriptomic comparison between cholangiocytes and hepatoblasts and facilitate understanding the fate-decision stage for differentiation into cholangiocytes. We obtained 52 Epcam$^+$ single cells from P3.25 fetal livers and sequenced them.

We compared the expression of LSPC-related markers between hepatoblasts and the P3.25 Epcam$^+$ cells (Fig. 4a-b). Some genes such as *Hnf4a*, *Dlk1*, *Anpep* and *Prom1* (*Cd133*) were highly expressed in hepatoblasts, but not in the Epcam$^+$ cells. Significantly, *Gpc3*

expression was maintained in all Epcam$^+$ cells, whilst *Afp* was only expressed in about half of the Epcam$^+$ cells. Another interesting phenomenon is that the expression level of *Cdh1* was increased in the Epcam$^+$ cells, suggesting high E-cadherin level could be a putative marker for cholangiocyte isolation. The genes highly and specifically expressed in the Epcam$^+$ cells included *Sox9* and *Spp1*, which are well-documented markers for cholangiocytes. The expression of *Krt7* or *Krt19* emerged in some of the Epcam$^+$ cells, but not in E11.5 ~ E16.5 hepatoblasts. We also checked the expression pattern of marker genes from Fig. 4a in P3.25 Epcam$^+$ cells and hepatoblasts to hESC-derived cholangiocytes and hepatoblasts (hESC-Chol and hESC-HB) [24]. Seven genes from Fig. 4a were found to be differently expressed between hESC-Chol and hESC-HB, and similar changing expression patterns were also observed between mouse cholangiocytes and hepatoblasts (Additional file 1: Figure S6a). The expression of *DLK1* was decreased in hESC-Chol in comparison with hESC-HB, also consistent with disappearing expression of *Dlk1* in single cells of mouse cholangiocytes (Additional file 1:

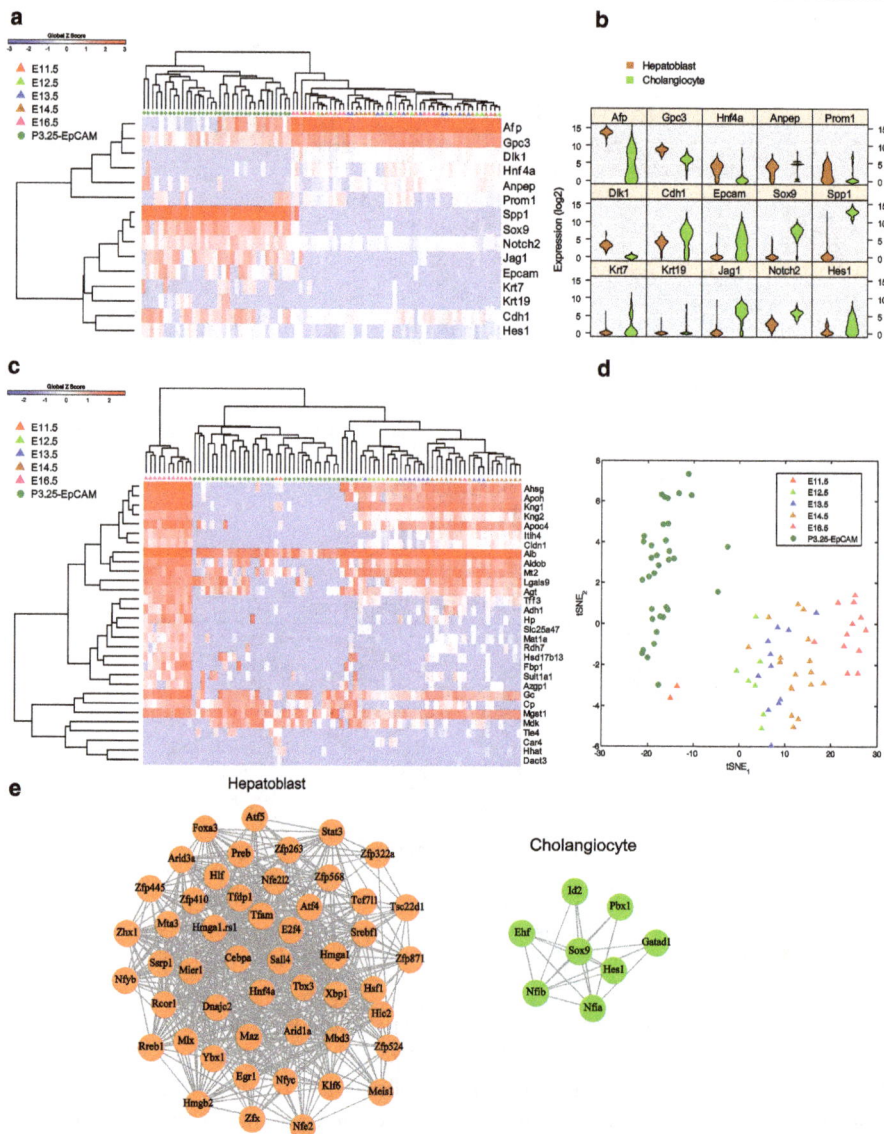

Fig. 4 Distinct transcriptomic features between hepatoblasts and cholangiocytes. **a** HC showing the heterogeneity of gene expression of some selected marker genes in hepatoblasts and cholangiocytes. **b** Violin plot of selected marker genes in hepatoblasts and cholangiocytes. Comparison of the gene expression profiles of P3.25 cholangiocytes and hepatoblasts from different stages by HC (**c**) and t-SNE plot (**d**) are shown. **e** Transcription factors covariance networks of hepatoblasts and cholangiocytes. Each node represents a TF, and each edge represents correlation coefficient higher than 0.35. The two networks are colored to discriminate TFs specifically related to hepatoblasts and cholangiocytes

Figure S6b). There was one exception that *AFP* expression was increased in hESC-Chol, while single cells of mouse cholangiocytes showed heterogeneous *Afp* expression pattern (Additional file 1: Figure S6b). This is probably because the hESC derived cholangiocytes are not exactly the same as their in vivo counterparts. The collective data indicated that these Epcam⁺ cells are cholangiocytes.

Currently it is still not clear when LSPCs are fated to hepatocytes or cholangiocytes. To decipher the issue, we compared transcriptomic feature of these P3.25 Epcam⁺ cholangiocytes with hepatoblasts from E11.5 ~ E16.5

using the differentially expressed transcripts among these stages (see Fig. 3). Interestingly, the gene expression profiles of P3.25 cholangiocytes were more closely related to that of E11.5 hepatoblasts rather than other hepatoblasts from later developmental stages (Fig. 4c-d). The resembled gene expression profiles between cholangiocytes and early hepatoblasts hint that hepatoblasts may be fated to cholangiocytes at an early stage. However, this single-cell genomics-derived model still needs further validation, especially solid evidence from well-designed lineage tracing experiments employing reliable markers.

We further focused on the signaling pathways related to liver development, including the Wnt/β-catenin, Notch, TGF-β and Hedgehog pathways. The results showed that *Jag1*, *Notch2* and *Hes1* were significantly ($p < 0.01$) higher expressed in cholangiocytes, suggesting that Notch activation was associated with cholangiocytes differentiation (Fig. 4a-b), consistent with previous results [18].

As known, Transcription factors (TFs) play important roles in liver development. Therefore, we screened TFs that were differentially expressed between hepatoblasts and cholangiocytes and found more TFs were specifically or highly expressed in hepatoblasts (Additional file 4: Table S3), suggesting stem/progenitor cells with higher plasticity could maintain a high self-renewal and differentiation capability under complex regulatory conditions. We then performed expression covariance network analysis and constructed two TF covariance networks specific to hepatoblasts and cholangiocytes (Fig. 4e). Within the hepatoblast-specific network, the TFs that had significant correlations with more TFs included Atf4, Tfam, Hmga1, Maz, Sall4, Mbd3, Tfdp1, Hnf4a, Ssrp1 and E2f4, some of which are required for liver functions or regulation of cell self-renewal and differentiation. For examples, zinc finger transcription factor Sall4 controls the cell fate decision of hepatoblasts and serves as a marker for progenitor subclass of hepatocellular carcinoma [25, 26]. For the cholangiocyte-specific network, such TFs included Sox9, Nfib, Nfia, Ehf, Id2, Gatad1, Pbx1 and Hes1, some of which are known to be involved in various stem cell-related signaling pathways. For example, Hes1 is closely related to Notch pathway, while the HMG-box transcription factor Sox9 has been shown to be related to progenitor status and differentiation of cholangiocytes [27, 28]. The TF covariance analysis not only supports that some well-studied TFs may play critical roles in LSPC differentiation but also provides candidate TFs for future investigation.

Assessment and prediction of LSPC biomarkers

The isolation of LSPCs in previous studies generally relied on limited surface markers. The gene expression patterns of commonly used markers confirmed the heterogeneity within LSPCs (Fig. 5a and Additional file 1: Figure S7a). We checked the combined expression profiles of some representative genes. As an example, the marker-positive cells with the gene pairs such as *Dlk1* vs. *Lgr5* and *Cd24a* vs. *Igdcc4* (*Nope*) were not identical in hepatoblasts at E13.5 although overlapped (Fig. 5a). The results indicated that LSPCs isolated with different markers may represent overlapping but not identical LSPC pool. As our single cells were randomly selected from fetal livers, systematic assessment of the sensitivity

and specificity of isolation markers for LSPCs could be performed according to the proposed approach (Fig. 5b).

As the phenotypes of LSPCs are dynamically changing during the developmental process, we separately analyzed 11 marker genes commonly used for cell isolation in hepatoblasts at each stage (Fig. 5c and Additional file 1: Figure S7b). *Cdh1* exhibited the best sensitivity and specificity for hepatoblast isolation from E12.5 to E16.5 livers (Fig. 5d), followed by *Anpep* and *Prom1* (Additional file 1: Figure S7c). *Dlk1* was expressed in both hepatoblasts and mesenchymal cells at early stage, where its specificity gradually increased from E11.5 to E16.5 for hepatoblasts and decreased from E11.5 to E14.5 for mesenchymal cells (Fig. 5d). The specificity of *Igdcc4* for hepatoblasts gradually increased from E11.5 to E16.5, despite its decreased sensitivity (Additional file 1: Figure S7c). We also checked *Gpc3*, and the specificity decreased in mesenchymal cells but increased in hepatoblasts over time (Additional file 1: Figure S7c). Our data provide a systematic evaluation of the isolation markers for LSPCs, which is helpful for evaluating the reliability of previously isolated LSPCs and for predicting isolation markers for further validation. For example, *Gcgr* and *Cdhr2* are predicted to be isolation markers for hepatoblasts (Additional file 1: Figure S7c).

To validate the results, we performed flow cytometric analysis of mouse fetal liver cells from different stages with antibodies against E-cadherin, Anpep, Dlk1 and Prom1. The co-expression of E-cadherin, Anpep and Dlk1 was observed in E12.5 ~ E16.5 fetal liver cells (Fig. 5e and Additional file 1: Figure S8a-b), consistent with our single-cell RNA-Seq data. However, there was inconsistency in regard to Dlk1 expression in E12.5 fetal livers. Flow cytometry indicated that Dlk1$^+$ cells almost perfectly overlapped with E-cadherin$^+$ cells (Additional file 1: Figure S8a), whereas RNA-Seq data showed that *Dlk1* expression could not distinguish between hepatoblasts and mesenchymal cells at E12.5 stage and that *Cdh1* is a specific marker for hepatoblasts (Fig. 5d). This finding suggested that protein levels are not always consistent with transcript levels. Flow cytometric analysis also revealed that only a portion of the Prom1$^+$ cells from E14.5 and E16.5 fetal livers were Dlk1-positive (Fig. 5f and Additional file 1: Figure S8c). In general, both single-cell RNA-Seq and flow cytometry supported E-cadherin, Anpep and Dlk1 as appropriate biomarkers for isolation of LSPCs, while Prom1 may slightly differ from the above three markers (Figs. 2d, 5e-f and Additional file 1: Figs. S7c, S8).

Discussion

There have been DNA microarrays or RNA-Seq analyses of fetal livers from different stages [29, 30], but analysis based on average gene expression signals from different

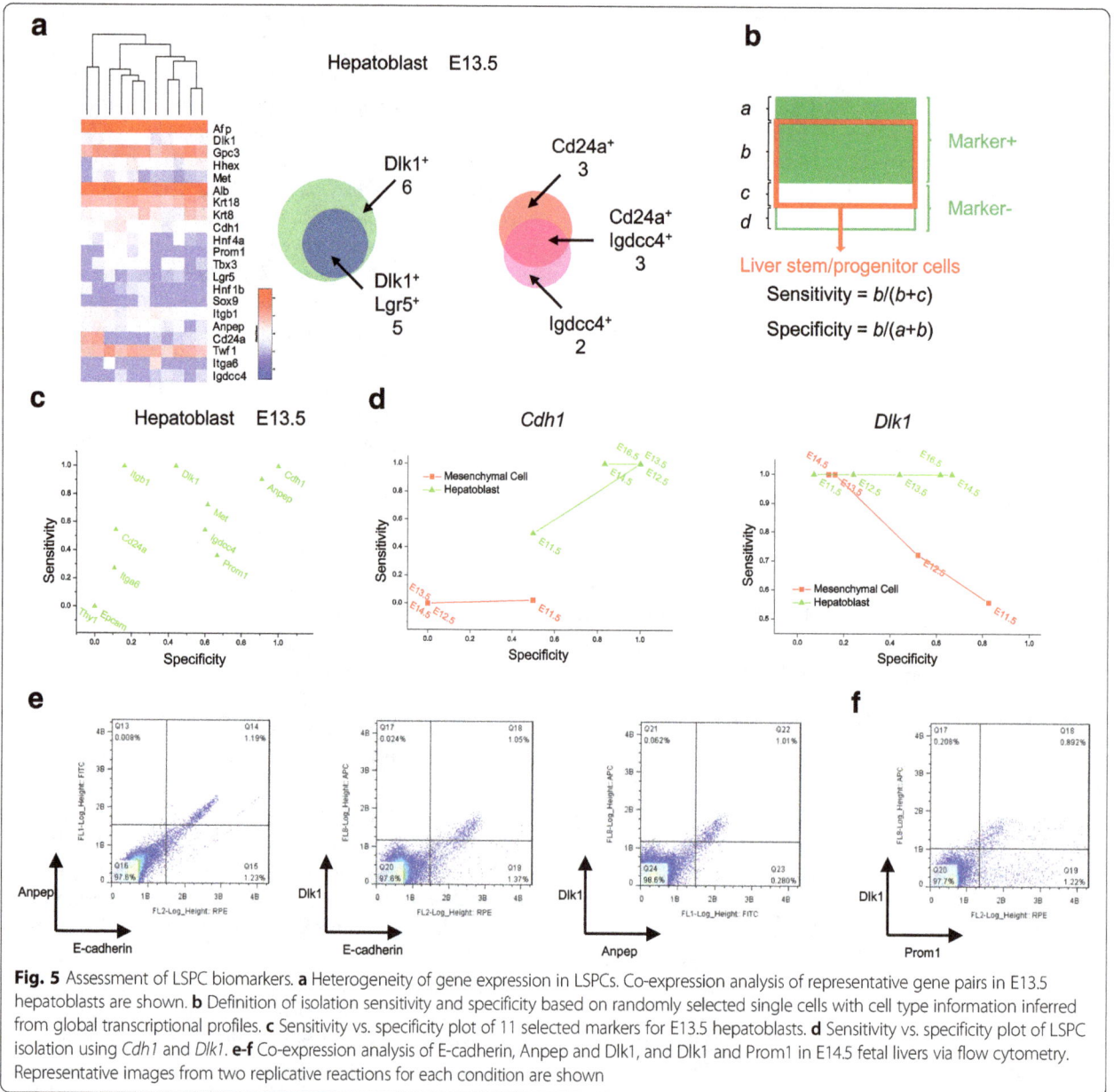

Fig. 5 Assessment of LSPC biomarkers. **a** Heterogeneity of gene expression in LSPCs. Co-expression analysis of representative gene pairs in E13.5 hepatoblasts are shown. **b** Definition of isolation sensitivity and specificity based on randomly selected single cells with cell type information inferred from global transcriptional profiles. **c** Sensitivity vs. specificity plot of 11 selected markers for E13.5 hepatoblasts. **d** Sensitivity vs. specificity plot of LSPC isolation using *Cdh1* and *Dlk1*. **e-f** Co-expression analysis of E-cadherin, Anpep and Dlk1, and Dlk1 and Prom1 in E14.5 fetal livers via flow cytometry. Representative images from two replicative reactions for each condition are shown

constituent cell types makes it difficult to ascribe the gene expression changes to a specific cell type. Single-cell analysis thus facilitates a more accurate identification of gene expression changes related to LSPCs. In this study, we employed single-cell qPCR and RNA-Seq to systematically re-visit the developmental process of mouse fetal livers, and the marker-free approach based on global transcriptional profiles enabled more reliable identification of the constituent cell types in fetal livers. Our data support the systematic assessment and prediction of markers for LSPC isolation, and reconstructed the developmental track of hepatoblasts at single-cell resolution.

Hepatoblasts will theoretically differentiate into both hepatocytes and cholangiocytes, and when they are fated to or differentiate into the two lineages is an important question still unanswered. Currently, it is found that, at around E13.5 ~ E16.5, hepatoblasts adjacent to portal veins are induced by Notch, TGF-β or other signals from nearby mesenchymal or endothelial cells to form the ductal plate and later differentiate into intra-hepatic ductal cells [18, 27]. This implies that hepatoblasts are not pre-fated to hepatic or biliary lineage but that their locations will decide their fates. Our single-cell data showed that the changes in hepatoblasts during this period are mainly related to stepwise increased hepatic functional gene expression (Figs. 3, 6). The cell cycle- and mitosis-related genes were significantly enriched in E14.5 hepatoblasts (Additional file 1: Figure S5a), suggesting that these cells may be induced by an inherited program or exterior environment to initiate differentiation via cell division (Fig. 6), consistent with the previous model.

It was found that the gene expression profiles of single Epcam$^+$ cholangiocytes from P3.25 livers were more closely related to E11.5 hepatoblasts rather than other later hepatoblasts, implying that LSPCs may have their fate decision for cholangiocytes differentiation at an early stage (Fig. 4). The differentiation of hepatoblasts into hepatocytes and cholangiocytes is asymmetrical, and the proportion of hepatic-fated hepatoblasts appeared to be much higher than ductal-fated cells, which could explain the reason why these hepatoblasts derived from our random approach only showed differentiation directions towards hepatocytes. Our single-cell transcriptomic analysis thus provides insights about the fate decision stage of hepatoblasts.

Our data showed the existence of some cells co-expressing both hepatic and mesenchymal markers during mouse liver development. In this study, the immunofluorescence assay confirmed there were cells co-expressing both Dlk1 and vimentin in E12.5 fetal livers. This finding is consistent with previous observations of liver cells that co-express epithelial and mesenchymal markers [31, 32]. The results suggested the possible involvement of epithelial-mesenchymal transition (EMT) or mesenchymal-epithelial transition (MET), consistent with recent observation of EMT-MET during the differentiation of hESC into hepatocytes [33], but this needs further evidences such as lineage tracing experiments.

Conclusions

In summary, our data provides a useful resource describing LSPCs at single-cell resolution during mouse liver development. Our marker-free approach provides insights into the reliable isolation of LSPCs and their developmental track, and analysis of single cholangiocytes further facilitates understanding of cell fate decision, differentiation and regulatory mechanisms of LSPCs.

Methods
Animals
C57BL/6 mice were purchased from Shanghai Laboratory Animal Center to provide fetal livers. No randomization method was used for group allocation, and animals were randomly selected and no animals were excluded from experiments.

Preparation of mouse liver cells
Liver cells were prepared from E11.5, E12.5, E13.5, E14.5, E16.5 and E18.5 mouse embryos and P2.5 and P3.25 neonatal mice. After adult or neonatal animals were euthanized by CO_2, fetal livers from siblings of littermates were isolated, pooled, minced and digested in 0.05% collagenase IV (Sigma-Aldrich) at 37 °C for 30 min. After the cells were dissociated by repeated pipetting, the suspensions were filtered through a 40-μm cell strainer to remove undigested tissues. The cells were washed twice with RPMI 1640 medium containing 10% FBS after centrifugation at 300×g for 5 min. The cell densities and sizes were determined to select suitable C1 RNA-Seq IFC.

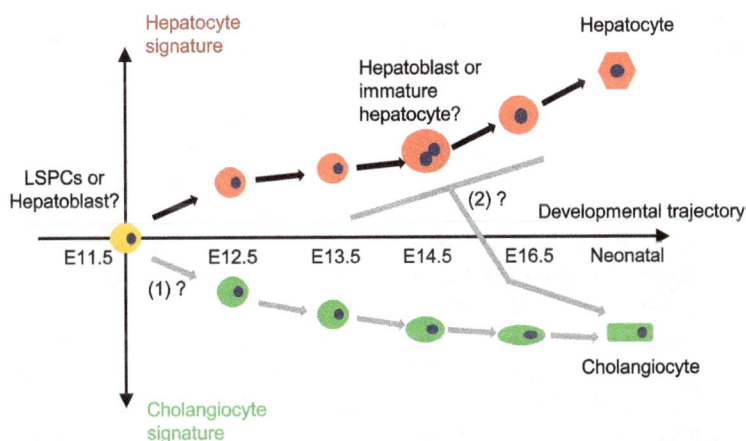

Fig. 6 The proposed schematic diagram of the fate decision and differentiation of LSPCs. From E11.5 to neonatal, the increased expression of hepatic- or biliary-related genes indicates an elevated hepatocyte or cholangiocyte signature in LSPCs. Black arrows indicate the observed developmental processes in our data, while grey arrows indicate the putative developmental steps. "(1)" and "(2)" denote two possible stages where the fate decisions of LSPCs occur

Single-cell qPCR and RNA-Seq

Single cells with diameters of 10 ~ 17 μm were captured randomly on a C1 RNA-Seq IFC (Fluidigm). A SMAR-Ter Ultra Low RNA kit for Illumina (Clontech) was used for on-chip cell lysis, reverse transcription and cDNA amplification. An ERCC Spike-in Mix (Ambion) was used as the technical control. The quality of cDNA libraries was checked by single-cell qPCR using selected genes on a BioMark HD system (Fluidigm), and the qPCR primers are shown in Additional file 5: Table S4. cDNA were fragmented and prepared using Nextera XT Kit and Index Kit (Illumina). Single-cell libraries were pooled and sequenced by NextSeq 500 (Illumina), with 2 × 151 bp or 2 × 76 bp sequencing modes. To check the expression levels of *Epcam*, single-cell qPCR was carried out using SYBR® Premix Ex Taq™ (Clontech) on the StepOnePlus system (Applied Biosystems).

Immunofluorescence assay

Fetal livers from E12.5 mouse were embedded in OCT compound, and 5-μm cryostat sections were fixed with 4% paraformaldehyde and then permeabilized with 0.1% Triton X-100. The sections were blocked with 1% BSA in Tris-buffered saline for 1 h at room temperature. To detect co-expression of Dlk1/Hnf4a or Dlk1/vimentin, the sections were incubated with the following primary antibodies overnight at 4 °C: Goat anti-Dlk1 antibody (C-19) (Santa Cruz, PN sc-8624), 1/100; Rabbit anti-HNF-4α antibody (H-171) (Santa Cruz, PN sc-8987), 1/100; Rabbit anti-Vimentin antibody [EPR3776] (Abcam, PN ab92547), 1/250. The sections were washed 3 times with Tris-buffered saline with 0.1% Tween 20 for 5 min each time, and then incubated with the following secondary antibodies for 1 h at room temperature: Donkey anti-Goat IgG (H + L) Cross-Adsorbed Secondary Antibody, Alexa Fluor® 488 (Invitrgen, PN A11055), 1/1000; Donkey anti-Rabbit IgG (H + L) Highly Cross-Adsorbed Secondary Antibody, Alexa Fluor 647 (Life Technologies, PN A31573), 1/1000. The nuclei were counterstained with DAPI, and immunofluorescence was detected with a fluorescent confocal microscope (Nikon).

Flow cytometric analysis and cell sorting

The following antibodies were used for flow cytometric analysis: E-cadherin, PE-conjugated anti-mouse/human CD324 (E-cadherin) antibody (BioLegend, PN 147304); Anpep, FITC-conjugated anti-CD13 antibody (Abcam, PN ab33486); Dlk1, APC-conjugated mouse Pref-1/DLK1/FA1 antibody (R&D Systems, PN FAB8634A-100); Prom1, PE-conjugated anti-Prominin-1 mouse antibody (Miltenyi, PN 130-102-210); and Epcam, PE-labeled anti-mouse CD326 (Epcam) antibodies (BioLegend, PN 118205). The incubation of fetal liver cells and antibodies were carried out per the vendors' suggestions.

Co-expression analyses of E-cadherin and Anpep, E-cadherin and Dlk1, Anpep and Dlk1, Dlk1 and Prom1 and E-cadherin, and Anpep and Dlk1 were performed using a MoFlo™ XDP cell sorting system (Beckman Coulter). Epcam+ cells were sorted from P3.25 fetal liver for single-cell RNA-Seq.

Sequencing data processing

bcl2fastq2 Conversion Software (Illumina) was used for de-multiplexing, adaptor trimming and generation of FASTQ files for each single cell according to the unique barcode combinations of Nextera XT Index kit. The raw reads in FASTQ format were mapped via Tophat onto the mouse reference genome GRCm38/mm10 downloaded from UCSC Genome Browser, using the default parameters. The transcript levels were then quantified as FPKM values generated by Cufflinks with default parameters. The FPKM values of ERCC Spike-ins were obtained in the same way with ERCC92 reference sequences. The FPKM values of genes and ERCC Spike-ins were then adjusted according the relative proportions of sequencing reads mapped to mouse genome or ERCC92 reference sequences.

Gene filtering to reduce technical noise

To reduce the interferences in the gene expression-based cell clustering analyses, the transcripts severely corrupted by noise should be removed. Here, we employed a novel approach to identify the noise corrupted transcripts. Specifically, we first constructed a regression model between noise variance and mean transcript RNA-seq FPKMs [34]:

$$CV^2 = a_1/\mu + a_0$$

where $CV^2 = \sigma_0^2/\mu^2$, σ_0^2 denotes the noise variance and μ denotes the average transcript reads, a_0 and a_1 are constants. The regression parameters a_1 and a_0 were estimated by constructing a generalized linear model regression between CV^2 and $1/\mu$. In the model construction, CV^2 is approximated by σ^2/μ^2, where σ^2 is the variance of a combination of noise and biological component. The noise variance for a given transcript can be estimated by $\sigma_0^2 = a_0\mu^2 + a_1\mu$. To reduce the uncertainty in model construction, only the transcripts with $\mu > \mu_{th}$ were applied to build the regression model. The μ_{th} was chosen such that the transcripts with $\mu > \mu_{th}$ only have 5% with $CV^2 > 0.3$. With the estimated noise variance σ_0^2, the reliability of each transcript was evaluated by the following criteria: (1) the 95% confidence interval calculated using μ and σ_0^2 should not include zero, which guarantees the overall expression of the given transcript is large enough to surpass noise interference; (2) σ^2 of the transcript should be larger than the corresponding σ_0^2, which filters out the transcripts which are unlikely to be differentially expressed over the dataset. Only those transcripts fulfilled the two criteria at the same time were retained for further analyses.

Single-cell data analysis

Outlier identification, HC, principal component analysis (PCA), violin plot and one-way ANOVA of single-cell qPCR and RNA-Seq data were performed using the SINGuLAR™ Analysis Toolset R package (Fluidigm). After gene filtering, the gene expression data of 507 single cells were fed into the identifyOutliers () function of SINGuLAR for outlier filtering with default threshold, and 456 single cells (89.9%) passed the filtering. For the 52 Epcam⁺ cells from P3.25 livers, 10 cells were identified as outliers, and 7 cells were further removed as they showed different gene expression pattern compared with the remaining 35 typical cholangiocytes.

For grouping of all fetal liver cells from E11.5 to E16.5, un-supervised HC of all the liver cells with the top 400 genes ranked by PCA scores was performed and then identified two groups. Among them, one group is hepatoblasts which express both hepatic-related and stem/progenitor-related genes. After removing cells belonging to hepatoblasts, un-supervised HC assigns the remaining cells into six major sample clusters (Sc-1, 2, 3, 4, 5 and 6). Erythrocyte-related genes were not included in the analysis as the cells in the embryonic stages express such genes at relatively low expression level compared with erythrocytes for undetermined reasons. Expression of known markers facilitated the cell type identification of each cluster, where both Sc-1 and Sc-5 cells expressing *Ptprc* (*Cd45*), *Cd68* and *Cd52* are likely to be macrophages, and thus merged as one group. After annotation of putative cell type identification of each single cell, one-way ANOVA was conducted and the top 400 ANOVA-ranked genes were used for the HC plot. Cell clusters were also visualized with t-SNE [35].

For the construction of developmental track of hepatoblasts, the developmental stage information of each single cell was annotated, and only these genes expressed in at least 3 single cells were included in one-way ANOVA to make sure that these genes are indeed related to the liver developmental process.

Gene ontology (GO) categories and gene set enrichment analysis (GSEA)

In the cell type identification process, the gene sets specifically expressed in each type (Fig. 2a) were analyzed to find out functional terms enriched based on GO Biological Process datasets, and Bonferroni correction for multiple testing was used [36, 37]. The statistically significantly ($p < 0.05$) enriched GO terms are helpful for cell type identification. For pairwise comparison of single cell groups, GSEA [38] was employed to identify the gene sets that are enriched in either group, with the following criteria: $p < 0.05$ and FDR $q < 0.25$. GSEA was carried out for comparison of hepatoblast groups across developmental stage E11.5 ~ E16.5. The detailed definitions of p and q values and corrections can be found in the references on GO and GSEA [36–38].

Transcription factor covariance network construction

Transcription factor list of *Mus musculus* from Transcription Factor Database [39] was used to screen TFs that show statistically significantly ($p < 0.01$) differential expression between hepatoblasts and cholangiocytes. Pairwise Pearson correlation coefficients between these selected TFs expressed in single cells were calculated to identify TFs that correlate with at least three other TFs with correlation coefficient higher than 0.35. The matrix was used to construct a weighted network using graph.adjacency () function of igraph package implemented in R as previously described [40], where vertices represent TFs and edges represent high correlation. The networks were visualized with Fruchterman-Reingold layout.

Performance assessment of the genes for isolation of hepatoblasts

Whether a marker is suitable for isolation of LSPCs can be assessed by two factors: (1) sensitivity, which refers to whether the marker-positive cells include most of the LSPCs; and (2) specificity, which refers to whether the purity of the marker-positive cells excludes other types of cells. The isolation sensitivity and specificity could be defined by randomly selected single cells and cell type information inferred from global transcriptional profiles. For each gene, we calculated its isolation sensitivity and specificity for hepatoblasts, and mesenchymal cells as a comparison. In addition to common isolation markers for hepatoblasts, we also attempted to predict new markers for isolation of hepatoblasts. We first used a cutoff of 0.5 for both sensitivity and selectivity for each group at each stage, found the genes that are shared by different stages, and then selected membrane receptors for further analysis. Plotting the sensitivity vs. specificity across different stages allowed us to analyze whether a given gene marker is suitable for the isolation of hepatoblasts.

Additional files

Additional file 1: Supplemental figures. **Figure S1.** Single-cell qPCR analysis of mouse fetal liver cells. **Figure S2.** Quality control of single-cell RNA-Seq analysis of mouse fetal liver cells. **Figure S3.** Grouping of fetal liver cells from E11.5 to E16.5. **Figure S4.** Validation of single-cell RNA-Seq results by immunofluorescence and qPCR. **Figure S5.** Dynamic developmental process of mouse LSPCs at single-cell resolution. **Figure S6.** Comparison of the gene expression patterns of some marker genes between mouse and human for cholangiocyte and hepatoblast. **Figure S7.** Assessment and prediction of LSPC biomarkers. **Figure S8.** Validation of some markers for LSPC isolation via flow cytometric analysis. (PDF 2895 kb)

Additional file 2: Table S1. Statistics of RNA sequencing reads of single cells. (XLSX 38 kb)

Additional file 3: Table S2. Top 30 ANOVA ranked genes for E11.5 ~ E16.5 hepatoblast. (XLSX 12 kb)

Additional file 4: Tables S3. Transcription factors differentially expressed between hepatoblast and cholangiocyte for covariance networks. (XLSX 12 kb)

Additional file 5: Table S4. Primers used for single-cell qPCR. (XLSX 11 kb)

Abbreviations
ANOVA: Analysis of variance; EMT: Epithelial-mesenchymal transition; FPKM: Fragments per kilobase of exon model per million; GO: Gene Ontology; GSEA: Gene set enrichment analysis; HC: Hierarchical clustering; LSPCs: Liver stem/progenitor cells; MET: Mesenchymal-epithelial transition; PCA: Principal component analysis; TFs: Transcription factors; t-SNE: t-distributed stochastic neighbor embedding

Acknowledgements
We would like to thank Professor Jian Huang for critical discussions, Professor Ying-Xin Qi, Professor Yue-Ying Wang and Dr. Yin Huang for providing BioMark and C1 systems, and Dr. Jia-Zhu Fang for technical assistance.

Funding
This work is supported in part by the National Natural Science Foundation of China (81472621, 81672772 and 81402329), Shanghai Sailing Program (15YF1405800), National Program on Key Research Project of China (2016YFC0902701, Precision Medicine), Start Funding for Young Teachers of SJTU (14X100040063), Medical and Engineering Crossover Fund of SJTU (YG2014QN14, YG2016QN71 and YG2016MS05), Shanghai Pujiang Program (16PJ1405200), and Shanghai Natural Science Foundation (16ZR1417900). The funding bodies didn't involve in the design of the study, collection, analysis, interpretation of data, or writing the manuscript.

Authors' contributions
ZH conceived and designed the study. XS, ZL, CW, XC, KH, YQ, NW and LW carried out experiments. XS, YS, XZ, GX, JY, and QL analyzed data. ZH and XS interpreted the data and drafted the manuscript. All authors read and approved final version of the manuscript.

Competing interests
The authors declare that they have no competing interests.

Author details
[1]Key Laboratory of Systems Biomedicine (Ministry of Education) and Collaborative Innovation Center of Systems Biomedicine, Shanghai Center for Systems Biomedicine, Shanghai Jiao Tong University, 800 Dongchuan Road, Shanghai 200240, China. [2]Key Laboratory of Computational Biology, CAS-MPG Partner Institute for Computational Biology, 320 Yueyang Road, Shanghai, China. [3]School of Mathematics and Statistics, The University of Sydney, Sydney, Australia. [4]Shanghai-MOST Key Laboratory for Disease and Health Genomics, Chinese National Human Genome Center at Shanghai, Shanghai, China.

References
1. Zaret KS. Regulatory phases of early liver development: paradigms of organogenesis. Nat Rev Genet. 2002;3(7):499–512.
2. Zorn AM: Liver development. StemBook, ed the stem cell research community, StemBook 2008.
3. Lemaigre FP. Mechanisms of liver development: concepts for understanding liver disorders and design of novel therapies. Gastroenterology. 2009;137(1):62–79.
4. Miyajima A, Tanaka M, Itoh T. Stem/progenitor cells in liver development, homeostasis, regeneration, and reprogramming. Cell Stem Cell. 2014;14(5):561–74.
5. Si-Tayeb K, Lemaigre FP, Duncan SA. Organogenesis and development of the liver. Dev Cell. 2010;18(2):175–89.
6. Okabe M, Tsukahara Y, Tanaka M, Suzuki K, Saito S, Kamiya Y, Tsujimura T, Nakamura K, Miyajima A. Potential hepatic stem cells reside in EpCAM+ cells of normal and injured mouse liver. Development. 2009;136(11):1951–60.
7. Tanaka M, Okabe M, Suzuki K, Kamiya Y, Tsukahara Y, Saito S, Miyajima A. Mouse hepatoblasts at distinct developmental stages are characterized by expression of EpCAM and DLK1: drastic change of EpCAM expression during liver development. Mech Dev. 2009;126(8-9):665–76.
8. Tanimizu N, Nishikawa M, Saito H, Tsujimura T, Miyajima A. Isolation of hepatoblasts based on the expression of Dlk/Pref-1. J Cell Sci. 2003;116(9):1775–86.
9. Nitou M, Sugiyama Y, Ishikawa K, Shiojiri N. Purification of fetal mouse hepatoblasts by magnetic beads coated with monoclonal anti-E-cadherin antibodies and their in vitro culture. Exp Cell Res. 2002;279(2):330–43.
10. Kakinuma S, Ohta H, Kamiya A, Yamazaki Y, Oikawa T, Okada K, Nakauchi H. Analyses of cell surface molecules on hepatic stem/progenitor cells in mouse fetal liver. J Hepatol. 2009;51(1):127–38.
11. Kamiya A, Kakinuma S, Yamazaki Y, Nakauchi H. Enrichment and clonal culture of progenitor cells during mouse postnatal liver development in mice. Gastroenterology. 2009;137(3):1114–26.
12. Tang F, Barbacioru C, Wang Y, Nordman E, Lee C, Xu N, Wang X, Bodeau J, Tuch BB, Siddiqui A, et al. mRNA-Seq whole-transcriptome analysis of a single cell. Nat Methods. 2009;6(5):377–82.
13. Treutlein B, Brownfield DG, Wu AR, Neff NF, Mantalas GL, Espinoza FH, Desai TJ, Krasnow MA, Quake SR. Reconstructing lineage hierarchies of the distal lung epithelium using single-cell RNA-seq. Nature. 2014;509(7500):371–5.
14. Zhou F, Li X, Wang W, Zhu P, Zhou J, He W, Ding M, Xiong F, Zheng X, Li Z, et al. Tracing haematopoietic stem cell formation at single-cell resolution. Nature. 2016;533(7604):487–92.
15. Macosko EZ, Basu A, Satija R, Nemesh J, Shekhar K, Goldman M, Tirosh I, Bialas AR, Kamitaki N, Martersteck EM, et al. Highly parallel genome-wide expression profiling of individual cells using Nanoliter droplets. Cell. 2015;161(5):1202–14.
16. Halpern KB, Shenhav R, Matcovitch-Natan O, Toth B, Lemze D, Golan M, Massasa EE, Baydatch S, Landen S, Moor AE, et al. Single-cell spatial reconstruction reveals global division of labour in the mammalian liver. Nature. 2017;542(7641):352–6.
17. Camp JG, Sekine K, Gerber T, Loeffler-Wirth H, Binder H, Gac M, Kanton S, Kageyama J, Damm G, Seehofer D, et al. Multilineage communication regulates human liver bud development from pluripotency. Nature. 2017;10.1038/nature22796.
18. Zong Y, Panikkar A, Xu J, Antoniou A, Raynaud P, Lemaigre F, Stanger BZ. Notch signaling controls liver development by regulating biliary differentiation. Development. 2009;136(10):1727–39.
19. Kadomatsu K, Muramatsu T. Midkine and pleiotrophin in neural development and cancer. Cancer Lett. 2004;204(2):127–43.
20. Buglino JA, Resh MD. Hhat is a palmitoylacyltransferase with specificity for N-palmitoylation of sonic hedgehog. J Biol Chem. 2008;283(32):22076–88.
21. Jacob L, Lum L. Deconstructing the hedgehog pathway in development and disease. Science. 2007;318(5847):66–8.
22. Jeng KS, Sheen IS, Jeng WJ, Yu MC, Hsiau HI, Chang FY, Tsai HH. Activation of the sonic hedgehog signaling pathway occurs in the CD133 positive cells of mouse liver cancer Hepa 1-6 cells. Onco Targets Ther. 2013;6:1047–55.
23. Omenetti A, Choi S, Michelotti G, Diehl AM. Hedgehog signaling in the liver. J Hepatol. 2011;54(2):366–73.

24. Dianat N, Dubois-Pot-Schneider H, Steichen C, Desterke C, Leclerc P, Raveux A, Combettes L, Weber A, Corlu A, Dubart-Kupperschmitt A. Generation of functional cholangiocyte-like cells from human pluripotent stem cells and HepaRG cells. Hepatology. 2014;60(2):700–14.

25. Oikawa T, Kamiya A, Kakinuma S, Zeniya M, Nishinakamura R, Tajiri H, Nakauchi H. Sall4 regulates cell fate decision in fetal hepatic stem/progenitor cells. Gastroenterology. 2009;136(3):1000–11.

26. Yong KJ, Gao C, Lim JS, Yan B, Yang H, Dimitrov T, Kawasaki A, Ong CW, Wong KF, Lee S, et al. Oncofetal gene SALL4 in aggressive hepatocellular carcinoma. N Engl J Med. 2013;368(24):2266–76.

27. Antoniou A, Raynaud P, Cordi S, Zong Y, Tronche F, Stanger BZ, Jacquemin P, Pierreux CE, Clotman F, Lemaigre FP. Intrahepatic bile ducts develop according to a new mode of tubulogenesis regulated by the transcription factor SOX9. Gastroenterology. 2009;136(7):2325–33.

28. Furuyama K, Kawaguchi Y, Akiyama H, Horiguchi M, Kodama S, Kuhara T, Hosokawa S, Elbahrawy A, Soeda T, Koizumi M, et al. Continuous cell supply from a Sox9-expressing progenitor zone in adult liver, exocrine pancreas and intestine. Nat Genet. 2011;43(1):34–41.

29. Li T, Huang J, Jiang Y, Zeng Y, He F, Zhang MQ, Han Z, Zhang X. Multi-stage analysis of gene expression and transcription regulation in C57/B6 mouse liver development. Genomics. 2009;93(3):235–42.

30. Renaud HJ, Cui YJ, Lu H, Zhong XB, Klaassen CD. Ontogeny of hepatic energy metabolism genes in mice as revealed by RNA-sequencing. PLoS One. 2014;9(8):e104560.

31. Najimi M, Khuu DN, Lysy PA, Jazouli N, Abarca J, Sempoux C, Sokal EM. Adult-derived human liver mesenchymal-like cells as a potential progenitor reservoir of hepatocytes? Cell Transplant. 2007;16(7):717–28.

32. Li B, Zheng YW, Sano Y, Taniguchi H. Evidence for mesenchymal-epithelial transition associated with mouse hepatic stem cell differentiation. PLoS One. 2011;6(2):e17092.

33. Li Q, Hutchins AP, Chen Y, Li S, Shan Y, Liao B, Zheng D, Shi X, Li Y, Chan W, et al. A sequential EMT-MET mechanism drives the differentiation of human embryonic stem cells towards hepatocytes. Nat Commun. 2017;8:15166.

34. Brennecke P, Anders S, Kim JK, Kolodziejczyk AA, Zhang X, Proserpio V, Baying B, Benes V, Teichmann SA, Marioni JC, et al. Accounting for technical noise in single-cell RNA-seq experiments. Nat Methods. 2013;10(11):1093–5.

35. Der Maaten LV, Hinton G. Visualizing Data using t-SNE. J Mach Learn Res. 2008;9:2579–605.

36. Ashburner M, Ball CA, Blake JA, Botstein D, Butler H, Cherry JM, Davis AP, Dolinski K, Dwight SS, Eppig JT, et al. Gene ontology: tool for the unification of biology. The Gene Ontology Consortium. Nat Genet. 2000;25(1):25–9.

37. Mi H, Muruganujan A, Casagrande JT, Thomas PD. Large-scale gene function analysis with the PANTHER classification system. Nat Protoc. 2013; 8(8):1551–66.

38. Subramanian A, Tamayo P, Mootha VK, Mukherjee S, Ebert BL, Gillette MA, Paulovich A, Pomeroy SL, Golub TR, Lander ES, et al. Gene set enrichment analysis: a knowledge-based approach for interpreting genome-wide expression profiles. Proc Natl Acad Sci U S A. 2005;102(43):15545–50.

39. Zhang HM, Chen H, Liu W, Liu H, Gong J, Wang H, Guo AY. AnimalTFDB: a comprehensive animal transcription factor database. Nucleic Acids Res. 2012;40(Database issue):D144–9.

40. Treutlein B, Lee QY, Camp JG, Mall M, Koh W, Shariati SA, Sim S, Neff NF, Skotheim JM, Wernig M, et al. Dissecting direct reprogramming from fibroblast to neuron using single-cell RNA-seq. Nature. 2016;534(7607):391–5.

11

Effect of two non-synonymous ecto-5′-nucleotidase variants on the genetic architecture of inosine 5′-monophosphate (IMP) and its degradation products in Japanese Black beef

Yoshinobu Uemoto[1,2], Tsuyoshi Ohtake[1], Nanae Sasago[1], Masayuki Takeda[1], Tsuyoshi Abe[1], Hironori Sakuma[1], Takatoshi Kojima[1] and Shinji Sasaki[1*]

Abstract

Background: *Umami* is a Japanese term for the fifth basic taste and is an important sensory property of beef palatability. Inosine 5′-monophosphate (IMP) contributes to *umami* taste in beef. Thus, the overall change in concentration of IMP and its degradation products can potentially affect the beef palatability. In this study, we investigated the genetic architecture of IMP and its degradation products in Japanese Black beef. First, we performed genome-wide association study (GWAS), candidate gene analysis, and functional analysis to detect the causal variants that affect IMP, inosine, and hypoxanthine. Second, we evaluated the allele frequencies in the different breeds, the contribution of genetic variance, and the effect on other economical traits using the detected variants.

Results: A total of 574 Japanese Black cattle were genotyped using the Illumina BovineSNP50 BeadChip and were then used for GWAS. The results of GWAS showed that the genome-wide significant single nucleotide polymorphisms (SNPs) on BTA9 were detected for IMP, inosine, and hypoxanthine. The *ecto-5′-nucleotidase (NT5E)* gene, which encodes the enzyme NT5E for the extracellular degradation of IMP to inosine, was located near the significant region on BTA9. The results of candidate gene analysis and functional analysis showed that two non-synonymous SNPs (c.1318C > T and c.1475 T > A) in *NT5E* affected the amount of IMP and its degradation products in beef by regulating the enzymatic activity of NT5E. The *Q* haplotype showed a positive effect on IMP and a negative effect on the enzymatic activity of NT5E in IMP degradation. The two SNPs were under perfect linkage disequilibrium in five different breeds, and different haplotype frequencies were seen among breeds. The two SNPs contribute to about half of the total genetic variance in IMP, and the results of genetic relationship between IMP and its degradation products showed that *NT5E* affected the overall concentration balance of IMP and its degradation products. In addition, the SNPs in *NT5E* did not have an unfavorable effect on the other economical traits.

Conclusion: Based on all the above findings taken together, two non-synonymous SNPs in *NT5E* would be useful for improving IMP and its degradation products by marker-assisted selection in Japanese Black cattle.

Keywords: GWAS, IMP, Japanese Black cattle, Meat quality, NT5E

* Correspondence: s0sasaki@nlbc.go.jp
[1]National Livestock Breeding Center, Nishigo, Fukushima 961-8511, Japan
Full list of author information is available at the end of the article

Background

Beef palatability is one of the most economically important objectives for the breeding of Japanese Black cattle, whose beef has a unique characteristic of intense marbling. The palatability of beef is primarily evaluated based on sensory characteristics such as taste, tenderness, juiciness, aroma, and so on. However, sensory characteristics are difficult to measure, are largely subjective, and possess low heritability [1, 2]. Therefore, it is difficult to genetically improve the quality of meat by relying on sensory characteristics alone. Therefore, another approach to incorporate indicators of these sensory characteristics is necessary.

Umami is a Japanese term for the fifth basic taste and is an important sensory property of foods, along with many other characteristics, including texture and flavor [3]. Inosine 5′-monophosphate (IMP) is a major nucleotide in postmortem muscle and contributes to the taste and flavor in meat [4]. The combination of IMP and glutamic acid or aspartic acid enhances the *umami* taste, and is known as *umami* intensity [5]. Recently, Suzuki et al. [6] reported that the 'strength aroma' and '*umami* intensity' based on the panel test contributed to the overall taste evaluation in seven beef brands of Japanese Black beef, and the amount of IMP was significantly correlated with the '*umami* intensity' of the panel test. The degradation products of IMP (inosine and hypoxanthine) are also important indicators of beef palatability. Inosine and hypoxanthine do not contribute to *umami* taste in

beef, but hypoxanthine, in combination with some amino acids and peptides, may contribute to bitterness in meat [7]. Therefore, the overall change in concentration of IMP and its degradation products can potentially affect the beef palatability. Therefore, a genetic understanding of IMP and its degradation products is important in beef cattle breeding to enhance the meat quality. The heritability estimates of glutamic acid and aspartic acid in beef were low (0.17 and 0.00, respectively) but those of IMP, inosine, and hypoxanthine were low to moderate (0.48, 0.33, and 0.23, respectively) in Japanese Black cattle population [8].

During the pathway of the formation and degradation of IMP in muscles, Adenosine triphosphate (ATP) is rapidly degraded to adenosine diphosphate (ADP) and adenosine monophosphate (AMP), which is then degraded to IMP. The IMP is further hydrolyzed to inosine by the enzymes of 5′-nucleotidase in intracellular and ecto-5′-nucleotidase (NT5E) in extracellular, which, in turn, is degraded to hypoxanthine (Fig. 1) [9–12]. The aging period after slaughter induces structural changes in the cell membrane, leading to intracellular water efflux [13]. Some of the enzymes involved in the degradation of IMP are gradually inactivated during the aging period of beef [14, 15]. Therefore, there is a strong biochemical relationship between IMP and its degradation products in beef, and detailed information on the genetic architecture of these traits is necessary to improve meat quality by using these traits as indicators. However, detailed studies on the genetic architecture of these traits

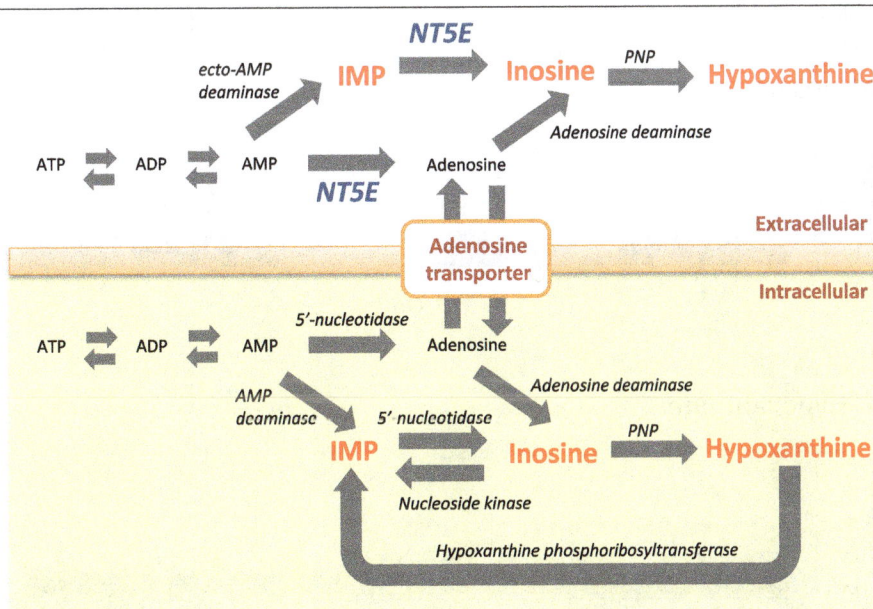

Fig. 1 Schematic representation of the pathway of formation and degradation of inosine 5′-monophosphate (IMP) in muscle. Each abbreviation is adenosine triphosphate (ATP), adenosine diphosphate (ADP), adenosine 5′-monophosphate (AMP), ecto-5′-nucleotidase (NT5E), and purine nucleoside phosphorylase (PNP). Black and red characters indicate products, and blue and black characters in italics indicate enzymes. This schematic representation is based on previous reports [9–12]

and their genetic relationship have, to our knowledge, not been performed.

To investigate the genetic architecture of IMP and its degradation products in Japanese Black beef, we performed genome-wide association studies (GWAS) and functional analysis to detect the causal variants that affect IMP and its degradation products at first. Second, we evaluated the allele frequencies in the different breed populations, the contribution of genetic variance, and the effect on other economical traits using the detected variants. This study identified quantitative trait nucleotides (QTNs) in *ecto-5′-nucleotidase (NT5E)*, which serves as a key enzyme for IMP and it degradation products.

Results
GWAS
A total of 574 Japanese Black cattle with records reported by Sakuma et al. [8] and BovineSNP50 genotypes reported by Sasago et al. [16] were used. The 5% genome-wide significant threshold was accounted for in multiple testing via Bonferroni correction with p-value = 1.37×10^{-6}. The results of GWAS are shown in Fig. 2

and Additional file 1: Table S1. The significant SNPs associated with the three phenotypes were detected on BTA 9. Specifically, the rs42865669 SNP on BTA 9 had the highest significance in IMP (p-value = 2.8×10^{-29}), inosine (p-value = 6.7×10^{-14}), and hypoxanthine (p-value = 1.8×10^{-12}). The rs42865669 SNP was not located within any genes, but was located about 500 kb from *NT5E* (Fig. 3a). *NT5E* encodes a membrane-bound enzyme for extracellular degradation of AMP to adenosine and degradation of IMP to inosine (Fig. 1). Thus, the *NT5E* could be regarded as the positional candidate gene for IMP and its degradation products. Based on the results from our GWAS, we performed candidate gene analysis on detecting the variants in *NT5E* and on testing the association between the variants in *NT5E* and these traits.

NT5E sequencing and association test
To detect variants in *NT5E*, we determined the nucleotide sequences of all of the exons and the proximal promoter region by direct sequencing. The variants detected are shown in Table 1. Eight SNPs and an insertions and deletions variant (indel) were detected. These variants were

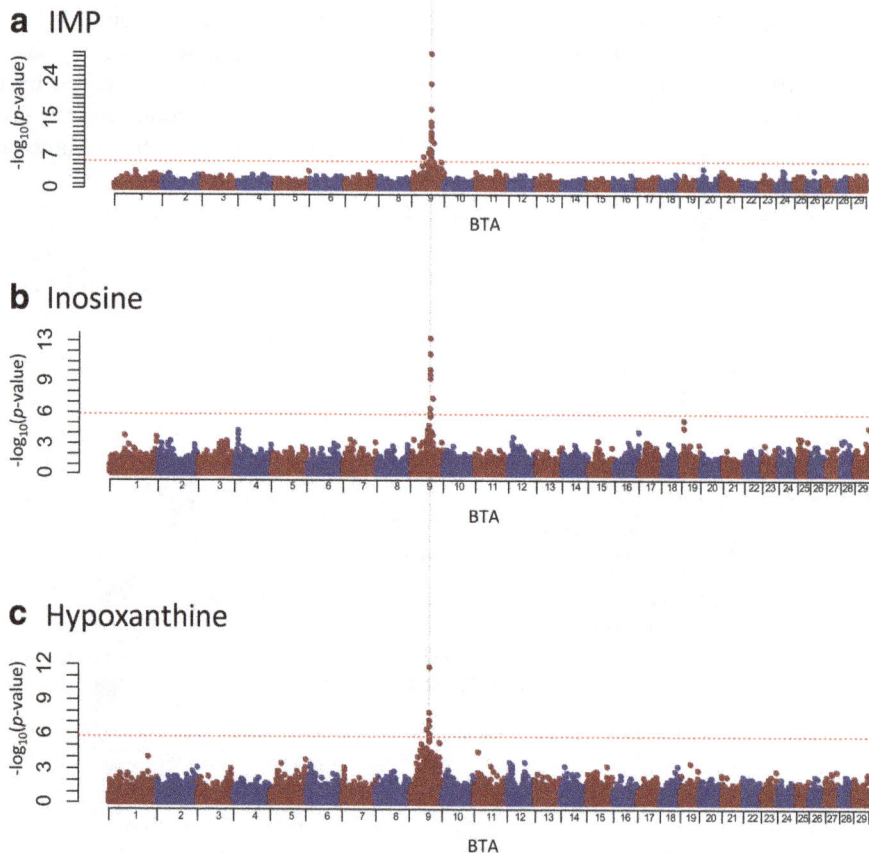

Fig. 2 Genome-wide plots of p-values (-log$_{10}$) for significantly associated loci. **a** Inosine 5′-monophosphate (IMP), **b** Inosine, and **c** Hypoxanthine. The x-axis indicates the chromosome number, and the y-axis indicates p-values (-log$_{10}$). Dashed red line indicates the threshold of the Bonferroni 5% significance level

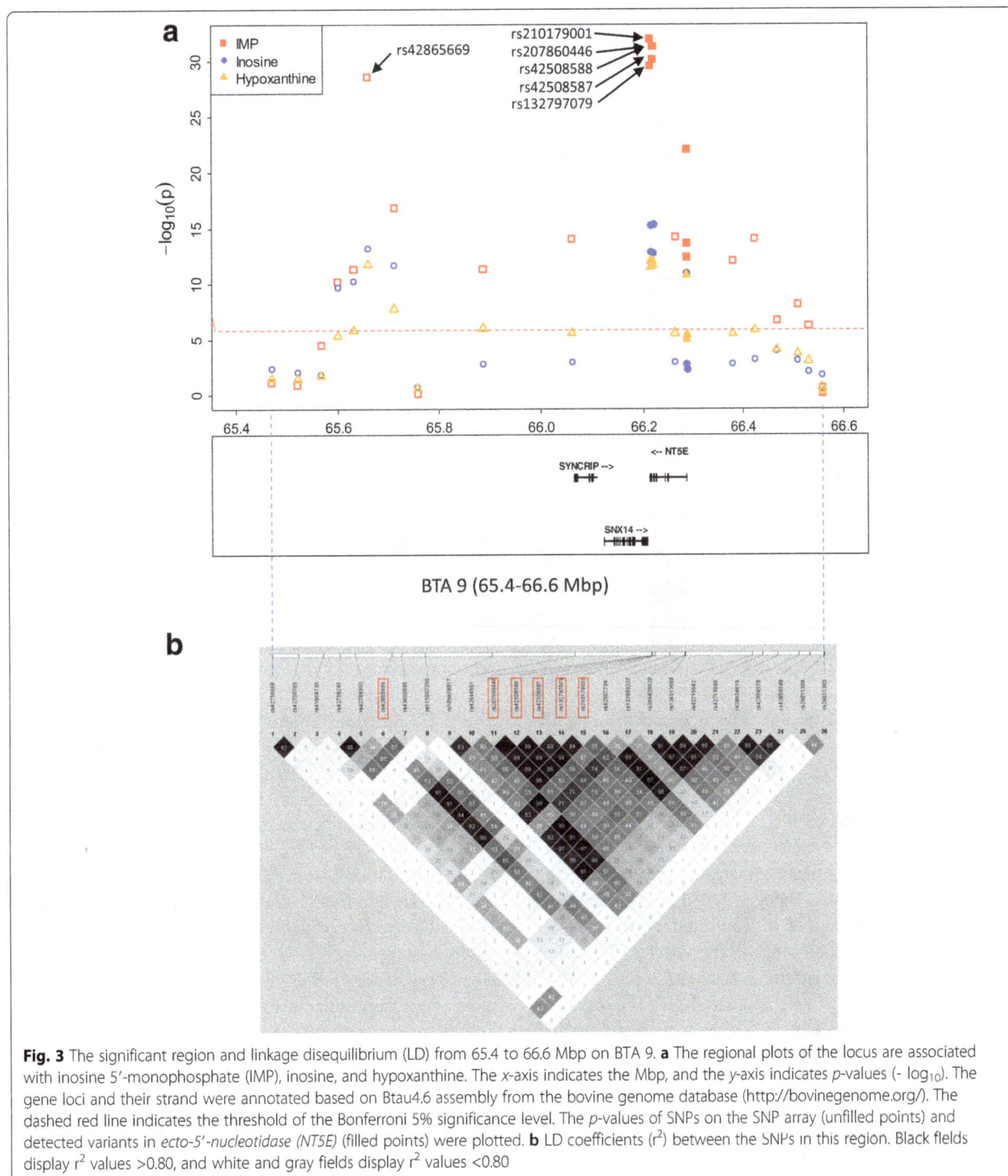

Fig. 3 The significant region and linkage disequilibrium (LD) from 65.4 to 66.6 Mbp on BTA 9. **a** The regional plots of the locus are associated with inosine 5'-monophosphate (IMP), inosine, and hypoxanthine. The *x*-axis indicates the Mbp, and the *y*-axis indicates *p*-values (- log$_{10}$). The gene loci and their strand were annotated based on Btau4.6 assembly from the bovine genome database (http://bovinegenome.org/). The dashed red line indicates the threshold of the Bonferroni 5% significance level. The *p*-values of SNPs on the SNP array (unfilled points) and detected variants in *ecto-5'-nucleotidase (NT5E)* (filled points) were plotted. **b** LD coefficients (r^2) between the SNPs in this region. Black fields display r^2 values >0.80, and white and gray fields display r^2 values <0.80

present in the dbSNP database. Three non-synonymous SNPs, an indel, two SNPs in 5'-upstream/5'-untransrated region (UTR), and two SNPs in 3'-UTR were genotyped in all animals. The association tests of these variants with three traits were performed, and the results are shown in Table 1 and Fig. 3a. The results showed that the three non-synonymous SNPs (c.1318C > T in the exon 7,

c.1475 T > A, and c.1526A > G in the exon 8) and the two SNPs in 3'-UTR of exon 9 (c.3060C > T and c.3098A > G) had high significance in IMP (*p*-value = 3.0×10^{-30} to 1.1×10^{-32}), inosine (*p*-value = 1.5×10^{-13} to 4.6×10^{-16}), and hypoxanthine (*p*-value = 3.2×10^{-12} to 9.6×10^{-13}). These SNPs were under high linkage disequilibrium (LD) in this population, and the LD coefficients (r^2) for the five

Table 1 The variants information in *ecto-5'-nucleotidase (NT5E)* and its association test with inosine 5'-monophosphate (IMP), inosine, and hypoxanthine in meats

Locus name[a]	refSNP variation ID	Position (bp)		Allele[a,b]		Locus	Amino Acid[c]	RAF[d,e]	*p*-value[e]		
		UMD3.1	Btau4.6	Ref	Alt				IMP	Inosine	Hypoxanthine
g.-622_-621insTTA	rs136113688	64,930,537	66,286,784	TTA	–	5'-upstream		0.44	4.2×10^{-13}	4.1×10^{-3}	3.9×10^{-6}
g.-82G > A	rs384428520	64,929,995	66,286,242	G	A	5'-upstream		0.42	2.2×10^{-14}	1.5×10^{-3}	9.4×10^{-6}
g.-17G > T	rs132959237	64,929,930	66,286,177	G	T	Exon 1 (5'-UTR)		0.65	8.8×10^{-23}	9.7×10^{-12}	1.7×10^{-11}
c.1044C > T	rs135498711	64,860,563	66,224,147	C	T	Exon 5	p.Gly348Gly	–	–	–	–
c.1318C > T	rs207860446	64,864,545	66,220,165	C	T	Exon 7	p.His440Tyr	0.58	5.4×10^{-32}	4.6×10^{-16}	2.3×10^{-12}
c.1475 T > A	rs42508588	64,866,290	66,218,420	T	A	Exon 8	p.Val492Glu	0.58	5.4×10^{-32}	4.6×10^{-16}	2.3×10^{-12}
c.1526A > G	rs42508587	64,866,341	66,218,369	A	G	Exon 8	p.Gln509Arg	0.56	8.2×10^{-31}	1.5×10^{-13}	1.0×10^{-12}
c.3060C > T	rs132797079	64,871,310	66,213,400	C	T	Exon 9 (3'-UTR)		0.60	3.0×10^{-30}	5.4×10^{-16}	3.2×10^{-12}
c.3098A > G	rs210179001	64,871,348	66,213,362	A	G	Exon 9 (3'-UTR)		0.56	1.1×10^{-32}	1.2×10^{-13}	9.6×10^{-13}

[a]Variants information in exon are based on mRNA reference sequence (GeneBank accession no. NM_174129.3). Variants information in 5'-upstream regions are based on DNA reference sequence of Btau4.6 assembly (GeneBank accession no. NC_007307)
[b]The Reference (Ref) and Altenative (Alt) alleles
[c]The protein information is based on GeneBank accession no. NP_776554.2
[d]RAF: Reference allele frequency
[e]The c.1044C > T SNP was not genotyped in all animals because of its synonymous SNP

SNPs ranged from 0.80 to 1.00 (Fig. 3b). The c.1318C > T and c.1475 T > A were under perfect LD in this population. The r^2 values of the five SNPs and the most significant SNP (rs42865669) in the SNP array were very high (r^2 = 0.82 to 0.91) (Fig. 3b). Among the haplotypes of five SNPs, we defined the haplotype with positive effect on IMP as *Q* haplotype and those with negative effect on IMP as *q* haplotype (Fig. 4a). The *q* haplotype had the same alleles of the *NT5E* mRNA reference sequence (NM_174129.3).

Functional effect of five SNPs

NT5E is the enzyme that causes the degradation of IMP and AMP on the plasma membrane to inosine and adenosine, respectively (Fig. 1). The three non-synonymous SNPs in *NT5E* could affect the protein structure and thus their enzymatic activity. To determine whether the enzymatic activity of NT5E in each haplotype is different, we transfected *NT5E*-expression plasmid into COS-7 cells. The ability to degrade IMP was examined using malachite green for the detection of the released inorganic phosphate (Pi). The results showed that the enzymatic activity of NT5E was significantly higher (*p*-value = 3.6 × 10^{-4}) in constructs with *q* haplotype (q-q-q) compared to those with *Q* haplotype (Q-Q-Q) (Fig. 4b). Next, the three non-synonymous SNPs in exon 7 and 8 were mutated from *q* haplotype to *Q* allele in each locus to determine their effect on the enzymatic activity. The results showed that the activity in the construct with c.1318C > T (Q-q-q) and c.1475 T > A (q-Q-q) were detected equally well as that of Q-Q-Q (*p*-values were 0.30 and 0.88, respectively). In contrast, the activity in the construct with c.1526A > G (q-q-Q) was similar to that of q-q-q (*p*-value = 0.99),

but was significantly different from those of Q-Q-Q (*p*-value = 3.7 × 10^{-3}), Q-q-q (*p*-value = 8.7 × 10^{-5}), and q-Q-q (*p*-value = 4.3 × 10^{-2}). These results showed that c.1318C > T and c.1475 T > A in *NT5*E are the QTNs for the degradation of IMP as they affect its enzymatic activity.

The two SNPs in 3'-UTR of *NT5E* (c.1318C > T and c.1475 T > A) could potentially affect the stability of the *NT5E* transcripts. Thus, we compared the relative abundances of *Q*- versus *q*-derived *NT5E* transcripts in skeletal muscles (*n* = 14) of heterozygotes (Fig. 4c). We isolated samples of genomic DNA and complementary DNA (cDNA) from heterozygotes and then compared their allelic ratios using PeakPicker2 software [17]. The results showed that *Q*-derived *NT5E* cDNA and genomic DNA were not significantly detected as *q*-derived cDNA and genomic DNA, respectively (*p*-value = 0.51). Therefore, the two SNPs in 3'-UTR of *NT5E* did not affect the allelic imbalances of *NT5E* mRNA expression, and thus did not affect the stability of the *NT5E* transcripts.

Bioinformatics analysis to investigate effect of non-synonymous SNPs

The multiple sequence alignments of the three amino acids based on the non-synonymous SNPs are shown in Fig. 4d. Val492 by *q* haplotype is highly conserved in mammals, and His440 and Gln509 by *q* haplotype are not conserved in some ruminants (goat and sheep), but the regions flanking His440 are highly conserved in mammals.

To locate the three amino acids based on the non-synonymous SNPs at the protein structure of NT5E, the

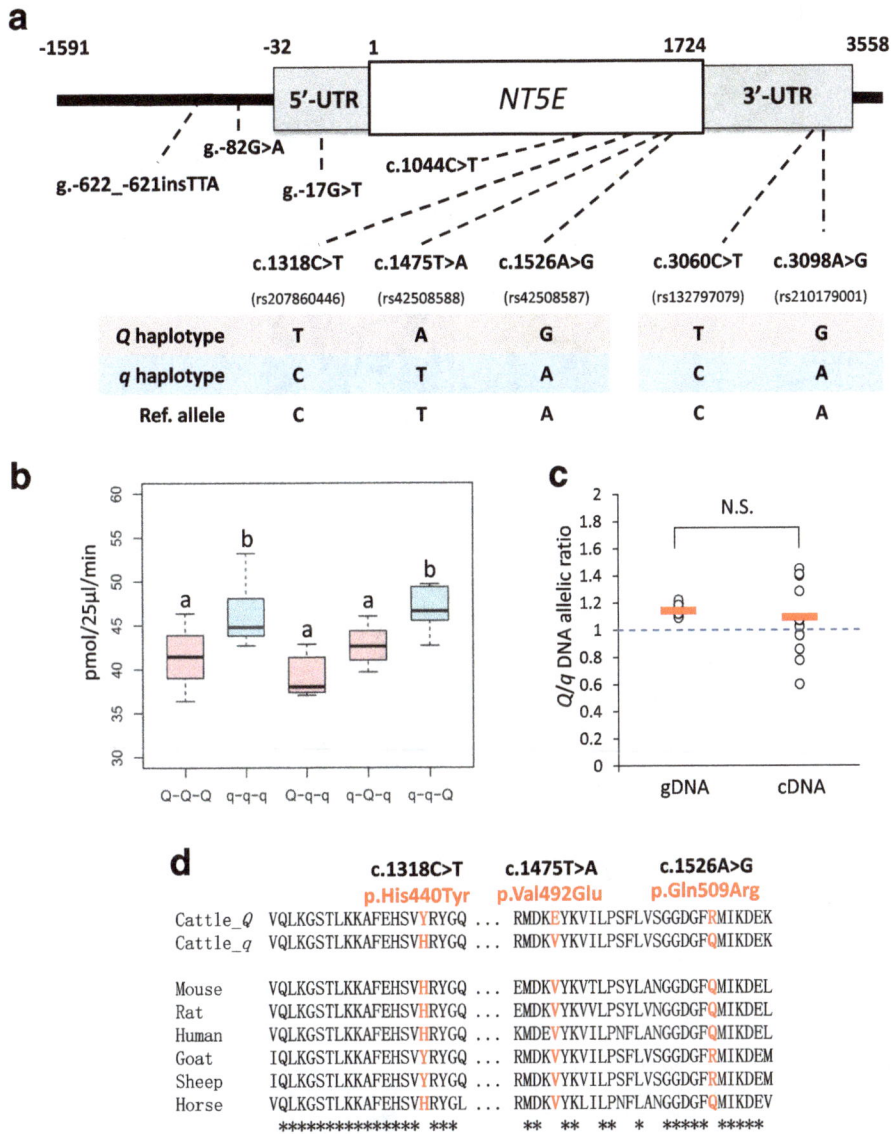

Fig. 4 The schematic structure and the SNP features of *ecto-5'-nucleotidase* (*NT5E*). **a** Schematic representation of the positions of variants from the 5'-upstream region to the 3'-UTR in *NT5E*. The detailed positions and names of the variants are shown in Table 1. The *Q* and *q* haplotypes are defined by the genotypes of the three non-synonymous SNPs on exon 7 and exon 8 and two SNPs on 3'-UTR of exon 9. The *Q* haplotype has a positive effect on inosine 5'-monophosphate (IMP). Bovine reference (Ref) allele from Gene Bank accession no. NM_174129.3 is also shown. **b** The IMPase activity of ecto-5'-nucleotidase in COS-7 cells. Q-Q-Q, the construct with *Q* haplotype; q-q-q, the construct with *q* haplotype; Q-q-q, the construct mutated from *q* haplotype to *Q* allele in c.1318C > T; q-Q-q, the construct mutated from *q* haplotype to *Q* allele in c.1475 T > A; q-q-Q, the construct mutated from *q* haplotype to *Q* allele in c.1526A > G. The superscript letters indicate significant differences among five constructs tested by analysis of variance followed by a Tukey HSD (honestly significant difference) multiple comparison test (*p*-value <0.05). **c** The allelic imbalance test for levels of *NT5E* mRNA in the heterozygotes. The y-axis shows the ratio of peak height of the *Q* allele over the *q* allele in the genomic DNA (gDNA) and the complementary DNA (cDNA) from the same animal. Red bars indicate the mean expression. No significant (N.S.) difference was shown between them. **d** Multiple sequence alignment of the regions flanking p.His440Tyr, p.Val492Glu, and p.Gln509Arg. *Q* haplotype and *q* haplotype sequences of the cattle are shown on the top and other mammalian sequences are shown below

3D protein structure of NT5E predicted by SWISS-MODEL server [18, 19] is illustrated in Fig. 5a. The crystal structure of human NT5E, which is known as CD73, is described a non-covalent homodimer [20], and the predicted structure of bovine NT5E using template human NT5E (4h2g) was illustrated. The docking of the

predicted bovine NT5E with the IMP ligand was carried out using SWISSDOCK server [21]. One of the top five clusters for IMP ligand was located on the β-sheet where p.Val492Glu was located (Fig. 5b). In contrast, p.His440-Tyr and p.Gln509Arg were not located near the binding site of the IMP ligands.

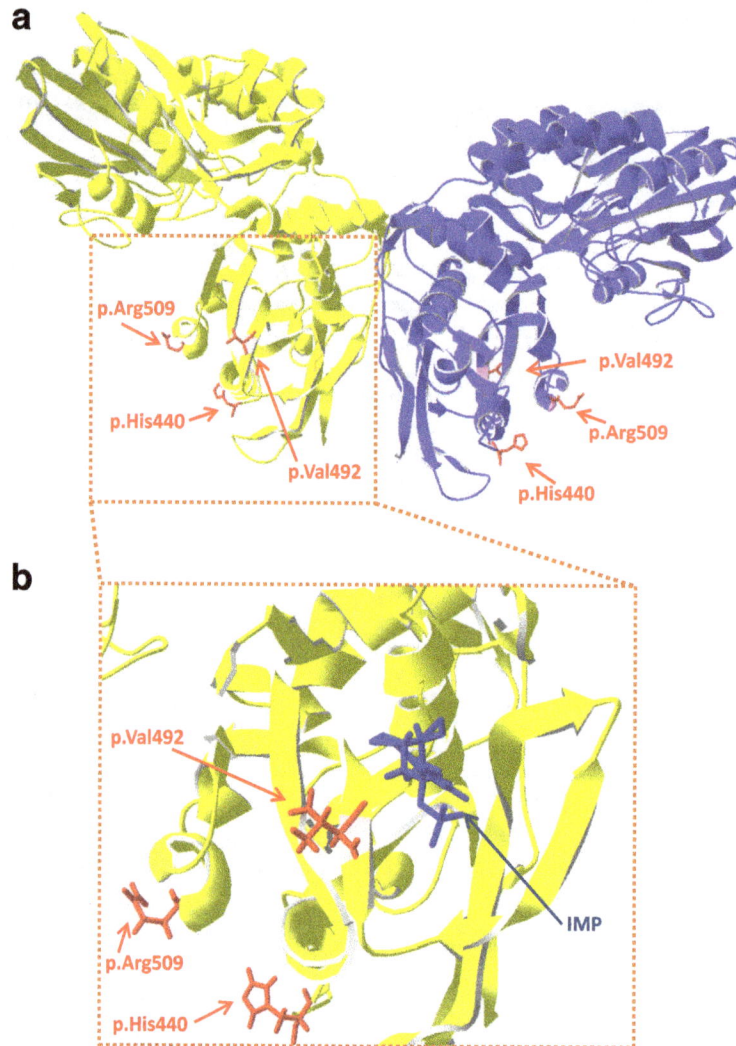

Fig. 5 Predicted structure of ecto-5′-nucleotidase (NT5E). **a** Predicted structure of NT5E modeled by the SWISS-MODEL server [18, 19]. **b** Docking of the predicted NT5E structure with inosine 5′-monophosphate (IMP) ligand. Three amino acids based on the three non-synonymous SNPs (p.His440, p.Val492, and p.Arg509) are shown as red sticks, and IMP ligand is shown as a blue stick

Comparison of Q and q haplotype frequencies among five different breeds

In the functional analysis, both the SNPs (c.1318C > T and c.1475 T > A) affect the enzymatic activity and were under perfect LD in this population. To determine whether the two SNPs segregate in other breeds, they were genotyped in another 1079 animals, which were from Japanese Black cattle, Japanese Shorthorn cattle, Japanese Brown cattle, Angus cattle, and Holstein cattle. The two SNPs were under perfect LD in all the breeds. The genotype (QQ, Qq, and qq) and haplotype (Q and q) frequencies of each breed are shown in Table 2. The genotype and haplotype frequencies of the Japanese Black cattle population used in GWAS (named as Japanese Black cattle 1) were similar to those of the Japanese Black cattle population composed of unrelated animals (named

as Japanese Black cattle 2). The different genotype and haplotype frequencies among these five breeds are also shown. For example, the frequencies of QQ genotype were 0.23 in Japanese Black cattle 2, 0.01 in Japanese Shorthorn cattle, 0.25 in Japanese Brown cattle, 0.09 in Angus cattle, and 0.04 in Holstein cattle.

Genetic architecture of IMP and its degradation products

In this population, the effects of the two non-synonymous SNPs in *NT5E* on IMP, inosine, and hypoxanthine were evaluated. The Q haplotype substitution effect and the proportions of genetic and phenotypic variances explained by the Q haplotype substitution effect are shown in Table 3. The genetic and phenotypic variances in each trait were estimated by single-trait animal model. The Q haplotype had negative effects on inosine (-0.07) and hypoxanthine (-0.12).

Table 2 Genotype and haplotype frequencies of two non-synonymous SNPs in *ecto-5'-nucleotidase (NT5E)* in five different breeds

Breeds[a]	N	Genotype frequency			Haplotype frequency	
		QQ	Qq	qq	Q	q
Japanese Black cattle 1	574	0.17	0.49	0.34	0.42	0.58
Japanese Black cattle 2	542	0.23	0.46	0.31	0.46	0.54
Japanese Shorthorn cattle	109	0.01	0.21	0.78	0.11	0.89
Japanese Brown cattle	106	0.25	0.55	0.20	0.53	0.47
Angus cattle	118	0.09	0.44	0.47	0.31	0.69
Holstein cattle	204	0.04	0.28	0.67	0.19	0.81

[a]Japanese Black cattle 1, the population used in genome-wide association study; Japanese Black cattle 2, the population composed of unrelated animals and used for calculating genotype and haplotype frequencies

The proportions of genetic and phenotypic variances for IMP, inosine, and hypoxanthine ranged from 0.30 to 0.46 and from 0.08 to 0.22, respectively. Moreover, the proportion of genetic variance for IMP was almost half of total genetic variance (0.46). In this study, we also evaluated the association of the two non-synonymous SNPs with other economically important traits, which were five carcass traits and 13 fatty acid compositions previously reported by Sasago et al. [16]. The results showed that no significant associations with *p*-value <0.01 were detected in these traits (Additional file 2: Table S2).

The genetic and phenotypic correlations among IMP, inosine, and hypoxanthine using the model (1) (see Materials and Methods) with and without *NT5E* effect as covariate are shown in Table 4. When the *NT5E* effect was included in the model (1), the genetic correlations of IMP with inosine and hypoxanthine increased (from -0.16 to 0.66 and from -0.72 to -0.49, respectively), and the genetic correlation of inosine with hypoxanthine decreased (from 0.53 to 0.19). The phenotypic correlations among the three traits did not show a large difference in the model (1) with and without *NT5E* effect, except for the phenotypic correlation of IMP with inosine (from 0.24 to 0.55).

Discussion

GWAS, candidate gene analysis, and functional analysis
In this study, we investigated the genetic architecture of IMP and its degradation products in Japanese Black beef.

First, we performed GWAS, candidate gene analysis, and functional analysis to detect the causal variants affecting IMP and its degradation products. The results of GWAS and candidate gene analysis showed that the three non-synonymous SNPs and the two SNPs in 3'-UTR in *NT5E* had high significance in these traits. In addition, no significant association of these traits was detected in loci outside of the *NT5E* locus. In functional analysis, the different enzymatic activity of NT5E was shown between *Q* and *q* allele of the two non-synonymous SNPs under in vitro conditions, when IMP is used as a substrate. In addition, the SNPs in 3'-UTR of *NT5E* did not affect the level of *NT5E* mRNA expression, which could not lead to an allelic imbalance. These results indicated that the two non-synonymous SNPs (c.1318C > T and c.1475 T > A) in *NT5E* affect the amount of IMP, inosine, and hypoxanthine in beef by regulating enzymatic activity.

The detailed studies on the quantitative trait loci (QTL) affecting IMP and its degradation products have not been reported in livestock population. On the other hand, significant associations between *NT5E* and some traits have been reported recently in humans. For example, the serum inosine concentration is the biomarker of metabolic traits involved in purine metabolic pathways [12]. Recently, the metabolome-wide GWAS was performed to evaluate the genetic variance in comprehensive human metabolisms [22]. The result showed

Table 3 Descriptive statistics, the results of genetic analysis, and the effect of *ecto-5'-nucleotidase (NT5E)* for inosine 5'-monophosphate (IMP) and its degradation products in beef

Traits[a]	Descriptive statistics			Genetic analysis			NT5E effect		
				Variance components[b]		Heritability	Q haplotype substitution effect	Proportion[c]	
	N	Mean	SD	Vg	Vp			Vg	Vp
IMP	571	0.40	0.21	0.018 (0.007)	0.038 (0.003)	0.48 (0.15)	0.13 (0.01)	0.46	0.22
Inosine	573	0.81	0.15	0.008 (0.003)	0.023 (0.002)	0.34 (0.14)	−0.07 (0.01)	0.30	0.10
Hypoxanthine	570	2.03	0.30	0.021 (0.011)	0.085 (0.006)	0.25 (0.13)	−0.12 (0.02)	0.33	0.08

[a]Unit is μmol/g meat. Standard errors are shown in parentheses
[b]V_g, Genetic variance; V_p, Phenotypic variance
[c]The proportion of V_g and V_p explained by the *Q* haplotype substitution effect

Table 4 Genetic and phenotypic correlations estimated by the models with and without *ecto-5′-nucleotidase (NT5E)* effect

Trait	Model without NT5E effect[a]			Model with NT5E effect[a]		
	IMP	Inosine	Hx	IMP	Inosine	Hx
Inosine 5′-monophosphate (IMP)		−0.16	−0.72		0.67	−0.48
		(0.28)	(0.24)		(0.21)	(0.37)
Inosine	0.24		0.53	0.55		0.18
	(0.05)		(0.30)	(0.03)		(0.45)
Hypoxanthine (Hx)	−0.30	0.35		−0.17	0.27	
	(0.04)	(0.04)		(0.05)	(0.04)	

[a]Upper diagonal is genetic correlation and lower diagonal is phenotypic correlation. Standard errors are shown in parentheses

that a significant association exists between an SNP near *NT5E* and inosine concentration in human serum. Another example is the arterial and joint calcifications, an extremely rare mendelian disorder associated with increased cardiovascular risk, the genetic architecture of which was unclear [23]. Hilaire et al. [23] performed linkage analysis and showed that rare mutations in *NT5E* affect the arterial and joint calcifications due to loss of NT5E function. Zhang et al. [24] also performed candidate gene analysis and showed the association between calcification of joints and arteries and another non-synonymous SNP in *NT5E* (p.Gly454Arg), which was far away from the binding site of substrate AMP but next to the lacked locus of shorter NT5E isoform known as CD73S [25]. These results indicated that the mutations in *NT5E* would affect the enzymatic activity of NT5E and thus were associated with these traits. The objective of finding variants associated with these traits are different in between human (for health) and cattle (for breeding), but our results could contribute to the genetic understanding the effect of *NT5E* variants on both human and cattle nucleotide metabolisms.

Bioinformatics analysis of two-synonymous SNPs
The two non-synonymous SNPs (c.1318C > T and c.1475 T > A) were under complete LD in five different breeds. The result indicated the difficulty in evaluating the effect of SNP by separating the two SNPs and obtaining more than two types of haplotypes in vivo. The c.1475 T > A SNP encodes a highly conserved amino acid in mammals, and the predicted 3D protein structure showed that the SNP (p.Val492Glu) is located on the β-sheet for binding IMP ligand. Therefore, c.1475 T > A SNP could have a direct effect on enzyme–substrate binding. As for c.1318C > T, it encodes amino acid which is not conserved in some ruminants (goat and sheep), but the region flanking the amino acid is highly conserved in mammals. In addition, the SNP (p.His440Tyr) is far away from the binding site of the IMP ligand. Thus, c.1318C > T may not have a direct effect on enzyme–substrate binding. However, it could affect the enzymatic activity of NT5E from the non-binding site.

For example, human *NT5E* has two splice variants, which encode full-length canonical NT5E protein and shorter NT5E isoform known as CD73S [25]. Human CD73S lacks 5′-nucleotidase activity, because this enzyme lacks amino acids 404–453 in exon 7 located in the C-terminus. The C-terminus contains the important interface for forming the functional NT5E homodimer expressed on the cell surface [20]. The c.1318C > T is located on the absent locus of exon 7 (corresponding to 440th amino acid in human NT5E). Therefore, c.1318C > T could also affect the formation of the functional NT5E homodimer, and thus affect the enzymatic activity of NT5E.

Genetic architecture of IMP and its degradation products
In this study, two QTNs (c.1318C > T and c.1475 T > A) in *NT5E* contributed to about half of the total genetic variance in IMP. The *Q* haplotype with positive effect on IMP has a negative effect on its enzymatic activity in IMP degradation, and thus contributes to increase *umami* taste (Fig. 6). In addition, the genetic correlation of IMP with inosine largely increased when *NT5E* effect was included as a covariate in the model (1). These results indicated that the QTNs in *NT5E* strongly affected the enzymatic activity of NT5E and thus strongly affects the overall balance in concentration of IMP and its degradation products in beef.

The changes in the amount of IMP and its degradation products in beef are influenced by postmortem conditioning and the aging period [14, 15]. The overall change in concentration of IMP and its degradation products strongly depend on the aging period until 20 days after slaughter [14, 15]. After 20 days, the amount of IMP remains almost unchanged [15]. Some of the enzymes involved in IMP degradation are gradually inactivated after 20 days, since the proteolytic activity in postmortem muscle affects other enzymatic processes, such as glycolysis [26]. The IMP in muscle is degraded by 5′-nucleotidase in intracellular and NT5E in extracellular (Fig. 1). The aging period after slaughter of cattle induces structural changes in the cell membrane and leads to intracellular water efflux [13]. The intracellular fluid could lead to an increase the amount of IMP in extracellular during aging,

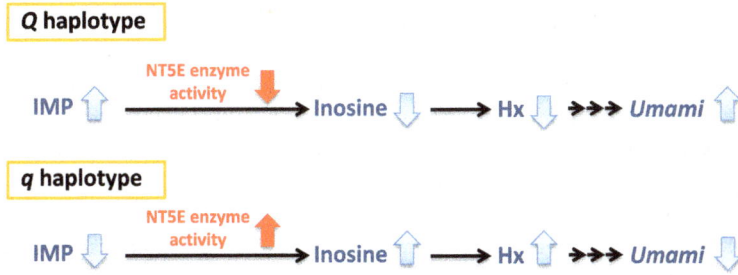

Fig. 6 Proposed model for the difference of ecto-5'-nucleotidase (NT5E) enzyme activity between *Q* and *q* haplotypes in beef. IMP, inosine 5'-monophosphate; Hx, hypoxanthine

and the NT5E then involves the degradation of IMP in extracellular. In this study, no significant association of IMP and its degradation products was detected in loci outside of the *NT5E* locus by GWAS. The result indicated that the enzymatic activity of NT5E strongly affects the difference in IMP degradation in beef.

Some of the major QTLs affecting the economically important traits are still segregated in a population, even if cattle populations go through intensive artificial selection for the traits. However, such QTLs with a favorable effect on the traits may have an unfavorable effect on other economical traits. For example, the allele in the *Acyl-CoA:diacylglycerol acyltransferase 1 (DGAT1)* increases milk fat content and decreases milk yield in dairy cattle [27], and the risk allele *FYVE, RhoGEF and PH domain-containing protein 3 (FGD3)* of skeletal dysplasia increases carcass weight [28]. Therefore, it is difficult to apply marker-assisted selection (MAS) using these QTLs. On the other hand, unselected traits, which have the potential for future breeding objectives, such as meat quality traits, could retain high genetic variance. A major QTL affecting the traits might not have an unfavorable effect on other selection traits. IMP is currently not under selection in Japanese Black cattle breeding, but the *NT5E* is a major QTL affecting the IMP in beef and is not associated with carcass traits and fatty acid compositions. In addition, the detected QTNs in *NT5E* were common variants and different haplotype frequency was seen among different breeds. The frequency of favorable *Q* haplotype was greater in Japanese Black cattle than in other breeds such as Angus cattle and Holstein cattle. Therefore, IMP could be a future objective trait in Japanese Black cattle, and *NT5E* could be useful as a means of improving meat quality by MAS.

Conclusion

The overall change in concentration of IMP and its degradation products in meat can potentially affect the beef palatability and especially *umami* taste. We investigated the genetic architecture of these traits to utilize them in Japanese Black cattle breeding to enhance meat quality.

Our study demonstrated that two non-synonymous SNPs in *NT5E* affect the amount of IMP and its degradation products in meat by regulating NT5E enzymatic activity. The genetic architecture of IMP and its degradation products included the *NT5E* with a very large effect, and the QTNs in *NT5E* affected the overall concentration balance of IMP and its degradation products. In addition, the QTNs in *NT5E* were common variants in different breed populations and did not have an unfavorable effect on the carcass traits and fatty acid composition. Based on all the above findings taken together, IMP could be a breeding target in the future for increasing *umami* taste in beef, and the QTNs in *NT5E* could be useful for improving meat quality by MAS in Japanese Black cattle.

Methods

Animals, phenotypes, and genotyping for GWAS

The complete description of the population was reported in the previous study [8, 16]. A commercial Japanese Black cattle population, produced in Yamagata Prefecture, Japan, was used in this study. *Longissimus thoracis* muscles were collected from 574 Japanese Black cattle slaughtered at a meat processing plant in Yamagata Prefecture from 2011 to 2013. The muscles located at the 7th thoracic vertebra (frozen 16–19 d after slaughter) were purchased from a distributor. The muscles were thawed at 2 °C for 16 h and were then cut for nucleotide measurements. The samples were then stored at -30 °C until analysis.

IMP, inosine, and hypoxanthine were measured in this study. Phenotypes that were not within the mean ± 3 standard deviation (SD) range for each trait were considered outliers and were deleted. The descriptive statistics of these traits are shown in Table 3. The details on the procedure for measuring the traits in this study have been described by Sakuma et al. [8]. The extractions of nucleotides were performed using approximately 0.10 g samples of meat taken from the minced raw meat (about 15 g in total). The sample was homogenized with ultra-pure water and N-hexane, with cytidine solution as an internal standard. The underlayer was obtained after the removal of fat and protein using hexane and acetonitrile, respectively. The supernatant was filtered through a 0.45-μm microfilter (Millex-LH, Merck Millipore, Billerica, USA), and the filtrate was then mixed with 45% acetonitrile solution. Approximately 20 μL of the filtrate was mixed with 180 μL of ultrapure water, and the resulting solution was analyzed using the high performance liquid chromatography (Waters 2695, Waters, Milford, USA), along with an Atlantis T3 column (4.6 × 150 mm, 5 μm, Waters) and a UV detector (Waters 2487, Waters). The nucleotides were identified by comparing their retention times with those of the established standards. The concentrations of each were calculated using internal and external standard solutions, expressed as μmol per gram of meat.

The complete description of the SNP array genotyping was reported in a previous study by Sasago et al. [16]. The genomic DNA of 574 animals was briefly extracted from muscle samples using phenol-chloroform extraction. The DNA samples were genotyped using the Illumina BovineSNP50 v2 BeadChip (Illumina, CA, USA) and the GenomeStudio software (Illumina, CA, USA). The SNP maps were updated according to the SNPchiMpv.3 database [29] and the UMD3.1 assembly, and autosomal chromosomes were used in this study. The SNP quality control was assessed using PLINK software [30]. The exclusion criteria for SNPs included a minor allele frequency (MAF) < 0.01, a call rate < 0.95, and a Hardy-Weinberg equilibrium test with p-value <0.001. A total of 40,657 SNPs on the array were used in the present study.

Genetic analysis

The genetic parameters of IMP, inosine, and hypoxanthine in meat were estimated by the following animal model

$$y_{ikjlmno} = \mu_i + sex_{ij} + year_{ik} + month_{il} + aging_{im} + farm_{in} + b_1 x_{ijklmno} + b_2 x_{ijklmno}^2 + u_{ijklmno} + e_{ijklmno}$$

(1)

where $y_{ijklmno}$ is the observation of the animal o for trait i; μ_i is the total mean for trait i; sex_{ij} is the fixed effect of sex j (2 classes) for trait i; $year_{ik}$ is the fixed effect of the slaughter year k (3 classes, 2011–2013) for trait i; $month_{il}$ is the fixed effect of the slaughter month l (12 classes) for trait i; $aging_{im}$ is the fixed effect of aging period m (4 classes, 16–19 days) for trait i; $farm_{in}$ is the fixed effect of the farm n (13 classes) for trait i; $b_1 x_{ijklmno} + b_2 x_{ijklmno}^2$ are the linear (b_1) and quadratic (b_2) regression coefficients on slaughter age ($x_{ijklmno}$) for trait i; $u_{ijklmno}$ is the random additive genetic effect of animal o for trait i; $e_{ijklmno}$ is the random residual effect for trait i. The pedigrees were traced back to five generations,

and a total of 3513 animals were used in this study. Single-trait animal model for estimating heritability and multi-trait animal model for estimating genetic and phenotypic correlations were applied using model (1). The ASReml 3.0 software [31] was used to estimate (co)variance components with standard errors and best linear unbiased estimators (BLUEs) of all fixed effects. The results of estimated variance components and heritability for three traits by single-trait animal model are shown in Table 3.

GWAS

Firstly, the phenotypic values were adjusted using any fixed non-genetic effects in model (1). The BLUEs for all fixed effects were obtained by single-trait animal model, and the adjusted phenotype was calculated by subtracting the BLUEs from raw phenotypic values. Secondly, the vectors of adjusted phenotypes ($\mathbf{y_{adj}}$) were used as dependent traits in a linear mixed model approach for each SNP as follows:

$$\mathbf{y_{adj}} = \beta_i \mathbf{w}_i + \mathbf{a} + \mathbf{e}'_i$$

(2)

where β_i is the allele substitution effect of the effect allele, $\mathbf{w}_i$ is a vector of the SNP genotypes (coded as 0, 1, or 2 for the homozygote, heterozygote, and the other homozygote, respectively), and $\mathbf{e}'_i$ is a vector of the random residual effect at the i-th SNP distributed as $N(0, \mathbf{I}\sigma_{e'}^2)$, where $\mathbf{I}$ and $\sigma_{e'}^2$ are the identity matrix and residual variance, respectively. $\mathbf{a}$ is a vector of random genetic effects ($\mathbf{a} \sim N(0, \mathbf{G}\sigma_a^2)$), where $\mathbf{G}$ and σ_a^2 are the genomic relationship matrix proposed by VanRaden [32] and the SNP genetic variance, respectively. The regression coefficient and p-values tested by Wald test were obtained using GEMMA software [33]. The proportion of phenotypic variance explained by the i-th SNP effect was calculated using the formula

$$Proportion_i = \frac{2p_i(1-p_i)\beta_i^2}{V}$$

where p_i is MAF of the i-th SNP [34], and V is the genotypic (V_g) or phenotypic (V_p) variances obtained in a single-trait animal model in genetic analysis (Table 3). The Bonferroni correction was applied to determine the 5% genome-wide significance thresholds ($p = 1.37 \times 10^{-6}$). The genes within ±50 kb of the significant SNPs were scanned using the NCBI2R R-package. The values of r^2 between individual SNPs were calculated, and the haplotype block pattern was visualized using Haploview 4.0 software [35].

NT5E sequencing and its effect

We used the genomic DNA to determine the nucleotide sequences of NT5E to detect its variants. Each of the eight animals was selected from the most extreme upper and lower residuals of IMP in the population. The residuals in this analysis were calculated by subtracting the additive genetic effect and all the non-genetic effects in model (1) from phenotypic values. For the analysis of polymorphisms in *NT5E*, the fragments of the full-length exon region and proximal promoter region (upstream region from the start codon to 1591 bp upper region) were directly sequenced with 15 primer sets (Additional file 3: Table S3). The primer sets were designed by PRIMER3 software [36] according to the information about bovine *NT5E* mRNA and Btau4.6 genome sequences (GenBank accession no. NM_174129.3 and NC_007307, respectively), because the *NT5E* was not correctly assigned in UMD3.1 assembly (GeneBank accession no. AC_000166). Polymerase chain reaction (PCR) was performed in a 15-µL volume of solution containing 20 ng of genomic DNA, 6.25 pmol of each primer, 0.2 mM of each deoxynucleoside triphosphate, 10 mM Tris-HCl (pH 8.3), 50 mM KCl, 1.5 mM MgCl$_2$, and 0.375 U of KOD-FX DNA polymerase (Toyobo, Osaka, Japan). The PCR was performed as follows: 94°C for 2 min followed by 30 cycles of 98°C for 10 s, 55°C for 30 s, and 68°C for 40 s. The PCR products were purified using the ExoSAP-IT PCR Product Cleanup (USB Corporation, Cleveland, OH, USA) and then directly sequenced using the ABI PRISM 3130 DNA Sequencer and Sequencing Analysis 3.4 software (Applied Biosystems Japan, Tokyo, Japan). The polymorphisms were then checked, and the reference SNP ID number was obtained through the dbSNP database (http://www.ncbi.nlm.nih.gov/SNP). A total of eight SNPs and one indel variant, including one synonymous SNP, were identified in this population, and all variants were present in dbSNP database. The eight variants except for one synonymous SNP were then genotyped in all animals by direct sequencing as shown above.

The association tests of all variants in *NT5E* with three traits were performed using the model (2). In addition, estimation of genetic and phenotypic correlations was performed using the model (1), which included the c.1475 T > A SNP as a covariate in the model (1). As for the most significant SNP, the association test of c.1475 T > A SNP with other economically important traits was also performed using the model (2). The economically important traits used in this study were five carcass traits and 13 fatty acid compositions described by Sasago et al. [16] (Additional file 2: Table S2).

Expression construct for NT5E
To measure the enzymatic activity of NT5E, the coding region of *NT5E* (NM_174129.3) of each haplotype (*Q* and *q* haplotype) was PCR amplified using PrimeSTAR Max DNA Polymerase (Takara, Cat. #R045A) from cDNA derived from the dermal primary fibroblasts using

a forward primer (5'- <u>atgaatcccggagcggctcgcaccccggcgctga ggatcctcgcgctgggcgcgttgctgtggcccgcgggcgcgcccc</u>ATGGGAT CCTACCCTTACGACGTTCCTGATTACGCTAGCCTC GAATTCtgggagctcaccatcttgcacacc-3'; lowercase letters + underline indicate the signal sequence, and uppercase letters indicate hemagglutinin [HA] tag), and a reverse primer (5'- ATAAGAATGCGGCCGCctattggtata aaataatgatc-3'; uppercase letters indicate the *Not*I linker). The PCR product was cloned into the blunted *EcoR*I and *Not*I sites of the pCAGGS vector [37]. The sequence and orientation of the insert were confirmed by sequencing. The expression of NT5E was confirmed by western blotting with an anti-HA antibody 3F10 (Roche, Cat. #11867423001, 100 ng/mL). Immunoreactivity was detected with a horseradish peroxidase-conjugated donkey anti-rat IgG antibody (Jackson ImmunoResearch, Cat. #712–035-153) and the ECL Prime Western Blotting Detection Reagent (GE Healthcare, Cat. #RPN2232). Chemiluminescence was detected with an ImageQuant LAS 4000 (GE Healthcare) and quantified using the ImageQuant TL Analysis Toolbox.

To determine whether non-synonymous SNPs affect the enzymatic activity of NT5E, pCAGGS-*NT5E* (*q* haplotype) was subjected to site-direct mutation using the following primers: c.1318C > T (forward primer: 5'-gagcacagcgtgTaccgctatggccaggccac-3', reverse primer: 5'- tagcggtAcacgctgtgctcgaaggccttcttc-3'); c.1475 T > A (forward primer: 5'- gaatggdataaggAgtacaaggtgatcctccc-3', reverse primer: 5'- cttgtacTccttatccattctaagaggctc-3'); and c.1526A > G (forward primer: 5'- gagacggattccGgatgataaaag atgaaaag-3', reverse primer: 5'- tatcatcCggaatccgtctcca ccactgacaag-3'). Uppercase letters indicate the mutation allele (*Q* allele). After the PCR amplification with PrimeSTAR Max DNA Polymerase, the PCR products were digested with *Dpn*I (Toyobo, Cat. #DPN-101) and were then transformed into One Shot TOP10 Competent *E. coli* (ThermoFisher, Cat. #C404003). The sequence were confirmed by sequencing.

IMPase activity assay
To determine whether the haplotype and non-synonymous SNPs in the exon 7 and exon 8 affect the enzymatic activity of NT5E in COS-7 cells, we transfected 0.4×10^5 cells per well into a 24-well plate with a mixture of 0.4 ng of the pCAGGS-NT5E. COS-7 cells were used in this study as it exhibits negligible endogenous nucleotide-metabolizing enzyme activities [38, 39]. About 48 h after the transfection, the cells were washed twice with saline. The IMP activity of the intact cells was determined by measuring the Pi liberated as a result of the degradation of 1 mM IMP (Sigma, Cat. # I4625) for 30 min using Malachite Green Phosphate Assay Kit (echelon, Cat. #K-1500) and plate reader (Bio-Rad, Cat. #iMark). The results were tested by analysis of variance

(one-way ANOVA) followed by the Tukey HSD (honestly significant difference) multiple comparisons test for significant difference among the five constructs (Q-Q-Q, q-q-q, Q-q-q, q-Q-q, and q-q-Q).

Allelic imbalance test

To quantify the potential allelic imbalance of *NT5E* transcripts, we designed PCR primers to c.1475 T > A in exon 8 of *NT5E*. The forward primer was 5'-gggacagggtggtcaagtta –3', and the reverse primer was 5'- cagagtcatgttttatcttttcatctt –3'. A total of 14 skeletal muscles from heterozygotes in Japanese Black cattle in NLBC were collected. We used 50 ng of template cDNA from skeletal muscles or 10 ng of genomic DNA from heterozygous animals for PCR amplification with TaKaRa Ex Taq HS DNA Polymerase (TaKaRa, Cat. #RR006). The PCR product was directly sequenced and purified using the CleanSEQ system (Agencourt, Cat. #A29154). The peak heights at the polymorphic sites were quantified using PeakPicker 2 software [17]. Allelic imbalances were estimated as the ratio of the peak height of the Q allele to that of the q allele in cDNA and in genomic DNA from the same animal. Calibration curves were generated using data obtained by mixing varying amounts of genomic DNA from Q and q homozygotes. Welch's t-test was performed to compare the differences between the ratios of the peak height of the Q allele over the q allele in genomic DNA and cDNA.

Bioinformatics analysis

For the three non-synonymous SNPs detected in the exon 7 and exon 8, multiple sequence alignments for *NT5E* were conducted using Clustal W [40] using cow, mouse, rat, human, goat, sheep, and horse genomic sequences. To illustrate the effects of the three non-synonymous SNPs, the 3D structure of NT5E reference sequence (NP_776554.2) was modeled by the homology modeling using the SWISS-MODEL server [18, 19]. The template structure of human NT5E, which is known as CD73, crystal form I (4h2f), crystal form II (4h2g), and crystal form III (4h2i) [20] from protein data bank (PDB) [41] were applied for the modeling. We used GMQE and QMEAN4 scores to discriminate the good model from all other models (higher numbers indicate higher reliability). The template structure of NT5E crystal form II (4h2g) had the highest values of GMQE (0.95) and QMEAN4 (0.53) in the SWISS-MODEL server (Additional file 4: Table S4). The structural model based on the template structure of NT5E crystal form II was then visualized by Swiss-PdbViewer software [42]. The docking of the obtained structural model with the 3D structure of IMP ligand (ZINC database ID: 14,951,284) [43] was carried out with SWISSDOCK server based on EADock DSS [21], because the NT5E

crystals in PDB were not obtained in the presence of IMP ligand. A total of 42 binding clusters were generated in the vicinity of all target cavities, and the top five most favorable clusters (Additional file 5: Table S5) were then visualized by the Swiss-PdbViewer software.

Genotype and haplotype frequencies of two non-synonymous SNPs in five different breeds

We obtained blood samples from Japanese Black cattle ($n = 542$), Japanese Shorthorn cattle ($n = 109$), Japanese Brown cattle ($n = 106$), Angus cattle ($n = 118$), and Holstein cattle ($n = 204$) raised at Tokachi, Ohu, Iwate, Tottori, Miyazaki, and Kumamoto stations of NLBC, respectively, in order to screen for the two non-synonymous SNPs. Japanese Black cattle, Japanese Shorthorn cattle, and Japanese Brown cattle are beef cattle called "Wagyu" [44], with different genetic background [45]. For Japanese Black cattle and Holstein cattle, the samples were selected using the criteria of at most 5 progenies in each sire, and these animals were low relatives with the progeny of 164 sires and 110 sires, respectively. The genomic DNA was extracted using phenol-chloroform extraction. The two non-synonymous SNPs were genotyped using the Cycleave PCR system (Takara Bio Inc., Shiga, Japan), and the PCR reaction mixture was prepared using Premix Ex Taq™ (Probe qPCR; TaKaRa Bio Inc.). Genotyping was done using an ABI StepOnePlus Real-Time PCR system (Applied Biosystems), and the primer and the probe information are shown in Additional file 6: Table S6. The reaction conditions were 95 °C for 20 s; 40 cycles at 95 °C for 1 s and 60 °C for 20 s; and 60 °C for 10 s.

Additional files

Additional file 1: Table S1. Significant genome-wide single nucleotide polymorphisms (SNPs) for inosine 5'-monophosphate (IMP), inosine, and hypoxanthine in meat. (XLSX 16 kb)

Additional file 2: Table S2. Association test of *ecto-5'-nucleotidase* (*NT5E*) with carcass traits and fatty acid compositions. (XLSX 12 kb)

Additional file 3: Table S3. Sequences of PCR primers used to amplify and sequence the genomic DNA. (XLSX 10 kb)

Additional file 4: Table S4. Details of the parameters for three selected models of human ecto-5'-nucleotidase in SWISS-MODEL server. (XLSX 9 kb)

Additional file 5: Table S5. Top five clustering results obtained from the docking of inosine 5'-monophosphate into ecto-5'-nucleotidase by SWISSDOCK server. (XLSX 9 kb)

Additional file 6: Table S6. Primer and reporter information for the two non-synonymous SNP genotyping. (XLSX 15 kb)

Abbreviations

ADP: Adenosine diphosphate; AMP: Adenosine monophosphate; ATP: Adenosine triphosphate; BLUE: Best linear unbiased estimator; DGAT1: Acyl-CoA:diacylglycerol acyltransferase 1; FGD3: FYVE, RhoGEF and PH domain-containing protein 3; GWAS: Genome-wide association study; IMP: Inosine 5'-monophosphate; LD: Linkage disequilibrium; MAS: Marker-assisted selection; NT5E: Ecto-5'-nucleotidase; PCR: Polymerase chain reaction; PDB: Protein data bank; QTL: Quantitative trait locus;

QTN: Quantitative trait nucleotide; SD: Standard deviation; SNP: Single nucleotide polymorphism; UTR: Untransrated region

Acknowledgments
The authors thank Mr. Noriaki Shoji and Mr. Kunihiko Saito for the collection of samples, and the other staffs of each branch of NLBC for generously providing the samples. We are grateful to Jun-ichi Miyazaki for providing the pCAGGS plasmid.

Funding
The work was supported by the National Livestock Breeding Center, Japan. The funders had no role in study design, data collection and analysis, decision to publish, or preparation of the manuscript.

Authors' contributions
YU conceived and designed the experiments, performed the statistical and bioinformatics analyses, and contributed to writing and improving the manuscript. TO, NS, MT, and TA performed DNA extraction, identified the polymorphisms, and performed genotyping. HS collected samples and phenotypes. TK designed the experiment and managed the entire project. SS conceived and designed the experiments, performed functional experiments, and contributed to writing and improving the manuscript. All authors read and approved the final manuscript.

Competing interests
The authors declare that they have no competing interests to National Livestock Breeding Center.

Author details
[1]National Livestock Breeding Center, Nishigo, Fukushima 961-8511, Japan.
[2]Present address: Graduate School of Agricultural Science, Tohoku University, Sendai, Miyagi 980-0845, Japan.

References
1. Gill JL, Matika O, Williams JL, Worton H, Wiener P, Bishop SC. Consistency statistics and genetic parameter for taste panel associated meat quality traits and their relationship with carcass quality traits in a commercial population of Angus-sired beef cattle. Animal. 2010;4:1–8.
2. Mateescu RG, Garrick DJ, Garmyn AJ, VanOverbeke DL, Mafi GG, Reecy JM. Genetic parameters for sensory traits in longissimus muscle and their associations with tenderness, marbling score, and intramuscular fat in Angus cattle. J Anim Sci. 2015;93:21–7.
3. Pegg RB, Shahidi F. Heat effects on meat/warmed-oven flavour. In: Jensen WK, Devine C, Dikeman M, editors. Encyclopedia of meat sciences Oxford: Elsevier ltd, vol. 2; 2004. p. 592–9.
4. Maga JA. Flavour potentiator. CRC Crit Rev Food Sci Nutr. 1983;18:231–312.
5. Nishimura T, Rhue MR, Okitani A, Kato H. Components contributing to the improvement of meat taste during storage. Agric Biol Chem. 1988;52:2323–30.
6. Suzuki K, Shioura H, Yokota S, Katoh K, Roh SG, Iida F, et al. Search for an index for the taste of Japanese Black cattle beef by panel testing and chemical composition analysis. Anim Sci J. 2017;88:421–32.
7. MacLeod G. The flavour of beef. In: Flavor of meat and meat products. Springer: US; 1994. p. 4–37.
8. Sakuma H, Saito K, Kohira K, Ohashi H, Shoji N, Uemoto Y. Estimates of genetic parameters for chemical traits of meat quality in Japanese Black cattle. Anim Sci J. 2017;88:203–12.
9. Surette ME, Gill TA, LeBlanc PJ. Biochemical basis of postmortem nucleotide catabolism in cod (Gadus morhua) and its relationship to spoilage. J Agr Food Chem. 1988;36:19–22.
10. Ribeiro JA, Cunha RA, Correia-de-Sá P, Sebastião AM. Purinergic regulation of acetylcholine release. Prog Brain Res. 1996;109:231–42.
11. Schulte G. Adenosine receptor signaling and the activation of mitogen-activated protein kinases. Dissertation. Stockholm: Repro Print AB; 2002. p. 7.
12. Bogan KL, Brenner C. 5'-nucleotidases and their new roles in NAD+ and phosphate metabolism. New J Chem. 2010;34:845–53.
13. Damez JL, Clerjon S, Abouelkaram S, Lepetit J. Dielectric behavior of beef meat in the 1–1500kHz range: simulation with the Fricke/Cole–Cole model. Meat Sci. 2007;77:512–9.
14. Koutsidis G, Elmore JS, Oruna-Concha MJ, Campo MM, Wood JD, Mottram DS. Water-soluble precursors of beef flavor. Part II: effect of post-mortem conditioning. Meat Sci. 2008;79:270–7.
15. Iida F, Miyazaki Y, Tsuyuki R, Kato K, Egusa A, Ogoshi H, et al. Changes in taste compounds, breaking properties, and sensory attributes during dry aging of beef from Japanese black cattle. Meat Sci. 2016;112:46–51.
16. Sasago N, Abe T, Sakuma H, Kojima T, Uemoto Y. Genome-wide association study for carcass traits, fatty acid composition, chemical composition, sugar, and the effect of their candidate genes in Japanese Black cattle. Anim Sci J. 2017;88:33–44.
17. Ge B, Gurd S, Gaudin T, Dore C, Lepage P, Harmsen E, et al. Survey of allelic expression using EST mining. Genome Res. 2005;15:1584–91.
18. Arnold K, Bordoli L, Kopp J, Schwede T. The SWISS-MODEL workspace: a web-based environment for protein structure homology modelling. Bioinformatics. 2006;22:195–201.
19. Biasini M, Bienert S, Waterhouse A, Arnold K, Studer G, Schmidt T, et al. SWISS-MODEL: modelling protein tertiary and quaternary structure using evolutionary information. Nucleic Acids Res. 2014;42:W252–8.
20. Knapp K, Zebisch M, Pippel J, El-Tayeb A, Müller CE, Sträter N. Crystal structure of the human ecto-5'-nucleotidase (CD73): insights into the regulation of purinergic signaling. Structure. 2012;20:2161–73.
21. Grosdidier A, Zoete V, Michielin O. SwissDock, a protein-small module docking web server based on EADock DSS. Nucleic Acids Res. 2011;39:270–7.
22. Suhre K, Shin SY, Petersen AK, Mohney RP, Meredith D, Wägele B, et al. Human metabolic individuality in biomedical and pharmaceutical research. Nature. 2011;477:54–60.
23. St. Hilaire C, Ziegler SG, Brusco A, Groden C, Gill F, et al. NT5E mutations and arterial calcifications. N Engl J Med. 2011;364:432–42.
24. Zhang Z, He JW, WZ F, Zhang CQ, Zhang ZL. Calcification of joints and arteries: second report with novel NT5E mutations and expansion of the phenotype. J Hum Genet. 2015;60:561–4.
25. Snider NT, Altshuler PJ, Wan S, Welling TH, Cavalcoli J, Omary MB. Alternative splicing of human NT5E in cirrhosis and hepatocellular carcinoma produces a negative regulator of ecto-5'-nucleotidase (CD73). Mol Biol Cell. 2014;25:4024–33.
26. Lametsch R, Roepstorff P, Bendixen E. Identification of protein degradation during post-mortem storage of pig meat. J Agr Food Chem. 2002;50:5508–12.
27. Grisart B, Coppieters W, Farnir F, Karim L, Ford C, Berzi P, et al. Positional candidate cloning of a QTL in dairy cattle: identification of a missense mutation in the bovine DGAT1 gene with major effect on milk yield and composition. Genome Res. 2002;12:222–31.
28. Takasuga A, Sato K, Nakamura R, Saito Y, Sasaki S, Tsuji T, et al. Non-synonymous FGD3 variant as positional candidate for disproportional tall stature accounting for a carcass weight QTL (CW-3) and skeletal dysplasia in Japanese Black cattle. PLoS Genet. 2015;11:e1005433.
29. Nicolazzi EL, Caprera A, Nazzicari N, Cozzi P, Strozzi F, Lawley C, et al. SNPchiMp v.3: integrating and standardizing single nucleotide polymorphism data for livestock species. BMC Genomics. 2015;16:283.
30. Purcell S, Neale B, Todd-Brown K, Thomas L, Ferreira MA, Bender D, et al. PLINK: a tool set for whole-genome association and population-based linkage analyses. Am J Hum Genet. 2007;81:559–75.
31. Gilmour AR, Gogel BJ, Cullis BR, Thompsion R. Asreml. User Guide Release 3. 0. Hemel Hempstead: VSN International Ltd; 2009.
32. VanRaden PM. Efficient methods to compute genomic predictions. J Dairy Sci. 2008;91:4414–23.

33. Zhou X, Stephens M. Genome-wide efficient mixed-model analysis for association studies. Nature Genet. 2012;44:821–4.

34. Falconer DS, Mackay TFC. Introduction to quantitative genetics. 4th ed. London: Longman Group; 1996.

35. Barrett JC, Fry B, Maller J, Daly MJ. Haploview: analysis and visualization of LD and haplotype maps. Bioinformatics. 2005;21:263–5.

36. Untergrasser A, Cutcutache I, Koressaar T, Ye J, Faircloth BC, Remm M, et al. Primer3 - new capabilities and interfaces. Nucleic Acids Res. 2012;40:e115.

37. Niwa H, Yamamura K, Miyazaki J. Efficient selection for high-expression transfectants with a novel eukaryotic vector. Gene. 1991;108:193–9.

38. Tkacz K, Cioroch M, Skladanowski AC, Makarewicz W. The cytotoxic effect of purine riboside on COS-7 cells. Adv Exp Med Biol. 2000;486:355–9.

39. Fausther M, Lavoie EG, Goree JR, Baldini G, Dranoff JA. NT5E mutations that cause human disease are associated with intracellular mistrafficking of NT5E protein. PLoS One. 2014;9:e98568.

40. Larkin MA, Blackshields G, Brown NP, Chenna R, McGettigan PA, McWilliam H, et al. Clustal W and Clustal X version 2.0. Bioinformatics. 2007;23:2947–8.

41. Rose PW, Prlić A, Bi C, Bluhm WF, Christie CH, Dutta S, et al. The RCSB protein data bank: views of structural biology for basic and applied research and education. Nucleic Acids Res. 2015;43:D345–56.

42. Guex N, Peitsch MC. SWISS-MODEL and the Swiss-PdbViewer: an environment for comparative protein modeling. Electrophoresis. 1997;18:2714–23.

43. Irwin JJ, Sterling T, Mysinger MM, Bolstad ES, Coleman RG. ZINC: a free tool to discover chemistry for biology. J Chem Inf Model. 2012;52:1757–68.

44. Oyama K. Genetic variability of Wagyu cattle estimated by statistical approaches. Anim Sci J. 2011;82:367–73.

45. Yonesaka R, Sasazaki S, Yasue H, Niwata S, Inayoshi Y, Mukai F, et al. Genetic structure and relationships of 16 Asian and European cattle populations using DigiTag2 assay. Anim Sci J. 2016;87:190–6.

12

Assembling large genomes: analysis of the stick insect (*Clitarchus hookeri*) genome reveals a high repeat content and sex-biased genes associated with reproduction

Chen Wu[1,2,3]* (iD), Victoria G. Twort[1,2,4], Ross N. Crowhurst[3], Richard D. Newcomb[1,3] and Thomas R. Buckley[1,2]

Abstract

Background: Stick insects (Phasmatodea) have a high incidence of parthenogenesis and other alternative reproductive strategies, yet the genetic basis of reproduction is poorly understood. Phasmatodea includes nearly 3000 species, yet only the genome of *Timema cristinae* has been published to date. *Clitarchus hookeri* is a geographical parthenogenetic stick insect distributed across New Zealand. Sexual reproduction dominates in northern habitats but is replaced by parthenogenesis in the south. Here, we present a de novo genome assembly of a female *C. hookeri* and use it to detect candidate genes associated with gamete production and development in females and males. We also explore the factors underlying large genome size in stick insects.

Results: The *C. hookeri* genome assembly was 4.2 Gb, similar to the flow cytometry estimate, making it the second largest insect genome sequenced and assembled to date. Like the large genome of *Locusta migratoria*, the genome of *C. hookeri* is also highly repetitive and the predicted gene models are much longer than those from most other sequenced insect genomes, largely due to longer introns. Miniature inverted repeat transposable elements (MITEs), absent in the much smaller *T. cristinae* genome, is the most abundant repeat type in the *C. hookeri* genome assembly. Mapping RNA-Seq reads from female and male gonadal transcriptomes onto the genome assembly resulted in the identification of 39,940 gene loci, 15.8% and 37.6% of which showed female-biased and male-biased expression, respectively. The genes that were over-expressed in females were mostly associated with molecular transportation, developmental process, oocyte growth and reproductive process; whereas, the male-biased genes were enriched in rhythmic process, molecular transducer activity and synapse. Several genes involved in the juvenile hormone synthesis pathway were also identified.

Conclusions: The evolution of large insect genomes such as *L. migratoria* and *C. hookeri* genomes is most likely due to the accumulation of repetitive regions and intron elongation. MITEs contributed significantly to the growth of *C. hookeri* genome size yet are surprisingly absent from the *T. cristinae* genome. Sex-biased genes identified from gonadal tissues, including genes involved in juvenile hormone synthesis, provide interesting candidates for the further study of flexible reproduction in stick insects.

Keywords: Phasmatodea, Genome assembly, RNA-Seq, Reproduction, Parthenogenesis, *Clitarchus hookeri*

* Correspondence: chen.wu@plantandfood.co.nz
[1]School of Biological Sciences, The University of Auckland, Auckland, New Zealand
[2]Landcare Research, Auckland, New Zealand
Full list of author information is available at the end of the article

Background

The insect order Phasmatodea, commonly known as stick insects or walking sticks, contains approximately 3000 species distributed worldwide [1]. At least 10% of stick insect species can reproduce parthenogenetically and for this reason have attracted much attention [1]. *Clitarchus hookeri* is one of the most common New Zealand stick insect species and is distributed across a wide range of habitats on both the North and South Islands with a lower population density at higher latitudes and altitudes [2]. Its reproductive biology is interesting as it displays extreme sexual dimorphism and geographical parthenogenesis [2, 3]. On the upper half of the North Island sexual reproduction dominates, which produces offspring with relatively equal numbers of both sexes, whereas, obligate parthenogenesis is widespread on the lower North Island and South Island, forming all-female populations [2, 3]. In addition, *C. hookeri* is also thought to have hybridized with the obligate parthenogenetic genus *Acanthoxyla* [4–7]. All these features make *C. hookeri* an ideal species for the study of geographical parthenogenesis, hybridisation, and mating behaviour [2–5, 7–13]. However, at the molecular level, little is known about these processes, impeding further understanding of their reproductive biology. Currently, except for some genes encoding male accessory gland proteins [12] from *C. hookeri*, the genes involved in other female and male reproductive traits, such as oogenesis, spermatogenesis, egg and sperm maturation are largely unknown. It is critical that these genes are characterised to advance our understanding of the evolution of parthenogenesis and other alternative reproductive processes within the Phasmatodea.

RNA-Seq based transcriptome profiling has been successfully employed to identify genes expressed in the germline tissues from both model and non-model insect species, such as *Aedes aegypti* (mosquito) [14], *Nasonia vitripennis* (wasp) [15] and *Periplaneta americana* (cockroach) [16]. By sampling gonadal tissues from males and females, a digital gene expression profile can be obtained from the RNA-Seq data for the identification of the genes predominantly expressed in one sex (sex-biased genes). These genes frequently encode proteins essential to the sex-linked characteristics that are critical to the study of reproduction and its evolution [17]. Coupling genome and transcriptome sequencing allows the use of a reference-guided approach to studying patterns of gene expression. This is thought to be more accurate than the de novo method, because it can reduce transcript redundancy, reveal strand orientation to tease apart overlapping transcripts and enhance the representation of lowly expressed genes [18].

Another interesting feature of the stick insects is their large genome size. Currently, only the genome of *Timema cristinae*, sister lineage to all other stick insects (Euphasmatodea), has been sequenced and assembled [19]. However, compared with the 1.5 Gb genome of *T. cristinae*, even larger genomes occur in Euphasmatodea. Multiple species from the European stick insect genus *Bacillus* are ~2 Gb [20], while the South American species *Anisomorpha buprestoides*, Lord Howe Island species *Dryococelus australis* and the Australian species *Extatosoma tiaratum* have even larger genomes, up to ~3 [21], ~4 [22] and ~8 Gb (reported from the Animal genome size database: http://www.genomesize.com/), respectively. Without genome sequences from euphasmids, we are unable to determine the main causes for the formation of these large genomes. High-throughput sequencing has been used to sequence the whole genomes of some model and non-model species, especially those from the more derived Holometabola with small or moderate genome sizes [23–27]. Recently, the *Locusta migratoria* (Orthoptera) genome sequence comprising 6.5 Gb has become available. The analysis of this genome revealed a large numbers of repetitive elements (~58.9% of the genome assembly) and gene copy expansion within some gene families (e.g. detoxification) [28]. This resource has enhanced our understanding of the causes and consequences of large insect genomes, but whether similar patterns also occur in other insect lineages with large genomes is unknown.

Here we have sequenced, assembled and analysed the genome of *C. hookeri*, and through comparison with other insects, provide insights into the evolution of stick insect genomes. We particularly focused on repetitive elements and predicted gene models, both potential factors underlying large euphasmid genome sizes. We also performed an analysis of the RNA-Seq data produced from sequencing the female reproductive tract and the male testis to identify genes that are essential to gamete production and maturation in *C. hookeri*. The candidate genes predominantly expressed in the gonad of one sex and the genes involved in the juvenile hormone pathway provide candidates for the further study of stick insect reproductive flexibility.

Methods

Sample collection and preparation

All samples used for sequencing in this study were collected from Totara Park, Auckland, New Zealand (37°0.111 S, 174° 55.039 E). A female *C. hookeri* (CLI739) was collected in 2013 and leg tissues were used for the estimation of genome size. Three female insects (CLI525, 600 and 654) were collected in 2012 for genome sequencing and leg tissues were used. Three males (CLI755, CLI757 and CLI760) and three females (CLI765, CLI767 and CLI768) were collected in 2014 for gonadal transcriptome sequencing. Females were nymphs when collected and separately reared from males until they laid the first egg (reached maturation). Live insects were snap frozen in liquid nitrogen on capture and stored at −80 °C after collection. Female reproductive tract (approximately 18 to 20 ovarioles, early developing eggs and oviducts), and male testicle pairs were dissected in ethanol (100%) for RNA extraction.

Genome size estimation

The genome size of *C. hookeri* was estimated using flow cytometry following the Otto two-step method with the substitution of propidium iodine for DAPI [29]. A female *L. migratoria* was used as an internal standard. Approximately 15 mm^2 of leg tissue with a standard was co-chopped in a few drops of ice cold Otto buffer 1 with a stainless steel razor blade and then incubated for approximately 2 min. The sample was then filtered through a 20 μm Celltrics filter (PARTEC GmbH) before adding 2.5 ml of Otto buffer 2 with 1 mg/ml propidium iodine. The sample was run on a PARTEC CyFlow Space with a 488 nm laser as the excitation source. The 2C content of the *L. migratoria* standard was determined to be 18.31 pg, using *Pisum sativum* Citrad (9.09 pg 2C content; 4445 Mbp) [30] internal standard, and the gain adjusted as required. The total amount of DNA in the sample was determined as the ratio of the average channel number of the sample 2 N to the average channel number of the standard 2 N times the 1C amount of DNA in the standard.

DNA extraction, library preparation and sequencing

Genomic DNA was extracted from *C. hookeri* leg tissue using the DNeasy Plant Mini kit (Qiagen) with the following modifications: frozen leg tissue was chopped and incubated for 1 h with digestion buffer before DNA elution columns were incubated with digestion buffer for 1 h, and after the addition of elution buffer columns were incubated at room temperature for 30 min prior to centrifugation. DNA was quantified by a Nanodrop 2000 spectrophotometer (Thermo Fisher Scientific) and quality checked by running a sample on 0.5% *w/v* agarose gel stained with 1X GelRed (Huntingtree Ltd.). The resulting nuclear genomic DNA was sent to New Zealand Genomics Limited (NZGL: http://www.nzgenomics.co.nz/) Otago, Dunedin, for library construction and sequencing. Seven paired-end (PE) sequencing libraries with average insert sizes of 200 bp (2 libraries), 350 bp, 500 bp and 720 bp (3 libraries) (Table 1) were prepared using the TruSeq® DNA LT Sample Prep Kit v2 (catalogue ID: FC-121-2001) and sequenced with 11 lanes on an Illumina HiSeq2000™. The three mate-pair (MP) libraries with insert sizes of 5 k bp and 8 k bp (2 libraries) (Table 1) were constructed using the Nextera® Mate Pair Sample Prep Kit (catalogue ID: FC-132-1001) and sequenced with eight

Table 1 Illumina sequencing output for *Clitarchus hookeri* whole genome assembly

Insert size (bp)	Sample × library × lane	Sequencing output (Gb)	Estimated Genome coverage (×)
200	CLI525 × 1 × 5, CLI600 × 1 × 1	261.2	59.4
350	CLI525 × 1 × 2	88.5	20.1
500	CLI600 × 1 × 1	39.7	9.0
720	CLI525 × 1 × 1, CLI600 × 2 × 2	72.3	16.4
5000	CLI654 × 1 × 4	126.4	28.7
8000	CLI525 × 1 × 2, CLI654 × 1 × 2	145.2	33.0
	Total	733.3	166.6

Genome coverage estimates were obtained using the flow cytometry genome size

lanes on the same platform. The number of libraries per insert size and the number of lanes sequenced per library are described in Table 1. All reads were 101 bases in length. For RNA-Seq, total RNA extraction and library preparation were performed as described in [12]. The extractions were barcoded and then pooled together for sequencing on the HiSeq2000 platform for one lane to generate 100 bp PE reads at NZGL.

De novo genome assembly and quality assessment

The raw PE reads were preprocessed to remove duplicate sequence pairs, possible contaminants, and low-quality bases as follows: 1) reads containing ambiguities (Ns) and duplicates were filtered using PRINSEQ (v0.20.3) [31] and FastUniq (v1.1) [32], respectively; 2) reads with adapters and low quality ends (Phred < 30) were trimmed using Cutadapt (v1.3) [33]; 3) a read pair with overlapping ends more than 10 bp was merged into a single read using "abyss-mergepairs" from ABySS (v1.5.1) [34]; 4) reads shorter than 50 bp and orphan reads (single pair) were discarded; 5) remaining reads (83.3%) were error corrected using "ErrorCorrectReads.pl" from ALLPATH-LG (v46436) [35]. The raw MP reads were preprocessed following 1), 2), 4), and then trimmed to retain the first 36 bases of the 5 prime end in order to minimise inclusion of Nextera® adapters resulting from library preparation.

De bruijn graph and initial contigs were constructed and assembled using "pregraph" and "contig" commands from SOAPdenovo2 (vR223) [36] with Kmer 75 on the PE data derived from a single insect (CLI525; Table 1). The resulting contigs were used to construct scaffolds using "map" and "scaff" commands from SOAPdenovo2 based on their relationships implied by mapping all PE reads to the contigs from the shortest to the largest insect size libraries. Sequence gaps were then filled by GapCloser (v1.12-r6) [36] with PE reads from CLI525, and the resulting sequences were re-scaffolded again by non-CLI525 PE reads using SSPACE-basic (v2.0) [37] with default options (−z 0, −k 5, −a 0.7 and −n 15) and CLI525 PE reads with additional option "−X 1" for extending scaffolds. SSPACE-basic was also used to post-scaffold the resulting scaffolds using MP reads from the shortest to the largest insert size libraries with default options, followed by a step of gap filling using CLI525 reads, as described above, to produce the final genome assembly.

Approximately 1.0 billion genomic PE reads and 54.7 million RNA-Seq reads [12] were mapped to the genome assembly using Bowtie2 (v2.2.0) [38] with paired-end mode (−q −1 R1.fastq −2 R2.fastq) to produce a proportion of the coverage of the assembly from the short reads. The assembly was then quality evaluated with Core Eukaryotic Genes Mapping Approach (CEGMA; v2.4) [39] with large genome size mode (−mam) and Benchmarking Universal Single-Copy Orthologs (BUSCO; v2.0.1) [39] with searching database "arthropoda_odb9" to detect the presence of a core protein set of 248 highly conserved eukaryotic genes and 1066 highly conserved arthropoda genes. Scaffolds were also searched against GenBank *nt* database (Release 212) to estimate the percentage of archea, bacteria and virus sequences.

Repeat identification

RepeatModeler (v1.0.8) [40] with default options was used to predict and classify repetitive elements. It employs two de novo repeat finding programs, RECON and RepeatScout, to identify repeat element boundaries, followed by an assignment to the repeat classes based on the sequence feature.The resulting repeat models were searched against GenBank non-redundant (*nr*) protein database (evalue $<10^{-5}$) using Blastx (v2.2.28) [41] to exclude potential protein-coding genes. An additional repeat classification was conducted using PASTEClassifier (v1.0) [42] to assign miniature inverted repeat transposable elements (MITEs). The abundances of all predicted repeats were estimated in the genome assembly with RepeatMasker (v4.0.5) [43] (−no_is, −gff and −lib RepeatModels.gff).

Gene model annotation

Structural gene annotation was performed using MAKER2 (v2.31.3) [44] on scaffolds longer than 2000 bp. Before annotation, meta parameters of *C. hookeri* protein-coding genes, including those determining intron and exon length distributions, splice site patterns, and translation start codon patterns were generated using AUGUSTUS (v3.0.2) [45] (optimize_augustus.pl). Spliced alignments of protein sequences as inputs for AUGUSTUS were generated by aligning *C. hookeri* protein sequences including CEGMA predicted proteins and those that were identified from the head and prothorax transcriptome [9] to the assembly using Scipio (v1.4.1) [46] with default options. The workflow of MAKER2 involves: 1) producing *ab initio*

gene predictions using trained *C. hookeri* meta parameters, 2) aligning de novo transcripts collected from [9], [12] and Arthropoda conserved protein sequences (OrthoDB: v7) [47] to the assembly followed by the identification of intron-exon boundaries and splice forms as evidence, 3) producing evidence-informed gene predictions, computing quality scores and selecting the gene models best supported by the evidence. The resulted *C. hookeri* gene models were searched for the presence of core proteins (BUSCO) and homology matches against *nr* (evalue $<10^{-5}$) for gene annotation. These gene models were also identified with *T. cristinae* orthologues from a reciprocal blast method using a custom python script. The *T. cristinae* gene models (v0.2) were downloaded from: http://nosil-lab.group.shef.ac.uk/?page_id=25.

Transcript construction and annotation

Raw reads were trimmed 5′ end (8 bases), adapter sequences, low quality 3′ ends and filtered reads containing ambiguous bases (Ns) using PRINSEQ (v0.20.3) [31] and CUTADAPT (v1.3) [33]. The program STAR (v2.5) [48] with options "−−outFilterType BySJout −−outFilterIntronMotifs RemoveNoncanonical −−outSAMstrandField intronMotif −−outSAMtype BAM SortedByCoordinate −−outReadsUnmapped Fastx" was used to align reads to the genome assembly (scaffolds longer than 10 kbp). We did not include the putative gene models generated from the in silico gene annotation above as the reference gene set for annotating the gonad-expressed genes; instead, the regions mapped by all the RNA-Seq reads were counted as the gene loci. Cufflinks (v2.2.1) [49] with default options was used to generate these gene loci on the genome assembly according to the alignments, followed by producing merged gene loci using an embedded command "cuff-merge". The gene loci present with intron-exon boundaries were stranded on the genomic scaffolds. The transcript set was constructed by extracting sequences from annotated gene loci using "gffread". For multiple isoforms detected from a single gene locus, the first (longest) isoform present was chosen as the representative transcript that was subjected to a search against the SwissProt (release-2015_12) [50], UniProt (release-2015_12) [51] and Flybase (*Drosophila melanogaster*: dmel_r6.08) [52] protein databases using BLASTx (v2.2.28, E-value cut-off: 10^{-5}, keeping the top hit). These isoforms were also searched against GenBank *nt* (release14) using Blastn (evalue $< 10^{-10}$) to screen for contamination from bacteria, fungi and virus.

Read quantification and differential expression analysis

The RNA-Seq reads were aligned to the annotated gene loci followed by quantification. The number of read pairs aligned to the stranded loci was calculated using htseq-count from HTSeq (v0.6.1) [53] with default options and the unstranded gene loci with an additional option "−s no". Differential expression comparison was performed in R (v3.1.1) [54] using the DESeq2 Bioconductor package [55]. This program takes read counts to estimate sample size factors, followed by estimating dispersions with expected mean values from the maximum likelihood estimate of log2 fold changes, and then fits a negative binomial distribution [55]. The principle component analysis from DESeq2 and an R package pheatmap were used to visualise global similarities and differences. Transcripts with an adjusted p value less than 0.05 and a minimum fold change (FC) of 2 were reported as differently expressed.

Gene ontology and pathways analysis

The predicted gene loci showing significantly sex-biased expression were used to detect enriched gene ontology (GO) terms and Kyoto Encyclopedia of Genes and Genomes (KEGG) pathways. The matched *D. melanogaster* FlyBase gene IDs were imported into the Database for Annotation, Visualization, and Integrated Discovery (DAVID, v6.8) [56, 57] for functional annotation and enrichment tests. The default FlyBase database was used as the background. The level one GOs (GOTERM_1) and the GOs directly mapped (GOTERM_DIRECT) from the source database (without parental terms) were generated with p values below 0.05, and the significantly enriched ones were defined according to the p values adjusted by the Benjamini and Hochberg (BH) procedure (<0.05). They were visualised using the R package GOplot [58]. The enriched KEGG pathways (KEGG_PATHWAY) were generated with a p value less than 0.1.

Results

Genome sequence and analysis

The genome size of *C. hookeri* was estimated at approximately 4.4 Gb using flow cytometry. Illumina reads derived from ten libraries with various insert sizes shown in Table 1 were subject to de novo assembly. Sequencing of multiple libraries derived from DNA of a single female (CLI525) yielded 400.4 Gb of PE reads for contig construction with an estimated coverage of 90.9 × the estimated genome size and the initial contigs constructed from these data had a N50 of 3715 bp. The final assembly had a total size of 4.2Gb with a N50 of 255.7 kb (Table 2). Nearly all genomic reads (>99%) were mapped back to the assembly while the map back rate of RNA-Seq reads was ~86%. BUSCO and CEGMA analysis suggested the whole genome sequence was reasonably complete with 91.6% complete and 3.6% partial BUSCO proteins present and 76.6% complete and 22.3% partial CEGMA genes present. The "scaffold13492" comprising 19,561 bp was identified as the *C. hookeri* mitochondrial genome. The total number of estimated

Table 2 *Clitarchus hookeri* genome assembly statistics

	Size (bp)	Number of scaffolds
N90	188	859,481
N80	5446	48,684
N70	30,351	11,639
N60	148,642	5894
N50	255,691	3749
Longest	4,944,527	–
Total (>100 bp)	4,244,875,252	4,114,148
Total (>2 kb)	3,503,002,174	78,458

archea, bacteria and virus scaffolds is only 1387 (Additional file 1).

Repetitive elements

The genome assembly of *C. hookeri* is highly repetitive, with approximately half the genome (51.6%) predicted as repeats. Among these, a total of 3210 repeat models were determined, 1404 (43.7%) of which were classified as different groups of interspersed transposon elements (TEs). To compare with *T. cristinae*, we used the same repeat identification method to identify repeats from the published genome assembly. In *T. cristinae*, 1288 repeat models were detected, including 433 (33.6%) assigned to the known repeat groups. The proportions of different repeat groups from the genomes of the two stick insects and the other two Polyneoptera species (*Zootermopsis nevadensis* and *L. migratoria*) [28, 59] are shown in Table 3.

In the *C. hookeri* genome, miniature inverted repeat transposable element (MITE) was the most abundant repeat type, which was identified with 87 putative models and 1,214,018 copies, covering 5.79% of the assembly. The short sequence "rnd-1_family-8" was detected as the most frequent MITE repeat in the genome. Similarly, MITEs were also reported to be highly abundant in the stick insect *Bacillus rossius*, *B. grandii* and *B. atticus* partial genomes [20]. However, they were absent from the *T. cristinae* genome. In comparison, Maverick was the most abundant repeat type in the *T. cristinae* genome, predicted with 12 repeats comprising 145,675 copies, which cover 3.13% of the genome, and the most frequent repeat copy was "rnd-2_family-6". Maverick was ranked as the third most abundant in the *C. hookeri* genome. The second most abundant DNA transposon in *C. hookeri* was TcMar-Tc1, the most abundant repeat type in the *L. migratoria* genome.

Class I TEs containing long terminal repeat retrotransposon (LTR), short interspersed element (SINE) and LINE were much less abundant than Class II TEs in the two stick insect genomes. Class I TEs constitute 5.74% and 8.08% of the *C. hookeri* and the *T. cristinae* genomes, respectively. LTR gypsy was the most abundant repeat

Table 3 Comparisons of repeats among four Polyneoptera genomes

	Zootermopsis nevadensis		Timema cristinae		Clitarchus hookeri		Locusta migratoria	
Genome size (Mb)	490		1030		4240		6500	
Repeat types	Length (Mb)	P%	Lengh (Mb)	P%	Lengh (Mb)	P%	Length (Mb)	P%
DNA	12.78	2.60	75.13	7.30	612.35	14.43	1480.54	22.69
LINE	22.13	4.50	34.34	3.34	55.84	1.32	1332.72	20.42
LTR	0.72	0.15	5.89	0.57	103.59	2.44	508.68	7.80
SINE	9.78	2.00	42.88	4.17	84.08	1.98	141.18	2.16
Simple repeat	1.39	0.28	9.00	0.87	96.82	2.28	13.03	0.20
Other	0.00	0.00	1.00	0.10	0.74	0.02	0.03	0.00
Unknown	81.32	16.50	243.05	23.61	1237.41	29.15	406.10	6.22
Total	128.41	26.00	411.29	39.96	2190.84	51.60	3840.81	58.86

type in the *C. hookeri* genome, comprising 194,787 copies, covering 1.44% of the genome. The short sequence "rnd-1_family-196" was detected with the highest copy number. In comparison, the LINE repeat RTE-BovB was the most abundant repeat type in the *T. cristinae* genome and the most frequent model was "rnd-3_family-716". This repeat type was predicted to occupy 2.33% of the *T. cristinae* genome, comprising 75,215 copies. It is also one of the most frequent repeat types in the *L. migratoria* genome. There were only a few repeat types that were uniquely present in one of the stick insect genomes when compared with the other. The repeats RTE-RTE (LINE), ERV1 (LTR) and TRIM (LTR) were detected only from the *C. hookeri* genome, whereas Ngaro (LTR) was only present in the *T. cristinae* genome. The SINE repeats appear to be slightly more abundant in the *C. hookeri* than the *T. cristinae* genome.

Protein-coding genes

The current predicted 66,470 *C. hookeri* gene models include 10,266 models revised from transcript and protein sequences after in silico prediction. This gene set includes 779 (73.1%) complete predicted BUSCO proteins with 747 (95.9%) single-copy genes, and 154 (14.4%) partial proteins. The proportion of gene models that have Blast matches with *nr* database proteins was 36.2% (24,085), the majority of which hit the eusocial termite *Z. nevadensis* (42.3%), followed by the red flour beetle *Tribolium castaneum* (5.7%) (Additional file 2). There were 8478 predicted gene orthologues between the two sets of stick insect gene models using a reciprocal blast approach. We compared the average values of a variety of gene model features across *C. hookeri*, *T. cristinae*, *Holacanthella duospinosa* (New Zealand giant collembolan) [60] and a wide range of arthropods with these values available [28, 59] (Fig. 1). This analysis demonstrates that insects from Polyneoptera frequently have genome sizes

in the gigabase range and that increases in genome size are positively correlated with increasing transcript and intron sizes.

Gonadal transcriptome assembly and annotation

Approximately 95.8% of the 600 million raw RNA-Seq reads passed the cleaning criteria and were mapped to the genome assembly. The mapping ratios ranged from 89.8% to 94.9% across individuals and the ratios of uniquely mapped reads were all above 70% (Table 4). We then annotated only those gene loci with mapped RNA-seq reads, which can maximise the number of genes predicted to be expressed in the tissues compared with reusing the in silico produced gene models. In total, the mapped reads generated 39,940 putative gene loci, including 23,778 (59.5%) genes containing more than one exon, with strand orientation determined according to the intron-exon boundaries. Their transcribed sequences show an N50 length of 3828 bp, and a minimum and maximum length of 64 and 37,398 bp, respectively. A total of 36,072 (90.3%) transcripts have lengths longer than 500 bp. Only 45 sequences had the best homologous matches from bacteria and virus sequences within the Genbank *nt* database. This suggests an extremely low level (0.1%) of xenobiotic RNA contamination in this predicted gene set compared with our previous dataset generated from the same laboratory procedure [12].

The number of transcripts having matches from the UniProt, SwissProt and *D. melanogaster* protein databases were 20,841 (52.24%), 14,270 (35.77%), and 12,443 (31.19%), respectively (Additional file 3). Among the UniProt blast hits, 37% were from the eusocial termite *Z. nevadensis*, followed by 7.5% and 4.5% from the pea aphid *Acyrthosiphon pisum* and the araneomorph spider *Stegodyphus mimosarum*, respectively. Notably, the proportion of single-exon transcripts having homologous matches was much lower than the multi-exon transcripts (Fig. 2), likely due to the fact that many of them

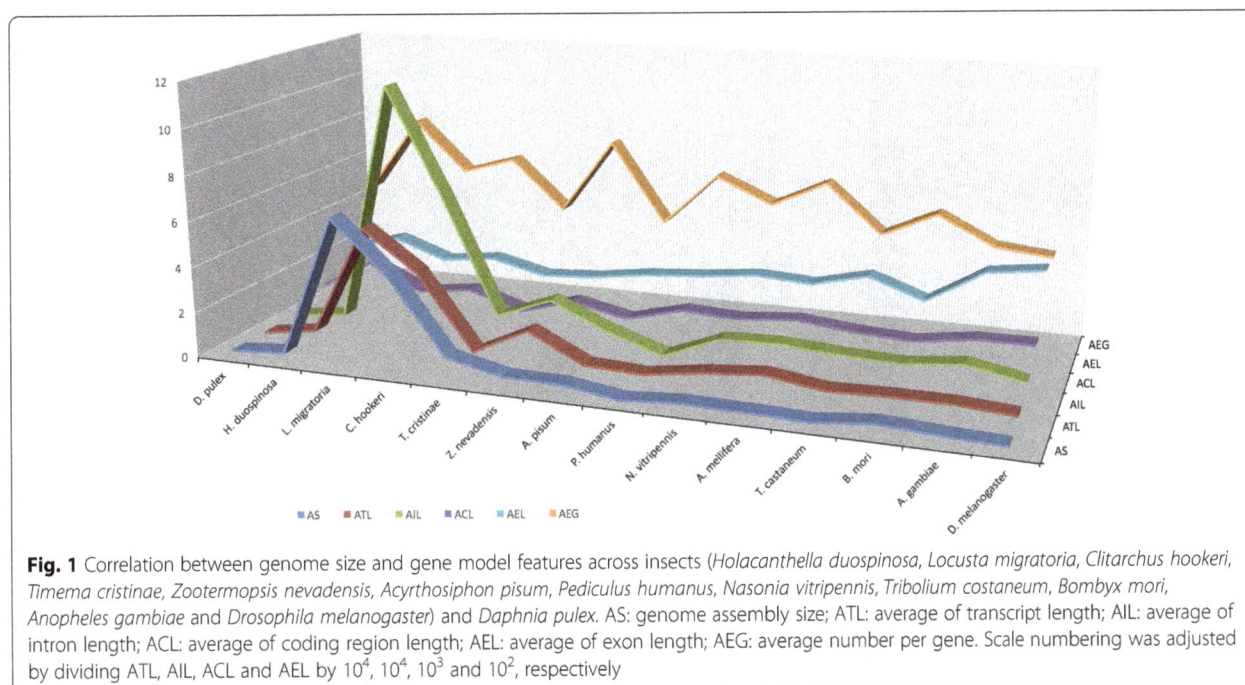

Fig. 1 Correlation between genome size and gene model features across insects (*Holacanthella duospinosa, Locusta migratoria, Clitarchus hookeri, Timema cristinae, Zootermopsis nevadensis, Acyrthosiphon pisum, Pediculus humanus, Nasonia vitripennis, Tribolium costaneum, Bombyx mori, Anopheles gambiae* and *Drosophila melanogaster*) and *Daphnia pulex*. AS: genome assembly size; ATL: average of transcript length; AIL: average of intron length; ACL: average of coding region length; AEL: average of exon length; AEG: average number per gene. Scale numbering was adjusted by dividing ATL, AIL, ACL and AEL by 10^4, 10^4, 10^3 and 10^2, respectively

were derived from non-coding regions, thus lacking similarity with sequences from protein databases.

Gonadal sex-biased genes

The female reproductive tract had 3262 gene loci that were uniquely expressed in this tissue, whereas approximately four times of this number (12,516) were detected in testis (ECS > 5). Comparative gene expression analysis between female and male gonadal samples revealed a large set of transcripts that were significantly differentially expressed (FC > 2 and BH adjusted $p < 0.05$). There were 6308 genes significantly overexpressed in the female, whereas the genes displaying male-biased expression had more than two-fold abundance (15,889) (Fig. 3 and Additional file 3). Notably, compared with female-biased transcripts, a much larger proportion of transcripts over-expressed in testis were lacking matches, especially when compared with the *D. melanogaster* protein database (Fig. 3).

The top 20 significantly over-expressed genes having *D. melanogaster* protein matches from each of the sexes are shown in Table 5. The genes with orthologues that were also highly expressed in *D. melanogaster* ovary were *CG1077, yellow-g, yellow-g2* and *Acph-1*, and testis were *CG12020, CG5458, CG32392, CG14838, CG13442, CG31068* and *CG17377*, respectively (data retrieved from the FlyAtlas Anatomy Microarray and modENCODE Anatomy RNA-Seq data on http://flybase.org).

Enriched gene ontology terms and pathways

The matched *D. melanogaster* protein hits from the *C. hookeri* sex-biased genes were used to detect enriched gene ontology (GO) terms. In summary, a total of 2405 *D. melanogaster* FlyBase IDs were subjected to multiple tests of GO enrichment using DAVID. Within level one GOs, most of the significantly enriched terms were female-biased. Some of these terms closely represent female features of reproduction, such as developmental

Table 4 Genomic DNA mapping statistics

Sample ID	Read pairs	Uniquely mapped	No. of splices	Multiply mapped	Total mapped (%)
CLI765	47,699,036	37,219,558 (78.0%)	12,199,348	7,414,350 (15.5%)	93.6
CLI767	50,546,345	35,708,416 (70.6%)	11,153,462	12,243,359 (24.2%)	94.9
CLI768	45,978,461	34,815,789 (75.7%)	10,547,208	8,859,828 (19.3%)	95.0
CLI755	44,052,337	34,808,276 (79.0%)	10,253,623	6,884,797 (15.6%)	94.6
CLI757	45,000,013	32,997,927 (73.3%)	10,140,927	8,889,222 (19.8%)	93.1
CLI760	52,261,126	40,301,287 (77.1%)	13,724,663	6,631,109 (12.7%)	89.9

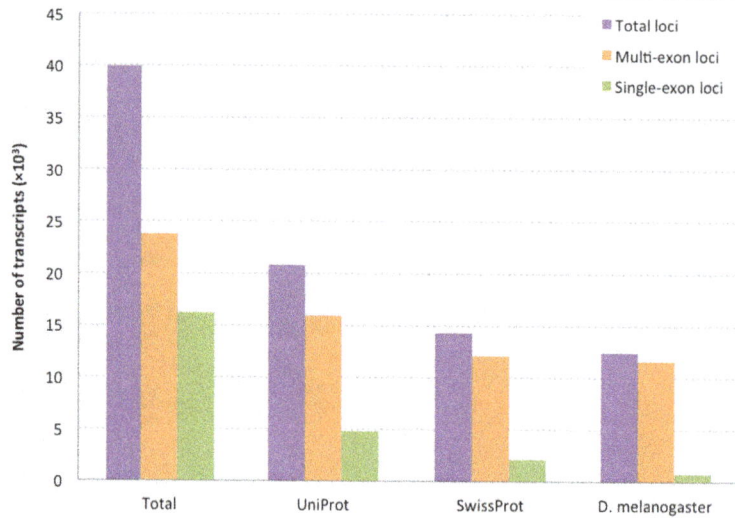

Fig. 2 Distribution of homologous sequence matches from the Uniprot, SwissProt and *Drosophila melanogaster* protein databases, respectively

process, signaling, reproductive process and immune system process from the biological process category (BP). In comparison, rhythmic process (BP), molecular transducer activity (MF) and synapse (CC) were significantly enriched among male-biased genes (Fig. 4a).

After removing all the parental terms, the more specific GOs were revealed (Fig. 4b). The significantly enriched terms for the male-biased genes include ATP binding, motor activity, microtubule binding and ATPase activity (MF), corresponding to sperm maturation and movement; whereas, the terms for the female-biased genes include calcium ion, protein and actin binding and ATPase activity (MF). In addition, females also over-expressed genes involved in glucose transmembrane transporter activity, oxidoreductase activity and carboxylic

ester hydrolase activity (MF) (Fig. 4b). Furthermore, we also found GOs associated with oocyte maturation and development, such as imaginal disc-derived wing morphogenesis, open tracheal system development (BP), myofibril assembly, neuron projection morphogenesis and dorsal closure (CC) that were enriched within female-biased genes.

The enriched KEGG pathways include many related to carbohydrate metabolism, including pentose and glucuronate interconversions and starch and sucrose metabolism enriched within female-biased genes, and glycolysis/gluconeogenesis, glycan degradation, glycosaminoglycan degradation and galactose metabolism pathways enriched within male-biased genes (Fig. 5). Other enriched pathways within female-biased genes were ECM-receptor

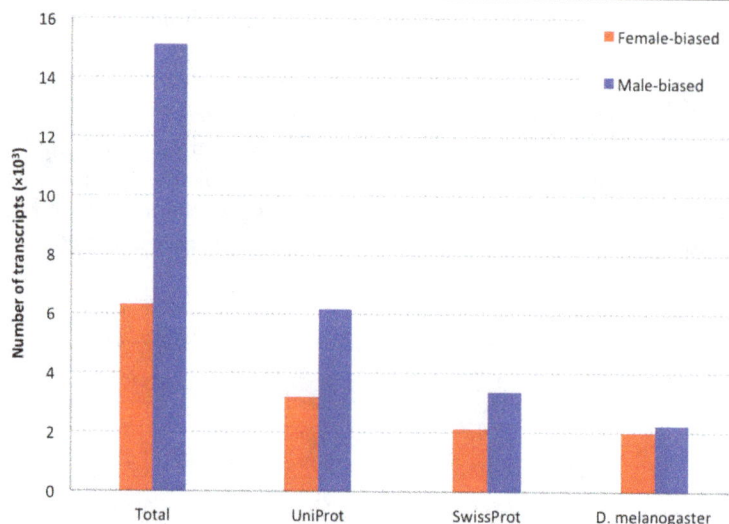

Fig. 3 Distribution of homologous sequence matches from the Uniprot, SwissProt and *Drosophila melanogaster* protein databases for female-biased and male-biased genes, respectively

Table 5 Summary of top 20 differentially expressed genes from female reproductive tract and testis, respectively

	Gene locus	Hit ID	Hit Name	E-value	Log2 FC	Adjusted p
Female-biased	XLOC_018823	FBgn0037405	CG1077[a]	2.38E-06	−13.06	1.58E-12
	XLOC_015696	FBgn0041709	yellow-g[a]	5.04E-19	−12.65	3.91E-11
	XLOC_032619	FBgn0266435	CG45065[a]	1.72E-151	−12.36	1.64E-10
	XLOC_030196	FBgn0035328	yellow-g2[a]	4.20E-65	−12.06	2.09E-36
	XLOC_007922	FBgn0035089	Phk-3	1.59E-08	−12.01	1.45E-91
	XLOC_009044	FBgn0034145	CG5065	1.75E-13	−11.93	2.21E-45
	XLOC_007550	FBgn0039084	CG10175	8.89E-79	−11.9	1.73E-50
	XLOC_021021	FBgn0038799	MFS9	1.96E-09	−11.89	1.59E-09
	XLOC_021022	FBgn0038799	MFS9	3.49E-45	−11.85	1.88E-21
	XLOC_025391	FBgn0013680	mt:ND2	6.59E-09	−11.81	1.74E-283
	XLOC_000499	FBgn0032433	Oatp33Ea	3.76E-32	−11.54	2.71E-09
	XLOC_003735	FBgn0034145	CG5065	3.59E-14	−11.52	6.18E-32
	XLOC_026733	FBgn0000032	Acph-1[a]	4.73E-69	−11.46	2.48E-41
	XLOC_002774	FBgn0000261	Cat	1.66E-34	−11.28	2.37E-08
	XLOC_005024	FBgn0039896	yellow-h	7.54E-111	−11.26	1.75E-18
	XLOC_024658	FBgn0030452	MFS10	5.42E-19	−11.23	2.92E-08
	XLOC_016572	FBgn0038799	MFS9	3.93E-29	−11.12	4.62E-08
	XLOC_032950	FBgn0000032	Acph-1[a]	3.86E-21	−10.99	6.21E-18
	XLOC_036184	FBgn0259247	laccase2	2.23E-09	−10.94	3.44E-09
	XLOC_018465	FBgn0026314	Ugt35b	6.37E-96	−10.91	4.59E-18
Male-biased	XLOC_037165	FBgn0014869	Pglym78	7.44E-70	13.61	1.81E-13
	XLOC_034828	FBgn0035273	CG12020[a]	9.64E-35	13.59	2.02E-13
	XLOC_033633	FBgn0020412	JIL-1	1.21E-19	13.47	4.23E-13
	XLOC_003842	FBgn0032478	CG5458[a]	3.75E-45	13.31	1.07E-12
	XLOC_039688	FBgn0038385	Fbxl7	6.21E-07	13.29	1.15E-12
	XLOC_035079	FBgn0019982	Gs1l	2.23E-53	13.29	1.19E-12
	XLOC_019177	FBgn0020412	JIL-1	2.99E-16	13.25	1.45E-12
	XLOC_039094	FBgn0052392	CG32392[a]	2.03E-08	13.12	3.14E-12
	XLOC_010029	FBgn0005612	Sox14	3.45E-08	13.05	4.59E-12
	XLOC_016670	FBgn0033635	CG7777	2.72E-26	12.98	2.59E-27
	XLOC_035177	FBgn0004380	Klp64D	1.96E-09	12.94	1.30E-35
	XLOC_024216	FBgn0264574	Glut1	1.89E-43	12.88	8.30E-25
	XLOC_004823	FBgn0035799	CG14838[a]	1.28E-72	12.84	1.01E-25
	XLOC_010324	FBgn0034546	CG13442[a]	2.57E-10	12.84	1.45E-11
	XLOC_019341	FBgn0036211	CG5946	9.38E-16	12.77	2.02E-11
	XLOC_002845	FBgn0020412	JIL-1	6.64E-13	12.75	2.79E-24
	XLOC_035152	FBgn0051068	CG31068[a]	1.24E-33	12.73	2.62E-26
	XLOC_038804	FBgn0031988	CG8668	1.07E-63	12.65	7.43E-25
	XLOC_020898	FBgn0031859	CG17377[a]	2.27E-10	12.58	1.47E-24
	XLOC_038151	FBgn0039396	CCAP-R	6.78E-28	12.57	5.83E-11

"[a]" indicates the transcript is also highly expressed in the *Drosophila melanogaster* ovary and testicle respectively

a

ID	Description
GO:0044699	single-organism process
GO:0051179	localization
GO:0065007	biological regulation
GO:0032502	developmental process
GO:0023052	signaling
GO:0050896	response to stimulus
GO:0040011	locomotion
GO:0009987	cellular process
GO:0040007	growth
GO:0022610	biological adhesion
GO:0071840	cellular component organization or biogenesis
GO:0032501	multicellular organismal process
GO:0007610	behavior
GO:0048511	rhythmic process
GO:0024414	reproductive process
GO:0051704	multi-organism process
GO:0002376	immune system process
GO:0016020	membrane
GO:0030054	cell junction
GO:0044425	membrane part
GO:0045202	synapse
GO:0044456	synapse part
GO:0005215	transporter activity
GO:0003824	catalytic activity
GO:0060089	molecular transducer activity

b

ID	Description
GO:0055085	transmembrane transport
GO:0007476	imaginal disc-derived wing morphogenesis
GO:0007155	cell adhesion
GO:0006468	protein phosphorylation
GO:0008045	motor neuron axon guidance
GO:0007605	sensory perception of sound
GO:0040011	locomotion
GO:0007424	open tracheal system development
GO:0005975	carbohydrate metabolic process
GO:0005886	plasma membrane
GO:0005737	cytoplasm
GO:0016021	integral component of membrane
GO:0005912	adherens junction
GO:0005811	lipid particle
GO:0030018	Z disc
GO:0016324	apical plasma membrane
GO:0043190	ATP-binding cassette (ABC) transporter complex
GO:0005925	focal adhesion
GO:0005524	ATP binding
GO:0003774	motor activity
GO:0005509	calcium ion binding
GO:0005515	protein binding

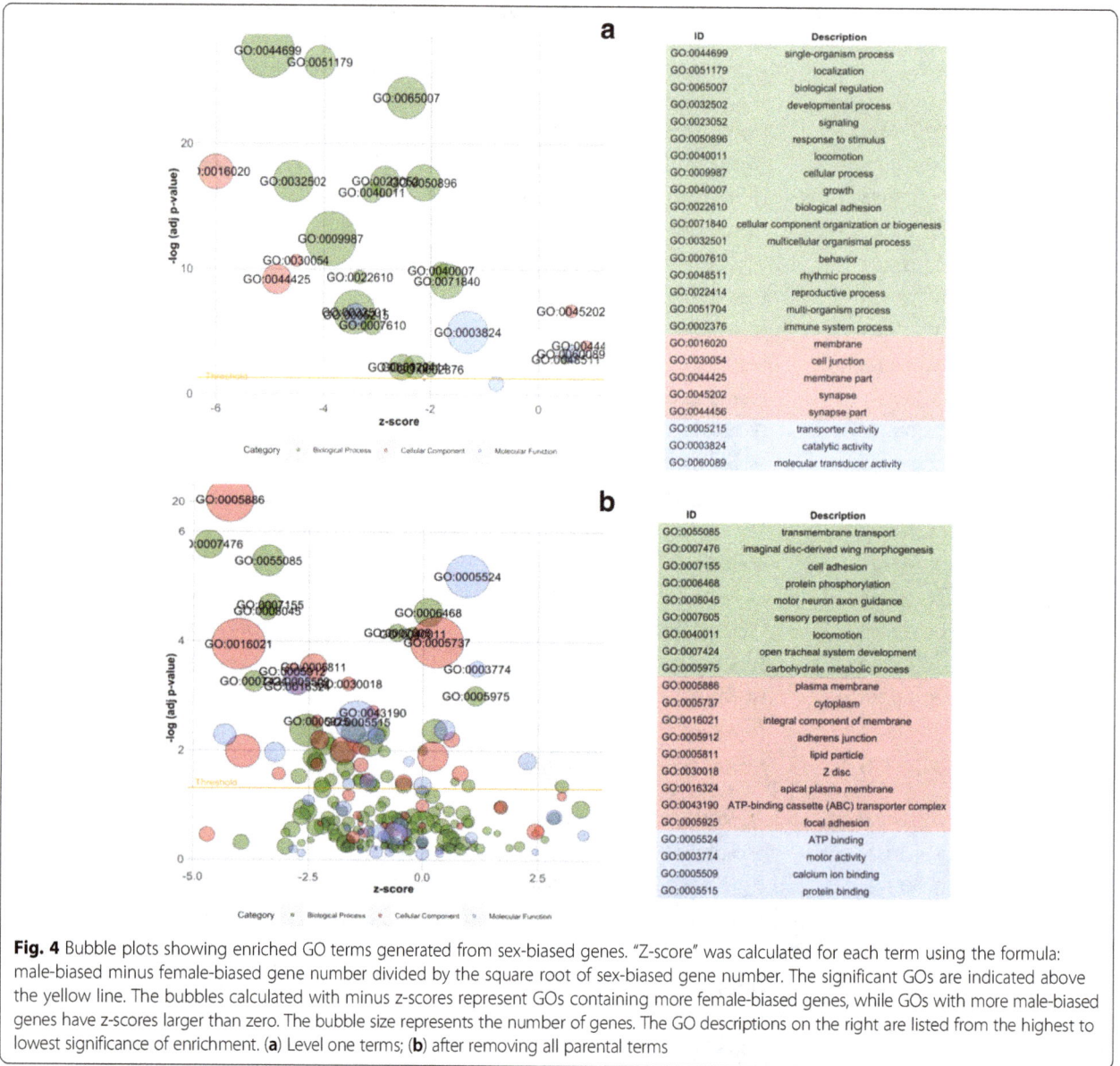

Fig. 4 Bubble plots showing enriched GO terms generated from sex-biased genes. "Z-score" was calculated for each term using the formula: male-biased minus female-biased gene number divided by the square root of sex-biased gene number. The significant GOs are indicated above the yellow line. The bubbles calculated with minus z-scores represent GOs containing more female-biased genes, while GOs with more male-biased genes have z-scores larger than zero. The bubble size represents the number of genes. The GO descriptions on the right are listed from the highest to lowest significance of enrichment. (**a**) Level one terms; (**b**) after removing all parental terms

interaction, hippo signaling pathway, phototransduction, insect hormone biosynthesis and retinol metabolism, and within male-biased genes were lysosome, phosphatidylinositol signaling system, glycosphingolipid biosynthesis, foxO signaling pathway and ABC transporters (Fig. 5). Within the juvenile hormone (JH) biosynthesis pathway (belonging to the insect hormone biosynthesis pathway), JH epoxide hydrolase 2 (*Jheh2*) and *CG9360* were over-expressed in females.

Discussion

The *C. hookeri* genome is the second largest insect genome published to date after *L. migratoria* [28]. It is also the second stick insect genome to be reported and the only one sequenced from the Euphasmatodea. Thus, the availability of this genome sequence provides valuable information to investigate the evolution of large insect genomes and the diversification of the Phasmatodea. In particular, parthenogenesis and other alternative reproductive strategies have played an important role in phasmid evolution. However, the molecular mechanisms underlying parthenogenesis have been poorly studied. Characterisation of genes essential to stick insect reproduction, especially from species with different reproductive modes, is the first step towards understanding reproductive flexibility. In this study, we constructed a *C. hookeri* genome assembly, which was then used as a reference genome to detect genes that are present in the female reproductive tract and testicle transcriptomes.

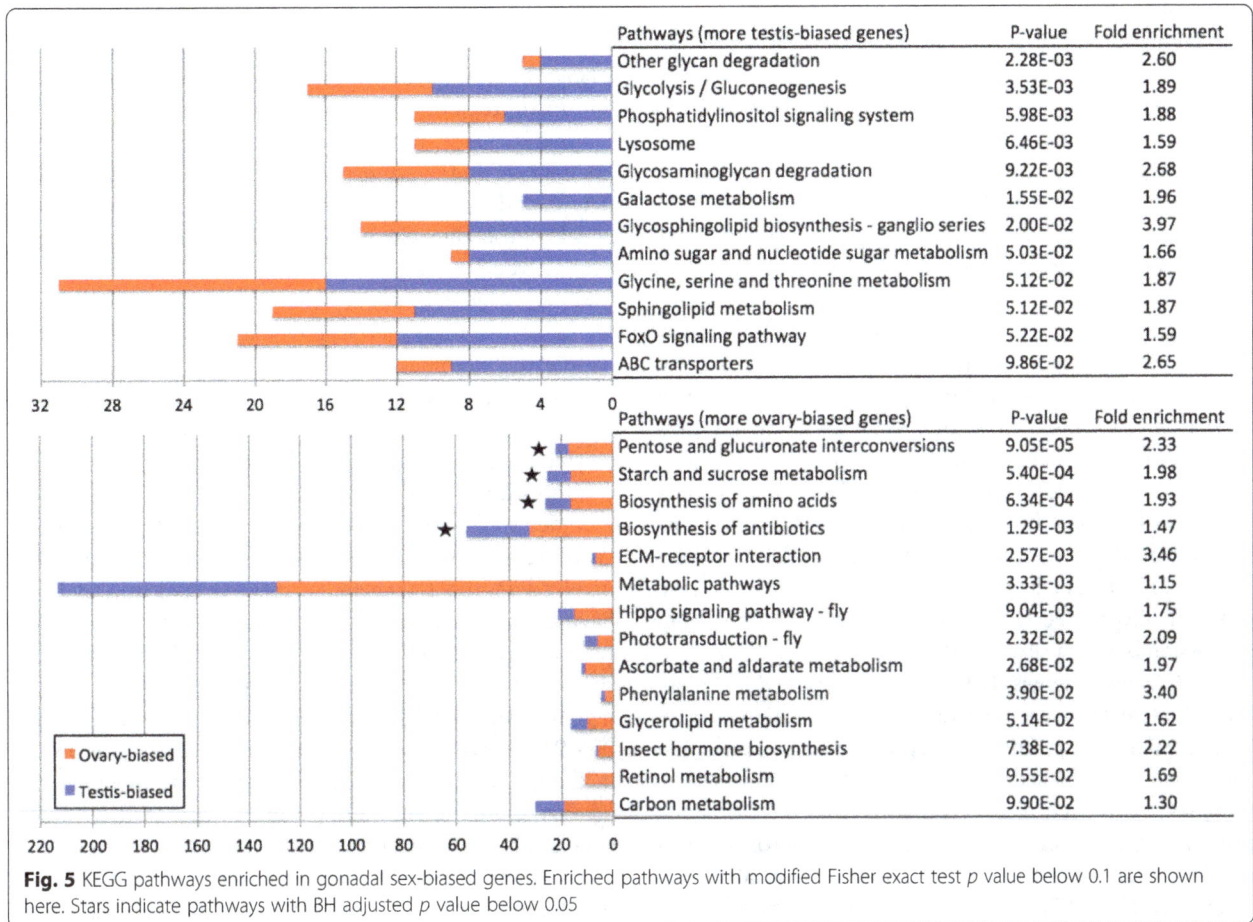

Pathways (more testis-biased genes)	P-value	Fold enrichment
Other glycan degradation	2.28E-03	2.60
Glycolysis / Gluconeogenesis	3.53E-03	1.89
Phosphatidylinositol signaling system	5.98E-03	1.88
Lysosome	6.46E-03	1.59
Glycosaminoglycan degradation	9.22E-03	2.68
Galactose metabolism	1.55E-02	1.96
Glycosphingolipid biosynthesis - ganglio series	2.00E-02	3.97
Amino sugar and nucleotide sugar metabolism	5.03E-02	1.66
Glycine, serine and threonine metabolism	5.12E-02	1.87
Sphingolipid metabolism	5.12E-02	1.87
FoxO signaling pathway	5.22E-02	1.59
ABC transporters	9.86E-02	2.65

Pathways (more ovary-biased genes)	P-value	Fold enrichment
★ Pentose and glucuronate interconversions	9.05E-05	2.33
★ Starch and sucrose metabolism	5.40E-04	1.98
★ Biosynthesis of amino acids	6.34E-04	1.93
★ Biosynthesis of antibiotics	1.29E-03	1.47
ECM-receptor interaction	2.57E-03	3.46
Metabolic pathways	3.33E-03	1.15
Hippo signaling pathway - fly	9.04E-03	1.75
Phototransduction - fly	2.32E-02	2.09
Ascorbate and aldarate metabolism	2.68E-02	1.97
Phenylalanine metabolism	3.90E-02	3.40
Glycerolipid metabolism	5.14E-02	1.62
Insect hormone biosynthesis	7.38E-02	2.22
Retinol metabolism	9.55E-02	1.69
Carbon metabolism	9.90E-02	1.30

Fig. 5 KEGG pathways enriched in gonadal sex-biased genes. Enriched pathways with modified Fisher exact test *p* value below 0.1 are shown here. Stars indicate pathways with BH adjusted *p* value below 0.05

Evolution of repeats and gene length expansion

The *C. hookeri* genome assembly comprising 4.2 Gb, was very similar to the flow cytometry estimate, and approximately four times larger than *T. cristinae* [19]. Similar to *L. migratoria*, the analysis of the *C. hookeri* genome also revealed a large volume of repetitive elements. DNA transposon (Class II) was the most dominant repeat group in the stick insect and the *L. migratoria* genomes; whereas, the genome of *Z. nevadensis*, only around one third of *T. cristinae* genome size, was mostly enriched with Non-LTR retroid long interspersed element (LINE). The comparison also shows that the increasing genome size from *Z. nevadensis* to *L. migratoria* is consistent with the growing absolute amount and proportion of DNA transposons. Relative to *T. cristinae*, the *C. hookeri* genome is expanded with a variety of repeat types. Notably, MITE, the most abundant group of DNA transposon in the *C. hookeri* genome assembly, was also reported to be present at [61] high frequency in the partial *Bacillus* genome sequences [20], but absent in the *T. cristinae* genome. Either the invasion and proliferation of MITEs occurred along the evolution of euphasmids or the loss of MITEs occurred along the *Timema* lineage. A survey of MITEs across a wide range of stick insect lineages may shed light on the evolution of this repeat group in stick insects. A relatively large proportion of un-classified repeats were also revealed in the two stick insect genomes, whether they indeed represent novel repeat families requires further analysis.

The annotated gene models from *L. migratoria* and *C. hookeri* were roughly 6 and 5 times longer than other insects with moderate to small genome sizes. This suggests there was an intron length expansion during the formation of large insect genomes. There are several consequences to an increase in intron length. First, long introns are associated with an increased metabolic cost as introns are transcribed together with exons [61]. Second, they are thought to be associated with the large genome chromosomal compactness [61]. Third, they may be negatively correlated with recombination [62]. The *C. hookeri* genome provides a novel resource to test these three hypotheses in insects. In addition to longer introns, the *C. hookeri* genes also exhibited a higher number of introns and exons per gene and a slightly longer coding sequence length compared with *T. cristinae*. However, whether these patterns hold across the Euphasmatodea requires the availability of more stick insect genome sequences.

Sex-biased genes

The comparative transcriptome analysis between female and male gonads revealed a large set of genes with sex-biased expression. The male-biased genes were more than twice as abundant as those displaying female-biased expression, and a much greater proportion of male-biased genes were lacking blast matches from the known proteins. Also, *Clitarchus hookeri* females sometimes reproduce by parthenogenesis, which might lead to less selective pressure on males because the male-biased genes are used less frequently than the female-biased genes [63, 64]. It has also been reported that the sexually-derived *Timema* parthenogenetic lineages have experienced sexual trait decline, such as shrunken spermatotheca and the loss of male attractiveness [65]. This has also likely occurred in the *C. hookeri* parthenogens. It is possible that this sexual trait decay correlates with female-biased gene expression change. All these hypotheses require further investigation of the sequence and expression divergence across euphasmids and the gene expression between sexual and parthenogenetic *C. hookeri*.

The *C. hookeri* genes that are over-expressed in female reproductive tract are enriched in development, signalling, growth, behaviour and reproductive process. The top five female-biased DE genes had *D. melanogaster* matches with a chorion-containing eggshell formation protein (*CG1077*) predicted to have anti-microbial activity [66], *yellow-g* and *yellow-g2* proteins essential to eggshell integrity [67], *CG45065* protein responding to mating [68], and *Phk-3* protein associated with metamorphosis [69]. Stick insects show interesting traits relating to egg morphology [70, 71]. Eggs have in a variety of shapes, often mimicking plant seeds, many of which contain a knob-like capitulum that resembles an elaiosome to attract ants for burial [72]. Recently, a species (*Korinninae* sp.) was found to produce an ootheca containing numerous eggs in a highly ordered arrangement, which is distinct from other stick insects that produce eggs singly by dropping them to the ground or inserting them into crevices or soil [73]. The genes identified in our study may be essential to eggshell formation and maturation and are candidates to further investigate egg variation and adaptations in stick insects.

In comparison, the male-biased genes were significantly enriched for rhythmic process, molecular transducer activity and synapse. Rhythmic processes play important roles in temporally coordinating the release of sequential sperm, the acidification of the vas deferens, and contractile activity [74–77]. The enrichment of the molecular transducer activity was also found in crab and a sex-changing fish testis [78, 79]. The top five male-biased genes with blast matches include a glycolysis protein *Pglym78* that is present in the semen and seminal vesicle tissue of a honey bee [80], *CG12020* enriched in

the sperm proteome with a role in protein folding [81, 82], *JIL-1* essential for chromosomal organisation [83], *CG5458* involved in sperm axoneme assembly [81, 82] and *Fbxl7* regulating mitosis through Aurora A [84].

Notably, a large number of sex-biased transcripts showed no matches with any of the databases. These transcripts may include: 1) highly diverged genes; 2) unknown genes; and 3) non-coding elements. Stick insects contain panoistic ovaries where oogonia eventually differentiate into oocytes [85]. However, insect ovary-biased genes have mostly been identified from the meroistic ovaries frequently present in the derived Holometabola, where oogonia differentiate into an oocyte and several nurse cells [86, 87]. Thus, some of the novel female-biased genes found in this study may be playing unique functions in the panoistic ovary.

Interestingly, we found some *C. hookeri* genes with matches to *D. melanogaster* proteins involved in the JH synthesis pathway. These genes include *Jheh1* and *Jheh2* involved in the JH catabolic process [88] and *CG9360* having an oxidoreductase activity [89], *Jheh2* and *CG9360* over-expressed in female *C. hookeri*. It has been suggested that in the cyclical parthenogenetic aphids, higher levels of JH induce parthenogenetic reproduction [90–93]. However, whether the level of JH has an impact in the differentiation of the southern parthenogenetic *C. hookeri* is unknown. This could be examined by assessing levels of JH or indirectly by measuring expression levels of the genes involved in JH hormone biosynthesis. In addition, we also found some enriched pathways were related in carbohydrate metabolism (e.g. starch and sucrose metabolism and glycan degradation), which are likely to contain genes playing important roles in nutrition and energy support for the production of gametes and the maintenance of reproductive organs.

Conclusions

The analysis of the *C. hookeri* genome assembly revealed a large, repetitive genome, likely resulting from the accumulation of DNA transposons along with an increase of intron length. MITE, the most abundant repeat type in the *C. hookeri* genome assembly, contributed significantly to the growth in genome size. Using a reference-guided approach coupled with differential expression analysis, a large number of sex-biased genes were identified by comparing gonadal transcriptomes between females and males. Female reproductive tract over-expressed genes were involved in development, signalling, growth, behaviour and reproductive process, whereas, testicle over-expressed genes were involved in rhythmic process, transducer activity and synapse. We also identified several genes involved in JH synthesis that were over-expressed in the female. These genes are an important resource for furthering understanding of the evolution of reproductive strategies within Phasmatodea.

Abbreviations
Bp: Base pair; BP: Biological process; CC: Cellular component; Gb: Gigabases; GO: Gene ontology; JH: Juvenile hormone; Jheh2: JH epoxide hydrolase; LTR: Long terminal repeat retrotransposon; MF: Molecular function; MITEs: Miniature inverted repeat transposable elements; MP: Mate pair; PE: Paired end; SINE: Short interspersed element; TEs: Interspersed transposon elements

Acknowledgments
We thank New Zealand eScience Infrastructure (NeSI) for high-performance computing facilities; Elena Hilario, Gary Houliston, Dagmar Goeke, Robyn Howitt, Julia Allwood, Ana Ramon-Laca, and Duckchul Park for laboratory support. We thank John Parsch and two anonymous reviewers for comments that improved the manuscript.

Funding
This research was funded by the Alan Wilson Centre (RM 13790/24947), the University of Auckland, and Strategic Science Investment Fund to Landcare Research from the Ministry of Business, Innovation and Employment's Science and Innovation Group.

Authors' contributions
TRB and RDN conceived and designed the project and advised on analyses. CW performed the genome and transcriptome assembly, analysis. RNC provided valuable suggestions and participated in the bioinformatics analysis. VGT performed flow cytometry experiment. VGT and CW performed DNA and RNA extractions. CW prepared the manuscript draft. TRB, RDN and RNC edited and commented extensively on the draft manuscript. All authors reviewed and approved the final manuscript.

Competing interests
The authors declare that they have no competing interests.

Author details
[1]School of Biological Sciences, The University of Auckland, Auckland, New Zealand. [2]Landcare Research, Auckland, New Zealand. [3]New Zealand Institute for Plant & Food Research Ltd, Auckland, New Zealand. [4]Department of Biology, Lund University, Lund, Sweden.

References
1. Scali V: Metasexual stick insects: model pathways to losing sex and bringing it back. In: Schön I, Martens K, Dijk PV, editors. Lost sex. Springer; 2009:317–345.
2. Buckley TR, Marske K, Attanayake D. Phylogeography and ecological niche modelling of the New Zealand stick insect Clitarchus hookeri (white) support survival in multiple coastal refugia. J Biogeogr. 2010;37(4):682–95.
3. Morgan-Richards M, Trewick SA, Stringer IA. Geographic parthenogenesis and the common tea-tree stick insect of New Zealand. Mol Ecol. 2010;19(6):1227–38.
4. Buckley TR, Attanayake D, Park D, Ravindran S, Jewell TR, Normark BB. Investigating hybridization in the parthenogenetic New Zealand stick insect Acanthoxyla (Phasmatodea) using single-copy nuclear loci. Mol Phylogenet Evol. 2008;48(1):335–49.
5. Morgan-Richards M, Trewick SA. Hybrid origin of a parthenogenetic genus? Mol Ecol. 2005;14(7):2133–42.
6. Trewick SA, Morgan-Richards M, Collins LJ. Are you my mother? Phylogenetic analysis reveals orphan hybrid stick insect genus is part of a monophyletic New Zealand clade. Mol Phylogenet Evol. 2008; 48(3):799–808.
7. Morgan-Richards M, Hills SF, Biggs PJ, Trewick SA. Sticky genomes: using NGS evidence to test hybrid speciation hypotheses. PLoS One. 2016;11(5):e0154911.
8. Myers SS, Trewick SA, Morgan-Richards M. Multiple lines of evidence suggest mosaic polyploidy in the hybrid parthenogenetic stick insect lineage Acanthoxyla. Insect Conserv Diver. 2013;6(4):537–48.
9. Dennis AB, Dunning LT, Sinclair BJ, Buckley TR. Parallel molecular routes to cold adaptation in eight genera of New Zealand stick insects. Sci Rep. 2015;5:13965.
10. Myers SS, Buckley TR, Holwell GI. Mate detection and seasonal variation in stick insect mating behaviour (Phamatodea: Clitarchus hookeri). Behaviour. 2015;152(10):1325–48.
11. Myers SS, Buckley TR, Holwell GI. Male genital claspers influence female mate acceptance in the stick insect Clitarchus hookeri. Behav Ecol Sociobiol. 2016;70(9):1547–56.
12. Wu C, Crowhurst RN, Dennis AB, Twort VG, Liu S, Newcomb RD, Ross HA, Buckley TR. De novo Transcriptome analysis of the common New Zealand stick insect Clitarchus hookeri (Phasmatodea) reveals genes involved in olfaction, digestion and sexual reproduction. PLoS One. 2016;11(6): e0157783.
13. Myers SS, Holwell GI, Buckley TR. Genetic and morphometric data demonstrate alternative consequences of secondary contact in Clitarchus stick insects. J Biogeogr. 2017;44(9):2069–81.
14. Akbari OS, Antoshechkin I, Amrhein H, Williams B, Diloreto R, Sandler J, Hay BA. The developmental transcriptome of the mosquito Aedes aegypti, an invasive species and major arbovirus vector. G3: Genes Genomes Genetics. 2013;3(9):1493–509.
15. Ferree PM, Fang C, Mastrodimos M, Hay BA, Amrhein H, Akbari OS. Identification of genes uniquely expressed in the germ-line tissues of the Jewel wasp Nasonia vitripennis. G3: Genes Genomes Genetics. 2015; 5(12):2647–53.
16. Chen W, Liu Y-X, Jiang G-F. De novo assembly and characterization of the testis Transcriptome and development of EST-SSR markers in the cockroach \textitPeriplaneta americana. Sci Rep. 2015;5:11144.
17. Parsch J, Ellegren H. The evolutionary causes and consequences of sex-biased gene expression. Nat Rev Genet. 2013;14(2):83–7.
18. Lu B, Zeng Z, Shi T. Comparative study of de novo assembly and genome-guided assembly strategies for transcriptome reconstruction based on RNA-Seq. Sci China Life Sci. 2013;56(2):143–55.
19. Soria-Carrasco V, Gompert Z, Comeault AA, Farkas TE, Parchman TL, Johnston JS, Buerkle CA, Feder JL, Bast J, Schwander T, et al. Stick insect genomes reveal natural selection's role in parallel speciation. Science. 2014; 344(6185):738–42.
20. Ricci M, Luchetti A, Bonandin L, Mantovani B. Random DNA libraries from three species of the stick insect genus bacillus (Insecta: Phasmida): repetitive DNA characterization and first observation of polyneopteran MITEs. Genome. 2013;56(12):729–35.
21. Hanrahan SJ, Johnston JS. New genome size estimates of 134 species of arthropods. Chromosom Res. 2011;19(6):809–23.
22. Mikheyev AS, Zwick A, Magrath MJL, Grau ML, Qiu L, Su YN, Yeates D. Museum genomics confirms that the Lord Howe Island stick insect survived extinction. Curr Biol. 2017; In press
23. Consortium HGS, et al. Insights into social insects from the genome of the honeybee Apis mellifera. Nature. 2006;443(7114):931.
24. Werren JH, Richards S, Desjardins CA, Niehuis O, Gadau J, Colbourne JK, Group NGW, et al. Functional and evolutionary insights from the genomes of three parasitoid Nasonia species. Science. 2010;327(5963):343–8.
25. Smith CD, Zimin A, Holt C, Abouheif E, Benton R, Cash E, Croset V, Currie CR, Elhaik E, Elsik CG, et al. Draft genome of the globally widespread and invasive argentine ant (Linepithema humile). Proc Natl Acad Sci U S A. 2011; 108(14):5673–8.
26. Zhan S, Merlin C, Boore JL, Reppert SM. The monarch butterfly genome yields insights into long-distance migration. Cell. 2011;147(5):1171–85.
27. You M, Yue Z, He W, Yang X, Yang G, Xie M, Zhan D, Baxter SW, Vasseur L, Gurr GM, et al. A heterozygous moth genome provides insights into herbivory and detoxification. Nat Genet. 2013;45(2):220–5.

28. Wang X, Fang X, Yang P, Jiang X, Jiang F, Zhao D, Li B, Cui F, Wei J, Ma C, et al. The locust genome provides insight into swarm formation and long-distance flight. Nat Commun. 2014;5:2957.

29. Otto F. DAPI staining of fixed cells for high-resolution flow cytometry of nuclear DNA. Methods Cell Biol. 1990;33:105–10.

30. Lysak MA, Dolezel J. Estimation of nuclear DNA content in *Sesleria* (Poaceae). Caryologia. 1998;51(2):123–32.

31. Schmieder R, Edwards R. Quality control and preprocessing of metagenomic datasets. Bioinformatics. 2011;27(6):863–4.

32. Xu H, Luo X, Qian J, Pang X, Song J, Qian G, Chen J, Chen S. FastUniq: a fast *de novo* duplicates removal tool for paired short reads. PLoS One. 2012;7(12):e52249.

33. Martin M: Cutadapt removes adapter sequences from high-throughput sequencing reads. EMBnet J. 2011;17(1):10–12.

34. Birol I, Jackman SD, Nielsen CB, Qian JQ, Varhol R, Stazyk G, Morin RD, Zhao Y, Hirst M, Schein JE, et al. *De novo* transcriptome assembly with ABySS. Bioinformatics. 2009;25(21):2872–7.

35. Gnerre S, Maccallum I, Przybylski D, Ribeiro FJ, Burton JN, Walker BJ, Sharpe T, Hall G, Shea TP, Sykes S, et al. High-quality draft assemblies of mammalian genomes from massively parallel sequence data. Proc Natl Acad Sci U S A. 2011;108(4):1513–8.

36. Luo R, Liu B, Xie Y, Li Z, Huang W, Yuan J, He G, Chen Y, Pan Q, Liu Y, et al. SOAPdenovo2: an empirically improved memory-efficient short-read de novo assembler. GigaScience. 2012;1(1):18.

37. Boetzer M, Henkel CV, Jansen HJ, Butler D, Pirovano W. Scaffolding pre-assembled contigs using SSPACE. Bioinformatics. 2011;27(4):578–9.

38. Langmead B, Salzberg SL. Fast gapped-read alignment with bowtie 2. Nat Methods. 2012;9(4):357–9.

39. Parra G, Bradnam K, Korf I. CEGMA: a pipeline to accurately annotate core genes in eukaryotic genomes. Bioinformatics. 2007;23(9):1061–7.

40. Smit A, Hubley R: RepeatModeler Open-1.0. <http://wwwrepeatmaskerorg/> 2008-2015.

41. Altschul SF, Madden TL, Schaffer AA, Zhang J, Zhang Z, Miller W, Lipman DJ. Gapped BLAST and PSI-BLAST: a new generation of protein database search programs. Nucleic Acids Res. 1997;25(17):3389–402.

42. Hoede C, Arnoux S, Moisset M, Chaumier T, Inizan O, Jamilloux V, Quesneville H. PASTEC: an automatic transposable element classification tool. PLoS One. 2014;9(5):e91929.

43. Smit A, Hubley R, Green P: RepeatMasker Open-4.0. <http://wwwrepeatmaskerorg/> 2013-2015.

44. Holt C, Yandell M. MAKER2: an annotation pipeline and genome-database management tool for second-generation genome projects. BMC Bioinformatics. 2011;12:491.

45. Stanke M, Waack S. Gene prediction with a hidden Markov model and a new intron submodel. Bioinformatics. 2003;19(Suppl 2):ii215–25.

46. Keller O, Odronitz F, Stanke M, Kollmar M, Waack S. Scipio: using protein sequences to determine the precise exon/intron structures of genes and their orthologs in closely related species. BMC Bioinformatics. 2008;9:278.

47. Waterhouse RM, Zdobnov EM, Tegenfeldt F, Li J, Kriventseva EV. OrthoDB: the hierarchical catalog of eukaryotic orthologs in 2011. Nucleic Acids Res. 2011;39(Database issue):D283–8.

48. Dobin A, Davis CA, Schlesinger F, Drenkow J, Zaleski C, Jha S, Batut P, Chaisson M, Gingeras TR. STAR: ultrafast universal RNA-seq aligner. Bioinformatics. 2013;29(1):15–21.

49. Trapnell C, Roberts A, Goff L, Pertea G, Kim D, Kelley DR, Pimentel H, Salzberg SL, Rinn JL, Pachter L. Differential gene and transcript expression analysis of RNA-seq experiments with TopHat and cufflinks. Nat Protoc. 2012;7(3):562–78.

50. O'Donovan C, Martin MJ, Gattiker A, Gasteiger E, Bairoch A, Apweiler R. High-quality protein knowledge resource: SWISS-PROT and TrEMBL. Brief Bioinform. 2002;3(3):275–84.

51. UniProt C. The universal protein resource (UniProt). Nucleic Acids Res. 2008; 36(Database issue):D190–5.

52. Drysdale RA, Crosby MA, FlyBase C. FlyBase: genes and gene models. Nucleic Acids Res. 2005;33(Database issue):D390–5.

53. Anders S, Pyl PT, Huber W. HTSeq–a python framework to work with high-throughput sequencing data. Bioinformatics. 2015;31(2):166–9.

54. Ihaka R, Gentleman R. R: a language for data analysis and graphics. J Comput Graph Stat. 1996;5(3):299–314.

55. Love MI, Huber W, Anders S. Moderated estimation of fold change and dispersion for RNA-seq data with DESeq2. Genome Biol. 2014;15(12):550.

56. Huang DW, Sherman BT, Lempicki RA. Bioinformatics enrichment tools: paths toward the comprehensive functional analysis of large gene lists. Nucleic Acids Res. 2009;37(1):1–13.

57. Huang DW, Sherman BT, Lempicki RA. Systematic and integrative analysis of large gene lists using DAVID bioinformatics resources. Nat Protoc. 2009;4(1):44–57.

58. Walter W, Sánchez-Cabo F, Ricote M. GOplot: an R package for visually combining expression data with functional analysis. Bioinformatics. 2015; 31(17):2912–4.

59. Terrapon N, Li C, Robertson HM, Ji L, Meng X, Booth W, Chen Z, Childers CP, Glastad KM, Gokhale K, et al. Molecular traces of alternative social organization in a termite genome. Nat Commun. 2014;5:3636.

60. Wu C, Jordan MD, Newcomb RD, Gemmell NJ, Bank S, Meusemann K, Dearden PK, Duncan EJ, Grosser S, Rutherford K, et al. Analysis of the genome of the New Zealand giant Collembola (*Holacanthella duospinosa*) sheds light on hexapod evolution. BMC Genomics. 2017; In press

61. Vinogradov AE. Intron-genome size relationship on a large evolutionary scale. J Mol Evol. 1999;49(3):376–84.

62. Comeron JM, Kreitman M. The correlation between intron length and recombination in *Drosophila*: dynamic equilibrium between mutational and selective forces. Genetics. 2000;156(3):1175–90.

63. Huylmans AK, Ezquerra AL, Parsch J, Cordellier M. *De novo* transcriptome assembly and sex-biased gene expression in the cyclical parthenogenetic *Daphnia galeata*. Genome Biol Evol. 2016;8(10):3120–39.

64. Purandare SR, Bickel RD, Jaquiery J, Rispe C, Brisson JA. Accelerated evolution of morph-biased genes in pea aphids. Mol Biol Evol. 2014; 31(8):2073–83.

65. Schwander T, Crespi BJ, Gries R, Gries G. Neutral and selection-driven decay of sexual traits in asexual stick insects. Proc Biol Sci. 2013; 280(1764):20130823.

66. Tootle TL, Williams D, Hubb A, Frederick R, Spradling A. *Drosophila* eggshell production: identification of new genes and coordination by Pxt. PLoS One. 2011;6(5):e19943.

67. Claycomb JM, Benasutti M, Bosco G, Fenger DD, Orr-Weaver TL. Gene amplification as a developmental strategy: isolation of two developmental amplicons in *Drosophila*. Dev Cell. 2004;6(1):145–55.

68. McGraw LA, Gibson G, Clark AG, Wolfner MF. Genes regulated by mating, sperm, or seminal proteins in mated female *Drosophila melanogaster*. Curr Biol. 2004;14(16):1509–14.

69. Sabatier L, Jouanguy E, Dostert C, Zachary D, Dimarcq JL, Bulet P, Imler JL. Pherokine-2 and -3. Eur J Biochem. 2003;270(16):3398–407.

70. Buckley TR, Bradler S. *Tepakiphasma ngatikuri*, a new genus and species of stick insect (Phasmatodea) from the far north of New Zealand. New Zealand Entomologist. 2010;33(1):118–26.

71. Sellick JC. Descriptive terminology of the phasmid egg capsule, with an extended key to the phasmid genera based on egg structure. Syst Entomol. 1997;22(2):97–122.

72. Hughes L, Westoby M. Capitula on stick insect eggs and elaiosomes on seeds: convergent adaptations for burial by ants. Funct Ecol. 1992;6(6): 642–8.

73. Goldberg J, Bresseel J, Constant J, Kneubühler B, Leubner F, Michalik P, Bradler S. Extreme convergence in egg-laying strategy across insect orders. Sci Rep. 2015;5:7825.

74. Beaver L, Gvakharia B, Vollintine T, Hege D, Stanewsky R, Giebultowicz J. Loss of circadian clock function decreases reproductive fitness in males of *Drosophila melanogaster*. Proc Natl Acad Sci. 2002;99(4):2134–9.

75. Bebas P, Cymborowski B, Giebultowicz J. Circadian rhythm of sperm release in males of the cotton leafworm, *Spodoptera littoralis*: in vivo and in vitro studies. J Insect Physiol. 2001;47(8):859–66.

76. Giebultowicz J, Riemann J, Raina A, Ridgway R. Circadian system controlling release of sperm in the insect testes. Science. 1989;245(4922):1098–100.

77. Kotwica-Rolinska J, Gvakharia BO, Kedzierska U, Giebultowicz JM, Bebas P: Effects of period RNAi on V-ATPase expression and rhythmic pH changes in the vas deferens of *Spodoptera littoralis* (Lepidoptera: Noctuidae). Insect Biochem Mol Biol 2013, 43(6):522-532.

78. He L, Wang Q, Jin X, Wang Y, Chen L, Liu L, Wang Y. Transcriptome profiling of testis during sexual maturation stages in \textitEriocheir sinensis using Illumina sequencing. PLoS One. 2012;7(3):e33735.

79. Liu H, Lamm MS, Rutherford K, Black MA, Godwin JR, Gemmell NJ. Large-scale transcriptome sequencing reveals novel expression patterns for key sex-related genes in a sex-changing fish. Biol Sex Differ. 2015;6(1):1.

80. Collins AM, Caperna TJ, Williams V, Garrett WM, Evans JD. Proteomic analyses of male contributions to honey bee sperm storage and mating. Insect Mol Biol. 2006;15(5):541–9.

81. Dorus S, Busby SA, Gerike U, Shabanowitz J, Hunt DF, Karr TL. Genomic and functional evolution of the *Drosophila melanogaster* sperm proteome. Nat Genet. 2006;38(12):1440–5.

82. Wasbrough ER, Dorus S, Hester S, Howard-Murkin J, Lilley K, Wilkin E, Polpitiya A, Petritis K, Karr TL: The *Drosophila melanogaster* sperm proteome-II (DmSP-II). J Proteome 2010, 73(11):2171-2185.

83. Ivaldi MS, Karam CS, Corces VG. Phosphorylation of histone H3 at Ser10 facilitates RNA polymerase II release from promoter-proximal pausing in *Drosophila*. Genes Dev. 2007;21(21):2818–31.

84. Coon TA, Glasser JR, Mallampalli RK, Chen BB. Novel E3 ligase component FBXL7 ubiquitinates and degrades aurora a, causing mitotic arrest. Cell Cycle. 2012;11(4):721–9.

85. Taddei C, Chicca MV, Maurizii MG, Scali V. The germarium of panoistic ovarioles of *Bacillus rossius* (Insecta Phasmatodea): larval differentiation. Invertebr Reprod Dev. 1992;21(1):47–56.

86. Büning J: The insect ovary: ultrastructure, previtellogenic growth and evolution: Springer Science & Business Media; 1994.

87. Irles P, Bellés X, Piulachs MD. Identifying genes related to choriogenesis in insect panoistic ovaries by suppression subtractive hybridization. BMC Genomics. 2009;10(1):1.

88. Campbell PM, Healy MJ, Oakeshott JG. Characterisation of juvenile hormone esterase in *Drosophila melanogaster*. Insect Biochem Mol Biol. 1992;22(7):665–77.

89. Mayoral JG, Nouzova M, Navare A, Noriega FG. NADP+–dependent farnesol dehydrogenase, a corpora allata enzyme involved in juvenile hormone synthesis. Proc Natl Acad Sci. 2009;106(50):21091–6.

90. Mittler T, Nassar S, Staal G. Wing development and parthenogenesis induced in progenies of kinoprene-treated gynoparae of *Aphis fabae* and *Myzus persicae*. J Insect Physiol. 1976;22(12):1717–25.

91. Hardie J, Baker FC, Jamieson GC, Lees AD, Schooley DA. The identification of an aphid juvenile hormone, and its titre in relation to photoperiod. Physiol Entomol. 1985;10(3):297–302.

92. Corbitt TS, Hardie J. Juvenile hormone effects on polymorphism in the pea aphid, *Acyrthosiphon pisum*. Entomol Exp Appl. 1985;38(2):131–5.

93. Liu L-J, Zheng H-Y, Jiang F, Guo W, Zhou S-T. Comparative transcriptional analysis of asexual and sexual morphs reveals possible mechanisms in reproductive polyphenism of the cotton aphid. PLoS One. 2014;9(6):e99506.

13

Identification of genome-wide SNP-SNP interactions associated with important traits in chicken

Hui Zhang[1,2,3], Jia-Qiang Yu[1,2,3], Li-Li Yang[1,2,3], Luke M. Kramer[4], Xin-Yang Zhang[1,2,3], Wei Na[1,2,3], James M. Reecy[4*] and Hui Li[1,2,3*]

Abstract

Background: In addition to additive genetic effects, epistatic interactions can play key roles in the control of phenotypic variation of traits of interest. In the current study, 475 male birds from lean and fat chicken lines were utilized as a resource population to detect significant epistatic effects associated with growth and carcass traits.

Results: A total of 421 significant epistatic effects were associated with testis weight (TeW), from which 11 sub-networks (Sub-network1 to Sub-network11) were constructed. In Sub-network1, which was the biggest network, there was an interaction between GGA21 and GGAZ. Three genes on GGA21 (*SDHB*, *PARK7* and *VAMP3*) and nine genes (*AGTPBP1*, *CAMK4*, *CDC14B*, *FANCC*, *FBP1*, *GNAQ*, *PTCH1*, *ROR2* and *STARD4*) on GGAZ that might be potentially important candidate genes for testis growth and development were detected based on the annotated gene function. In Sub-network2, there was a SNP on GGA19 that interacted with 8 SNPs located on GGA10. The SNP (Gga_rs15834332) on GGA19 was located between C-C motif chemokine ligand 5 (*CCL5*) and *MIR142*. There were 32 Refgenes on GGA10, including *TCF12* which is predicted to be a target gene of miR-142-5p. We hypothesize that miR-142-5p and *TCF12* may interact with one another to regulate testis growth and development. Two genes (*CDH12* and *WNT8A*) in the same cadherin signaling pathway were implicated as potentially important genes in the control of metatarsus circumference (MeC). There were no significant epistatic effects identified for the other carcass and growth traits, e.g. heart weight (HW), liver weight (LW), spleen weight (SW), muscular and glandular stomach weight (MGSW), carcass weight (CW), body weight (BW1, BW3, BW5, BW7), chest width (ChWi), metatarsus length (MeL).

Conclusions: The results of the current study are helpful to better understand the genetic basis of carcass and growth traits, especially for testis growth and development in broilers.

Keywords: Carcass and growth traits, Testis, Epistasis, SNP-SNP interaction, Chicken

Background

Epistasis arises due to interactions, either between single nucleotide polymorphisms (SNPs), genes or quantitative trait loci (QTLs), which result in non-linear effects that control variation in phenotypes. Epistasis can have a large influence on phenotypic variation of traits such as starvation resistance, startle response, and chill coma recovery in Drosophila [1]. Identification of epistatic effects associated with quantitative traits will help us to better understand the genetic architecture that underlies complex variation of phenotypes for both humans and animals [1, 2]. Therefore, more and more attention has been placed on epistasis, which has resulted in some valuable insights [3–8].

With the advent of SNP arrays and genomic resequencing, it is relatively easy to genotype a wide array of individuals. As a result, many genome–wide association studies (GWAS) have been carried out in the past several years. Most of these studies have focused on single locus additive genetic tests. However, this is not

* Correspondence: jreecy@iastate.edu; lihui@neau.edu.cn
[4]Department of Animal Science, Iowa State University, 2255 Kildee Hall, Ames, IA 50011, USA
[1]Key Laboratory of Chicken Genetics and Breeding, Ministry of Agriculture, Harbin 150030, People's Republic of China
Full list of author information is available at the end of the article

that only type of genetic association. A genome wide SNP-SNP interaction analysis should provide new insights into the genetic architecture that underlies variation in complex traits.

Recently, genome wide SNP-SNP interaction analysis have been conducted in humans [4, 9] and domestic livestock species [10–12]. In chicken (*Gallus gallus*), it has been suggested that epistatic interactions between genes (or QTLs) are important for variation in quantitative traits [8, 13, 14]. However, the study of epistatic interactions at the whole genome level have been limited [15].

Two Northeast Agricultural University broiler lines that have been divergently selected for abdominal fat content (named as NEAUHLF) for more than 10 years were used in the current study. Previously, we reported that 52 pairs of SNPs had significant epistatic interactions that were associated with abdominal fat weight [15]. In the current study, the significant epistatic interactions for carcass and growth traits were identified. The results of this study may provide some helpful information to better understand the genetic basis of carcass and growth traits in broilers.

Methods
Experimental populations
Two Northeast Agricultural University broiler lines that have been divergently selected for abdominal fat content (NEAUHLF) were used to identify epistatic interactions. The NEAUHLF lines have been selected since 1996 using abdominal fat percentage (AFP = abdominal fat weight/body weight at 7 weeks of age) and plasma very low-density lipoprotein (VLDL) concentration as selection criteria. The G0 generation of NEAUHLF came from the same grandsire line, which originated from the Arbor Acres broiler, which was then divided into two lines according to their plasma VLDL concentration at 7 weeks of age. The G0 birds were mated (one sire: four dams) to produce 25 half-sib families for each line, with an average of 70 G1 offspring per family in two hatches. From G1 to G11, the birds of each line were raised in two hatches with five birds per cage. Plasma VLDL concentrations were measured for all male birds, which had free access to feed and water at 7 weeks, and the AFP of the male birds in the first hatch was measured after slaughter at 7 weeks. Sib birds from the families with lower (lean line) or higher (fat line) AFP than the average value for the population were selected as candidates for breeding, considering the plasma VLDL concentration and the body weights of male birds in the second hatch and the egg production of female birds in both hatches. These birds were kept under the same environmental conditions and had free access to feed and water. Commercial corn-soybean-based diets that met all National Research Council (NRC) requirements were provided. From hatch to 3 weeks of age, the birds received a starter feed (3,000kal ME = kg and 210 g = kg CP) and from 4 weeks of age to slaughter the birds were fed a grower diet (3100 kal ME = kg and 190 g = kg CP). The birds used in the current study included 475 male individuals from the 11th generation of NEAUHLF [15]. The birds were weighed at 0, 1, 3, 5 and 7 weeks of age (BW0, BW1, BW3, BW5 and BW7). At 7 weeks of age, the metatarsus length (MeL), metatarsus circumference (MeC) and chest width (ChWi) were measured prior to slaughter as described previously [16]. Carcass weight (CW), testis weight (TeW), heart weight (HW), liver weight (LW), spleen weight (SW), muscular and glandular stomach weight (MGSW) were obtained after the birds were slaughtered.

SNP genotyping
Genotyping was carried out using the Illumina Inc. (San Diego, CA, USA) chicken 60 K SNP chip, which contained 57,636 SNPs. After quality control, 48,824 SNPs in 475 individuals were used in the epistatic interaction analyses. The quality control of the SNP genotypes was described previously by Zhang et al. [17].

Genome-wide Pairwise interaction analysis
The EPISNP3 module in epiSNP_v4.2_Windows software package was used to identify significant epistatic effects [18]. The statistical model used to test for epistatic effects associated with carcass and growth traits was as follows: y = Xg + Zb + e, where y is the column vector of phenotypic values, g is the effects of SNP genotypes, X is the design matrix of g, b is the fixed effects of Line and BW7 (or BW0), Z is the model matrix of b, and e is the random error.

The P-values of the epistatic effects were Bonferroni corrected for multiple testing (5.96×10^9 independent tests, with a significance threshold of $P < 0.05$), which resulted in $P < 8.39 \times 10^{-12}$ as a significance threshold. Significant interactions, including additive by additive (AA), additive by dominance (AD) or dominance by additive (DA) and dominance by dominance (DD), between two SNPs on the same chromosome were deleted because these interactions may potentially be markers for a haplotype effect that contains a single QTL [12]. The remaining significant SNP interactions were further filtered with the criterion that only those interaction with at least 10 animals in every genotype combination were considered [12], which is roughly equivalent to a 15% minor allele frequency for each variant in an additive by additive epistatic interaction.

SNP-SNP network
The figures that illustrate the SNP-SNP networks with the significant epistatic effects for carcass and growth

traits were drawn using the epiNet option within the epiSNP_v4.2_Windows software package [18].

Linkage disequilibrium (LD) analysis

The linkage disequilibrium (LD) between SNPs was calculated using Haploview software (version 4.2). The solid spine method within the package was used to define the LD block.

Annotation of SNP-SNP network

Genes within 1 Mb (upstream and downstream) of the SNPs that had significant interactions with another SNP for carcass and growth traits were retrieved from UCSC (https://genome.ucsc.edu/) (Galgal4). Functional annotation of genes was performed using DAVID bioinformatics resources 6.8 (http://david.abcc.ncifcrf.gov/summary.jsp) for Gene Ontology (GO) terms and Kyoto Encyclopedia of Genes and Genomes (KEGG) pathway analysis. Statistical significance was set at the nominal P-value < 0.05.

Results

Phenotypic and SNPs information

Phenotypic summary statistics for carcass and growth traits are shown in Table 1 and the phenotypic distributions for the traits in the lean and fat lines respectively are shown in Additional file 1: Figure S1. There were extremely significant differences ($P < 0.01$) in Chwi, MeL, MeC, LW, SW, TeW and significant difference ($P < 0.05$) in HW between the lean and fat lines. After quality control, 48,824 SNPs were utilized for epistatic interaction analyses (Table 2). These SNPs were distributed on 28 autosomes, Z chromosome, two linkage groups, and SNPs not assigned to any chromosomes in

chickens. These markers covered about 1027.01 Mb of the chicken genome, with an average SNP density of 16.08 kb/SNP.

Epistatic analysis of carcass trait

The pairwise interaction effects between every two SNPs across the whole chicken genome for TeW were calculated using EPISNP3 [18]. After filtering, 421 pairs of SNPs were significantly associated with testis weight ($P < 8.39 \times 10^{-12}$). Of these 421 significant SNP by SNP interactions, 403 (95.72%) exhibited an additive by additive interaction, 18 (4.28%) exhibited an additive by dominance (or dominance by additive) interaction, and no dominance by dominance interactions were detected (Additional file 2: Table S1). The most significant additive by additive effect detected occurred between GGA3 (GGaluGA22768) and GGA10 (Gga_rs14722408). The phenotypic distributions of the four different genotype classes of this additive by additive effect are showed in Additional file 3: Figure S2.

To investigate the complex mechanism of epistatic effects on TeW, networks were constructed using the 421 significant SNP by SNP interactions. The epistatic interaction sub-networks that contained more than three nodes are shown in Additional file 4: Figure S3. Eleven sub-networks were detected (Additional file 4: Figure S3). Sub-network1 was the biggest and contained 372 pairs of SNP by SNP interaction effects. Based on LD information, a simpler sub-network, which was derived from Sub-network1, was obtained, in which numerous SNPs are represented by a single LD block (Fig. 1, Additional file 5: Figure S4). The blocks in the simpler sub-network represented the SNPs that contained in the LD blocks.

Table 1 The Mean ± Standard deviation (SD) of the carcass and growth traits in lean and fat lines, respectively, and in the combined population

Traits	Combined population (475 birds)	Lean line (203 birds)	Fat line (272 birds)
BW1 (g)	121.97 ± 12.34	121.05 ± 12.80	122.68 ± 11.95
BW3 (g)	615.22 ± 65.97	617.35 ± 71.98	613.65 ± 61.23
BW5 (g)	1491.19 ± 142.53	1487.53 ± 159.13	1493.91 ± 129.10
BW7 (g)	2400.97 ± 221.65	2419.53 ± 246.45	2387.11 ± 200.51
ChWi (cm)	9.23 ± 0.74	9.54 ± 0.70^A	9.00 ± 0.68^B
MeL (cm)	9.25 ± 0.46	9.56 ± 0.37^A	9.02 ± 0.38^B
MeC (cm)	5.10 ± 0.39	5.46 ± 0.27^A	4.84 ± 0.20^B
CW (g)	2170.03 ± 203.31	2164.95 ± 225.19	2173.84 ± 185.59
LW (g)	57.54 ± 9.08	55.35 ± 8.44^B	59.18 ± 9.21^A
HW (g)	10.68 ± 1.72	10.86 ± 1.78^A	10.54 ± 1.67^a
SW (g)	3.25 ± 1.05	2.88 ± 0.86^B	3.53 ± 1.09^A
MGSW (g)	31.10 ± 5.57	31.46 ± 6.07	30.83 ± 5.15
TeW (g)	1.03 ± 0.85	1.39 ± 1.03^A	0.77 ± 0.55^B

Note: Different letters indicate significant differences between the lean and fat lines. Uppercase ($P < 0.01$) and lowercase ($P < 0.05$) letters indicate significant differences

Table 2 Summary information of the genome-wide SNP markers

GGA[1]	SNPs number	GGA length (Mb)	Mean distance (kb)
1	7538	200.95	26.66
2	5652	154.79	27.39
3	4322	113.65	26.30
4	3518	94.16	26.77
5	2295	62.23	27.11
6	1814	35.84	19.76
7	1907	38.17	20.01
8	1486	30.62	20.61
9	1240	24.02	19.37
10	1379	22.42	16.26
11	1312	21.87	16.67
12	1425	20.46	14.36
13	1204	18.32	15.21
14	1062	15.76	14.84
15	1082	12.93	11.95
16	16	0.42	26.12
17	922	10.61	11.51
18	917	10.89	11.87
19	880	9.90	11.25
20	1574	13.92	8.84
21	796	6.95	8.73
22	327	3.89	11.90
23	643	6.02	9.37
24	758	6.37	8.40
25	181	2.02	11.17
26	670	5.03	7.51
27	506	4.84	9.56
28	607	4.47	7.37
LGE22C19W28_E50C23	115	0.88	7.67
LEG64	3	0.02	6.80
Z	2001	74.59	37.28
UN[a]	672	/	/
Total/Mean value	48,824	1027.01	16.08

[a]These SNPs were not assigned to any chromosomes
[1]GGA is an abbreviation for Gallus gallus

Two hundred and fifty-five of the interactions in Sub-network1 were between GGA21 and GGAZ, which indicated an interaction between the two chromosomes. The 255 interactions detected between GGA21 and GGAZ involved 24 SNPs on GGA21 and 19 SNPs on GGAZ, which spanned 572 kb (from 76,023 bp to 647,587 bp) and 9.5 Mb (from 37,246,321 bp to 46,745,968 bp), respectively. There were 13 Refgenes in the 572 kb region on GGA21 and 41 Refgenes in the 9.5 Mb region on GGAZ

(Table 3). Six GO terms, including protein-arginine deiminase activity, protein citrullination, positive regulation of collateral sprouting, cytoplasm, cell fate determination and protein autophosphorylation, were significantly ($P < 0.05$) enriched. No significant KEGG pathways were detected. Three genes on GGA21 (*SDHB*, *PARK7* and *VAMP3*) and nine genes (*AGTPBP1*, *CAMK4*, *CDC14B*, *FANCC*, *FBP1*, *GNAQ*, *PTCH1*, *ROR2* and *STARD4*) on GGAZ might be important for testis growth and development based on their basic functions.

Sub-networks 2, 3, 4, 5, 8 and 11 each had several SNPs in the same LD block on one chromosome, which interacted with a single SNP on another chromosome. For example, in Sub-network2 eight SNPs on GGA10 interacted with the SNP (Gga_rs15834332) on GGA19 (Fig. 2). The eight SNPs on GGA10 were spread across a 4.3 Mb region that contained 32 chicken Refgenes (Fig. 2). The SNP Gga_rs15834332 on GGA19 was located between two genes, C-C motif chemokine ligand 5 (*CCL5*) and *MIR142*, which is related to two miRNAs, miR-142-5p and miR-142-3p. Interestingly, five genes in the 4.3 Mb region on GGA10 are predicted to be target genes of miR-142-5p and miR-142-3p (Table 4). Among these five target genes, transcription factor 12 (*TCF12*) was the only gene that was predicted to be the target gene of miR-142-5p by three different packages, including TargetScan (http://www.targetscan.org), miRDB (http://mirdb.org) and PicTar (http://pictar.mdc-berlin.de). Unfortunately, it was impossible to predict which genes in the other regions of the genome may be good candidates to control testis growth and development.

For CW, HW, LW, SW and MGSW, no significant epistatic interactions were detected.

Epistatic analysis of growth trait

For BW1, BW3, BW5, BW7, ChWi and MeL, no significant epistatic interactions were detected. Fifteen pairs of SNPs with significant interaction effects on MeC were detected (Table 5). These 15 interactions were all additive by additive interactions, which implicated an interaction between GGA2 and GGA13 (Table 5). There was a single network that contained all 15 interactions (Fig. 3). The fifteen interactions occurred between five SNPs in a single LD block on GGA2, and three SNPs in a single LD block on GGA13 (Fig. 3). The genes inside the two LD blocks and within 0.5 Mb 5′ and 3′ of the LD blocks were found. There was only a single Refgene (Cadherin-12, *CDH12*) in the region on GGA2. There were fifteen Refgenes (*CDX1*, *CSF1R*, *NPY7R*, *FLT4*, *CANX*, *HNRNPH1*, *DGUOK*, *RUFY1*, *MAPK9*, *RASGEF1C*, *HNRNPAB*, *NME5*, *WNT8A*, *FAM13B* and *NPY6R*) located in the region on GGA13.

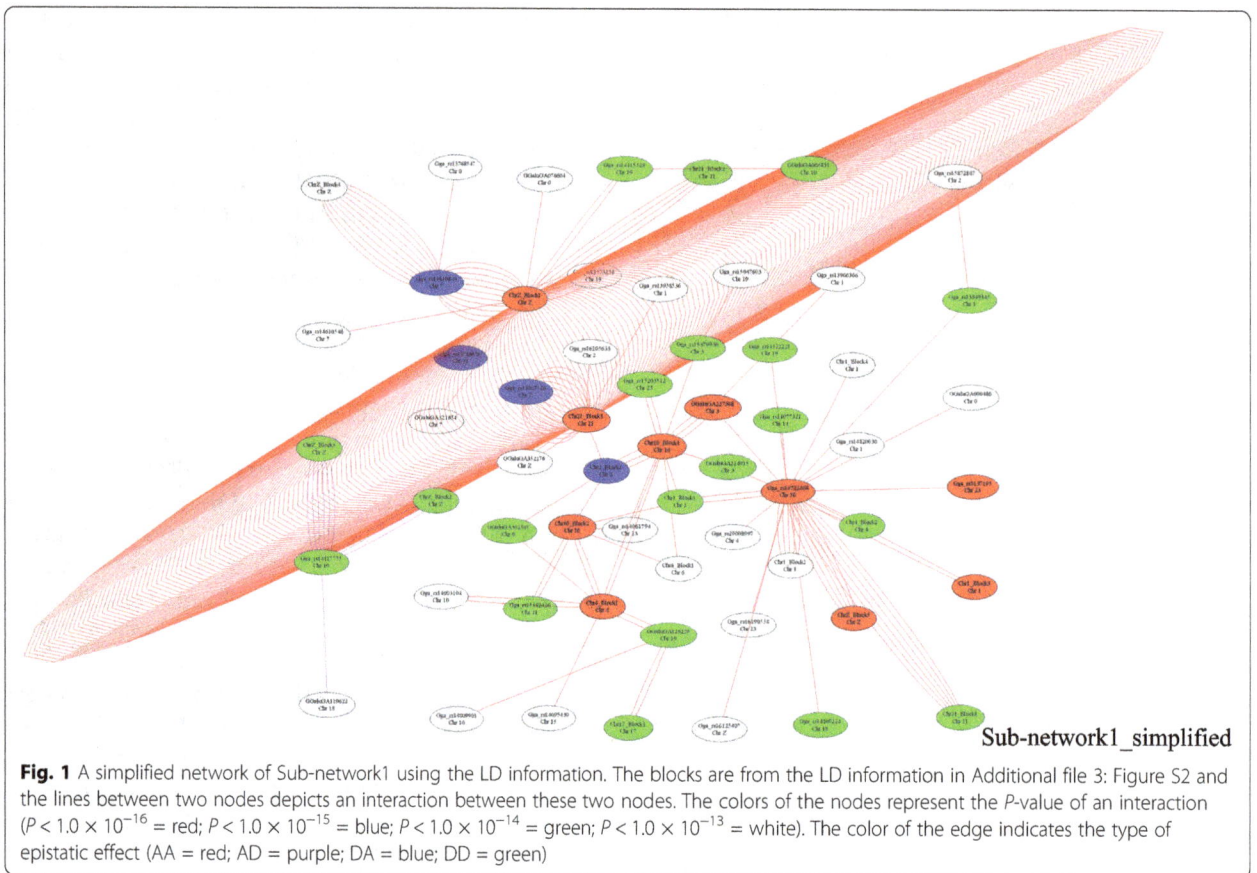

Fig. 1 A simplified network of Sub-network1 using the LD information. The blocks are from the LD information in Additional file 3: Figure S2 and the lines between two nodes depicts an interaction between these two nodes. The colors of the nodes represent the P-value of an interaction ($P < 1.0 \times 10^{-16}$ = red; $P < 1.0 \times 10^{-15}$ = blue; $P < 1.0 \times 10^{-14}$ = green; $P < 1.0 \times 10^{-13}$ = white). The color of the edge indicates the type of epistatic effect (AA = red; AD = purple; DA = blue; DD = green)

Discussion

The birds from lean and fat chicken lines used in the current study had significantly different amounts of abdominal fat content and significantly different testis weight, which was an ideal population to study the genetic architecture of abdominal fat deposition and testis growth and development [19]. The birds used in the current study were 7-week-old, therefore, the results of the current study could reflect early testis development. Testis weight has been reported to be controlled by genetic factors in both chickens and mice [20–24]. Roosters with small testes often have poor fertility [25]. There have been previously reported studies on the genetics of testis development in chicken [19]. However, most of the previous studies focused on additive genetic effects, while, epistatic effects were ignored. It is unknown how important epistatic interactions are for development of the testis. In the current study, we used 475 birds from lean and fat lines to conduct epistasis analysis for testis weight and other carcass and growth traits in chickens.

The two lines used in the current study were selected for 11 years and some regions of the genome may be fixed because of selection pressure. If this occurred, it

would not be possible to identify epistatic interactions in these regions of the genome. In contrast, if the two lines were crossed, it would be possible to detect epistatic effects in this intercross population. Despite this problem, some epistatic effects were detected in these two lines using epiSNP. Due to the selection applied to these population, it is possible that genomic stratification has occurred. epiSNP, which was used to conduct these analyses is capable of adjusting for family structure [18].

In this study, the SNP by SNP interaction effects for carcass and growth traits were filtered by three criteria. First, to correct for multiple testing comparisons, a Bonferroni correction was used instead of false discovering rate (FDR) method to minimize false positives. Utilization of the Bonferroni correction method will however also decrease the discovery of true interactions compared to the use of a false discovery rate correction. Therefore, the SNP X SNP interaction detected in this study may be fraction of the interactions that affect these. Second, when considering the linkage disequilibrium (LD) between the SNPs and the QTL for traits of interest, an interaction between two SNPs on the same chromosome may detect a haplotype effect but not an interaction. In other words, it is not possible to separate

Table 3 The SNPs on GGAZ interacted with the SNPs on GGA21, and the Refgenes in the two important regions

Chr	Position	Locus	RS#	Refgenes
Z	37,246,321	Gga_rs14765324	rs14765324	*VPS13A, AGTPBP1, AUH, CAMK4, CDC14B, CDC42SE2, CFC1B, CKS2,*
	37,440,770	Gga_rs16768474	rs16768474	*CTSL2, DAPK1, FANCC, FBP1, GADD45B, GNAQ, HABP4, HINT1, ISCA1,*
	37,562,582	Gga_rs14765605	rs14765605	*LOC427470, LOC770548, MIR1456, MIR23B, MIR24-2, MIR27B, MIR7-1,*
	38,066,346	Gga_rs16768723	rs16768723	*MIR7439, NAA35, NFIL3, NREP, NTRK2, PTCH1, REEP5, RMI1, ROR2,*
	38,424,448	Gga_rs14766107	rs14766107	*SEMA4D, SLC25A46, SPINZ, STARD4, SYK, TLE4, WDR36, ZCCHC6*
	38,545,044	GGaluGA351567	rs317567123	
	38,795,339	Gga_rs16047676	rs16047676	
	39,120,198	Gga_rs16781643	rs16781643	
	39,189,946	Gga_rs14745723	rs14745723	
	39,223,668	Gga_rs16131986	rs16131986	
	39,256,770	Gga_rs14787078	rs14787078	
	39,415,378	Gga_rs16781713	rs16781713	
	40,393,058	Gga_rs14787751	rs14787751	
	40,411,585	Gga_rs16782083	rs16782083	
	40,514,085	Gga_rs16754179	rs16754179	
	41,304,561	Gga_rs16132921	rs16132921	
	45,300,030	Gga_rs14015526	rs14015526	
	45,695,927	GGaluGA352176	rs316100592	
	46,745,968	Gga_rs14016510	rs14016510	
21	76,023	Gga_rs15179992	rs15179992	*PADI3, PADI1, PADI2, SDHB, MRPS16, PARK7, UTS2, PER3, VAMP3,*
	87,719	GGaluGA181809	rs312621287	*PHF13, ZBTB48, ICMT, RPL22*
	113,769	Gga_rs15179999	rs15179999	
	125,790	Gga_rs10732124	rs10732124	
	140,660	Gga_rs15180005	rs15180005	
	145,617	Gga_rs15180007	rs15180007	
	153,135	GGaluGA181823	rs315641745	
	165,390	Gga_rs16176404	rs16176404	
	197,067	Gga_rs15180023	rs15180023	
	219,311	Gga_rs15180012	rs15180012	
	277,332	Gga_rs15180032	rs15180032	
	291,087	Gga_rs15180041	rs15180041	
	297,203	Gga_rs16176409	rs16176409	
	299,152	Gga_rs16176412	rs16176412	
	304,979	GGaluGA181852	rs314825899	
	320,390	Gga_rs16176425	rs16176425	
	332,352	GGaluGA181865	rs314965954	
	332,687	GGaluGA181868	rs313888034	
	362,742	GGaluGA181877	rs316891294	
	402,941	Gga_rs14281175	rs14281175	
	423,299	Gga_rs13602346	rs13602346	
	493,436	Gga_rs14281291	rs14281291	
	631,537	Gga_rs16176824	rs16176824	
	647,587	GGaluGA182048	rs312963281	

true interaction effects from haplotype effects if the two SNPs are located on the same chromosome. Therefore, in order to increase the power to detect the true interaction effects for the growth and carcass traits, we deleted interactions that occurred on the same chromosome [12]. Third, the smaller the number of birds in any given genotype class, the less likely it is to get a good estimate of the genotype effect. Thus, in order to increase the power to detect the true interactions, only those interactions that contained at least 10 animals in every genotype combination were considered [12]. We carried out the filter of the interactions according to these criteria in order to reduce the chances of obtaining false-positive results (type I errors).

For testis weight, a total of 421 pairs of SNP-SNP epistatic interactions were detected. These pairs of SNPs comprised 211 single SNPs, and none of these individual SNPs were identified in our previous GWAS analysis for testis weight [19]. A similar phenomenon was also detected by Wu et al. for psoriasis in human [26]. These results indicated that all SNPs on the chip should be tested for identifying potential interaction effects. In contrast, testing for interactions between SNPs that have been previously identified in GWAS is not enough.

There were 11 sub-networks with at least three nodes that were identify by using the 421 pairs of interaction effects. Sub-network1 was the biggest one and most of the interactions in Sub-network1 occurred between GGAZ and GGA21. The two regions on GGA21 and GGAZ that may harbor genes important for testis growth and development spanned 572 kb and 9.5 Mb, respectively. Three genes on GGA21 (*SDHB*, *PARK7* and

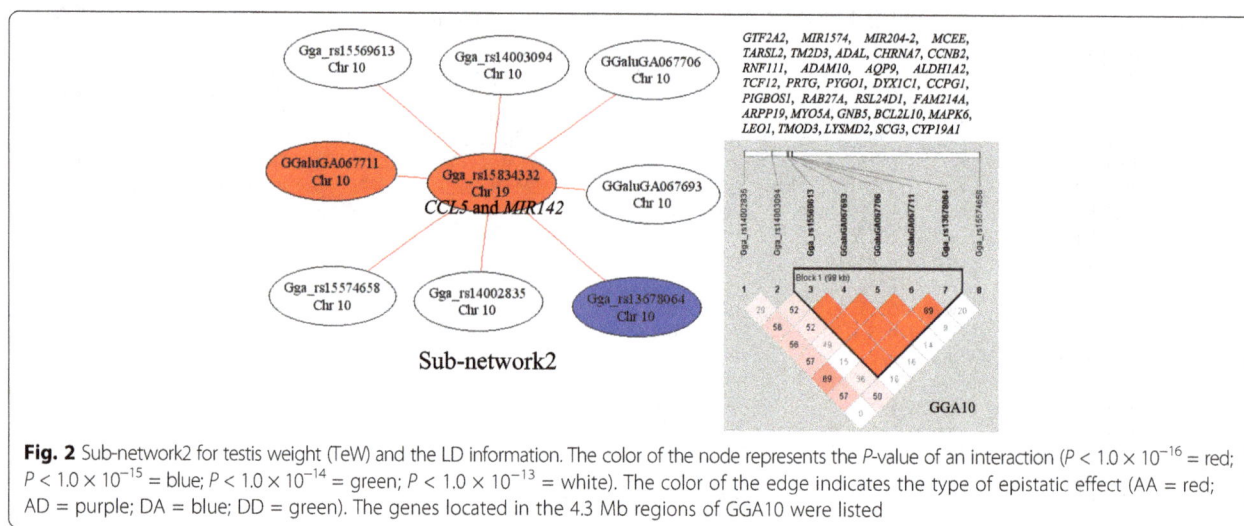

Fig. 2 Sub-network2 for testis weight (TeW) and the LD information. The color of the node represents the P-value of an interaction ($P < 1.0 \times 10^{-16}$ = red; $P < 1.0 \times 10^{-15}$ = blue; $P < 1.0 \times 10^{-14}$ = green; $P < 1.0 \times 10^{-13}$ = white). The color of the edge indicates the type of epistatic effect (AA = red; AD = purple; DA = blue; DD = green). The genes located in the 4.3 Mb regions of GGA10 were listed

VAMP3) and nine genes (AGTPBP1, CAMK4, CDC14B, FANCC, FBP1, GNAQ, PTCH1, ROR2 and STARD4) on GGAZ might be important for testis growth and development based on their annotated functions. The previous result indicated that the motility and viability of sperm were positively correlated with mitochondrial SDHB [27]. Therefore, SDHB may serve as a marker of sperm quality and male fertility [28]. PARK7 (DJ1) is highly expressed in human testes and has been shown to be essential for sperm maturation and fertilization [29–32]. VAMP3 has been shown to play an important role in the process of fertilization of sperms in pig [33]. In mice, AGTPBP1 was important for spermatogenesis, moreover, it was important for the survival of germ cells from the spermatocyte stage onward [34]. In mice and rats, the the CAMK4 gene encodes two proteins, Ca^2$^+$/calmodulin-dependent protein kinase IV (CaMKIV) and calspermin (CaS) [35–38]. CaMKIV is highly expressed in mouse testis and ovary and plays a essential role in male and female fertility [38–40]. CDC14B mutant mice were less fertile than the wild-type control [41]. Reduced fertility was reported for Fancc$^{-/-}$ mice [42]. The expression of some proteins, including FBP1, were altered and their functions may be damaged in infertile men with unilateral varicocele [43]. The results of

a previous study identified that Gnaq$^{d/d}$ male mice were subfertile [44]. The desert hedgehog (Dhh)-null mutant male mice had less mature sperm cells and lower numbers of Leydig cells (LCs), and Dhh played an important role in spermatogenesis by acting in a paracrine manner through the Ptch1 receptor component [45–47]. In mice, female Ror2$^{W749FLAG/W749FLAG}$ were fertile, however, Ror2$^{W749FLAG/W749FLAG}$ male mice showed a decreased in fertility [48]. StarD6 was testis-specific expressed which indicated that it may be important for fertility [49]. StarD6 is homology to StarD4, which indicated that StarD4 may have similar function as StarD6. In Sub-network2, eight SNPs on GGA10 all interacted with the SNP (Gga_rs15834332) on GGA19, which could be seen as the hub site of Sub-network2. Thus, Gga_rs15834332 may be the important node which interacts with a 4.3 Mb region on GGA10. The Gga_rs15834332 on GGA19 was located between CCL5 and MIR142. A total of 32 Refgenes were located in the 4.3 Mb region of GGA10. Among these genes, TCF12 was predicted as a target gene for miR-142-5p [50]. We also detected that TCF12 was the only gene that was predicted to be the target of miR-142-5p using three packages online. Therefore, it is proposed that miR-142-5p and TCF12 might work together to regulate the reproductive

Table 4 Target genes of miR-142-5p and miR-142-3p in the 4.3 Mb region on GGA10 in Sub-network2 for TeW predicted by three packages online

Gene symbol	Description	Position (Mb)	MiRNA	Packages
RNF111	Ring Finger Protein 111	6.32–6.36	miR-142-3p	Targetscan
TCF12	transcription factor 12	6.90–7.00	miR-142-5p	MIRDB, Targetscan PicTar
ARPP19	cAMP–regulated phosphoprotein, 19 kDa	8.23–8.24	miR-142-5p	Targentscan
MYO5A	myosin VA (heavy chain 12, myoxin)	8.24–8.33	miR-142-3p	Targentscan
MAPK6	Mitogen–Activated Protein Kinase 6	8.42– 8.45	miR-142-5p	Targentscan

Table 5 Significant epistatic effects on MeC

Chr1	Position1	Locus1	RS#	Chr2	Position2	Locus2	RS#	Test	P_value
2	73,383,598	Gga_rs14204534	rs14204534	13	13,041,051	Gga_rs14998703	rs14998703	AA	3.39×10^{-12}
2	73,383,598	Gga_rs14204534	rs14204534	13	13,361,236	Gga_rs14998801	rs14998801	AA	3.39×10^{-12}
2	73,383,598	Gga_rs14204534	rs14204534	13	13,343,436	Gga_rs15704596	rs15704596	AA	3.39×10^{-12}
2	73,420,901	Gga_rs14204566	rs14204566	13	13,041,051	Gga_rs14998703	rs14998703	AA	3.39×10^{-12}
2	73,420,901	Gga_rs14204566	rs14204566	13	13,361,236	Gga_rs14998801	rs14998801	AA	3.39×10^{-12}
2	73,420,901	Gga_rs14204566	rs14204566	13	13,343,436	Gga_rs15704596	rs15704596	AA	3.39×10^{-12}
2	73,507,376	Gga_rs16037701	rs16037701	13	13,041,051	Gga_rs14998703	rs14998703	AA	3.44×10^{-12}
2	73,507,376	Gga_rs16037701	rs16037701	13	13,361,236	Gga_rs14998801	rs14998801	AA	3.44×10^{-12}
2	73,507,376	Gga_rs16037701	rs16037701	13	13,343,436	Gga_rs15704596	rs15704596	AA	3.44×10^{-12}
2	73,559,761	GGaluGA153643	rs317095612	13	13,041,051	Gga_rs14998703	rs14998703	AA	3.44×10^{-12}
2	73,559,761	GGaluGA153643	rs317095612	13	13,361,236	Gga_rs14998801	rs14998801	AA	3.44×10^{-12}
2	73,559,761	GGaluGA153643	rs317095612	13	13,343,436	Gga_rs15704596	rs15704596	AA	3.44×10^{-12}
2	73,584,220	Gga_rs14204639	rs14204639	13	13,041,051	Gga_rs14998703	rs14998703	AA	4.47×10^{-12}
2	73,584,220	Gga_rs14204639	rs14204639	13	13,361,236	Gga_rs14998801	rs14998801	AA	4.47×10^{-12}
2	73,584,220	Gga_rs14204639	rs14204639	13	13,343,436	Gga_rs15704596	rs15704596	AA	4.47×10^{-12}

function of male broilers. Furthermore, *TCF12* was a partner of *TCF21*, which was detected as an important gene for testis growth and development in our previous GWAS result [19].

For MeC, 15 pairs of SNPs with significant epistatic effects were detected, which indicated an interaction between GGA2 and GGA13. It is proposed that *CDH12* on GGA2 and *WNT8A* on GGA13, which are both located in the same cadherin signaling pathway, may be important for bone growth. It had been shown that the pathway was involved in many biological processes, such as development, neurogenesis, cell adhesion, and inflammation, and also involved in many disease, such as cancer [51].

Conclusions

In the current study, a large number of epistatic interactions were found to be significantly associated with testis weight in chicken. It appears that miR-142-5p along with its target gene *TCF12*, and some other genes in GGA21 and GGAZ (*SDHB*, *PARK7*, *VAMP3*, *AGTPBP1*, *CAMK4*, *CDC14B*, *FANCC*, *FBP1*, *GNAQ*, *PTCH1*, *ROR2* and *STARD4*) might be important for testis growth and development. In contrast, very few significant epistatic interactions were identified for other carcass and growth traits. These results indicate that epistatic interaction may play very different roles in the control of phenotypic variation for different traits in chickens.

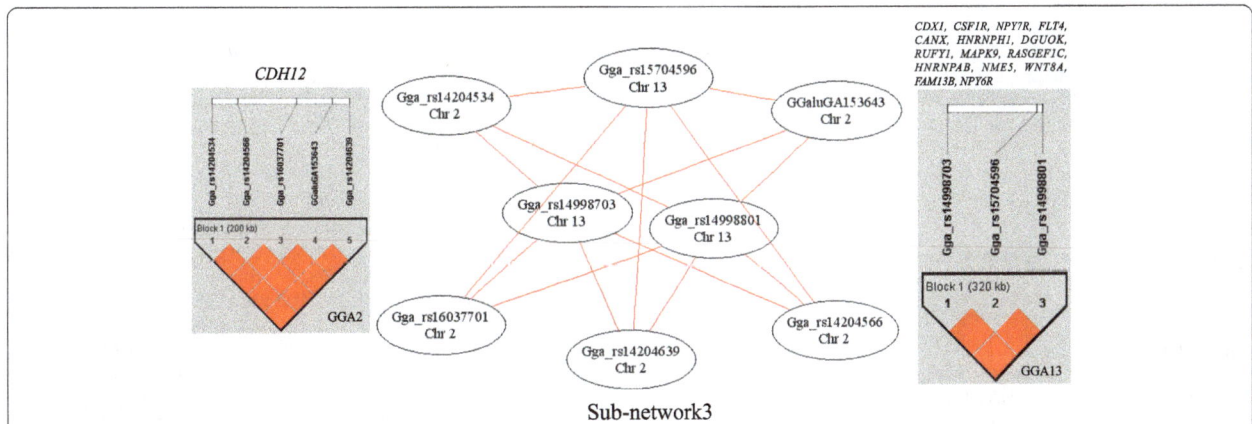

Fig. 3 Epistatic network among SNPs with significant epistatic effect on metatarsus circumference (MeC). A node represents a SNP. The chromosome in which the SNP is located is shown in the circle. A pair of SNPs connected by an edge had a significant interaction. The color of a node represent the P-value of the interaction ($P < 1.0 \times 10^{-12}$ = red; $P < 1.0 \times 10^{-11}$ = blue; $P < 1.0 \times 10^{-10}$ = green; $P < 1.0 \times 10^{-9}$ = white). The color of the edge indicates the type of epistatic effect (AA = red; AD = purple; DA = blue; DD = green). Genes in the LD blocks were listed

Additional files

> **Additional file 1: Figure S1.** Phenotypic distribution of the carcass and growth traits in the lean and fat lines, respectively. The red and light blue bars represent the traits in the lean and fat lines, respectively. The dark blue bars represent the overlap of the traits between the two lines. (PDF 322 kb)
>
> **Additional file 2: Table S1.** The 421 pairs of SNPs with significant interaction effects on testis weight (TeW). (XLSX 67 kb)
>
> **Additional file 3: Figure S2.** The phenotypic distributions of the four different genotype classes for the most significant AA effect between GGaluGA22768 on GGA3 and Gga_rs14722408 on GGA10. (PDF 122 kb)
>
> **Additional file 4: Figure S3.** Epistatic network among SNPs that affect testis weight (TeW). Each node represents a SNP. The chromosome in which a given SNP is located is shown within the circle. A pair of SNPs connected by an edge had a significant interaction. The colors of the nodes represent the P-value of an interaction ($P < 1.0 \times 10^{-16}$ = red; $P < 1.0 \times 10^{-15}$ = blue; $P < 1.0 \times 10^{-14}$ = green; $P < 1.0 \times 10^{-13}$ = white). The color of the edge indicates the type of epistatic effect (AA = red; AD = purple; DA = blue; DD = green). (PDF 1222 kb)
>
> **Additional file 5: Figure S4.** The LD blocks of the SNPs on every chromosome in Sub-network1 for testis weight (TeW). This information was used to simply Sub-network1. (PDF 619 kb)

Abbreviations
GO: Gene ontology; GWAS: Genome–wide association study; KEGG: Kyoto encyclopedia of genes and genomes; LD: Linkage disequilibrium; QTL: Quantitative trait loci; SNP: Single nucleotide polymorphism

Acknowledgements
The authors would like to acknowledge the members of the Poultry Breeding Group of the College of Animal Science and Technology at the Northeast Agricultural University for managing the birds and collecting the data.

Funding
This research was supported by the National 863 Project of China (No. 2013AA102501), the National Natural Science Foundation (No. 31301960), the China Postdoctoral Science Foundation (No. 2015 M581421), Heilongjiang Postdoctoral Financial Assistance (No. LBH-TZ0612), and the University Nursing Program for Young Scholars with Creative Talents in Heilongjiang Province (No. UNPYSCT-2015007).

Authors' contributions
HZ analyzed and interpreted the data, drafted, and wrote the manuscript. LLY, JQY and LK participated in the interpretation of data. XYZ and WN participated in the experiments. JR participated in the interpretation of data and helped write the manuscript. HL led the conception and design of the study and helped write the manuscript. All authors submitted comments on drafts, and read and approved the final manuscript.

Competing interests
The authors declare that they have no competing interests.

Author details
[1]Key Laboratory of Chicken Genetics and Breeding, Ministry of Agriculture, Harbin 150030, People's Republic of China. [2]Key Laboratory of Animal Genetics, Breeding and Reproduction, Education Department of Heilongjiang Province, Harbin 150030, People's Republic of China. [3]College of Animal Science and Technology, Northeast Agricultural University, Harbin 150030, People's Republic of China. [4]Department of Animal Science, Iowa State University, 2255 Kildee Hall, Ames, IA 50011, USA.

References
1. Huang W, Richards S, Carbone MA, Zhu D, Anholt RR, Ayroles JF, et al. Epistasis dominates the genetic architecture of drosophila quantitative traits. Proc Natl Acad Sci U S A. 2012;109(39):15553–9.
2. Rajon E, Plotkin JB. The evolution of genetic architectures underlying quantitative traits. Proc Biol Sci. 2013;280(1769):20131552.
3. Hibar DP, Stein JL, Jahanshad N, Kohannim O, Toga AW, McMahon KL, et al. Exhaustive search of the SNP-sNP interactome identifies epistatic effects on brain volume in two cohorts. Med Image Comput Comput Assist Interv. 2013;16(Pt 3):600–7.
4. Hibar DP, Stein JL, Jahanshad N, Kohannim O, Hua X, Toga AW, et al. Genome-wide interaction analysis reveals replicated epistatic effects on brain structure. Neurobiol Aging. 2015;36(Suppl 1):151–8.
5. Young KL, Graff M, North KE, Richardson AS, Bradfield JP, Grant SF, et al. Influence of SNP*SNP interaction on BMI in European American adolescents: findings from the National Longitudinal Study of adolescent health. Pediatr Obes. 2016;11(2):95–101.
6. Carlborg O, Kerje S, Schütz K, Jacobsson L, Jensen P, Andersson LA. Global search reveals epistatic interaction between QTL for early growth in the chicken. Genome Res. 2003;13(3):413–21.
7. Cheng HH, Zhang Y, Muir WM. Evidence for widespread epistatic interactions influencing Marek's disease virus viremia levels in chicken. Cytogenet Genome Res. 2007;117(1-4):313–8.
8. Carlborg O, Jacobsson L, Ahgren P, Siegel P, Andersson L. Epistasis and the release of genetic variation during long-term selection. Nat Genet. 2006; 38(4):418–20.
9. Li J, Zhang Q, Chen F, Yan J, Kim S, Wang L, et al. Genetic interactions explain variance in Cingulate Amyloid burden: an AV-45 PET genome-wide association and interaction study in the ADNI cohort. Biomed Res Int. 2015; 2015:647389.
10. Kogelman LJ, Kadarmideen HN. Weighted Interaction SNP Hub (WISH) network method for building genetic networks for complex diseases and traits using whole genome genotype data. BMC Syst Biol. 2014;8(Suppl 2):5.
11. Ali AA, Khatkar MS, Kadarmideen HN, Thomson PC. Additive and epistatic genome-wide association for growth and ultrasound scan measures of carcass-related traits in Brahman cattle. J Anim Breed Genet. 2015;132(2): 187–97.
12. Kramer LM, Ghaffar MA, Koltes JE, Fritz-Waters ER, Mayes MS, Sewell AD, et al. Epistatic interactions associated with fatty acid concentrations of beef from angus sired beef cattle. BMC Genomics. 2016;17(1):891.
13. Hu G, Wang SZ, Wang ZP, Li YM, Li H. Genetic epistasis analysis of 10 peroxisome proliferator-activated receptor γ-correlated genes in broiler lines divergently selected for abdominal fat content. Poult Sci. 2010;89(11):2341–50.
14. Ek W, Marklund S, Ragavendran A, Siegel P, Muir W, Carlborg O. Generation of a multi-locus chicken introgression line to study the effects of genetic interactions on metabolic phenotypes in chickens. Front Genet. 2012;3:29.
15. Li F, Hu G, Zhang H, Wang S, Wang Z, Li H. Epistatic effects on abdominal fat content in chickens: results from a genome-wide SNP-SNP interaction analysis. PLoS One. 2013;8(12):e81520.
16. Zhang H, Zhang YD, Wang SZ, Liu XF, Zhang Q, Tang ZQ, et al. Detection and fine mapping of quantitative trait loci for bone traits on chicken chromosome one. J Anim Breed Genet. 2010;127(6):462–8.
17. Zhang H, Hu X, Wang Z, Zhang Y, Wang S, Wang N, et al. Selection signature analysis implicates the PC1/PCSK1 region for chicken abdominal fat content. PLoS One. 2012;7(7):e40736.
18. Ma L, Runesha HB, Dvorkin D, Garbe JR, Da Y. Parallel and serial computing tools for testing single-locus and epistatic SNP effects of quantitative traits in genome-wide association studies. BMC Bioinformatics. 2008;9:315.
19. Zhang H, Na W, Zhang HL, Wang N, ZQ D, Wang SZ, et al. TCF21 is related to testis growth and development in broiler chickens. Genet Sel Evol. 2017;49(1):25.
20. Sarabia Fragoso J, Pizarro Díaz M, Abad Moreno JC, Casanovas Infesta P, Rodriguez-Bertos A, Barger K. Relationships between fertility and some parameters in male broiler breeders (body and testicular weight, histology and immunohistochemistry of testes, spermatogenesis and hormonal levels). Reprod Domest Anim. 2013;48(2):345–52.
21. Chubb C. Genes regulating testis size. Biol Reprod. 1992;47(1):29–36.
22. Lüpold S, Linz GM, Rivers JW, Westneat DF, Birkhead TR. Sperm competition selects beyond relative testes size in birds. Evolution. 2009;63(2):391–402.

23. Soulsbury CD. Genetic patterns of paternity and testes size in mammals. PLoS One. 2010;5(3):e9581.

24. Vizcarra JA, Kirby JD, Kreider DL. Testis development and gonadotropin secretion in broiler breeder males. Poult Sci. 2010;89(2):328–34.

25. John P. Testes development and fertility. Aviagen brief. 2008;0608-AVN-011.

26. Wu X, Dong H, Luo L, Zhu Y, Peng G, Reveille JD, et al. A novel statistic for genome-wide interaction analysis. PLoS Genet. 2010;6(9):e1001131.

27. Xue XP, Shang XJ, Fu J, Chen YG, Shi YC. Detection and significance of succinate dehydrogenase of sperm mitochondria. Zhonghua Nan Ke Xue. 2003;9(8):601–3.

28. Rahman MS, Kwon WS, Lee JS, Yoon SJ, Ryu BY, Pang MG. Bisphenol-a affects male fertility via fertility-related proteins in spermatozoa. Sci Rep. 2015;5:9169.

29. Nagakubo D, Taira T, Kitaura H, Ikeda M, Tamai K, Iguchi-Ariga SM, et al. DJ-1, a novel oncogene which transforms mouse NIH3T3 cells in cooperation with ras. Biochem Biophys Res Commun. 1997;231(2):509–13.

30. Klinefelter GR, Laskey JW, Ferrell J, Suarez JD, Roberts NL. Discriminant analysis indicates a single sperm protein (SP22) is predictive of fertility following exposure to epididymal toxicants. J Androl. 1997;18(2):139–50.

31. Ooe H, Taira T, Iguchi-Ariga SM, Ariga H. Induction of reactive oxygen species by bisphenol a and abrogation of bisphenol A-induced cell injury by DJ-1. Toxicol Sci. 2005;88(1):114–26.

32. Hao LY, Giasson BI, Bonini NM. DJ-1 is critical for mitochondrial function and rescues PINK1 loss of function. Proc Natl Acad Sci U S A. 2010;107(21):9747–52.

33. Tsai PS, Garcia-Gil N, van Haeften T, Gadella BM. How pig sperm prepares to fertilize: stable acrosome docking to the plasma membrane. PLoS One. 2010;5(6):e11204.

34. Kim N, Xiao R, Choi H, Jo H, Kim JH, et al. Abnormal sperm development in pcd(3J)–/– mice: the importance of Agtpbp1 in spermatogenesis. Mol Cells. 2011;31(1):39–48.

35. Ono T, Means AR. Calspermin is a testis specific calmodulin-binding protein closely related to Ca2+/calmodulin-dependent protein kinases. Adv Exp Med Biol. 1989;255:263–8.

36. Ono T, Slaughter GR, Cook RG, Means AR. Molecular cloning sequence and distribution of rat calspermin, a high affinity calmodulin-binding protein. J Biol Chem. 1989;264(4):2081–7.

37. Ohmstede CA, Bland MM, Merrill BM, Sahyoun N. Relationship of genes encoding Ca2+/calmodulin-dependent protein kinase Gr and calspermin: a gene within a gene. Proc Natl Acad Sci U S A. 1991;88(13):5784–8.

38. JY W, Ribar TJ, Cummings DE, Burton KA, McKnight GS, Means AR. Spermiogenesis and exchange of basic nuclear proteins are impaired in male germ cells lacking Camk4. Nat Genet. 2000;25(4):448–52.

39. JY W, Gonzalez-Robayna IJ, Richards JS, Means AR. Female fertility is reduced in mice lacking Ca2+/calmodulin-dependent protein kinase IV. Endocrinology. 2000;141(12):4777–83.

40. JY W, Ribar TJ, Means AR. Spermatogenesis and the regulation of ca(2+)-calmodulin-dependent protein kinase IV localization are not dependent on calspermin. Mol Cell Biol. 2001;21(17):6066–70.

41. Wei Z, Peddibhotla S, Lin H, Fang X, Li M, Rosen JM, et al. Early-onset aging and defective DNA damage response in Cdc14b-deficient mice. Mol Cell Biol. 2011;31(7):1470–7.

42. Whitney MA, Royle G, Low MJ, Kelly MA, Axthelm MK, Reifsteck C, et al. Germ cell defects and hematopoietic hypersensitivity to gamma-interferon in mice with a targeted disruption of the Fanconi anemia C gene. Blood. 1996;88(1):49–58.

43. Agarwal A, Sharma R, Durairajanayagam D, Ayaz A, Cui Z, Willard B, et al. Major protein alterations in spermatozoa from infertile men with unilateral varicocele. Reprod Biol Endocrinol. 2015;13:8.

44. Babwah AV, Navarro VM, Ahow M, Pampillo M, Nash C, Fayazi M, et al. GnRH neuron-specific ablation of Gαq/11 results in only partial inactivation of the Neuroendocrine-reproductive Axis in both male and female mice: in vivo evidence for Kiss1r-coupled Gαq/11-independent GnRH secretion. J Neurosci. 2015;35(37):12903–16.

45. Bitgood MJ, Shen L, McMahon AP. Sertoli cell signaling by desert hedgehog regulates the male germline. Curr Biol. 1996;6(3):298–304.

46. Yao HH, Whoriskey W, Capel B. Desert hedgehog/patched 1 signaling specifies fetal Leydig cell fate in testis organogenesis. Genes Dev. 2002;16(11):1433–40.

47. Morales CR, Fox A, El-Alfy M, Ni X, Argraves WS. Expression of Patched-1 and smoothened in testicular meiotic and post-meiotic cells. Microsc Res Tech. 2009;72(11):809–15.

48. Raz R, Stricker S, Gazzerro E, Clor JL, Witte F, Nistala H, et al. The mutation ROR2W749X, linked to human BDB, is a recessive mutation in the mouse, causing brachydactyly, mediating patterning of joints and modeling recessive Robinow syndrome. Development. 2008;135(9):1713–23.

49. Soccio RE, Adams RM, Romanowski MJ, Sehayek E, Burley SK, Breslow JL. The cholesterol-regulated StarD4 gene encodes a StAR-related lipid transfer protein with two closely related homologues, StarD5 and StarD6. Proc Natl Acad Sci U S A. 2002;99(10):6943–8.

50. Liao R, Sun J, Zhang L, Lou G, Chen M, Zhou D, et al. MicroRNAs play a role in the development of human hematopoietic stem cells. J Cell Biochem. 2008;104(3):805–17.

51. Nelson WJ, Nusse R. Convergence of Wnt, beta-catenin, and cadherin pathways. Science. 2004;303(5663):1483–7.

Artificial selection for odor-guided behavior in *Drosophila* reveals changes in food consumption

Elizabeth B. Brown[1], Cody Patterson[1], Rayanne Pancoast[1,2] and Stephanie M. Rollmann[1]*

Abstract

Background: The olfactory system enables organisms to detect chemical cues in the environment and can signal the availability of food or the presence of a predator. Appropriate behavioral responses to these chemical cues are therefore important for organismal survival and can influence traits such as organismal life span and food consumption. However, understanding the genetic mechanisms underlying odor-guided behavior, correlated responses in other traits, and how these constrain or promote their evolution, remain an important challenge. Here, we performed artificial selection for attractive and aversive behavioral responses to four chemical compounds, two aromatics (4-ethylguaiacol and 4-methylphenol) and two esters (methyl hexanoate and ethyl acetate), for thirty generations.

Results: Artificial selection for odor-guided behavior revealed symmetrical responses to selection for each of the four chemical compounds. We then investigated whether selection for odor-guided behavior resulted in correlated responses in life history traits and/or food consumption. We found changes in food consumption upon selection for behavioral responses to aromatics. In many cases, lines selected for increased attraction to aromatics showed an increase in food consumption. We then performed RNA sequencing of lines selected for responses to 4-ethylguaiacol to identify candidate genes associated with odor-guided behavior and its impact on food consumption. We identified 91 genes that were differentially expressed among lines, many of which were associated with metabolic processes. RNAi-mediated knockdown of select candidate genes further supports their role in odor-guided behavior and/or food consumption.

Conclusions: This study identifies novel genes underlying variation in odor-guided behavior and further elucidates the genetic mechanisms underlying the interrelationship between olfaction and feeding.

Keywords: *Drosophila*, Olfaction, Feeding, RNA sequencing, RNA interference

Background

Sensory systems enable organisms to interact with the environment. Whether avoiding a predator, seeking a mate, or searching for food, these behaviors are mediated through the sensory detection and subsequent processing of environmental cues. In the case of olfaction, aversive and attractive olfactory cues are used, for instance, by organisms to locate and evaluate food resources. Moreover, olfactory cues can also influence other traits, such as organismal life span and starvation resistance [1–5]. Given their importance to survival and reproduction, behavioral geneticists have long sought to understand the proximate mechanisms underlying olfactory behavior, its interrelationship with other traits, and the mechanisms governing or constraining its phenotypic evolution. Thus, remarkable progress has been made in uncovering the neural circuitry underlying the detection of chemical cues [6–8]. However, understanding how sensory input is processed to result in divergent behavioral responses (olfactory attraction or aversion) and the genetic mechanisms that underlie its association with traits such as feeding and life span remains an important challenge.

* Correspondence: stephanie.rollmann@uc.edu
[1]Department of Biological Sciences, University of Cincinnati, Cincinnati, OH 45221-0006, USA
Full list of author information is available at the end of the article

Drosophila melanogaster has emerged as a model system for investigating the genetic factors underlying the detection and discrimination of olfactory cues [6, 8]. Odorants are detected by odorant receptors expressed in olfactory sensory neurons (OSNs) located on either of two olfactory organs, the third segment of the antenna or the maxillary palp. Each OSN typically expresses a single odorant receptor (OR) type and projects its axons to a distinct glomerulus in the antennal lobe [9–13]. These odorant receptors comprise a family of sixty genes that together with the highly conserved co-receptor, *Orco*, are believed to form ligand-gated ion channels [10, 14–18]. First order OSNs form synaptic connections with second order projection neurons, which extend their axons to the mushroom body and lateral horn regions of the brain [19–22]. The resulting spatial and temporal representation of glomerular activity allows for the discrimination among the diverse odors present in the environment [10, 23, 24]. More recently, studies have also focused on the identification of genes associated with variation in olfactory behavior through the use of candidate gene association studies [25–29] as well as genome-wide association (GWA) mapping [30–32]. These analyses identified genes that form a pleiotropic network of interactions and are largely involved in nervous system development and function.

In addition, studies in *Drosophila* have focused on examining the interrelationship between olfaction, feeding, and life history traits, including longevity and starvation resistance [3, 4, 33, 34]. Changes in sensory perception, for example, through silencing of the co-receptor *Orco*, resulted in extended longevity and increased resistance to starvation. Altered metabolism was also observed, with sex-specific changes in triglyceride levels [3]. Additionally, the olfactory system is also associated with mediating feeding [35–38]. A starvation-dependent shift in OSN sensitivity via short neuropeptide F, a fly homolog of neuropeptide Y (NPY) [39], and insulin signaling regulates food search behavior [40]. Moreover, neurons expressing neuropeptide F (dNPF) respond to food odors, increasing in activity with increased food-odor attractiveness [41].

Here, we conduct artificial selection experiments to investigate the mechanisms underlying olfactory behavior and correlated responses with other traits. We independently selected for attractive and aversive behavioral responses to two aromatic compounds (4-ethylguaiacol and 4-methylphenol) and two esters (ethyl acetate and methyl hexanoate) for thirty generations. These odorants are a natural byproduct of yeast fermentation [42–45], a component of the feeding and breeding substrate of *Drosophila*. We then used these artificially selected lines to investigate potential relationships with other traits

and found odor- and sex-specific differences in food consumption. We subsequently performed RNA sequencing (RNA-seq) experiments to identify changes in gene expression using lines selected for differences in behavioral responses to the odorant 4-ethylguaiacol and examined the role of a subset of these candidate genes in mediating odor-guided behavior and food consumption.

Results

Selection for odor-guided behavior

To understand the genetic mechanisms underlying shifts in odor-guided behavior, we performed artificial selection experiments using the Flyland population, an outbred population derived from the *Drosophila* Genetic Reference Panel [46]. We measured its behavioral responses to four odorants, 4-ethylguaiacol (4EG), 4-methylphenol (4MP), methyl hexanoate (MH), and ethyl acetate (EA), at several different concentrations. We observed concentration-specific differences in behavioral responses to each odorant tested (Fig. 1a-d). Based on these results, we independently generated three replicate high and low lines selected for attractive or aversive behavioral responses to each odorant, as well as three replicate unselected (control) lines. After seven generations, mean high and low behavioral responses to all odorants significantly diverged relative to the controls, and plateaued at approximately generation ten (Fig. 1e-h, Additional file 1: Table S1). The response to selection was symmetrical, with responses of the controls intermediate between the high and low lines (Fig. 1i-l). No significant changes in locomotion were observed among the selection regimes (Additional file 2: Figure S1).

Correlated responses to selection for odor-guided behavior

Changes in olfactory perception have been shown to affect life history traits, such as longevity and starvation resistance [3–5]. To test whether artificial selection for odor-guided behavior resulted in correlated responses in these traits, we measured longevity and starvation resistance across all four selection regimes. Overall no significant differences in life span were observed (Additional file 3: Figure S2; Additional file 1: Table S3). Measurements of starvation resistance, a trait often positively correlated with life span [47], also revealed no differences among lines in all four selection regimes (Additional file 4: Figure S3; Additional file 1: Table S3). One exception to these results was found for measurements of life span among the MH selected lines. Females selected for increased attraction to MH had a significant increase in life span relative to the control and low lines (Additional file 3: Figure S2).

Alterations in chemosensory perception can also influence feeding behavior [35–38]. We therefore investigated

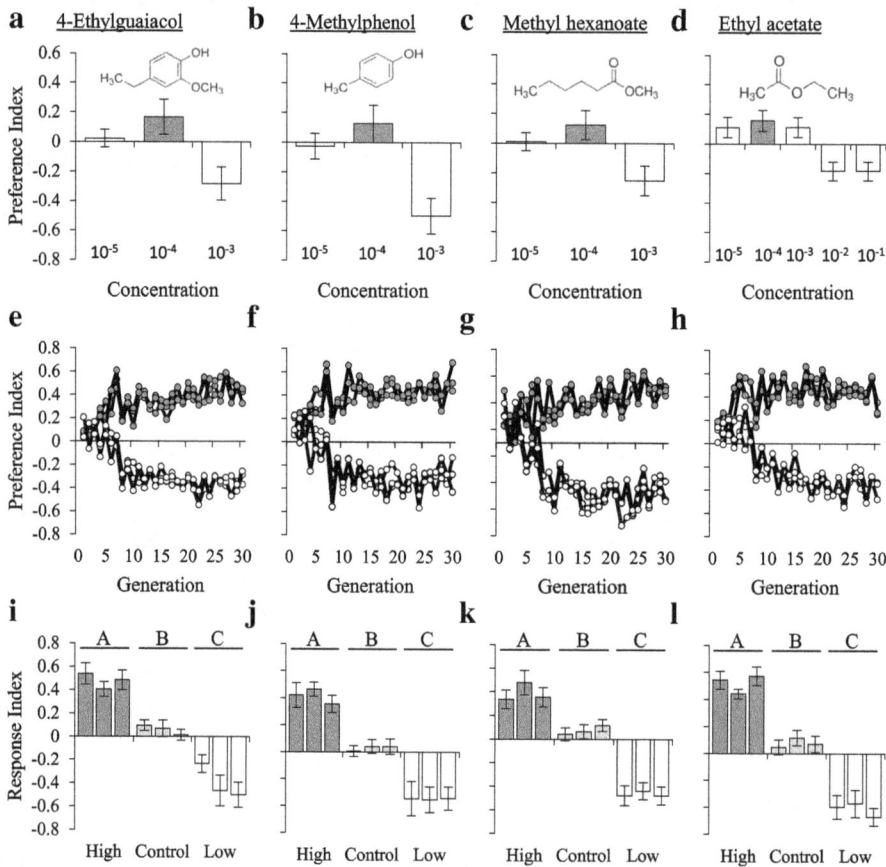

Fig. 1 Odor-guided behavioral responses to four odorants. Row 1: Dose response curves of the base population to four odorants. Preference indexes for (**a**) 4-ethylguaiacol, (**b**) 4-methylphenol, (**c**) methyl hexanoate, and (**d**) ethyl acetate are shown. Gray bars indicate the concentration used in the artificial selection experiments. Data shown are means ± SE. N = 20. Row 2: Phenotypic response to artificial selection for odor-guided behavioral responses to (**e**) 4-ethylguaiacol, (**f**) 4-methylphenol, (**g**) methyl hexanoate, and (**h**) ethyl acetate. Mean preference index at each generation for the three replicate high selected lines (gray filled dots) and three low selected lines (white filled dots) are shown. N = 6. Row 3: Symmetrical response to selection for (**i**) 4-ethylguaiacol, (**j**) 4-methylphenol, (**k**) methyl hexanoate, and (**l**) ethyl acetate. Data shown are means ± SE. Letters indicate P < 0.05 using Tukey's post hoc test. N = 20

whether changes in odor-guided behavior were associated with changes in food consumption using the CAFE assay [48]. When fed a standard liquid diet of sucrose and yeast extract, we observed significant sex-specific differences in food consumption among the 4EG and 4MP selected regimes in which high lines generally consumed significantly more than low and control lines (Fig. 2). In the case of lines selected for responses to 4EG, food consumption was significantly increased in males in the high selected lines, with the same trend apparent in females. For lines selected for responses to 4MP, food consumption was significantly increased in females in the high selected lines, with again a similar trend observed in males. This pattern of food consumption was not observed for lines selected for responses to the esters, MH and EA. In general, no significant changes in food consumption were observed among these lines, with the exception of males selected for responses to EA, in which both high and

low lines showed increased consumption relative to the control.

Food consumption of the 4EG and 4MP selected lines were again tested, and differences in consumption observed using two additional diets. First, we replaced yeast extract with live *Brettanomyces* yeast. Regardless of sex, all 4EG and 4MP selected lines increased consumption overall. Moreover, consistent with previous results, we generally observed increased food consumption in high lines relative to low and control lines (Additional file 5: Figure S4a, b). Next, we fed flies a diet of sucrose only. We again observed significant differences in food consumption for both sexes, with high lines consuming more food than the low and control lines (Additional file 5: Figure S4c, d). Also, food consumption in the control lines remained below that of the low lines. Finally, we used a binary-choice assay to investigate whether sucrose supplemented with either 4EG or 4MP conferred differences in food preference [45]. Flies were

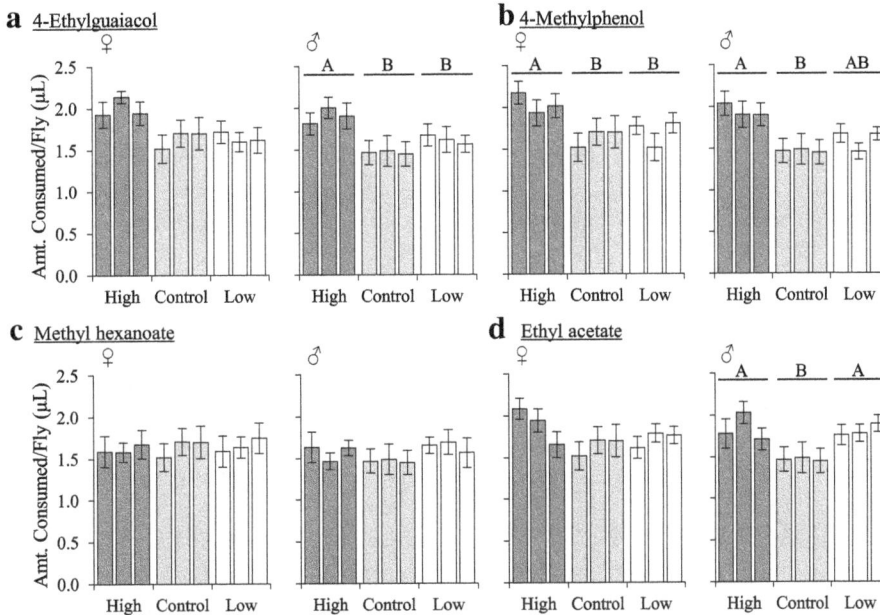

Fig. 2 Food consumption measurements of lines selected for increased and decreased behavioral responses to (**a**) 4-ethylguaiacol, (**b**) 4-methylphenol, (**c**) methyl hexanoate, and (**d**) ethyl acetate using the CAFE assay. The total amount of yeast extract consumed was measured. Data shown are means ± SE for females (left) and males (right). Letters indicate $P < 0.05$ using Tukey's post hoc test. $N = 12$

given a choice between food with or without supplementation of odorant. We hypothesized we would observe an increase in preference for food supplemented with odorant in the high lines and reduced consumption in the low lines, given their attraction/aversion to each odorant, respectively. Interestingly, we found no significant differences in food preference, such that all selected lines equally, albeit mildly, preferred food supplemented with either 4EG or 4MP (Additional file 5: Figure S4e, f). Together, these results suggest that selection for behavioral responses to these aromatics results in correlated responses in feeding behavior with the high lines typically consuming more, irrespective of diet, but importantly that selection for increased attraction and aversion does not directly translate into corresponding changes in food consumption.

The changes in food consumption may result from changes in metabolic processes or from general differences in body mass [49]. To investigate whether selection for odor-guided behavior was also associated with changes in these traits, we measured dry mass in the 4EG and 4MP selected lines. For 4EG selected lines, no significant differences in body mass were observed (Additional file 6: Figure S5a). However, for the 4MP selected lines, the high selected lines weighed significantly more than the other lines (Additional file 6: Figure S5b). Measurements of metabolic processes (triglycerides, glucose, and glycogen) revealed no significant differences after adjustment for dry mass (Additional file 6: Figure S5c-h). One exception being males selected for behavioral responses to 4MP,

in which the low lines had significantly lower glucose levels than the control lines, but neither of which were significantly different from the high lines. In short, for the 4EG and 4MP selection regimes, high lines consistently consumed more than low lines, however, these differences in food consumption cannot be clearly attributed to changes in the metabolic traits measured here.

Transcriptional response to selection for odor-guided behavior

To examine the genetic mechanisms underlying differences in odor-guided behavior and the correlated responses on feeding, we conducted RNA-seq analyses on lines selected for divergent responses to 4EG. This selection regime was chosen for further analysis because differences were observed between the high and low selected lines for both odor-guided behavior and food consumption, independent of mass. Differential gene expression was examined in whole heads to include in the analysis genes associated with the peripheral olfactory organs, the brain, as well as feeding organs. In total, we obtained 780,640,808 100 bp reads from 18 cDNA libraries. Of 753,144,236 reads that passed quality filtering, 95.32% could be aligned to the Drosophila genome, and 98.64% of the aligned reads mapped to uniquely (Additional file 1: Table S4, Additional file 1: Table S5). Of the 17,471 annotated genes [50], 9238 had at least one read per million in at least half the samples. This set of genes was used for subsequent statistical analyses.

Differential gene expression was evaluated between the different combinations of selection regimes, i.e., high vs. control, low vs. control, and high vs. low. A total of 43, 32, and 45 genes were significantly differentially expressed in the high vs. control, low vs. control, and high vs. low comparisons, respectively (Additional file 4: Fig. 3a; Additional file 1: Table S6–8). Eleven genes were shared between high vs. low and low vs. control comparisons, thereby highlighting genes that may contribute to aversive behavioral responses to 4EG. Nine genes were shared between high vs. control and low vs. control comparisons which may consist of genes that are responsible for generalized changes in behavior, regardless of hedonic value, as these genes were differentially expressed regardless of selection for high or low behavioral responses. Nine genes were also shared between the high vs. control and high vs. low comparisons and could give insights into the genes responsible for attraction to 4EG and its link with food consumption (Fig. 3b). Finally, we chose to focus on the high vs. low differentially expressed genes for subsequent analyses. We assessed whether there was overrepresentation of Gene Ontology (GO) terms. No significant overrepresentation for any GO terms were found (Additional file 7: Figure S6). However, of the 45

significantly differentially expressed genes, 13 were annotated for the GO term metabolic process and four for response to stimulus.

Functional tests of candidate genes

To further assess the candidate genes significantly differentially expressed between high and low selected lines for their contribution to changes in odor-guided behavior and/or food consumption, we used RNA interference [51, 52] and the GAL4/UAS system [53] to knockdown gene expression. We selected a subset of 16 genes based on their statistical level of significance, their identification across treatments, and their GO classification. Since OSNs and projection neurons are two primary relay stations at which olfactory information is processed, we knocked down gene expression in OSNs expressing *Or71a*, an odorant receptor shown to be tuned to 4EG using the *Or71a*-GAL4 driver line, and in projection neurons of the antennal lobe using the GH146-GAL4 driver line [54]. We observed significant differences in odor-guided behavior for seven of the genes tested (Fig. 4a.b; Additional file 1: Table S9; *CG6044*, *Cyp6a2*, *Egfr*, *grp*, *GstD2*, *tej*, *VepD*), and differences in food consumption for 10 genes (Fig. 4c, d; Additional file 1: Table S9; *Cdc6*, *CG6044*, *Cyp6a2*, *Egfr*, *grp*, *SoYb*, *Spn42Dc*, *tej*, *Tret1–2*, *VepD*). RNAi targeting of six of these genes

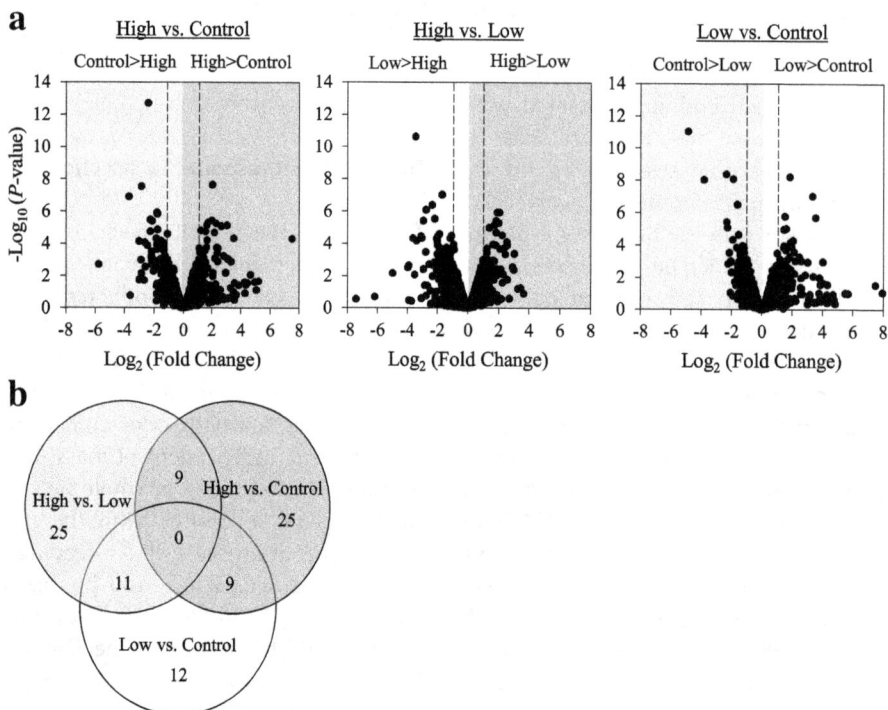

Fig. 3 RNA-seq analyses of genes differentially expressed among lines selected for 4-ethylguaiacol. (**a**) Volcano-plot of RNA-seq results for all pairwise comparisons. For each comparison, genes to the left of the > symbol are upregulated, while genes to the right are downregulated. Vertical dashed lines represent a two-fold cutoff. (**b**) Venn diagram illustrates the differentially expressed genes that overlap between each comparison

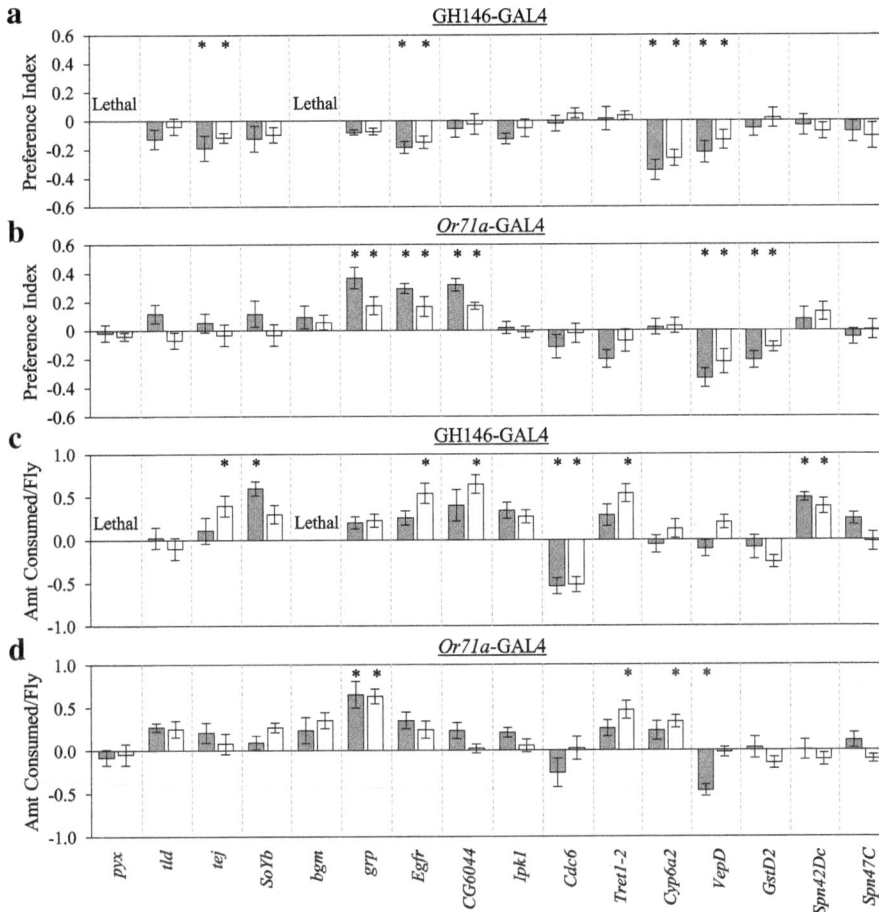

Fig. 4 Functional tests of candidate genes and their effects on (**a**), (**b**) odor-guided behavioral responses to 4-ethylguaiacol and (**c**), (**d**) food consumption. For each trait, the effects of RNAi-mediated knockdown of 16 candidate genes in either projection neurons of the antennal lobe (GH146-GAL4) or in Or71a-expressing neurons (Or71a-GAL4) were examined in both females (gray bars) and males (white bars). Values are deviations from the control. Error bars indicate SE. *: $P < 0.05$. $N = 10$

resulted in significant differences in both odor-guided behavior and food consumption (*CG6044, Cyp6a2, Egfr, grp, tej, VepD*). The effects on behavior depended on sex, and the gene and/or neuronal population targeted. For example, expression of *grp*-RNAi in *Or71a*-expressing OSNs resulted in increased behavioral responses to 4EG as well as an increase in food consumption. However, a similar positive response was not always observed, as was the case with RNAi-mediated knockdown of *Egfr* and *tej* in projection neurons.

Discussion
Selection for odor-guided behavior
Artificial selection is a powerful method that allows us to examine divergence in odor-guided behavioral responses and how selection for attraction and aversion may result in correlated responses with other traits. Using this approach, selection lines were generated for odor-guided behavioral responses to four odorants, two

aromatics and two esters. The response to selection was symmetric, such that both lines selected for attractive and aversive behavioral responses significantly differed relative to control. We then used these artificially selected lines to investigate correlated responses with odor-guided behavior and observed changes in food consumption associated with selection for behavioral responses to aromatics.

These aromatic compounds are detected by receptors expressed in olfactory sensory neurons of the maxillary palp. Detection of olfactory cues via the maxillary palp has been posited to mediate taste enhancement [55], with OSNs from the palp projecting to both the antennal lobe as well as the subesophageal ganglion, the primary taste center of the brain [56]. More specifically, these odorants (4EG and 4MP) bind to odorant receptor 71a (*Or71a*), which is expressed in palp OSNs [45, 57, 58]. Odor-evoked responses of *Or71a*-expressing neurons are of particular interest because previous studies suggest that detection of ethylphenols, including 4EG, by these neurons serves as a proxy for the detection of dietary antioxidants. Moreover, thermogenetic activation of these neurons via expression of dTRPA1 resulted in changes in feeding [45]. Thus, together with previous findings, our study sets the stage for future dissection of how changes in the maxillary palp may affect these behaviors.

In this study, the bidirectional response to selection for behavioral responses to aromatics resulted in an asymmetrical effect on food consumption. Significant increases in food consumption were typically observed in the high lines, relative to the low and control lines, irrespective of diet. Furthermore, we hypothesized that olfactory aversion would be associated with reduced food consumption, particularly when food was supplemented with the 'aversive' odorant. In point of fact, this was not the case. Previous studies have suggested that the context in which an odor is detected can be an important modulator of behavior [59]. For example, CO_2, a chemical compound emitted from stressed flies, results in robust aversive behavioral responses [60]. However, when this odorant is paired with food odors, the aversive response is suppressed [61, 62]. Such context dependent behavior is modulated at multiple levels of the olfactory circuitry, from the periphery to the brain. This study suggests that the mechanisms underlying olfactory behavior and food consumption are partially independent and again suggests that the behaviors may be context dependent.

Finally, correlated changes in food consumption may result from changes in metabolism or body mass [49]. Yet, for both selection regimes we observed little to no concomitant shift in traits associated with metabolism. It is possible that correlated changes in food consumption are indeed associated with changes in the metabolic traits measured here, but not at levels sufficient for detection. Alternatively, potential differences in other metabolic traits not measured in this study, such as metabolic rate, could underlie the observed differences in food consumption [63]. In the case of dry mass, a difference among lines was only observed in the 4MP selection regime despite changes in both food consumption and odor-guided behavioral responses being observed in both 4EG and 4MP selection regimes. This suggests that the mechanisms underlying variation in responses to these odorants may at least partially differ.

Transcriptional response to selection for odor-guided behavior

To further understand the mechanisms underlying the divergence in odor-guided behavior and its association with food consumption, we conducted an RNA-seq experiment using lines selected for attractive and aversive behavioral responses to 4EG. We hypothesized that we would identify odorant receptor and odorant binding protein genes, as these have been previously associated with variation in odor-guided behavior [25, 26, 28, 29, 64], most notably Or71a, because 4EG is a ligand of this receptor [45]. Contrary to our expectations, however, we did not observe differences in expression of Or71a, nor in any other odorant receptor or odorant binding protein genes. Interestingly, however, we did identify genes (Cyp6a8, Cyp6a20, Cyp6a2, Cyp4d1, GstD2) belonging to the cytochrome P450 and glutathione S-transferase families, which have been implicated in odor signal termination [65]. The rapid degradation of odors, and subsequent termination of the odor signal, enables an organism to appropriately behaviorally respond to volatile changes in the environment. Therefore, changes in the expression of these genes may alter odor perception. Moreover, of the 91 differentially expressed genes identified in our study, 19 were also identified in genome-wide association analyses (GWA) on odor-guided behavior (ACXD, bgm, Cdc6, CG5773, CG8785, CG9616, CG10064, CG10205, CG12374, CG14608, CG31809, CG34273, CG43448, Cpr66D, Cyp6a8, Egfr, Hel89B, luna, pwn) [30–32]. These overlapping candidate genes were identified despite using different odorants with varying valence, thereby suggesting that they indeed may function in odor-guided behavior regardless of odor intensity or hedonic value. Finally, our analyses revealed novel genes not previously implicated in mediating odor-guided behavior and afford new insights into the genetic mechanisms underlying these behavioral responses.

Given the observed differences in food consumption, we also hypothesized that we would identify genes involved in the insulin signaling and dNPF pathways, as both pathways are known to regulate multiple aspects of feeding behavior [66–69]. However, no differentially expressed genes were associated directly with these pathways. This can possibly be attributed to their high degree of conservation [39, 70], given that many of these genes are selectively constrained [71, 72]. However, our analyses did reveal differential expression of genes involved in metabolism (bgm, CG2469, CG10116, CG10962, CG33958, Cyp4d1, GstD2, HIP, Ipk1, phr, SP1029, tej, tld). Our analyses also revealed two genes identified in GWA analyses to be associated with variation in food consumption (CG43448, Egfr) [73], and three that overlap with the genomic response to selection for increased feeding (CG14696, Egfr, tld) [74]. The identification of these candidate genes in our study raises the possibility that they function at the interface between olfaction and feeding.

Functional assessment of candidate genes

We functionally assessed the effects of 16 candidate genes in Or71a-expressing neurons and projection neurons of the olfactory system. RNAi mediated targeting of eleven candidate genes resulted in changes in olfactory and/or food consumption. The remaining five genes did not significantly influence behavior. We cannot exclude the possibility, however, that these genes influence these traits through other mechanisms not tested in this study. Of those that did influence behavior in our study, several genes were of note. RNAi-targeting of Epidermal growth factor receptor (Egfr) resulted in a significant decrease in attraction when it was knocked down in projection neurons, but an opposite effect in Or71a-expressing neurons. This gene plays an extensive role in development, such as in cell fate specification [75–78]. Additionally, it has been implicated in GWA studies of Drosophila olfactory and feeding behaviors [32, 73] as well in mammalian obesity [79, 80]. Moreover, RNAi-mediated knockdown of Trehalose transporter 1–2 (Tret1–2) resulted in increased food consumption in males. A structurally similar gene, Tret1–1, regulates the release of trehalose, the primary sugar in insect hemolymph [81]. Although, Tret1–2 does not function in trehalose transport [81], our results suggest a role for Tret1–2 in mediating food consumption.

Finally, perhaps one of the most intriguing genes was grapes (grp). This gene had effects on both odor-guided behavior and food consumption. The directionality of grp expression is similar in both our RNA-seq analyses and RNAi experiments. High selected lines had significantly lower grp expression than low selected lines, and knockdown of grp expression in Or71a-expressing neurons recapitulated this pattern with a significant increase in attraction to 4EG and significantly higher amounts of food consumed relative to the control. This gene is expressed in the antenna [82], involved in sensory organ development [83], and has been previously identified in a GWA analysis on odor-guided behavior [32]. Moreover, grp belongs to the calcium/calmodulin-dependent protein kinase gene family [84–86]. Another member of this gene family, Ca^{2+}/calmodulin kinase II (CaMKII), is not differentially expressed in this study, but is well known for its role in olfaction. This gene regulates the termination of olfactory signaling by olfactory adenylyl cyclase [87] and has been implicated in olfactory memory formation [88]. Members of the calcium/calmodulin-dependent protein kinase family have also been implicated in the regulation of metabolism. In rats, administration of NPY to the hypothalamus, a region of the brain controlling appetite [89], increases CaMKII activity [90]. Two additional members of this family, CaMKI and CaMKIV, stimulate insulin biosynthesis in pancreatic β-cells in response to glucose stimulation

[91]. The involvement of this gene family in olfaction as well as the metabolic control of feeding sets the stage for future investigations into the molecular mechanisms by which grp affects olfactory perception and food consumption.

Conclusions

Artificial selection for attractive and aversive behavioral responses to four odorants revealed symmetrical responses to selection, such that the selection lines differed in their behavioral responses relative to the controls. Measurements of food consumption were associated with selection for behavioral responses to 4EG and 4MP. Differential expression analyses revealed genes involved in metabolism in lines selected for behavioral responses to 4EG. RNAi mediated knockdown of gene expression of several of these candidate genes revealed their role within specific neuronal populations of the olfactory system and sets the stage for future work on the functional mechanisms by which these genes affect both odor-guided behavior and food consumption. This work provides novel insights into the genetic architecture underlying olfactory behavior and its association with feeding behavior.

Methods

Drosophila maintenance and husbandry

Flies were reared on standard cornmeal/agar/molasses media at 25 °C under a 12 h light–dark cycle. Flyland was kindly provided by Dr. Trudy F. Mackay [46]. Transgenic RNAi lines obtained from the Drosophila Transgenic RNAi Project (Harvard Medical School) include: pyx (31297), tld (51507), tej (36879), SoYb (36881), bgm (55918), grp (36685), Egfr (36773), CG6044 (28610), Ipk1 (35250), Cdc6 (55734), as well as their co-isogenic controls attP2 (36303) and attP40 (36304) [51, 52]. RNAi lines were also obtained from the Vienna Drosophila Resource Center [92]. These lines include: Tret1–2 (v40980), Cyp6a2 (v48849), VepD (v103259), GstD2 (v109123), Spn42Dc (v13263), Spn47C (v100328) as well as the control lines w^{1118} (v60000) and y,w[1118];P{attP.y[+],w[3`] (v60100). Each RNAi line was crossed to the following drivers: GH146-GAL4 (gift from F. Hamada, University of Cincinnati) and Or71a-GAL4 (23122). Controls for genetic background were generated by crossing GAL4 driver lines to the appropriate host strain used to generate the RNAi line.

T-maze assay

Behavioral assays were conducted as previously described [93], with minor modification. Briefly, thirty flies were placed into the center of a T-maze apparatus and allowed to acclimate for one minute. Flies were then given a choice between the two arms of the maze, one arm containing the diluted odorant and the other containing the paraffin oil vehicle. Both arms had airflows of 500 mL/min. After one minute, the number of flies on each side were counted. The preference index (PI) was calculated using the formula: $PI = (O - N) / (O + N)$, where O is the total number of flies on the odor side, N is the total number of flies on the side not containing odor. A positive PI is indicative of attraction to the odor (with a maximal response of +1), whereas a negative PI indicates repulsion (with a maximal response of −1). All assays were conducted in the morning in the dark at 25 °C and 70% humidity. All flies were aged 3–7 days post-eclosion and were starved overnight on 1% agar (MoorAgar Inc.; Rocklin, CA). Each line and sex were tested separately and the arm from which odor was emitted was randomized each day of testing. RNA interference experiments were conducted using the same protocol as above.

Artificial selection

Artificial selection experiments were conducted for behavioral responses to four odorants: 4-ethylguaiacol (Sigma-Aldrich; St. Louis, MO), 4-methylphenol (Sigma-Aldrich), methyl hexanoate (Sigma-Aldrich), and ethyl acetate (Sigma-Aldrich), using Flyland as the base population [45]. All behavioral assays were performed at 0.01%. To commence selection, we measured behavioral responses of virgin females and males to each odorant. Upon completion of each assay, flies were collected from the odor- and non-odor sides of the T-maze. Flies collected from the odor side were used to establish the high lines, while files from the non-odor side were used to establish the low lines. Behavioral tests were performed until a minimum of 25 females and 25 males for each of three replicate high and low responding lines were obtained for each odorant. To establish the control lines, assays were conducted in which there was no odor present in the T-maze. Flies were then collected from one side of the T-maze, selected at random. This selection regime was repeated each generation for 30 generations. Symmetrical responses to selection were assessed at generation 18.

Locomotion

Locomotor reactivity was measured as described previously [94]. Briefly, single flies were placed into vials containing standard food media and acclimated overnight. Locomotion was quantified by recording the amount of time a single fly was active over a 45 s time period immediately following a mechanical disturbance. For each line and sex, 10 replicate measurements were taken at generation 18.

Correlated responses to selection

Food consumption was measured using the CAFE assay [48, 95]. Briefly, each chamber contained a calibrated glass micropipette (VWR, Radnor, PA) filled with 5 µl of liquid medium that was inserted through a foam plug and held in place with a pipette tip. At the bottom of each chamber, 1% agar was used as a water source. Flies were habituated to the chambers for 24 h with ad libitum food prior to testing. Food consumption was measured for 24 h. For feeding preference experiments, feeding preference was calculated using the formula: $(O - C) / (O + C)$, where O is the amount of food consumed with supplementation of odorant and C is the amount of food consumed without odorant [45]. Identical chambers without flies were maintained to assess evaporation and the total amount consumed was adjusted accordingly. Flies tested were 2–4 days post-eclosion. Unless otherwise specified, the liquid food was 5% sucrose (Sigma-Aldrich), 5% yeast extract (Fisher Scientific, Hampton, NH) or live *Brettanomyces* yeast (William's Brewing, San Leandro, CA), and 0.001% FD&C blue dye (Spectra Colors Corp., Kearny, NJ). For each line and sex, 12 replicates housing five flies each were tested.

For dry mass, longevity, and starvation resistance measurements, flies were reared as larvae at constant density of 50 larvae per vial. To measure dry mass, flies were separated by sex and placed on fresh media for 24 h after which they were dried for 24 h at 70 °C. For each line and sex, 10 replicate measurements were taken. Longevity was measured as previously described [96]. Briefly, 10 flies were placed into a vial and then scored every 24 h. Flies were transferred to fresh media every 2–3 days, during which dead flies were removed. For each line and sex, eight replicate vials were measured. For starvation resistance, flies were starved on 1% agar and survival measured every 8 h until death. For each line and sex, eight replicate vials of 10 flies each were measured. All experiments were performed on mated flies. Longevity and starvation resistance were measured at generation 18 and dry mass at 26.

Triglyceride (TAG), glucose, and glycogen levels were calculated as previously described [97]. For each line and sex, three replicate measurements consisting of five adult flies were taken at generation 28. Each sample was homogenized in 100 µl PBS, heat treated at 70 °C for 10 min, then flash frozen. Samples were subsequently thawed and assayed for glucose, glycogen and TAG content. Glucose and glycogen were quantified using the Glucose assay kit (Sigma-Aldrich). Glucose was measured directly from the homogenate, whereas glycogen was first digested to glucose during a 60 min incubation at 37 °C with 15 µl amyloglucosidase solution at 1.5 U ml^{-1} (Sigma-Aldrich), from which the previous glucose measurements were then subtracted. TAG levels were quantified by measuring glycerol content before and after digestion with Triglyceride Reagent (Sigma-Aldrich). TAG content was determined as the difference in glycerol content between the TAG-digested and TAG-undigested measurements.

Statistical analysis

A Shapiro-Wilk test was performed for each experiment to assess normality prior to subsequent statistical analysis (data not shown). For measurements of olfactory behavior, food consumption, and nutrient stores, we performed a nested mixed model analysis of variance (ANOVA) that accounted for the number of replicate lines within each selection regime (high, low and control): $Y = \mu + Selection + Line (Selection) + Sex + Selection \times Sex + Line (Selection) \times Sex + \varepsilon$. Where Selection is the fixed effect of selection treatment (high, control, or low behavioral responses), Line is the random effect of replicate within each selection regime, Sex is the fixed effect of sex, and ε indicates error. If no significant difference between sexes was observed, the data were pooled. *Post-hoc* analyses were conducted using Tukey's HSD test. For starvation resistance and longevity measurements, log-rank tests were performed for survivorship analyses. For RNAi experiments, significant differences between the knockdown and its corresponding isogenic control were assessed using Dunnett's tests for each sex. All data was analyzed using JMP 12.0 software (SAS Institute Inc., Cary, NC).

RNA isolation and sequencing

Whole heads from 100 female adult flies aged 3–7 days post-eclosion were hand dissected in the morning. Two independent biological replicates were collected for each of the three high, low, and control lines. Total RNA was isolated using an RNeasy Mini Kit (Qiagen, Valencia, CA, 74,104). Total RNA was provided to the Weill Cornell Medical College Genomics Resources Core Facility for subsequent RNA sequencing using standard protocols, during which cDNA libraries were generated from each sample, and then sequenced using Illumina HiSeq4000 to generate 100 bp reads.

RNA-seq processing and analysis

Adapters were removed from raw sequence reads using the program Trim Galore! (http://www.bioinformatics.babraham.ac.uk/projects/trim_galore), modifying the default parameters to allow a maximum error rate of zero. From there, the Cutadapt program was used to trim low quality sequences with Phred scores below 20 as well as remove reads shorter than 30 bp from the analysis [98]. The remaining RNA-seq reads were then aligned to the *Drosophila melanogaster* reference genome (version 6.10) [50] using STAR [99]. Differential expression between selection regimes was assessed using the Bioconductor EdgeR package [100]. Raw read counts were filtered to keep only genes that contain at least one read

per million in at least half the samples. The data were then normalized for library size using the *calcNormFactors* function. To identify the differentially expressed genes between high, low, and control selected lines, three comparisons were performed: (a) high vs low, (b) high vs control, (c) low vs control. For each comparison, the three replicate lines composing each high, low, and control treatments were pooled. To account for multiple testing, we applied a FDR of 0.10. The program Panther was used to assess whether there was overrepresentation of Gene Ontology (GO) terms [101, 102].

Additional files

Additional file 1: Table S1. Analyses of variance on odor-guided behavioral responses to artificial selection for each odorant at a given generation. **Table S2.** Analyses of variance on each trait measured. Traits include locomotor reactivity, longevity, starvation resistance, measurements of food consumption, dry mass, triglyceride levels, glucose, and glycogen. **Table S3.** Log-rank tests on longevity and starvation resistance measurements. **Table S4.** Summary of RNA-Seq datasets for control and 4-ethylguaiacol selected lines. Total reads are the number of reads that have passed quality filtering for each sample. Aligned reads and percent aligned are the number and percent of total reads that could be aligned to the reference genome. Uniquely mapped and percentage mapped reads are the number and percentage of reads that mapped uniquely. **Table S5.** RNA-seq results for control and 4-ethylguaiacol selected lines. For each gene, its FlyBase ID and the total number of raw reads in each sample is provided. **Table S6.** Differentially expressed genes between control and high lines selected for behavioral responses to 4-ethylguaiacol. For each gene, the log2 fold change (FC) and the log2 counts per million (CPM) are listed. Also listed is the corresponding likelihood ratio (LR), its *P*-value (*P*), and false discovery rate (FDR). **Table S7.** Differentially expressed genes between control and low lines selected for behavioral responses to 4-ethylguaiacol. For each gene, the log2 fold change (FC) and the log2 counts per million (CPM) are listed. Also listed is the corresponding likelihood ratio (LR), its *P*-value (*P*), and false discovery rate (FDR). **Table S8.** Differentially expressed genes between high and low lines selected for behavioral responses to 4-ethylguaiacol. For each gene, the log2 fold change (FC) and the log2 counts per million (CPM) are listed. Also listed is the corresponding likelihood ratio (LR), its *P*-value (*P*), and false discovery rate (FDR). **Table S9.** Dunnett's test on odor-guided behavior and food consumption measurements for genes silenced in distinct neuronal subpopulations using RNA interference. For both odor-guided behavior and food consumption, the GAL4 driver lines tested, gene names, and *P*-value (*P*) are listed. (XLSX 4068 kb)

Additional file 2: Figure S1. Locomotor reactivity of (**a**) 4-ethylguaiacol, (**b**) 4-methylphenol, (**c**) methyl hexanoate, and (**d**) ethyl acetate selected lines. Data shown are means ± SE. N = 20. (PDF 16 kb)

Additional file 3: Figure S2. Longevity of lines selected for (**a, b**) 4-ethylguaiacol, (**c, d**) 4-methylphenol, (**e, f**) methyl hexanoate, and (**g, h**) ethyl acetate. For each line and sex, survivorship curves (panels 1 and 2) and median survivorship (panels 3 and 4) are shown. Data shown are median ± SE for females and males (left and right columns, respectively). N = 70. Letters indicate P < 0.05 using Tukey's post hoc test. (PDF 126 kb)

Additional file 4: Figure S3. Starvation resistance of lines selected for (**a, b**) 4-ethylguaiacol, (**c, d**) 4-methylphenol, (**e, f**) methyl hexanoate, and (**g, h**) ethyl acetate. For each line and sex, survivorship curves (panels 1 and 2) and median survivorship (panels 3 and 4) are shown. Data shown are median ± SE for females and males (left and right columns, respectively). N = 70. (PDF 111 kb)

Additional file 5: Figure S4. Feeding measurements of lines selected for increased and decreased behavioral responses to 4-ethylguaiacol and 4-methylphenol using the CAFE assay. Row 1: Food consumption measurements of live *Brettanomyces* yeast for lines selected for (**a**)

4-ethylguaiacol and (**b**) 4-methylphenol. $N = 12$. Row 2: Food consumption measurements of sucrose for lines selected for (**c**) 4-ethylguaiacol and (**d**) 4-methylphenol. N = 12. Row 3: Binary preference assay for food with or without supplementation of either (**e**) 4-ethylguaiacol or (**f**) 4-methylphenol. Positive values indicate preference for food supplemented with odor. N = 12. Data shown are means ± SE for females (left) and males (right). Letters indicate $P < 0.05$ using Tukey's post hoc test. (PDF 104 kb)

Additional file 6: Figure S5. Dry mass and measurements of metabolism. (**a, b**) Dry mass ($N = 10$), (**c, d**) adjusted triglyceride levels ($N = 3$), (**e, f**) adjusted glucose (N = 3), and (**g, h**) adjusted glycogen measurements (N = 3) for lines selected for increased and decreased behavioral responses to 4-ethylguaiacol and 4-methylphenol. Data shown are means ± SE for females and males (left and right columns, respectively). Letters indicate $P < 0.05$ using Tukey's post hoc test. (PDF 207 kb)

Additional file 7: Figure S6. Categorization of differentially expressed genes among lines selected for differences in behavioral responses to 4-ethylguaiacol into (**a**) biological process, (**b**) molecular function, and (**c**) cellular component gene ontology terms. (PDF 18 kb)

Abbreviations
χ^2: Chi-square value; 4EG: 4-ethylguaiacol; 4MP : 4-methylphenol; CPM: Counts per million; Df: Degrees of freedom; EA: Ethyl acetate; F: F ratio; FC : Fold count; FDR: False discovery rate; GO: Gene ontology; GWA: Genome-wide association; LR: Likelihood ratio; MH: Methyl hexanoate; OSN: Olfactory sensory neuron; P: P-value; S.S.: Sum of squares; TAG: Triglyceride

Acknowledgements
We would like to thank Peter Andolfatto, John Layne, and Lu Yang for help with RNA-seq analyses and/or helpful discussions. We also thank Allie Elchert, Zachary Moore, Mary Shaw, and members of the Rollmann and Andolfatto labs for technical assistance. We thank TRiP at Harvard Medical School (NIH/NIGMS R01-GM084947) for transgenic RNAi stocks.

Funding
This work was supported by the National Institutes of Health (GM080592 to SMR). EB was supported by the University of Cincinnati Graduate School Dean's Fellowship, while the National Science Foundation REU (DBI-1262863) supported RP.

Authors' contributions
EB and SMR designed the study and wrote the manuscript. EB, RP, and CP conducted the experiments. EB analyzed the data. All authors read and approved of the manuscript.

Competing interests
The authors declare that they have no competing interests.

Author details
[1]Department of Biological Sciences, University of Cincinnati, Cincinnati, OH 45221-0006, USA. [2]Department of Biology, Xavier University, Cincinnati, OH 45207, USA.

References
1. Apfeld J, Kenyon C. Regulation of lifespan by sensory perception in *Caenorhabditis elegans*. Nature. 1999;402:804–9.
2. Alcedo J, Kenyon C. Regulation of *C. elegans* longevity by specific gustatory and olfactory neurons. Neuron. 2004;41:45–55.
3. Libert S, Zwiener J, Chu X, Vanvoorhies W, Roman G, Pletcher S. Regulation of *Drosophila* life span by olfaction and food-derived odors. Science. 2007; 315:1133–7.
4. Poon P, Kuo TH, Linford N, Roman G, Pletcher S. Carbon dioxide sensing modulates lifespan and physiology in *Drosophila*. PLoS Biol. 2010;8: e1000356.
5. Gendron CM, Kuo TH, Harvanek ZM, Chung BY, Yew JY, Dierick HA, Pletcher SD. *Drosophila* life span and physiology are modulated by sexual perception and reward. Science. 2014;343:544–8.
6. Vosshall LB, Stocker RF. Molecular architecture of smell and taste in *Drosophila*. Annu Rev Neurosci. 2007;30:505–33.
7. CY S, Menuz K, Calrson JR. Olfactory perception: receptors, cells, and circuits. Cell. 2009;139:45–59.
8. Wilson RI. Early olfactory processing in *Drosophila*: mechanisms and principles. Annu Rev Neurosci. 2013;36:217–41.
9. Laissue P, Reiter C, Hiesinger P, Halter S, Fischbach K, Stocker R. Three-dimensional reconstruction of the antennal lobe in *Drosophila melanogaster*. J Comp Neurol. 1999;405:543–52.
10. Vosshall L, Wong A, Axel R. An olfactory sensory map in the fly brain. Cell. 2000;102:147–59.
11. Dobritsa A, Naters W, Warr C, Steinbrecht R, Carlson J. Integrating the molecular and cellular basis of odor coding in the *Drosophila* antenna. Neuron. 2003;37:827–41.
12. Hallem E, Ho MG, Carlson JR. The molecular basis of odor coding in the *Drosophila* antenna. Cell. 2004;117:965–79.
13. Couto A, Alenius M, Dickson B. Molecular, anatomical, and functional organization of the *Drosophila* olfactory system. Curr Biol. 2005;15:1535–47.
14. Clyne P, Warr C, Freeman M, Lessing D, Kim J, Carlson JA. Novel family of divergent seven-transmembrane proteins candidate odorant receptors in *Drosophila*. Neuron. 1999;22:327–38.
15. Robertson H, Warr C, Carlson J. Molecular evolution of the insect chemoreceptor gene superfamily in *Drosophila melanogaster*. Proc Natl Acad Sci U S A. 2003;100(Suppl2):14537–42.
16. Larsson M, Domingos A, Jones W, Chiappe M, Amrein H, Vosshall L. *Or83b* encodes a broadly expressed odorant receptor essential for *Drosophila* olfaction. Neuron. 2004;43:703–14.
17. Sato K, Pellegrino M, Nakagawa T, Nakagawa T, Vosshall L, Touhara K. Insect olfactory receptors are heteromeric ligand-gated ion channels. Nature. 2008; 452:1002–6.
18. Wicher D, Schäfer R, Bauernfeind R, Stensmyr MC, Heller R, Heinemann SH, Hansson BS. *Drosophila* odorant receptors are both ligand-gated and cyclic-nucleotide-activated cation channels. Nature. 2008;452:1007–11.
19. Marin E, Jefferis G, Komiyama T, Zhu H, Luo L. Representation of the glomerular olfactory map in the *Drosophila* brain. Cell. 2001;109:243–55.
20. Wong A, Wang J, Axel R. Spatial representation of the glomerular map in the *Drosophila* protocerebrum. Cell. 2002;109:229–41.
21. Tanaka N, Awasaki T, Shimada T, Ito K. Integration of chemosensory pathways in the *Drosophila* second-order olfactory centers. Curr Biol. 2004;14:449–57.
22. Jefferis GS, Potter CJ, Chan AM, Marin EC, Rohlfing T, Maurer CR, Luo L. Comprehensive maps of *Drosophila* higher olfactory centers: spatially segregated fruit and pheromone representation. Cell. 2007;128:1187–203.
23. Gao Q, Yuan B, Chess A. Convergent projections of *Drosophila* olfactory neurons to specific glomeruli in the antennal lobe. Nat Neurosci. 2000; 3:780–5.
24. Wang J, Wong A, Flores J, Vosshall L, Axel R. Two-photon calcium imaging reveals an odor-evoked map of activity in the fly brain. Cell. 2003;112:271–82.
25. Wang P, Lyman R, Shabalina S, Mackay TF, Anholt RR. Association of polymorphisms in odorant-binding protein genes with variation in olfactory response to benzaldehyde in *Drosophila*. Genetics. 2007;177:1655–65.
26. Wang P, Lyman R, Mackay TF, Anholt RR. Natural variation in odorant recognition among odorant-binding proteins in *Drosophila melanogaster*. Genetics. 2010;184:759–67.

27. Arya G, Weber A, Wang P, Magwire M, Negron Y, Mackay TF, Anholt RR. Natural variation, functional pleiotropy and transcriptional contexts of odorant binding protein genes in Drosophila melanogaster. Genetics. 2010; 186:1475–85.

28. Rollmann SM, Wang P, Date P, West S, Mackay TF, Anholt RR. Odorant receptor polymorphisms and natural variation in olfactory behavior in Drosophila melanogaster. Genetics. 2010;186:687–97.

29. Richgels P, Rollmann S. Genetic variation in odorant receptors contributes to variation in olfactory behavior in a natural population of Drosophila melanogaster. Chem Senses. 2012;37:229–40.

30. Brown E, Layne J, Zhu C, Jegga A, Rollmann S. Genome-wide association mapping of natural variation in odour-guided behaviour in Drosophila. Genes Brain Behav. 2013;12:503–15.

31. Swarup S, Huang W, Mackay TF, Anholt RR. Analysis of natural variation reveals neurogenetic networks for Drosophila olfactory behavior. Proc Natl Acad Sci U S A. 2013;110:1017–22.

32. Arya G, Magwire M, Huang W, Serrano-Negron Y, Mackay TF, Anholt RR. The genetic basis for variation in olfactory behavior in Drosophila melanogaster. Chem Senses. 2015;40:233–43.

33. Rollmann SM, Magwire MM, Morgan TJ, Ozsoy ED, Yamamoto A, Mackay TF, Anholt RR. Pleiotropic fitness effects of the Tre1-Gr5a region in Drosophila melanogaster. Nature Genet. 2006;38:824–9.

34. Ostojic I, Boll W, Waterson MJ, Chan T, Chandra R, Pletcher SD, Alcedo J. Positive and negative gustatory inputs affect Drosophila lifespan partly in parallel to dFOXO signaling. Proc Natl Acad Sci U S A. 2014; 111(22):8143–8.

35. Pool AH, Scott K. Feeding regulation in Drosophila. Curr Opin Neurobiol. 2014;29:57–63.

36. Wright G. To feed or not to feed: circuits involved in the control of feeding in insects. Curr Opin Neurobiol. 2016;41:87–91.

37. Sachse S, Beshel J. The good, the bad, and the hungry: how the central brain codes odor valence to facilitate food approach in Drosophila. Curr Opin Neurobiol. 2016;40:53–8.

38. Kim SM, CY S, Wang JW. Neuromodulation of innate behaviors in Drosophila. Annu Rev Neurosci. 2017;40:327–48.

39. Brown MR, Crim JW, Arata RC, Cai HN, Chun C, Shen P. Identification of a Drosophila brain-gut peptide related to the neuropeptide Y family. Peptides. 1999;20:1035–42.

40. Root CM, Ko KI, Jafari A, Wang JW. Presynaptic facilitation by neuropeptide signaling mediates odor-driven food search. Cell. 2011;145:133–44.

41. Beshel J, Zhong Y. Graded encoding of food odor value in the Drosophila brain. J Neurosci. 2013;33:15693–704.

42. Nordström K. Formation of ethyl acetate in fermentation with brewer's yeast. J Inst Brew. 1960;67:173–81.

43. Tressel R, Drawert F. Biogenesis of banana volatiles. J Agr Food Chem. 1973; 21:560–5.

44. Comuzzo P, Tat L, Tonizzo A, Battistutta F. Yeast derivatives (extracts and autolysates) in winemaking: release of volatile compounds and effects on wine aroma volatility. Food Chem. 2006;99:217–30.

45. Dweck H, Ebrahim S, Farhan A, Hansson B, Stensmyr M. Olfactory proxy detection of dietary antioxidants in Drosophila. Curr Biol. 2015;25:455–66.

46. Huang W, Richards S, Carbone MA, Zhu D, Anholt RR, Ayroles JF, Duncan L, Jordan KW, Lawrence F, Magwire MM, Warner CB, Blankenburg K, Han Y, Javaid M, Jayaseelan J, Jhangiani SN, Muzny D, Ongeri F, Perales L, Wu YQ, Zhang Y, Zou X, Stone EA, Gibbs RA, Mackay TF. Epistasis dominates the genetic architecture of Drosophila quantitative traits. Proc Natl Acad Sci U S A. 2012;109:15553–9.

47. Schwasinger-Schmidt T, Kachman S, Harshman L. Evolution of starvation resistance in Drosophila melanogaster: measurement of direct and correlated responses to artificial selection. J Evol Biol. 2012;25:378–87.

48. Ja WW, Carvalho G, Mak EM, de las Rosa NN , Fang AY, Liong JC, Brummel T, Benzer S. Prandiology of Drosophila and the CAFE assay. Proc Natl Acad Sci U S A 2007;104:8253–8256.

49. Edgar B. How flies get their size: genetics meets physiology. Nature Rev Genet. 2006;7:907–16.

50. Attrill H, Falls K, Goodman JL, Millburn GH, Antonazzo G, Rey AJ, Marygold SJ. FlyBase: establishing a gene group resource for Drosophila melanogaster. Nucleic Acids Res. 2016;44:D786–92.

51. Ni JQ, Markstein M, Binari R, Pfeiffer BD, Liu LP, Villalta C, Booker M, Perkins LA, Perrimon N. Vector and parameters for targeted transgenic RNA interference in Drosophila melanogaster. Nat Methods. 2007;5:49–51.

52. Ni JQ, Liu LP, Binari R, Hardy R, Shim HS, Cavallaro A, Booker M, Pfeiffer BD, Markstein M, Wang H, Villalta C, Laverty TR, Perkins LA, Perrimon NA. Drosophila resource of transgenic RNAi lines for neurogenetics. Genetics. 2009;182:1089–100.

53. Brand A, Perrimon N. Targeted gene expression as a means of altering cell fates and generating dominant phenotypes. Dev Camb Engl. 1993; 118:401–15.

54. Jefferis GS, Marin EC, Stocker RF, Luo L. Target neuron prespecification in the olfactory map of Drosophila. Nature. 2001;414:204–8.

55. Shiraiwa T. Multimodal chemosensory integration through the maxillary palp in Drosophila. PLoS One. 2008;3(5):e2191.

56. Singh R, Nayak S. Fine structure and primary sensory projections of sensilla on the maxillary palp of Drosophila melanogaster Meigen (Diptera : Drosophilidae). Int J Insect Morphol Embryol. 1985;14:291–306.

57. de Bruyne M, Clyne PJ, Carlson JR. Odor coding in a model olfactory organ: the Drosophila maxillary palp. J Neurosci. 1999;19(1):4520–32.

58. Nowotny T, de Bruyne M, Berna AZ, Warr CG, Trowell SC. Drosophila olfactory receptors as classifiers for volatiles from disparate real world applications. Bioinspir Biomim. 2014;9:1–13.

59. CY S, Wang JW. Modulation of neural circuits: how stimulus context shapes innate behavior in Drosophila. Curr Opin Neurobiol. 2014;29:9–16.

60. Suh GS, Wong AM, Hergarden AC, Wang JW, Simon AF, Benzer S, Axel R, Anderson DJA. Single population of olfactory sensory neurons mediates an innate avoidance behavior in Drosophila. Nature. 2004;431:854–9.

61. Turner SL, Ray A. Modification of CO2 avoidance behavior in Drosophila by inhibitory odorants. Nature. 2009;461:277–82.

62. CY S, Menuz K, Reisert J, Carlson JR. Non-synaptic inhibition between grouped neurons in an olfactory circuit. Nature. 2012;492:66–72.

63. Clark RM, Zera AJ, Behmer ST. Metabolic rate is canalized in the face of variable life history and nutritional environment. Funct Ecol. 2016;30:922–31.

64. Swarup S, Williams T, Anholt RR. Functional dissection of odorant binding protein genes in Drosophila melanogaster. Genes Brain Behav. 2011;10:648–57.

65. Vogt RG. Molecular basis of pheromone detection in insects. In: Gilbert LI, Iatro K, Gill S, editors. Comprehensive insect physiology biochemistry, pharmacology and molecular biology. London: Elsevier; 2005. p. 753–804.

66. Shen P, Cai HN. Drosophila neuropeptide F mediates integration of chemosensory stimulation and conditioning of the nervous system by food. J Neurobiol. 2001;47(1):16–25.

67. Wu Q, Wen T, Lee G, Park JH, Cai HN, Shen P. Developmental control of foraging and social behavior by the Drosophila neuropeptide Y-like system. Neuron. 2003;39:147–61.

68. Erion R, Sehgal A. Regulation of insect behavior via the insulin-signaling pathway. Front Physiol. 2013;4:353.

69. Nassel DR, Liu Y, Luo J. Insulin/IGF signaling and its regulation in Drosophila. Gen Comp Endocrinol. 2015;221:255–66.

70. Porte D Jr, Baskin DG, Schwartz MW. Insulin signaling in the central nervous system: a critical role in metabolic homeostasis and disease from C. elegans to humans. Diabetes 2005;54(5):1264–1276.

71. Alvarez-Ponce D, Aguadé M, Rozas J. Network-level molecular evolutionary analysis of the insulin/TOR signal transduction pathway across 12 Drosophila genomes. Genome Res. 2009;19(2):234–42.

72. Wang M, Wang Q, Wang Z, Wang Q, Zhang X, Pan Y. The molecular evolutionary patterns of the insulin/FOXO signaling pathway. Evol Bioinformatics Online. 2013;9:1–16.

73. Garlapow ME, Huang W, Yarboro MT, Peterson KR. Quantitative genetics of food intake in Drosophila melanogaster. PLoS One. 2015;10(9):e0138129.

74. Garlapow ME, Everett LJ, Zhou S, Gearhart AW, Fay KA, Huang W, Morozoba TV, Arya GH, Turlapati L, St. Armour G, Hussain YN, McAdams SE, Fochler S, Mackay TF. Genetic and genomic response to selection for food consumption in Drosophila melanogaster. Behav Genet 2017;47(2):227–243.

75. Jhaveri D, Sen A, Reddy V, Rodrigues V. Sense organ identity in the Drosophila antenna is specified by the expression of the proneural gene atonal. Mech Dev. 2000;99:101–11.

76. Kumar JP, Moses KEGF. Receptor and notch signaling act upstream of eyeless/Pax6 to control eye specification. Cell. 2001;104:687–97.

77. Shilo B. Signaling by the Drosophila epidermal growth factor receptor pathway during development. Exp Cell Res. 2003;284:140–9.

78. Sen A, Kuruvilla D, Pinto L, Sarin A, Rodrigues V. Programmed cell death and context dependent activation of the EGF pathway regulate gliogenesis in the Drosophila olfactory system. Mech Dev. 2004;121:65–78.

79. Kurachi H, Adachi H, Ohtsuka S, Morishige K, Amemiya K, Keno Y, Shimomura I, Tokunaga K, Miyake A, Matsuzawa Y, Tanizawa O. Involvement of epidermal growth factor in inducing obesity in ovariectomized mice. Am J Phys. 1993;265:E323–31.

80. Adachi H, Kurachi H, Homma H, Adachi K, Imai T, Morishige K, Matsuzawa Y, Miyake A. Epidermal growth factor promotes adipogenesis of 3T3-L1 cell in vitro. Endocrinology. 1994;135(5):1824–30.

81. Kanamori Y, Saito A, Hagiwara-Komoda Y, Mitsumasu K, Kikuta S, Watanabe M, Cornette R, Kikawada T, Okuda T. The trehalose transporter 1 gene sequence is conserved in insects and encodes proteins with different kinetic properties involved in trehalose import into peripheral tissues. Inest Biochem Mol Biol. 2010;40:30–7.

82. Menuz K, Larter NK, Park J, Carlson JR. An RNA-seq screen of the *Drosophila* antenna identifies a transporter necessary for ammonia detection. PLoS Genet. 2014;10:e1004810.

83. Abdelilah-Seyfried S, Chan YM, Zeng C, Justice NJ, Younger-Shepherd S, Sharp LE, Barbel S, Meadows SA, Jan LY, Jan YNA. Gain-of-function screen for genes that affect the development of the *Drosophila* adult external sensory organ. Genetics. 2000;155:733–52.

84. Morrison DK, Murakami MS, Cleghon V. Protein kinases and phosphatases in the *Drosophila* genome. J Cell Biol. 2000;150(2):F57–62.

85. Manning G, Plowman GD, Hunter T, Sudarsanam S. Evolution of protein kinase signaling from yeast to man. Trends Biochem Sci. 2002;27(10):514–20.

86. Sopko R, Foos M, Vinayagam A, Zhai B, Binari R, Hu Y, Randklev S, Perkins LA, Gygi SP, Perrimon N. Combining genetic perturbations and proteomics to examine kinase-phosphatase networks in *Drosophila* embryos. Dev Cell. 2014;31:114–27.

87. Wei J, Zhao AZ, Chan GCK, Baker LP, Impey S, Beavo JA, Storm DR. Phosphorylation and inhibition of olfactory adenylyl cyclase by CaM kinase II neurons: a mechanism for attenuation of olfactory signals. Neuron. 1998;21:495–504.

88. Guven-Ozkan T, Davis R. Functional neuroanatomy of *Drosophila* olfactory memory formation. Learn Mem. 2014;21:519–26.

89. Berthoud HR. Metabolic and hedonic drives in the neural control of appetite: who is the boss? Curr Opin Neurobiol. 2011;21:888–96.

90. Sheriff S, Chance WT, Fischer JE, Balasubramaniam A, Neuropeptide Y. Treatment and food deprivation increase cyclic AMP response element-binding in rat hypothalamus. Mol Pharmacol. 1997;51:597–604.

91. Yu X, Murao K, Sayo Y, Imachi H, Cao WM, Ohtsuka S, Niimi M, Tokumitsu H, Inuzuka H, Wong NCW, Kobayashi R, Ishida T. The role of calcium/calmodulin-dependent protein kinase cascade in glucose upregulation of insulin gene expression. Diabetes. 2004;53:1475–81.

92. Dietzl G, Chen D, Schnorrer F, KC S, Barinova Y, Fellner M, Gasser B, Kinsey K, Oppel S, Scheiblauer S, Couto A, Marra V, Keleman K, Dickson BJA. Genome-wide transgenic RNAi library for conditional gene inactivation in *Drosophila*. Nature. 2007;448:151–6.

93. Helfand S, Carlson J. Isolation and characterization of an olfactory mutant in *Drosophila* with a chemically specific defect. Proc Natl Acad Sci. 1989;86: 2908–12.

94. Jordan K, Morgan T, Mackay TF. Quantitative trait loci for locomotor behavior in *Drosophila melanogaster*. Genetics. 2006;174:271–84.

95. Deshpande SA, Carvalho GB, Amador A, Phillips AM, Hoxha S, Lizotte KJ, Ja WW. Quantifying *Drosophila* food intake: comparative analysis of current methodology. Nat Methods. 2014;11:535–40.

96. Linford N, Bilgir C, Ro J, Pletcher S. Measurement of lifespan in *Drosophila melanogaster*. J Vis Exp. 2013;71:e50068.

97. Tennessen J, Barry W, Cox J, Thummel C. Methods for studying metabolism in *Drosophila*. Methods. 2014;68:105–15.

98. Martin M. *Cutadapt* removes adapter sequences from high-throughput sequencing reads. EMBnet. Journal. 2011;17:10–2.

99. Dobin A, Davis CA, Schlesinger F, Drenkow J, Zaleski C, Jha S, Batut P, Chaisson M, Gingeras TR. STAR: ultrafast universal RNA-seq aligner. Bioinformatics. 2013;29(1):15–21.

100. Robinson MD, McCarthy DJ, Smyth GK. EdgeR: a Bioconductor package for differential expression analysis of digital gene expression data. Bioinformatics. 2010;26(1):139–40.

101. Mi H, Muruganujan A, Casagrande JT, Thomas PD. Large-scale gene function analysis with the PANTHER classification system. Nat Protoc. 2013;8:1551–66.

102. Mi H, Sagar P, Muruganujan A, Casagrande JT, Thomas PDPANTHER. Version 10: expanded protein families and functions, and analysis tools. Nucleic Acids Res. 2016;44:D336–42.

Permissions

All chapters in this book were first published in GENOMICS, by BioMed Central; hereby published with permission under the Creative Commons Attribution License or equivalent. Every chapter published in this book has been scrutinized by our experts. Their significance has been extensively debated. The topics covered herein carry significant findings which will fuel the growth of the discipline. They may even be implemented as practical applications or may be referred to as a beginning point for another development.

The contributors of this book come from diverse backgrounds, making this book a truly international effort. This book will bring forth new frontiers with its revolutionizing research information and detailed analysis of the nascent developments around the world.

We would like to thank all the contributing authors for lending their expertise to make the book truly unique. They have played a crucial role in the development of this book. Without their invaluable contributions this book wouldn't have been possible. They have made vital efforts to compile up to date information on the varied aspects of this subject to make this book a valuable addition to the collection of many professionals and students.

This book was conceptualized with the vision of imparting up-to-date information and advanced data in this field. To ensure the same, a matchless editorial board was set up. Every individual on the board went through rigorous rounds of assessment to prove their worth. After which they invested a large part of their time researching and compiling the most relevant data for our readers.

The editorial board has been involved in producing this book since its inception. They have spent rigorous hours researching and exploring the diverse topics which have resulted in the successful publishing of this book. They have passed on their knowledge of decades through this book. To expedite this challenging task, the publisher supported the team at every step. A small team of assistant editors was also appointed to further simplify the editing procedure and attain best results for the readers.

Apart from the editorial board, the designing team has also invested a significant amount of their time in understanding the subject and creating the most relevant covers. They scrutinized every image to scout for the most suitable representation of the subject and create an appropriate cover for the book.

The publishing team has been an ardent support to the editorial, designing and production team. Their endless efforts to recruit the best for this project, has resulted in the accomplishment of this book. They are a veteran in the field of academics and their pool of knowledge is as vast as their experience in printing. Their expertise and guidance has proved useful at every step. Their uncompromising quality standards have made this book an exceptional effort. Their encouragement from time to time has been an inspiration for everyone.

The publisher and the editorial board hope that this book will prove to be a valuable piece of knowledge for researchers, students, practitioners and scholars across the globe.

List of Contributors

Jenelle R. Dunkelberger
Department of Animal Science, Iowa State University, Ames, IA 50011, USA
Topigs Norsvin USA, Burnsville, MN 55337, USA

Nick V. L. Serão, Jack C. M. Dekkers
Department of Animal Science, Iowa State University, Ames, IA 50011, USA

Ziqing Weng
Department of Animal Science, Iowa State University, Ames, IA 50011, USA
ABS Global Inc., DeForest,WI 53532, USA

Emily H. Waide
Department of Animal Science, Iowa State University, Ames, IA 50011, USA
The Seeing Eye Inc., Morristown, NJ 07960, USA

Megan C. Niederwerder, Maureen A. Kerrigan, Raymond R. R. Rowland
Department of Diagnostic Medicine/Pathobiology, College of Veterinary Medicine, Kansas State University, Manhattan, KS 66506, USA

Joan K. Lunney
USDA, ARS,BARC, APDL, Beltsville, MD 20705, USA

J. Michael Proffitt, Jeremy Glenn
Department of Genetics, Texas Biomedical Research Institute, San Antonio,Texas, USA

Anthony J. Cesnik, Michael R. Shortreed
Department of Chemistry, University of Wisconsin, Madison, Wisconsin, USA

Avinash Jadhav
Department of Genetics, Texas Biomedical Research Institute, San Antonio, Texas, USA
Department of Internal Medicine, Section of Molecular Medicine, Wake Forest School of Medicine, NRC Building, G-55, Winston-Salem, North Carolina 27157, USA

Lloyd M. Smith
Department of Chemistry, University of Wisconsin, Madison, Wisconsin, USA
Genome Center of Wisconsin, University of Wisconsin, Madison, Wisconsin, USA

Kylie Kavanagh
Department of Pathology and Comparative Medicine, Wake Forest School of Medicine, Winston-Salem, North Carolina,USA

Laura A. Cox
Department of Genetics, Texas Biomedical Research Institute, San Antonio,Texas, USA
Southwest National Primate Research Center, Texas Biomedical Research Institute, San Antonio, Texas, USA

Michael Olivier
Department of Genetics, Texas Biomedical Research Institute, San Antonio, Texas, USA
Southwest National Primate Research Center, Texas Biomedical Research Institute, San Antonio, Texas, USA
Department of Internal Medicine, Section of Molecular Medicine, Wake Forest School of Medicine, NRC Building, G-55, Winston-Salem, North Carolina 27157, USA

Hana Hall, Spencer E. Escobedo, Kaelan J. Brennan
Department of Biochemistry, Purdue University, West Lafayette, IN 47907,USA

Patrick Medina, Jeremiah Rounds, Christopher Vincent
Department of Statistics, Purdue University, West Lafayette, IN 47907,USA

Daphne A. Cooper, Pedro Miura
Department of Biology, University of Nevada, Reno, NV 89557, USA

Rebecca Doerge
Carnegie Mellon University, Pittsburgh, PA 15213, USA

Vikki M. Weake
Department of Biochemistry, Purdue University, West Lafayette, IN 47907, USA
Purdue University Center for Cancer Research, Purdue University, West Lafayette 47907, USA

Mirjam Frischknecht
Qualitas AG, Chamerstrasse 56a, 6300 Zug, Switzerland

School of Agricultural, Forest and Food Sciences HAFL, Bern University of Applied Sciences, Länggasse 85, 3052 Zollikofen, Switzerland

Beat Bapst, Franz R. Seefried, Birgit Gredler-Grandl
Qualitas AG, Chamerstrasse 56a, 6300 Zug, Switzerland

Heidi Signer-Hasler, Christine Flury
School of Agricultural, Forest and Food Sciences HAFL, Bern University of Applied Sciences, Länggasse 85, 3052 Zollikofen, Switzerland

Dorian Garrick
Institute of Veterinary, Animal & Biomedical Sciences, Massey University, Hamilton, New Zealand

Christian Stricker
agn Genetics GmbH, 8b Börtjistrasse, 7260 Davos, Switzerland

Intergenomics Consortium
Interbull center, SLU - Box 7023, S-75007 Uppsala, Sweden

Ruedi Fries
Technische Universität München, Liesel-Beckmann-Straße 1, 85354 Freising-Weihenstephan, Germany

Ingolf Russ
Tierzuchtforschung e.V, Senator-Gerauer-Str. 23, 85586 Poing, Germany

Johann Sölkner
University of Natural Resources and Life Sciences, Gregor-Mendel-Str 33, 1180 Wien, Austria

Anna Bieber
Research Institute of Organic Agriculture (FiBL), Ackerstrasse 113, 5070 Frick, Switzerland

Maria G. Strillacci
Department of Veterinary Medicine, University of Milan, Via Celoria 10, 20133 Milan, Italy

Xiaosai Niu, Yuyang Wang, Min Li, Xiaorong Zhang, Yantao Wu
Jiangsu Co-Innovation Center for Prevention of Animal Infectious Diseases and Zoonoses, College of Veterinary Medicine, Yangzhou University, 48 East Wenhui Road, Yangzhou, Jiangsu 225009, China

Il-Hwan Kim
Department of Entomology, University of Wisconsin-Madison, Madison, WI,USA

Laboratory of Malaria and Vector Research, National Institute of Allergy and Infectious Diseases, Rockville, MD, USA

Sudarshan K. Aryal
Department of Nematology, University of California, Riverside, CA, USA

Dariush T. Aghai, Ángel M. Casanova-Torres, Kai Hillman, Michael P. Kozuch, Jerald C. Ensign
Department of Bacteriology, University of Wisconsin-Madison, Madison, WI, USA

Erin J. Mans, Terra J. Mauer, Heidi Goodrich-Blair
Department of Bacteriology, University of Wisconsin-Madison, Madison, WI, USA
Department of Microbiology, University of Tennessee-Knoxville, Knoxville, TN, USA

Jean-Claude Ogier, Sophie Gaudriault
DGIMI, INRA, Université de Montpellier, 34095 Montpellier, France

Walter G. Goodman
Department of Entomology, University of Wisconsin-Madison, Madison, WI,USA

Adler R. Dillman
Department of Nematology, University of California, Riverside, CA, USA

Hui-Zeng Sun, Kai Shi, Xue-Hui Wu, Ming-Yuan Xue, Zi-Hai Wei, Jian-Xin Liu, Hong-Yun Liu
Institute of Dairy Science, MoE Key Laboratory of Molecular Animal Nutrition, College of Animal Sciences, Zhejiang University, Hangzhou 310058, People'sRepublic of China

Larissa Fernanda Simielli Fonseca, Daniele Fernanda Jovino Gimenez, Danielly Beraldo dos Santos Silva, Fernando Baldi, Jesus Aparecido Ferro, Lucia Galvão Albuquerque
Faculty of Agricultural and Veterinary Sciences, São Paulo State University, FCAV/UNESP, Jaboticabal, São Paulo, Brazil

Roger Barthelson
CyVerse, University of Arizona, Tucson, USA

Olivia Carulei, Nicola Douglass
Division of Medical Virology, Department of Pathology, Faculty of Health Sciences, University of Cape Town, Cape Town, South Africa

Anna-Lise Williamson
Division of Medical Virology, Department of Pathology, Faculty of Health Sciences, University of Cape Town, Cape Town, South Africa
Institute of Infectious Disease and Molecular Medicine, University of Cape Town, Cape Town, South Africa
National Health Laboratory Service, Cape Town, South Africa

Xianbin Su, Yi Shi, Xin Zou, Zhao-Ning Lu, Chong-Chao Wu, Xiao-Fang Cui, Kun-Yan He, Qing Luo, Yu- Lan Qu, Na Wang, Lan Wang
Key Laboratory of Systems Biomedicine (Ministry of Education) and Collaborative Innovation Center of Systems Biomedicine, Shanghai Center for Systems Biomedicine, Shanghai Jiao Tong University, 800 Dongchuan Road, Shanghai 200240, China

Gangcai Xie
Key Laboratory of Computational Biology, CAS-MPG Partner Institute for Computational Biology, 320 Yueyang Road, Shanghai, China

Jean Y. H. Yang
School of Mathematics and Statistics, The University of Sydney, Sydney, Australia.

Ze-Guang Han
Key Laboratory of Systems Biomedicine (Ministry of Education) and Collaborative Innovation Center of Systems Biomedicine, Shanghai Center for Systems Biomedicine, Shanghai Jiao Tong University, 800 Dongchuan Road, Shanghai 200240, China
Shanghai-MOST Key Laboratory for Disease and Health Genomics, Chinese National Human Genome Center at Shanghai, Shanghai, China

Yoshinobu Uemoto
National Livestock Breeding Center, Nishigo, Fukushima 961-8511, Japan
Graduate School of Agricultural Science, Tohoku University, Sendai, Miyagi 980- 0845, Japan

Tsuyoshi Ohtake, Nanae Sasago, Masayuki Takeda, Tsuyoshi Abe, Hironori Sakuma, Takatoshi Kojima, Shinji Sasaki
National Livestock Breeding Center, Nishigo, Fukushima 961-8511, Japan

Chen Wu
School of Biological Sciences, The University of Auckland, Auckland, New Zealand
Landcare Research, Auckland, New Zealand
New Zealand Institute for Plant & Food Research Ltd, Auckland, New Zealand

Victoria G. Twort
School of Biological Sciences, The University of Auckland, Auckland, New Zealand
Landcare Research, Auckland, New Zealand
Department of Biology, Lund University, Lund, Sweden

Ross N. Crowhurst
New Zealand Institute for Plant & Food Research Ltd, Auckland, New Zealand

Richard D. Newcomb
School of Biological Sciences, The University of Auckland, Auckland, New Zealand
New Zealand Institute for Plant & Food Research Ltd, Auckland, New Zealand

Thomas R. Buckley
School of Biological Sciences, The University of Auckland, Auckland, New Zealand
Landcare Research, Auckland, New Zealand

Hui Zhang, Jia-Qiang Yu, Li-Li Yang, Xin-Yang Zhang, Wei Na, Hui Li
Key Laboratory of Chicken Genetics and Breeding, Ministry of Agriculture, Harbin 150030, People's Republic of China
Key Laboratory of Animal Genetics, Breeding and Reproduction, Education Department of Heilongjiang Province, Harbin 150030, People's Republic of China
College of Animal Science and Technology, Northeast Agricultural University, Harbin 150030, People's Republic of China

Luke M. Kramer, James M. Reecy
Department of Animal Science, Iowa State University, 2255 Kildee Hall, Ames, IA 50011, USA

Elizabeth B. Brown, Cody Patterson, Stephanie M. Rollmann
Department of Biological Sciences, University of Cincinnati, Cincinnati, OH 45221-0006, USA

Rayanne Pancoast
Department of Biological Sciences, University of Cincinnati, Cincinnati, OH 45221-0006, USA
Department of Biology, Xavier University, Cincinnati, OH45207, USA

Index

www.ingramcontent.com/pod-product-compliance
Lightning Source LLC
Chambersburg PA
CBHW082024190326
41458CB00010B/3267